T0324845

Quantum Phase Transitions
Second Edition

This is the first book to describe the physical properties of quantum materials near critical points with long-range many-body quantum entanglement. Readers are introduced to the basic theory of quantum phases, their phase transitions, and their observable properties.

This second edition begins with nine chapters, six of them new, suitable for an introductory course on quantum phase transitions, assuming no prior knowledge of quantum field theory. There are several new chapters covering important recent advances, such as the Fermi gas near unitarity, Dirac fermions, Fermi liquids and their phase transitions, quantum magnetism, and solvable models obtained from string theory. After introducing the basic theory, it moves on to a detailed description of the canonical quantum-critical phase diagram at nonzero temperatures. Finally, a variety of more complex models is explored. This book is ideal for graduate students and researchers in condensed matter physics and particle and string theory.

Subir Sachdev is Professor of Physics at Harvard University and holds a Distinguished Research Chair at the Perimeter Institute for Theoretical Physics. His research has focused on a variety of quantum materials, and especially on their quantum phase transitions.

Quantum Phase Transitions

Second Edition

SUBIR SACHDEV

Harvard University

CAMBRIDGE
UNIVERSITY PRESS

CAMBRIDGE
UNIVERSITY PRESS

University Printing House, Cambridge CB2 8BS, United Kingdom

One Liberty Plaza, 20th Floor, New York, NY 10006, USA

477 Williamstown Road, Port Melbourne, VIC 3207, Australia

4843/24, 2nd Floor, Ansari Road, Daryaganj, Delhi - 110002, India

79 Anson Road, #06-04/06, Singapore 079906

Cambridge University Press is part of the University of Cambridge.

It furthers the University's mission by disseminating knowledge in the pursuit of education, learning and research at the highest international levels of excellence.

www.cambridge.org
Information on this title: www.cambridge.org/9780521514682

First published 2011
5th printing 2015

A catalogue record for this publication is available from the British Library

Library of Congress Cataloging in Publication data
Sachdev, Subir, 1961-
Quantum phase transitions / Subir Sachdev. – Second edition.
p. cm
Includes bibliographical references and index.
ISBN 978-0-521-51468-2 (Hardback)
1. Phase transformations (Statistical physics) 2. Quantum theory. I. Title.
QC175.16.P5S23 2011
530.4′74–dc22

2010050328

ISBN 978-0-521-51468-2 Hardback

To my parents and Menaka, Monisha, and Usha

Contents

From the Preface to the first edition

The past decade has seen a substantial rejuvenation of interest in the study of quantum phase transitions, driven by experiments on cuprate superconductors, heavy fermion materials, organic conductors, and related compounds. Although quantum phase transitions in simple spin systems, like the Ising model in a transverse field, were studied in the early 1970s, much of the subsequent theoretical work examined a particular example: the metal–insulator transition. While this is a subject of considerable experimental importance, the greatest theoretical progress was made for the case of the Anderson transition of non-interacting electrons, which is driven by localization of the electronic states in the presence of a random potential. The critical properties of this transition of noninteracting electrons constituted the primary basis upon which most condensed matter physicists have formed their intuition on the behavior of the systems near a quantum phase transition. However, it is clear that strong electronic interactions play a crucial role in the systems of current interest noted earlier, and simple paradigms for the behavior of such systems near quantum critical points are not widely known.

It is the purpose of this book to move interactions to center stage by describing and classifying the physical properties of the simplest interacting systems undergoing a quantum phase transition. The effects of disorder will be neglected for the most part but will be considered in the concluding chapters. Our focus will be on the dynamical properties of such systems at nonzero temperature, and it will become apparent that these differ substantially from the noninteracting case. We shall also be considering inelastic collision-dominated quantum dynamics and transport: our results will apply to clean physical systems whose inelastic scattering time is much shorter than their disorder-induced elastic scattering time. This is the converse of the usual theoretical situation in Anderson localization or mesoscopic system theory, where inelastic collision times are conventionally taken to be much larger than all other timescales.

One of the most interesting and significant regimes of the systems we shall study is one in which the inelastic scattering and phase coherence times are of order $\hbar/k_BT$, where T is the absolute temperature. The importance of such a regime was pointed out by Varma *et al.* [523,524] by an analysis of transport and optical data on the cuprate superconductors. Neutron scattering measurements of Hayden *et al.* [210] and Keimer *et al.* [263] also supported such an interpretation in the low doping region. It was subsequently realized [86,419,440] that the inelastic rates are in fact a *universal number* times $k_BT/\hbar$, and they are a robust property of the high-temperature limit of renormalizable, interacting quantum field theories that are not asymptotically free at high energies. In the Wilsonian picture, such a field theory is defined by renormalization group flows away from a critical point describing a second-order quantum phase transition. It is not essential for this critical point to be in an experimentally accessible regime of the phase diagram: the quantum field theory it defines may still be an appropriate description of the physics over a substantial intermediate energy and temperature scale. Among the implications of such an interpretation of the experiments was the requirement that response functions should have prefactors of anomalous powers of T and a singular dependence on the wavevector; recent observations of Aeppli *et al.* [5], at somewhat higher dopings, appear to be consistent with this. These recent experiments also suggest that the appropriate quantum critical points involve competition between phases with or without conventional superconducting, spin-, or charge-density-wave order. There is no global theory yet for such quantum transitions, but we shall discuss numerous simpler models here that capture some of the basic features.

It is also appropriate to note here theoretical studies [25, 93, 94, 336, 514] on the relevance of finite temperature crossovers near quantum critical points of Fermi liquids [218] to the physics of heavy fermion compounds.

A separate motivation for the study of quantum phase transitions is simply the value in having another perspective on the physics of an interacting many-body system. A traditional analysis of such a system would begin from either a weak-coupling Hamiltonian, and then build in interactions among the nearly free excitations, or a strong-coupling limit, where the local interactions are well accounted for, but their coherent propagation through the system is not fully described. In contrast, a quantum critical point begins from an intermediate coupling regime, which straddles these limiting cases. One can then use the powerful technology of scaling to set up a systematic expansion of physical properties away from the special critical point. For many low-dimensional strongly correlated systems, I believe that such an approach holds the most promise for a comprehensive understanding. Many of the vexing open problems are related to phenomena at intermediate temperatures, and this is precisely the region over which the influence of a quantum critical point is dominant. Related motivations for the study of quantum phase transitions appear in a recent discourse by Laughlin [286].

The particular quantum phase transitions that are examined in this book are undoubtedly heavily influenced by my own research. However, I believe that my choices can also be justified on pedagogical grounds and lead to a logical development of the main physical concepts in the simplest possible contexts. Throughout, I have also attempted to provide experimental motivations for the models considered; this is mainly in the form of a guide to the literature, rather than in-depth discussion of the experimental issues. I have highlighted some especially interesting experiments in a recent popular introduction to quantum phase transitions [428]. An experimentally oriented introduction to the subject of quantum phase transitions can also be found in the excellent review article of Sondhi, Girvin, Carini, and Shahar [481]. Readers may also be interested in a recent introductory article [533], intended for a general science audience.

Acknowledgments

Chapter 21 was co-authored with T. Senthil and adapted from his 1997 Yale University Ph.D. thesis; I am grateful to him for agreeing to this arrangement.

Some portions of this book grew out of lectures and write-ups I prepared for schools and conferences in Trieste, Italy [418], Xiamen, China [419], Madrid, Spain [421], Geilo, Norway [424], and Seoul, Korea [427]. I am obliged to Professors Yu Lu, S. Lundqvist, G. Morandi, Hao Bai-Lin, German Sierra, Miguel Martin-Delgado, Arne Skjeltorp, David Sherrington, Jisoon Ihm, Yunkyu Bang, and Jaejun Yu for the opportunities to present these lectures. I also taught two graduate courses at Yale University and a mini-course at the Université Joseph Fourier, Grenoble, France on topics discussed in this book; I thank both institutions for arranging and supporting these courses. I am indebted to the participants and students at these lectures for stimulating discussions, valuable feedback, and their interest. Part of this book was written during a sojourn at the Laboratoire des Champs Magnétiques Intenses in Grenoble, and I thank Professors Claude Berthier and Benoy Chakraverty for their hospitality. My research has been supported by grants from the Division of Materials Research of the U.S. National Science Foundation.

I have been fortunate in having the benefit of interactions and collaborations with numerous colleagues and students who have generously shared insights that appear in many of these pages. I would particularly like to thank my collaborators Chiranjeeb Buragohain, Andrey Chubukov, Kedar Damle, Sankar Das Sarma, Antoine Georges, Ilya Gruzberg, Satya Majumdar, Reinhold Oppermann, Nick Read, R. Shankar, T. Senthil, Sasha Sokol, Matthias Troyer, Jinwu Ye, Peter Young, and Lian Zheng.

The evolution of the book owes a great deal to comments of readers of earlier versions, who unselfishly donated their time in working through unpolished drafts; naturally, they bear no responsibility for the remaining errors and obscurities. I am most grateful to Sudip Chakravarty, Andrey Chubukov, Kedar Damle, Ilya Gruzberg, Sankar Das Sarma, Bert Halperin, T. Senthil, R. Shankar, Oleg Starykh, Chandra Varma, Peter Young, Jan Zaanen, and two anonymous referees. The detailed comments provided by Steve Girvin and Wim van Saarloos were especially valuable. My thanks to them, and the others, accompany an admiration for their generous collegial spirit. I also acknowledge salutary encouragement from Jan Zaanen.

My wife, Usha, and my daughters, Monisha and Menaka, patiently tolerated my mental and physical absences during the writing (and rewritings) of this book. Ultimately, it was their cheerful support that made the project possible and worthwhile.

Preface to the second edition

Research on quantum phase transitions has undergone a vast expansion since the publication of the first edition, over a decade ago. Many new theoretical ideas have emerged, and the arena of experimental systems has grown rapidly. The cuprates have been firmly established to be d-wave superconductors, with a massless Dirac spectrum for their electronic excitations; the latter spectrum has also been observed in graphene and on the surface of topological insulators. Such fermions play a key role in a variety of quantum phase transitions. The observation of quantum oscillations in the presence of strong magnetic fields in the underdoped cuprates has highlighted the relevance of competing orders, and their quantum critical points. Optical lattices of ultracold atoms now offer a realization of the boson Hubbard model, and exhibit the superfluid–insulator transition. And ideas on quantum criticality and entanglement have had an interesting interplay with developments in quantum information science.

The second edition does not present a fully comprehensive survey of these ongoing developments. I believe the core topics of the first edition had a certain coherence, and they continue to be central to the more modern developments; I did not wish to dilute the global perspective they offer in understanding both condensed matter and ultracold atom experiments. However, wherever possible, I have discussed important advances, or directed the reader to review articles.

Also, in the last few years, a remarkable connection has developed between ideas on quantum criticality and the string theory of quantum black holes. I briefly survey the initial developments in Section 15.5. The subject has advanced rapidly since then, with interesting applications to quantum critical states of fermions at nonzero density: this recent work is not discussed here. In any case, this book should be useful background reading for this emerging and growing field of research.

The primary change in the second edition is pedagogical. I have had the benefit of teaching a course on quantum phase transitions several times since the first edition, both at Yale and at Harvard. I am also grateful for the opportunity to lecture at various summer and winter schools (Altenberg, Boulder, Cargese, Goa, Groningen, Jerusalem, Les Houches, Mahabaleshwar, Milos, Prague, Trieste, Windsor). The content of these lectures is now in the new Part II of the book. Chapters 3–8 are new, although they do extract some material from the earlier chapters of the first edition. Part II, titled "A first course," is intended for a stand-alone course on the basic theory of quantum phase transitions, and for self-study. It should be accessible to students in both theory and experiment, after they have taken the core graduate courses on quantum mechanics and statistical mechanics. No prior knowledge of quantum field theory is assumed. Exercises are included at the ends of chapters, drawn from the problem sets of my courses.

After completing Part II, a course can choose from the more advanced topics in Parts III and IV. I recommend a basic survey of the nonzero temperature phase diagram from Chapters 10 and 11. This can be followed by a treatment of Fermi systems drawn from Chapters 17 and 18. Chapters 19 and 20 offer many possibilities for student presentations.

The chapters in the new Parts III and IV have been significantly updated from the first edition. Chapter 16 has a new section on the Fermi gas near unitarity: this was a simple and natural extension of the previous discussion on dilute quantum liquids. These results apply to ultracold atomic systems near a Feshbach resonance. Chapter 17, on Dirac fermions, is entirely new. I took this opportunity to introduce the basics of the theory of unconventional superconductivity induced by antiferromagnetism, as it applies to the cuprates and the pnictides. Dirac fermions also offer a gentle way of introducing non-trivial quantum phase transitions of Fermi systems. Chapter 18, on Fermi liquids and their phase transitions, has been almost completely re-written: this reflects advances in our understanding, and its relevance in many experimental contexts. Chapter 19, on quantum magnetism, has numerous updates to reflect our improved understanding of spin liquids, and a brief discussion of deconfined criticality. However, I have not attempted to cover the many modern developments in quantum magnetism: a more comprehensive starting point is offered by my Solvay lecture [430].

My web site, http://sachdev.physics.harvard.edu, will have updates and corrections.

Acknowledgments

I am very grateful to all the students in my courses for their interest and valuable feedback. The notes of Suzanne Pittman and Jihye Seo were invaluable in writing Chapters 3–8. Gilad Ben-Shach, Thiparat Chotibut, Debanjan Chowdhury, Sean Hartnoll, Yejin Huh, Max Metlitski, and Eun Gook Moon provided very useful feedback on the initial drafts. The treatment of Fermi liquids in Chapter 18 is based on the ideas of Max Metlitski [333, 334].

I thank Simon Capelin, from Cambridge University Press, for guiding both editions over many years.

I thank the Perimeter Institute, Waterloo for hospitality while I was working on the second edition. Finally, I remain grateful to the National Science Foundation for continued support of my research.

PART I

INTRODUCTION

1.1   What is a quantum phase transition?

Consider a Hamiltonian, $H(g)$, whose degrees of freedom reside on the sites of a lattice, and which varies as a function of a dimensionless coupling, g. Let us follow the evolution of the ground state energy of $H(g)$ as a function of g. For the case of a finite lattice, this ground state energy will generically be a smooth, analytic function of g. The main possibility of an exception comes from the case when g couples only to a conserved quantity (i.e. $H(g) = H_0 + gH_1$, where H_0 and H_1 commute). This means that H_0 and H_1 can be simultaneously diagonalized and so the eigenfunctions are independent of g even though the eigenvalues vary with g; then there can be a level-crossing where an excited level becomes the ground state at $g = g_c$ (say), creating a point of nonanalyticity of the ground state energy as a function of g (see Fig. 1.1). The possibilities for an *infinite* lattice are richer. An avoided level-crossing between the ground and an excited state in a finite lattice could become progressively sharper as the lattice size increases, leading to a nonanalyticity at $g = g_c$ in the infinite lattice limit. We shall identify any point of nonanalyticity in the ground state energy of the infinite lattice system as a quantum phase transition: The nonanalyticity could be either the limiting case of an avoided level-crossing or an actual level-crossing. The first kind is more common, but we shall also discuss transitions of the second kind in Chapters 16 and 19. The phase transition is usually accompanied by a qualitative change in the nature of the correlations in the ground state, and describing this change will clearly be one of our major interests.

Actually our focus will be on a limited class of quantum phase transitions – those that are *second order*. Loosely speaking, these are transitions at which the characteristic energy scale of fluctuations above the ground state vanishes as g approaches g_c. Let the energy Δ represent a scale characterizing some significant spectral density of fluctuations at zero temperature (T) for $g \neq g_c$. Thus Δ could be the energy of the lowest excitation above the ground state, if this is nonzero (i.e. there is an energy gap Δ), or if there are excitations at arbitrarily low energies in the infinite lattice limit (i.e. the energy spectrum is *gapless*), Δ is the scale at which there is a qualitative change in the nature of the frequency spectrum from its lowest frequency to its higher frequency behavior. In most cases, we will find that as g approaches g_c, Δ vanishes as

$$\Delta \sim J|g - g_c|^{z\nu}, \tag{1.1}$$

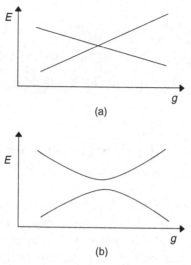

Fig. 1.1 Low eigenvalues, E, of a Hamiltonian $H(g)$ on a finite lattice, as a function of some dimensionless coupling, g. For the case where $H(g) = H_0 + gH_1$, where H_0 and H_1 commute and are independent of g, there can be an actual level-crossing, as in (a). More generally, however, there is an "avoided level-crossing," as in (b).

(exceptions to this behavior appear in Section 20.2.6). Here J is the energy scale of a characteristic microscopic coupling, and $z\nu$ is a *critical exponent*. The value of $z\nu$ is usually *universal*, that is, it is independent of most of the microscopic details of the Hamiltonian $H(g)$ (we shall have much more to say about the concept of universality below, and in the following chapters). The behavior (1.1) holds both for $g > g_c$ and for $g < g_c$ with the same value of the exponent $z\nu$, but with different nonuniversal constants of proportionality. We shall sometimes use the symbol Δ_+ (Δ_-) to represent the characteristic energy scale for $g > g_c$ ($g < g_c$).

In addition to a vanishing energy scale, second-order quantum phase transitions invariably have a diverging characteristic length scale ξ. This could be the length scale determining the exponential decay of equal-time correlations in the ground state or the length scale at which some characteristic crossover occurs to the correlations at the longest distances. This length diverges as

$$\xi^{-1} \sim \Lambda |g - g_c|^{\nu}, \tag{1.2}$$

where ν is a critical exponent, and Λ is an inverse length scale (a "momentum cutoff") of order the inverse lattice spacing. The ratio of the exponents in (1.1) and (1.2) is z, the dynamic critical exponent. The characteristic energy scale vanishes as the zth power of the characteristic inverse length scale

$$\Delta \sim \xi^{-z}. \tag{1.3}$$

It is important to note that the discussion above refers to singularities in the *ground state* of the system. So strictly speaking, quantum phase transitions occur only at zero temperature, $T = 0$. Because all experiments are necessarily at some nonzero, though possibly very small, temperature, a central task of the theory of quantum phase transitions is to describe the consequences of this $T = 0$ singularity on physical properties at $T > 0$. It turns out that working outward from the quantum critical point at $g = g_c$ and $T = 0$ is a powerful way of understanding and describing the thermodynamic and dynamic properties of numerous systems over a broad range of values of $|g - g_c|$ and T. Indeed, it is not even necessary that the system of interest ever have its microscopic couplings reach a value such that $g = g_c$: it can still be very useful to argue that there is a quantum critical point at a physically inaccessible coupling $g = g_c$ and to develop a description in the deviation $|g - g_c|$. It is one of the purposes of this book to describe the physical perspective that such an approach offers, and to contrast it with more conventional expansions about very weak (say $g \to 0$) or very strong couplings (say $g \to \infty$).

1.2 Nonzero temperature transitions and crossovers

Let us now discuss some basic aspects of the $T > 0$ phase diagram. First, let us ask only about the presence of phase transitions at nonzero T. With this limited criterion, there are two important possibilities for the $T > 0$ phase diagram of a system near a quantum critical point. These are shown in Fig. 1.2, and we will meet examples of both kinds in this book. In the first, shown in Fig. 1.2a, the thermodynamic singularity is present only at $T = 0$, and all $T > 0$ properties are analytic as a function of g near $g = g_c$. In the second, shown in

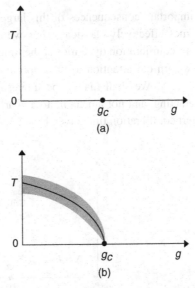

Fig. 1.2   Two possible phase diagrams of a system near a quantum phase transition. In both cases there is a quantum critical point at $g = g_c$ and $T = 0$. In (b), there is a line of $T > 0$ second-order phase transitions terminating at the quantum critical point. The theory of phase transitions in classical systems driven by thermal fluctuations can be applied within the shaded region of (b).

Fig. 1.2b, there is a line of $T > 0$ second-order phase transitions (this is a line at which the thermodynamic free energy is not analytic) that terminates at the $T = 0$ quantum critical point at $g = g_c$.

Moving beyond phase transitions, let us ask some basic questions about the dynamics of the system. A very general way to characterize the dynamics at $T > 0$ is in terms of the thermal equilibration time τ_{eq}. This is the characteristic time in which *local* thermal equilibrium is established after imposition of a weak external perturbation (say, a heat pulse). Here we are excluding equilibration with respect to globally conserved quantities (such as energy or charge) which will take a long time to equilibrate, dependent upon the length scale of the perturbation: hence the emphasis on local equilibration. Global equilibration is described by the equations of hydrodynamics, and we expect such equations to apply in all cases at times much larger than τ_{eq}. We focus here on the value of τ_{eq} as a function of $g - g_c$ and T. From the energy scales discussed in Section 1.1, we can immediately draw an important distinction between two regimes of the phase diagram. We characterized the ground state by the energy Δ in (1.1). At nonzero temperature, we have a second energy scale, $k_B T$. Comparing the values of Δ and $k_B T$, we are immediately led to the important phase diagram in Fig. 1.3. We will see that the two regimes, $\Delta > k_B T$ and $\Delta < k_B T$, are distinguished by different theories of thermal equilibration and of the values of τ_{eq}. In the regime where $\Delta > k_B T$, we will always find long equilibration times which satisfy

$$\tau_{\text{eq}} \gg \frac{\hbar}{k_B T}, \quad \Delta > k_B T. \tag{1.4}$$

One of the important consequences of this large value of τ_{eq} is that the dynamics of the system becomes effectively classical. Thus we can use classical equations of motion to describe the re-equilibration dynamics at the time scale τ_{eq}.

Let us now turn our attention to the important "Quantum Critical" region in Fig. 1.3, where $k_B T > \Delta$. We shall mainly be interested in quantum critical points which are strongly interacting, and not amenable to a nearly-free particle description. In such cases we find a short equilibration time given by

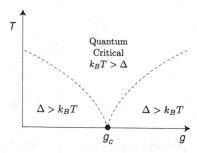

Fig. 1.3 Separation of the phase diagram into distinct regimes determined by the energy scale Δ, which characterizes the ground state, and $k_B T$. The dashed lines are not phase transitions, but smooth crossovers at $T \sim |g - g_c|^{z\nu}$. The phase transition in Fig. 1.2b lies within the $\Delta > k_B T$ region, and is not shown above.

$$\tau_{eq} \sim \frac{\hbar}{k_B T}, \quad k_B T > \Delta. \tag{1.5}$$

Now the equilibration occurs in a time which is actually independent of the microscopic energy scale J, and is determined by $k_B T$ alone. Moreover, and most interestingly, we cannot use an effectively classical description for the re-equilibration at times of order τ_{eq}. Quantum and thermal fluctuations are equally important in the dynamics in the quantum critical region, and developing a theory for this dynamics will be a central focus of Part III.

What about the $T > 0$ phase transition line in Fig. 1.2b? We have not shown this line in Fig. 1.3. Such a transition should be viewed as reflecting the physics of the $\Delta > k_B T$ region, and so the transition line lies in the corresponding region of Fig. 1.3. In other words, this transition is not really a property of the quantum critical point at $g = g_c$, but of the quantum phase at $g < g_c$. (There could also be a separate transition reflecting the physics of the $g > g_c$ phase, which we have not shown in our phase diagrams.) As we move closer to this phase transition line, we will show that not only does τ_{eq} become long, but so do all the time scales associated with long wavelength thermal fluctuations. Indeed we will find that the typical frequency at which the important long-distance degrees of freedom fluctuate, ω_{typ}, satisfies

$$\hbar \omega_{typ} \ll k_B T. \tag{1.6}$$

Under these conditions, it will be seen that a purely *classical* description can be applied to these important degrees of freedom – this classical description works in the shaded region of Fig. 1.2b. Consequently, the ultimate critical singularity along the line of $T > 0$ phase transitions in Fig. 1.2b is described by the theory of second-order phase transitions in classical systems. This theory was developed thoroughly in the past three decades and has been explained in many popular reviews and books [59, 172, 244, 312, 557]. We will discuss the needed basic features of this theory in Chapters 3 and 4. Note that the shaded region of classical behavior in Fig. 1.2b lies within the wider window of the phase diagram, with moderate values of $|g - g_c|$ and T, which we asserted above should be described as an expansion about the quantum critical point at $g = g_c$ and $T = 0$. So our study of quantum phase transitions will also apply to the shaded region of Fig. 1.2b, where it will yield information complementary to that available by directly thinking of the $T > 0$ phase transition in terms of purely classical models.

We note that phase transitions in classical models are driven only by thermal fluctuations, as classical systems usually freeze into a fluctuationless ground state at $T = 0$. In contrast, quantum systems have fluctuations driven by the Heisenberg uncertainty principle even in the ground state, and these can drive interesting phase transitions at $T = 0$. The $T > 0$ region in the vicinity of a quantum critical point therefore offers a fascinating interplay of effects driven by quantum and thermal fluctuations; sometimes, as in the shaded region of Fig. 1.2b, we can find some dominant, effective degrees of freedom whose fluctuations are purely classical and thermal, and then the classical theory will apply. However, as already noted, our attention will not be limited to such regions, and we shall be interested in a broader section of the phase diagram.

1.3 Experimental examples

To make the concepts of the previous sections less abstract, let us mention some experimental studies of simple second-order quantum phase transitions. We will meet numerous other examples in this book, but for now we focus on examples directly related to the canonical theoretical models of quantum phase transitions to be discussed in Section 1.4, and in Parts II and III.

- The low-lying magnetic excitations of the insulator $LiHoF_4$ consist of fluctuations of the Ho ions between two spin states that are aligned parallel and antiparallel to a particular crystalline axis. These states can be represented by a two-state "Ising" spin variable on each Ho ion. At $T = 0$, the magnetic dipolar interactions between the Ho ions cause all the Ising spins to align in the same orientation, and so the ground state is a ferromagnet. Bitko, Rosenbaum, and Aeppli [49] placed this material in a magnetic field transverse to the magnetic axis. Such a field induces quantum tunneling between the two states of each Ho ion, and a sufficiently strong tunneling rate can eventually destroy the long-range magnetic order. Such a quantum phase transition was indeed observed [49], with the ferromagnetic moment vanishing continuously at a quantum critical point. Note that such a transition can, in principle, occur precisely at $T = 0$, when it is driven entirely by quantum fluctuations. We shall call the $T = 0$ state without magnetic order a *quantum paramagnet*. However, we can also destroy the magnetic order at a fixed transverse magnetic field (possibly zero), simply by raising the temperature, enabling the material to undergo a conventional Curie transition to a high-temperature magnetically disordered state. Among the objectives of this book is to provide a description of the intricate crossover between the zero-temperature quantum transition and the finite-temperature transition driven partially by thermal fluctuations; we shall also delineate the important differences between the $T = 0$ quantum paramagnet and the high-temperature "thermal paramagnet;" see Chapters 11, 13, and 14.

  A more recent realization of an Ising model in a transverse field has appeared in experiments by Coldea and collaborators [90] on crystals of $CoNb_2O_6$, which belongs to the columbite family of minerals. In this case, the Ising spin resides on the Co^{++} ion, again aligned by the spin–orbit interaction to orient parallel or anti-parallel to a crystalline axis. An important difference from $LiHoF_4$ is that the interactions between the spins are essentially nearest-neighbor, and the long-range dipolar couplings are unimportant; the short-range interactions arise from the Heisenberg exchange process, and their energy scale is determined by the electrostatic Coulomb interactions. Thus $CoNb_2O_6$ provides a nearly ideal realization of the quantum Ising models which will be the focus of our study in Parts II and III. The dominant exchange couplings are along a particular crystalline axis, and so it is also a useful testing ground for exact results in one dimension.

- Experiments on ultracold atoms in optical lattices by Greiner, Bloch, and collaborators [175] have provided a celebrated example of the superfluid–insulator quantum phase transition. Atoms of ^{87}Rb are cooled to temperatures so low that their quantum statistics is important. These atoms are bosons and so they ultimately Bose condense into a superfluid state. Then, by applying a periodic potential on the atoms by an optical lattice, Greiner *et al.* localized the atoms in the minima of the periodic potential, leading to a quantum phase transition to an insulating state. At densities where the number of atoms is commensurate with the number of minima of the periodic potential, this transition is described by the O(2) quantum rotor model, which we introduce in Section 1.4 and discuss at length in Parts II and III.

- $TlCuCl_3$ is an insulator whose only low-lying electronic excitations are rotations of the $S = 1/2$ spins residing on the Cu^{++} ions. Unlike the case for the Co^{++} ions in $CoNb_2O_6$, the spin-orbit interactions are relatively weak on Cu^{++}, and a single spin can freely orient along any direction in spin space. A special feature of the crystal structure of $TlCuCl_3$ is that the Cu atoms are naturally dimerized, i.e. each Cu site has a single partner Cu site, and the exchange interactions are strongest between the partners in each pair. The exchange interaction has an antiferromagnetic sign, and consequently neighboring spins prefer to be oriented in anti-parallel directions. Under ambient pressure, each Cu spin forms a singlet valence bond with its partner, much like that between the two electrons in a hydrogen molecule. Thus although the neighboring spins within a dimer are always anti-parallel, they fluctuate along all directions in spin space in a rotationally invariant manner. We will refer to this state as a quantum paramagnet; it has an energy gap to all excitations above the ground state. Under applied pressure, $TlCuCl_3$ undergoes a quantum phase transition [414] to an ordered antiferromagnet: a Néel state. In this Néel state, the spins freeze into a definite orientation so that nearby spins are anti-parallel to each other. Such an arrangement is more nearly optimal when the exchange couplings between spins in different dimers are significant. As we discuss below in Section 1.4, this transition between the quantum paramagnet and the Néel state is described by the O(3) quantum rotor model, which will also be discussed in Parts II and III.

1.4 Theoretical models

Our strategy in this book will be to thoroughly analyze the physical properties of quantum phase transitions in two simple theoretical model systems in Parts II and III: the quantum Ising and rotor models. Fortunately, these simple models also have direct experimental realizations in the systems already surveyed in Section 1.3. Below, we introduce the quantum Ising and rotor models in turn, discussing the nature of the quantum phase transitions in them, and relating them to the experimental systems above. Other experimental connections will be discussed in subsequent chapters.

Part IV will survey some important quantum phase transitions in other models of physical interest. Our motivation in dividing the discussion in this manner is mainly pedagogical: the quantum transitions of the Ising/rotor models have an essential simplicity, but their behavior is rich enough to display most of the basic phenomena we wish to explore. It will therefore pay to meet the central physical ideas in this simple context first.

1.4.1 Quantum Ising model

We begin by writing down the Hamiltonian of the quantum Ising model. It is

$$H_I = -Jg \sum_i \hat{\sigma}_i^x - J \sum_{\langle ij \rangle} \hat{\sigma}_i^z \hat{\sigma}_j^z. \tag{1.7}$$

As in the general notation introduced above, $J > 0$ is an exchange constant, which sets the microscopic energy scale, and $g > 0$ is a dimensionless coupling, which will be used to tune H_I across a quantum phase transition. The quantum degrees of freedom are represented by operators $\hat{\sigma}_i^{z,x}$, which reside on the sites, i, of a hypercubic lattice in d dimensions; the sum $\langle ij \rangle$ is over pairs of nearest-neighbor sites i, j. The $\hat{\sigma}_i^{x,z}$ are the familiar Pauli matrices; the matrices on different sites i act on different spin states, and so matrices with $i \neq j$ commute with each other. In the basis where the $\hat{\sigma}_i^z$ are diagonal, these matrices have the well-known form

$$\hat{\sigma}^z = \begin{pmatrix} 1 & 0 \\ 0 & -1 \end{pmatrix}, \quad \hat{\sigma}^y = \begin{pmatrix} 0 & -i \\ i & 0 \end{pmatrix}, \quad \hat{\sigma}^x = \begin{pmatrix} 0 & 1 \\ 1 & 0 \end{pmatrix}, \tag{1.8}$$

on each site i. We will denote the eigenvalues of $\hat{\sigma}_i^z$ simply by σ_i^z, and so σ_i^z takes the values ± 1. We identify the two states with eigenvalues $\sigma_i^z = +1, -1$ as the two possible orientations of an "Ising spin," which can be oriented up or down in $|\uparrow\rangle_i, |\downarrow\rangle_i$. Consequently at $g = 0$, when H_I involves only the $\hat{\sigma}_i^z$, H_I will be diagonal in the basis of eigenvalues of $\hat{\sigma}_i^z$, and it reduces simply to the familiar classical Ising model. However, the $\hat{\sigma}_i^x$ are off-diagonal in the basis of these states, and therefore they induce quantum-mechanical tunneling events that flip the orientation of the Ising spin on a site. The physical significance of the two terms in H_I should be clear in the context of our earlier discussion in Section 1.3 for LiHoF$_4$ and CoNb$_2$O$_6$. The term proportional to J is the magnetic interaction between the spins, which prefers their global ferromagnetic alignment. While the interaction in LiHoF$_4$ has a long-range dipolar nature, that in CoNb$_2$O$_6$ has a nearest-neighbor form like that in (1.7). The term proportional to Jg is the applied external transverse magnetic field, which disrupts the magnetic order.

Let us make these qualitative considerations somewhat more precise. The ground state of H_I can depend only upon the value of the dimensionless coupling g, and so it pays to consider the two opposing limits $g \gg 1$ and $g \ll 1$.

First consider $g \gg 1$. In this case the first term in (1.7) dominates, and, to leading order in $1/g$, the ground state is simply

$$|0\rangle = \prod_i |\rightarrow\rangle_i, \tag{1.9}$$

where

$$|\rightarrow\rangle_i = (|\uparrow\rangle_i + |\downarrow\rangle_i)/\sqrt{2},$$
$$|\leftarrow\rangle_i = (|\uparrow\rangle_i - |\downarrow\rangle_i)/\sqrt{2}, \tag{1.10}$$

are the two eigenstates of $\hat{\sigma}_i^x$ with eigenvalues ± 1. The values of σ_i^z on different sites are totally uncorrelated in the state (1.9), and so $\langle 0|\hat{\sigma}_i^z\hat{\sigma}_j^z|0\rangle = \delta_{ij}$. Perturbative corrections in

$1/g$ will build in correlations in σ^z that increase in range at each order in $1/g$; for g large enough these correlations are expected to remain short-ranged, and we expect in general that

$$\langle 0|\hat{\sigma}_i^z \hat{\sigma}_j^z|0\rangle \sim e^{-|x_i - x_j|/\xi} \qquad (1.11)$$

for large $|x_i - x_j|$, where x_i is the spatial coordinate of site i, $|0\rangle$ is the exact ground state for large-g, and ξ is the "correlation length" introduced above (1.2).

Next we consider the opposing limit $g \ll 1$. We will find that the nature of the ground state is qualitatively different from the large-g limit above, and we shall use this to argue that there must be a quantum phase transition between the two limiting cases at a critical $g = g_c$ of order unity. For $g \ll 1$, the second term in (1.7) coupling neighboring sites dominates; at $g = 0$ the spins are either all up or all down (in eigenstates of σ^z):

$$|{\uparrow}\rangle = \prod_i |{\uparrow}\rangle_i \quad \text{or} \quad |{\downarrow}\rangle = \prod_i |{\downarrow}\rangle_i. \qquad (1.12)$$

Turning on a small-g will mix in a small fraction of spins of the opposite orientation, but in an infinite system the degeneracy will survive at any finite order in a perturbation theory in g. This is because there is an exact global Z_2 symmetry transformation (generated by the unitary operator $\prod_i \sigma_i^x$), which maps the two ground states into each other, under which H_I remains invariant:

$$\hat{\sigma}_i^z \to -\hat{\sigma}_i^z, \quad \hat{\sigma}_i^x \to -\hat{\sigma}_i^x, \qquad (1.13)$$

and there is no tunneling matrix element between the majority up and down spin sectors of the infinite system at any finite order in g. The mathematically alert reader will note that establishing the degeneracy to all orders in g, is not the same thing as establishing its existence for any small nonzero g, but more sophisticated considerations show that this is indeed the case. A thermodynamic system will always choose one or other of the states as its ground states (which may be preferred by some infinitesimal external perturbation), and this is commonly referred to as a "spontaneous breaking" of the Z_2 symmetry. As in the large-g limit, we can characterize the ground states by the behavior of correlations of $\hat{\sigma}_i^z$; the nature of the states (1.12) and small-g perturbation theory suggest that

$$\lim_{|x_i - x_j| \to \infty} \langle 0|\hat{\sigma}_i^z \hat{\sigma}_j^z|0\rangle = N_0^2, \qquad (1.14)$$

where $|0\rangle$ is either of the ground states obtained from $|{\uparrow}\rangle$ or $|{\downarrow}\rangle$ by perturbation theory in g, and $N_0 \neq 0$ is the "spontaneous magnetization" of the ground state. This identification is made clearer by the simpler statement

$$\langle 0|\hat{\sigma}_i^z|0\rangle = \pm N_0, \qquad (1.15)$$

which also follows from the perturbation theory in g. We have $N_0 = 1$ for $g = 0$, but quantum fluctuations at small-g reduce N_0 to a smaller, but nonzero, value.

Now we make the simple observation that it is not possible for states that obey (1.11) and (1.14) to transform into each other analytically as a function of g. There must be a critical value $g = g_c$ at which the large $|x_i - x_j|$ limit of the two-point correlator changes from (1.11) to (1.14) – this is the position of the quantum phase transition, which is the focus of intensive study in this book. Our arguments so far do not exclude the possibility that there could be more than one critical point, but this is known not to happen for H_I, and we will assume here that there is only one critical point at $g = g_c$. For $g > g_c$ the ground state is, as noted earlier, a *quantum paramagnet*, and (1.11) is obeyed. We will find that as g approaches g_c from above, the correlation length, ξ, diverges as in (1.2). Precisely at $g = g_c$, neither (1.11) nor (1.14) is obeyed, and we find instead a power-law dependence on $|x_i - x_j|$ at large distances. The result (1.14) holds for all $g < g_c$, when the ground state is *magnetically ordered*. The spontaneous magnetization of the ground state, N_0, vanishes as a power law as g approaches g_c from below.

Finally, we make a comment about the excited states of H_I. In a finite lattice, there is necessarily a nonzero energy separating the ground state and the first excited state. However, this energy spacing can either remain finite or approach zero in the infinite lattice limit, the two cases being identified as having a gapped or gapless energy spectrum, respectively. We will find that there is an energy gap Δ that is nonzero for all $g \neq g_c$, but that it vanishes upon approaching g_c as in (1.1), producing a gapless spectrum at $g = g_c$.

1.4.2 Quantum rotor model

We turn to the somewhat less familiar quantum rotor models. Elementary quantum rotors do not exist in nature; rather, each quantum rotor is an effective quantum degree of freedom for the low-energy states of a small number of electrons or atoms. We will first define the quantum mechanics of a single rotor and then turn to the lattice quantum rotor model. The connection to the experimental models introduced in Section 1.3 is described below in Section 1.4.3. Further details of this connection appear in Chapters 9 and 19.

Each rotor can be visualized as a particle constrained to move on the surface of a (fictitious) $(N > 1)$-dimensional sphere. The orientation of each rotor is represented by an N-component unit vector $\hat{\mathbf{n}}_i$ which satisfies

$$\hat{\mathbf{n}}_i^2 = 1. \tag{1.16}$$

The caret on $\hat{\mathbf{n}}_i$ reminds us that the orientation of the rotor is a quantum mechanical operator, while i represents the site on which the rotor resides; we will shortly consider an infinite number of such rotors residing on the sites of a d-dimensional lattice. Each rotor has a momentum $\hat{\mathbf{p}}_i$, and the constraint (1.16) implies that this must be tangential to the surface of the N-dimensional sphere. The rotor position and momentum satisfy the usual commutation relations

$$[\hat{n}_\alpha, \hat{p}_\beta] = i\delta_{\alpha\beta} \tag{1.17}$$

on each site i; here $\alpha, \beta = 1 \ldots N$. (Here, and in the remainder of the book, we will always measure time in units in which

$$\hbar = 1, \tag{1.18}$$

unless stated explicitly otherwise. This is also a good point to note that we will also set Boltzmann's constant

$$k_B = 1 \qquad (1.19)$$

by absorbing it into the units of temperature, T.) We will actually find it more convenient to work with the $N(N-1)/2$ components of the rotor angular momentum

$$\hat{L}_{\alpha\beta} = \hat{n}_\alpha \hat{p}_\beta - \hat{n}_\beta \hat{p}_\alpha. \qquad (1.20)$$

These operators are the generators of the group of rotations in N dimensions, denoted $O(N)$. Their commutation relations follow straightforwardly from (1.17) and (1.20). The case $N = 3$ will be of particular interest to us. For this we define $\hat{L}_\alpha = (1/2)\epsilon_{\alpha\beta\gamma}L_{\beta\gamma}$ (where $\epsilon_{\alpha\beta\gamma}$ is a totally antisymmetric tensor with $\epsilon_{123} = 1$), and then the commutation relations between the operators on each site are

$$[\hat{L}_\alpha, \hat{L}_\beta] = i\epsilon_{\alpha\beta\gamma}\hat{L}_\gamma,$$
$$[\hat{L}_\alpha, \hat{n}_\beta] = i\epsilon_{\alpha\beta\gamma}\hat{n}_\gamma,$$
$$[\hat{n}_\alpha, \hat{n}_\beta] = 0; \qquad (1.21)$$

the operators with different site labels all commute.

The dynamics of each rotor is governed simply by its kinetic energy term; interesting effects arise from potential energy terms that couple the rotors together, and these will be considered in a moment. Each rotor has the kinetic energy

$$H_K = \frac{J\tilde{g}}{2}\hat{\mathbf{L}}^2, \qquad (1.22)$$

where $1/J\tilde{g}$ is the rotor moment of inertia (we have put a tilde over g as we wish to reserve g for a different coupling to be introduced below). The Hamiltonian H_K can be readily diagonalized for general values of N by well-known group theoretical methods. We quote the results for the physically important cases of $N = 2$ and 3. For $N = 2$ the eigenvalues are

$$J\tilde{g}\ell^2/2 \quad \ell = 0, 1, 2, \ldots; \quad \text{degeneracy} = 2 - \delta_{\ell,0}. \qquad (1.23)$$

Note that there is a nondegenerate ground state with $\ell = 0$, while all excited states are two-fold degenerate, corresponding to a left- or right-moving rotor. This spectrum will be important in the mapping to physical models to be discussed in Section 1.4.3. For $N = 3$, the eigenvalues of H_K are

$$J\tilde{g}\ell(\ell + 1)/2 \quad \ell = 0, 1, 2, \ldots; \quad \text{degeneracy} = 2\ell + 1, \qquad (1.24)$$

corresponding to the familiar angular momentum states in three dimensions. These states can be viewed as representing the eigenstates of an even number of antiferromagnetically coupled Heisenberg spins, as discussed more explicitly in Section 1.4.3 and in Chapter 19, where we will see that there is a general and powerful correspondence between quantum antiferromagnets and $N = 3$ rotors.

We are ready to write down the full quantum rotor Hamiltonian, which will be the focus of intensive study in Parts II and III. We place a single quantum rotor on the sites, i, of a d-dimensional lattice, obeying the Hamiltonian

$$H_R = \frac{J\tilde{g}}{2} \sum_i \hat{\mathbf{L}}_i^2 - J \sum_{\langle ij \rangle} \hat{\mathbf{n}}_i \cdot \hat{\mathbf{n}}_j. \tag{1.25}$$

We have augmented the sum of kinetic energies of each site with a coupling, J, between rotor orientations on neighboring sites. This coupling energy is minimized by the simple "magnetically ordered" state in which all the rotors are oriented in the same direction. In contrast, the rotor kinetic energy is minimized when the orientation of the rotor is maximally uncertain (by the uncertainty principle), and so the first term in H_R prefers a quantum paramagnetic state in which the rotors do not have a definite orientation (i.e. $\langle \mathbf{n} \rangle = 0$). Thus the roles of the two terms in H_R closely parallel those of the terms in the Ising model H_I. As in Section 1.4.1, for $\tilde{g} \gg 1$, when the kinetic energy dominates, we expect a quantum paramagnet in which, following (1.11),

$$\langle 0 | \hat{\mathbf{n}}_i \cdot \hat{\mathbf{n}}_j | 0 \rangle \sim e^{-|x_i - x_j|/\xi}. \tag{1.26}$$

Similarly, for $\tilde{g} \ll 1$, when the coupling term dominates, we expect a magnetically ordered state in which, as in (1.14),

$$\lim_{|x_i - x_j| \to \infty} \langle 0 | \hat{\mathbf{n}}_i \cdot \hat{\mathbf{n}}_j | 0 \rangle = N_0^2. \tag{1.27}$$

Finally, we can anticipate a second-order quantum phase transition between the two phases at $\tilde{g} = \tilde{g}_c$, and the behavior of N_0 and ξ upon approaching this point will be similar to that in the Ising case. These expectations turn out to be correct for $d > 1$, but we will see that they need some modifications for $d = 1$. In one dimension, we will show that $\tilde{g}_c = 0$ for $N \geq 3$, and so the ground state is a quantum paramagnetic state for all nonzero $\tilde{g}$. The case $N = 2$, $d = 1$ is special: there is a transition at a finite $\tilde{g}_c$, but the divergence of the correlation length does not obey (1.2) and the long-distance behavior of the correlation function $\tilde{g} < \tilde{g}_c$ differs from (1.27). This case will not be considered until Section 20.3 in Part IV.

1.4.3 Physical realizations of quantum rotors

We will consider the $N = 3$ quantum rotors first, and expose a simple and important connection between O(3) quantum rotor models and a certain class of "dimerized" antiferromagnets, of which TlCuCl$_3$ is the example we highlighted in Section 1.3. Actually the connection between rotor models and antiferromagnets is far more general than the present discussion may suggest, as we see later in Chapter 19. However, this discussion should enable the reader to gain an intuitive feeling for the physical interpretation of the degrees of freedom of the rotor model.

Consider a dimerized system of "Heisenberg spins" $\hat{\mathbf{S}}_{1i}$ and $\hat{\mathbf{S}}_{2i}$, where i now labels a pair of spins (a "dimer"). Their Hamiltonian is

$$H_d = K \sum_i \hat{\mathbf{S}}_{1i} \cdot \hat{\mathbf{S}}_{2i} + J \sum_{\langle ij \rangle} \left(\hat{\mathbf{S}}_{1i} \cdot \hat{\mathbf{S}}_{1j} + \hat{\mathbf{S}}_{2i} \cdot \hat{\mathbf{S}}_{2j} \right). \tag{1.28}$$

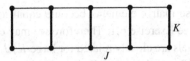

Fig. 1.4 A dimerized quantum spin system. Spins with angular momentum S reside on the circles, with antiferromagnetic exchange couplings as shown.

The $\hat{\mathbf{S}}_{ni}$ ($n = 1, 2$ labels the spins within a dimer) are spin operators usually representing the total spin of a set of electrons in some localized atomic states, see Fig. 1.4. On each site, the spins $\hat{\mathbf{S}}_{ni}$ obey the angular momentum commutation relations

$$\left[\hat{S}_\alpha, \hat{S}_\beta\right] = i\epsilon_{\alpha\beta\gamma}\hat{S}_\gamma \tag{1.29}$$

(the site index has been dropped above), while spin operators on different sites commute. These commutation relations are the same as those of the $\hat{\mathbf{L}}$ operators in (1.21). However, there is one crucial difference between the Hilbert spaces of states acted on by the quantum rotors and the Heisenberg spins. For the rotor models we allowed states with arbitrary total angular momentum ℓ on each site, as in (1.24), and so there were an infinite number of states on each site. For the present Heisenberg spins, however, we will only allow states with total spin S on each site, and we will permit S to be integer or half-integer. Thus there are precisely $2S + 1$ states on each site

$$|S, m\rangle \quad \text{with } m = -S \ldots S, \tag{1.30}$$

and the operator identity

$$\hat{\mathbf{S}}_{ni}^2 = S(S + 1) \tag{1.31}$$

holds for each i and n. In addition to describing TlCuCl$_3$, Hamiltonians like H_d describe spin-ladder compounds in $d = 1$ [33, 102] and "double layer" antiferromagnets in the family of the high-temperature superconductors in $d = 2$ [129,322,337,443,444,506,507].

Let us examine the properties of H_d in the limit $K \gg J$. As a first approximation, we can neglect the J couplings entirely, and then H_d splits into decoupled pairs of sites, each with a strong antiferromagnetic coupling K between two spins. The Hamiltonian for each pair can be diagonalized by noting that $\mathbf{S}_{1i}$ and $\mathbf{S}_{2i}$ couple into states with total angular momentum $0 \leq \ell \leq 2S$, and so we obtain the eigenenergies

$$(K/2)(\ell(\ell + 1) - 2S(S + 1)), \qquad \text{degeneracy } 2\ell + 1. \tag{1.32}$$

Note that these energies and degeneracies are in one-to-one correspondence with those of a single quantum rotor in (1.24), apart from the difference that the upper restriction on ℓ being smaller than $2S$ is absent in the rotor model case. If one is interested primarily in low-energy properties, then it appears reasonable to represent each pair of spins by a quantum rotor.

We have seen that the $K/J \to \infty$ limit of H_d closely resembles the $\tilde{g} \to \infty$ limit of H_R. To first order in $\tilde{g}$, we can compare the matrix elements of the term proportional to J in H_R among the low-lying states, with those of the J term in H_d; it is not difficult to see that these matrix elements become equal to each other for an appropriate choice of couplings: see Exercise 6.1. Therefore we may conclude that the low-energy properties of the two models are closely related for large K/J and $\tilde{g}$. Somewhat different considerations in Chapter 19 will show that the correspondence also applies to the quantum critical point and to the magnetically ordered phase.

The main lesson of the above analysis is that the O(3) quantum rotor model represents the low-energy properties of quantum antiferromagnets of Heisenberg spins, with each rotor being an effective representation of a *pair* of antiferromagnetically coupled spins. The strong-coupling spectra clearly indicate the operator correspondence $\hat{\mathbf{L}}_i = \hat{\mathbf{S}}_{1i} + \hat{\mathbf{S}}_{2i}$, and so the rotor angular momentum represents the total angular momentum of the underlying spin system. Examination of matrix elements in the large-S limit shows that $\hat{\mathbf{n}}_i \propto \hat{\mathbf{S}}_{1i} - \hat{\mathbf{S}}_{2i}$: the rotor coordinate $\hat{\mathbf{n}}_i$ is the antiferromagnetic order parameter of the spin system. Magnetically ordered states of the rotor model with $\langle \hat{\mathbf{n}}_i \rangle \neq 0$, which we will encounter below, are therefore spin states with long-range antiferromagnetic order and have a vanishing total ferromagnetic moment. Quantum Heisenberg spin systems with a net ferromagnetic moment are *not* modeled by the quantum rotor model (11.1) – these will be studied in Section 19.2 by a different approach.

Let us now consider the $N = 2$ quantum rotors, and introduce their connection to the superfluid–insulator transition of bosons. For $N = 2$, it is useful to introduce an angular variable θ_i on each site, so that

$$\mathbf{n}_i = (\cos\theta_i, \sin\theta_i). \tag{1.33}$$

The rotor angular momentum has only one component, which can be represented in the Schrödinger picture as the differential operator

$$\hat{L}_i = \frac{1}{i}\frac{\partial}{\partial\theta_i} \tag{1.34}$$

acting on a wavefunction which depends on all the θ_i. The rotor Hamiltonian is therefore

$$H_R = -\frac{J\tilde{g}}{2}\sum_i \frac{\partial^2}{\partial\theta_i^2} - J\sum_{\langle ij \rangle}\cos(\theta_i - \theta_j), \tag{1.35}$$

which is a form that has appeared in numerous studies in different physical contexts. For $\tilde{g} \to \infty$, the eigenstates of H_R are of the form $\prod_i |m_i\rangle$, where m_i is the integer angular momentum quantum number of site i; in the Schrödinger form, these states have the wavefunction $\exp\left(i\sum_i m_i\theta_i\right)$. Now we interpret m_i as the *change* in occupation number of a boson trapped in a potential which has its minimum at site i. The boson could be an ultracold ^{87}Rb atom, or a Cooper pair in a superconducting quantum dot, as illustrated in Fig. 1.5. The occupation number is measured with respect to a "background" number of bosons found in the insulator, and hence m_i can take negative values whose absolute value does not exceed this number. In the rotor model, m_i can run all the way to $-\infty$, but as in the $N = 3$ case, we do not expect these additional high-energy states to be important for low-energy physics.

From the wavefunction of these localized boson states, we see that the term proportional to J in (1.35) has the effect of shifting nearest-neighbor pairs of angular momenta

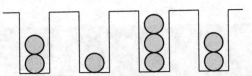

Bosons hopping on a lattice with potential minima at site i. Relative to an insulator with 2 bosons on each site, the state shown has boson numbers $m_i = (\ldots 0, -1, 1, 0 \ldots)$.

as $m_i \to m_i \pm 1$, $m_j \to m_j \mp 1$. In other words, bosons tunnel between sites i and j with matrix element $-J$. Such a tunneling event becomes more probable when the optical lattice potential is weak (*i.e.* $\tilde{g}$ is small), and strong tunneling eventually induces a transition to the superfluid state where (1.27) is obeyed. We will see later in Chapter 8 why (1.27) implies superfluidity of the bosons. More details on the connection between $N = 2$ quantum rotors and quantum boson models appear in Chapter 9.

Having introduced our key players, the quantum Ising model (1.7) and the quantum rotor model (1.25), we outline here the general strategy followed in describing their physical properties in Parts II and III. We introduce the idea of a continuum limit, and the classical and quantum field theories we will study. We also highlight some key questions in the theory of quantum phase transitions, towards which much of the subsequent discussion is directed.

A central concept which will play a fundamental role in our analysis is the connection between $(D > 1)$-dimensional classical statistical mechanical models and the d-dimensional quantum Ising and rotor models introduced in Chapter 1, where

$$D = d + 1. \tag{2.1}$$

This mapping is not an exact equivalence in general, but does become quantitatively precise in the vicinity of continuous phase transitions, as we discuss below. The nature of this general quantum–classical mapping will be discussed and its limitations and utility will be highlighted. As we noted at the beginning of Chapter 1, the present quantum–classical mapping should not be confused with the d-dimensional classical physics of d-dimensional quantum models in the vicinity of $T > 0$ phase transitions, as in the shaded region of Fig. 1.2.

We set the stage by simply writing down the D-dimensional classical statistical mechanical models. For the quantum Ising case (which we often refer to as the $N = 1$ case, because the order parameter has a single component), we consider the classical Ising partition function

$$\mathcal{Z} = \sum_{\{\sigma_i^z = \pm 1\}} \exp\left(K \sum_{\langle i,j \rangle} \sigma_i^z \sigma_j^z \right), \tag{2.2}$$

where K is a dimensionless coupling that characterizes the "temperature" of the classical problem, and the sum is over all 2^M possible configurations of Ising spins in a system of M sites in D dimensions. For $N > 1$, we have the classical O(N) spin model

$$\mathcal{Z} = \prod_i \int D\mathbf{n}_i \delta\left(\mathbf{n}_i^2 - 1 \right) \exp\left(K \sum_{\langle i,j \rangle} \mathbf{n}_i \cdot \mathbf{n}_j \right), \tag{2.3}$$

where $\mathbf{n}_i$ is a $N \geq 2$ component unit vector on the sites, i, of a hypercubic lattice in D dimensions.

Our claim is that the above classical partition functions are "equal" (in a sense to be made precise in Part II) to the partition functions of the quantum Ising and rotor models of Chapter 1:

$$\mathcal{Z} \sim \text{Tr} \exp \left(-\frac{H_{I,R}}{k_B T} \right). \tag{2.4}$$

It is important to note that the temperature, T, of the quantum models has no connection to the inverse temperature, K, of the classical models. Instead, as we will see, K determines the value of the dimensionless coupling g in the quantum Ising and rotor models.

Before we can explain the "classical" interpretation of T, we need to describe the quantum–classical mapping more precisely. We will show in Part II how the quantum partition function in (2.4) can be written in a Feynman "sum-over-histories." In this picture, we evolve the quantum states forward in imaginary time, τ, using the Heisenberg imaginary-time evolution operator, $\exp(-H_{I,R}\tau)$. We then see that it is useful to take a spacetime point of view, in which τ is viewed as another dimension, along with the d spacetime dimensions. In this manner, we obtain a partition function which is to be evaluated in D spacetime dimensions, which will turn out to be the models in (2.2) and (2.3). This connection is illustrated in Fig. 2.1.

Now we see from (2.4) that the quantum partition function is equivalent to an imaginary time evolution over a length L_τ, given by

$$L_\tau = \frac{\hbar}{k_B T} \tag{2.5}$$

(momentarily inserting factors of $\hbar$ and k_B). Thus the temperature T in the quantum model $H_{I,R}$ maps to a *finite size* in the classical models (2.2) and (2.3). Because the quantum trace in (2.4) involves the same initial and final traces, periodic boundary conditions are imposed in the classical model along the τ direction. More formally stated, a quantum model defined on a d-dimensional space R^d maps onto a classical model on $R^d \times S^1$, where the circle S^1 has circumference L_τ. In particular, the classical model in infinite D-dimensional spacetime maps onto a quantum model at *zero* temperature.

The above discussion gives a qualitative and intuitive picture of the mapping, but it is not numerically precise, as it glossed over the limit of temporal lattice spacing $a \to 0$ we will

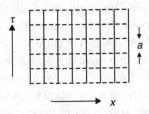

Fig. 2.1 D-dimensional lattice on which (2.2) and (2.3) are defined. The spatial coordinate x is a schematic for $d = D - 1$ directions. The vertical co-ordinate is imaginary time, τ, and the quantum model evolves forward by $\exp(-H_{I,R}a)$, where a is the "distance" between neighboring rows. The total length of the time coordinate is $L_\tau = \hbar/(k_B T)$, and periodic boundary conditions are imposed along the time coordinate.

need to take. As we outline below, and discuss in more detail in Part II, mapping becomes numerically precise in the vicinity of phase transitions.

The models (2.2) and (2.3) are central to the theory of finite-temperature phase transitions in classical statistical mechanics. We will review their basic properties in Chapters 3 and 4. For all values of N in $D > 2$, and for $N = 1, 2$ in $D = 2$, these models display a phase transition between a low "temperature" magnetically ordered phase for $K > K_c$ and a high "temperature" disordered phase for $K < K_c$. These phases are characterized by correlations of the order parameter σ^z, $\mathbf{n}$ in a manner closely analogous to the magnetically ordered and quantum paramagnetic phases of Chapter 1. So in the $K < K_c$ disordered phase we have, as in (1.26),

$$\langle \hat{\mathbf{n}}_i \cdot \hat{\mathbf{n}}_j \rangle \sim e^{-|x_i - x_j|/\xi}, \tag{2.6}$$

for large $|x_i - x_j|$, where the average is with respect to the classical partition function (3.2) and x_i is a D-dimensional coordinate. Similarly, for $K > K_c$ we have, in (1.27),

$$\lim_{|x_i - x_j| \to \infty} \langle 0 | \hat{\mathbf{n}}_i \cdot \hat{\mathbf{n}}_j | 0 \rangle = N_0^2, \tag{2.7}$$

where N_0 is the spontaneous magnetization (this does not apply to the special case $D = 2$, $N = 1$, where the behavior for $K > K_c$ will be discussed in Section 20.3). Similar results hold for the $N = 1$ case with the variable σ^z. Upon approaching K_c, N_0 vanishes as a power law, and ξ diverges as

$$\xi^{-1} \sim a|K - K_c|^\nu, \tag{2.8}$$

with ν a critical exponent. Again, an exception to this is the case $N = 2$, $D = 2$ where the divergence of ξ has a different form. Also for the cases $N > 2$, $D = 2$ there is no phase transition at any finite K, but there is a diverging correlation length for $K \to \infty$, and most of the considerations below apply to these cases as well.

An important consequence of the divergence of the correlation lengths (2.8) and (1.2) near the phase transition in both the classical and quantum models is that of *universality*. This is the claim that most microscopic details of the lattice models do not modify the essential structure of the corrections in the critical region at length scales of order ξ. With $\xi \gg a$, a lattice spacing, it seems reasonable that fluctuations of individual spins on the lattice scale do not matter in their details, and some "renormalized" theory is important at scales ξ and larger. This argument can also be made using energy scales of the quantum model, in which case the requirement of universality is that $\Delta \ll J$. We provide a specific justification of the hypothesis of universality using the renormalization group in Chapter 4.

We can now make a more precise statement of quantum–classical mapping. The universal properties of the d-dimensional quantum Ising and rotor models in their region of large correlation length are identical to those of the D-dimensional classical models (2.2) and (2.3). Further, correlators of the classical model in D dimensions map onto imaginary time correlators of the d-dimensional quantum model, where one of the classical D dimensions behaves like the quantum imaginary time direction, and the remaining $D - 1$ classical directions map onto the d spatial directions of the quantum model. The mapping has an immediate consequence: as the quantum imaginary time direction is simply one of

the spatial directions of the classical model, we compare (2.8) with (1.1) and (1.2) and conclude that we must have the dynamic exponent $z = 1$ for the quantum Ising/rotor models.

Having identified the appropriate universal limit of the quantum models, it is appropriate
to ask: what is the quantum theory that describes these universal properties? These turn out
to be continuum quantum field theories, which are introduced in the following section.

2.1 Quantum field theories

The following discussion will be carried out in the language of the quantum Ising and rotor
models. However, essentially the same arguments can also be made for the classical models
(2.2) and (2.3), as we will see in Chapters 3 and 4.

Let us consider the regime where $|g - g_c|$ is small, so that

$$\Delta \ll J \text{ and } \xi^{-1} \ll \Lambda. \tag{2.9}$$

Suppose, further, that we are observing the system at a temperature T, a length scale x,
and a frequency scale ω, and all of these are of the order of the temperature, length, and
energy scales that can be created out of Δ, ξ, and the fundamental constants. We will
then be particularly interested in dynamic response functions of the system near a quantum
critical point in the limit where the inequalities (2.9) are well satisfied. From a particle
theorist's perspective, this means we are taking the limits $\Lambda \to \infty$ and $J \to \infty$ while
keeping Δ, ξ, x, ω, and T fixed. In terms of dimensionless parameters, this means we
are sending $\Lambda\xi \to \infty$ and $J/\Delta \to \infty$, while keeping $\hbar\omega/\Delta$, x/ξ, and $k_B T/\Delta$ fixed. A
glance at (1.1) and (1.2) shows that these limits can only be taken while tuning g to become
progressively closer to g_c. The complementary condensed matter theorist's perspective
is that we are keeping Λ and J fixed and looking at the system's response at small Δ,
large ξ, and at long distances and times and low temperatures; the two approaches are
clearly equivalent as the limits of the dimensionless ratios are the same. The resulting
response functions can be considered to be correlators of a *quantum field theory*, which is
now associated with a Hamiltonian defined in the continuum and has no intrinsic short-
distance or high-energy cutoff. A quantum field theory shares many of the characteristics
of ordinary quantum mechanics, with a unitary time evolution operator defined by the
continuum Hamiltonian, except that it has an infinite number of degrees of freedom per
unit volume.

The physical utility of the quantum field theory relies mainly on its *universality*. As we
have sent $\Lambda \to \infty$ and $J \to \infty$, it appears plausible that changes in the structure of $H(g)$
at the lattice scale will not modify the nature of the quantum field theory that eventually
appears, and the only consequence is a change in the values of the dimensionful parameters
Δ and ξ (this change results from modifications of the prefactors in (1.1) and (1.2), which,
as we have already asserted, are nonuniversal). A general rule of thumb is that only essential qualitative features, such as the symmetry of the order parameter, the dimensionality
of space, and constraints placed by conservation laws, survive the continuum limit, and the
structure of the quantum field theory is severely constrained by these restrictions.

We have argued above that every second-order quantum phase transition defines a quantum field theory in the continuum. Our attack on the quantum phase transition problem in this book can be considered as consisting of two essential steps. First, we understand and classify the various quantum field theories that can arise out of quantum phase transitions in lattice Hamiltonians of physical interest. And second, we describe the dynamical properties of these quantum field theories at finite temperatures. The latter will then model the universal properties of the physical lattice Hamiltonians in the vicinity of the quantum critical point.

We can now answer the basic question: what are the quantum field theories associated with the second-order quantum phase transitions in the quantum rotor model H_R in (1.25) and the quantum Ising model H_I in (1.7)? It is possible to give a common treatment of H_I and H_R, with H_I simply being the $N = 1$ case of a general discussion for H_R. We attempt to write down a Feynman path integral for the quantum partition function (2.4). As we argued earlier, this is expressed in terms of a functional integral over all possible time histories (the "sum over histories" formulation of quantum mechanics) of the rotor coordinate $\mathbf{n}_i(\tau)$ over an imaginary time $0 \leq \tau \leq \hbar/k_B T$ (and similarly for σ_i^z for $N = 1$). Clearly, this time axis is the $(d + 1)$th dimension of the corresponding classical model. The final quantum field theory is conveniently expressed in terms of a coarse-grained field $\phi_\alpha(x, \tau)$ defined by

$$\phi_\alpha(x, \tau) \sim \sum_{i \in \mathcal{N}(x)} n_{i\alpha}(\tau), \tag{2.10}$$

where x is a point in d-dimensional space, $\mathcal{N}(x)$ is a coarse-graining neighborhood of x, the index $\alpha = 1 \ldots N$, and the overall normalization of ϕ_α can be chosen at our convenience. For the case $N = 1$, we simply replace $n_{i\alpha}$ by σ_i^z. Because the $\mathbf{n}_i$ can point in different directions at each i, the magnitude of ϕ_α can vary over a wide range. Indeed, it seems reasonable that instead of applying a "hard" constraint like $\mathbf{n}_i^2 = 1$, we can view ϕ_α as a "soft" spin whose magnitude can vary freely over all positive values. A remnant of the hard constraint on the microscopic degrees of freedom is that we have a local effective potential $V(\phi_\alpha^2)$, which controls fluctuations of ϕ_α^2 and prevents it from becoming too large. We can also make a polynomial expansion for V, and it turns out to be adequate to truncate it at terms of order $(\phi_\alpha^2)^2$. In this manner, the quantum field theory obtained by considering the vicinity of the quantum critical points in $H_{R,I}$ is defined by the following imaginary time Feynman path integral over all possible time histories of the field $\phi_\alpha(x, \tau)$ for the partition function $\mathcal{Z}$:

$$\mathcal{Z} = \int \mathcal{D}\phi_\alpha(x, \tau) \exp(-\mathcal{S}_\phi),$$

$$\mathcal{S}_\phi = \int d^d x \int_0^{\hbar/k_B T} d\tau \left\{ \frac{1}{2} \left[(\partial_\tau \phi_\alpha)^2 + c^2 (\nabla_x \phi_\alpha)^2 + r\phi_\alpha^2(x) \right] + \frac{u}{4!} \left(\phi_\alpha^2(x) \right)^2 \right\}, \tag{2.11}$$

where c is a velocity, r and u are coupling constants, and the functional integral is over fields periodic in τ with period $\hbar/k_B T$ (i.e. $\phi_\alpha(x, \tau) = \phi_\alpha(x, \tau + \hbar/k_B T)$). The two nongradient terms in (2.11) arise from the polynomial expansion of the potential $V(\phi_\alpha^2)$ noted above; the spatial gradient term represents the energy cost for the spatial variations in the orientation of the magnetic order. The time derivative term arises from the quantum-mechanical tunneling terms proportional to Jg ($J\tilde{g}$) in H_I (H_R), and we will see how they lead to second-order time derivatives in Chapters 5 and 6. This quantum field theory undergoes a quantum phase transition, from a phase with $\langle\phi_\alpha\rangle \neq 0$ to one with $\langle\phi_\alpha\rangle = 0$, by tuning the coupling r through a critical value r_c at $T = 0$.

An alternative formulation of this quantum field theory is sometimes useful for analyzing H_R at small $\tilde{g}$ and for low values of d; this formulation applies only for $N \geq 2$ and yields a field theory with precisely the same universal properties as the formulation in (2.11). The basic idea is that at small $\tilde{g}$, the predominant fluctuations will be variations in the orientation of the local direction of $\mathbf{n}_i$. Also, the orientation should not vary significantly from site to site, and we can therefore simply promote $\mathbf{n}_i(\tau)$ to a *unit-length* continuum field $\mathbf{n}(x, \tau)$ and obtain

$$\mathcal{Z} = \int \mathcal{D}\mathbf{n}(x, \tau)\delta\big(\mathbf{n}^2(x, \tau) - 1\big)\exp(-\mathcal{S}_n),$$

$$\mathcal{S}_n = \frac{N}{2cg}\int d^d x \int_0^{\hbar/k_B T} d\tau\big[(\partial_\tau\mathbf{n})^2 + c^2(\nabla_x\mathbf{n})^2\big], \qquad (2.12)$$

where the small $\tilde{g}$ expressions for g and c are given in (6.11) and (6.51), and $\mathbf{n}(x, \tau)$ satisfies a periodicity condition similar to that for ϕ_α. This field theory is often called the $O(N)$ quantum nonlinear sigma model in d dimensions, for obscure historical reasons. The action is only quadratic in the field $\mathbf{n}(x, \tau)$, but the model is not a free field theory because of the constraint $\mathbf{n}^2(x, \tau) = 1$ imposed at each point in spacetime. Note also that (2.12) is the obvious higher dimensional generalization of the $D = 1$ field theory (6.45) studied in Chapter 3: instead of having only one "quantum" τ direction, we also have d additional spatial directions labeled by x, along with the corresponding gradient squared term in the action.

We note one important property of the quantum field theories (2.11) and (2.12), which will not generalize to some of the other quantum phase transitions studied in Part IV. These field theories are clearly invariant under "relativistic" transformations in spacetime, with the velocity c playing the role of the velocity of light. Consequently spatial and temporal scales must behave equivalently near the quantum critical point; this implies that the dynamic critical exponent must be $z = 1$, a value which is implicitly assumed in some of the discussion in Parts II and III. Our discussion of transitions with $z \neq 1$ is deferred to Part IV.

The description of the universal dynamical properties of (2.11) and (2.12) will occupy a substantial portion of Part III. Formally, the imaginary time correlations of an infinite d-dimensional quantum system at a temperature T are simply related to the correlations of a D-dimensional classical system that is infinite in d directions and of finite extent L_τ in one direction.

2.2 What's different about quantum transitions?

The quantum–classical mapping discussed so far in Part I is in fact a very general result and not a specific property of the Ising/rotor models. One can always reinterpret the imaginary time functional integral of a d-dimensional quantum field theory as the finite "temperature" Gibbs ensemble of a D-dimensional classical field theory. We will often use this mapping between d-dimensional quantum mechanics and D-dimensional classical statistical mechanics, and we will refer to it as the $\mathcal{QC}$ mapping. However, in general, the resulting classical statistical mechanics problem will not be as simple as it was for the Ising/rotor models. Quantum critical points often have $z \neq 1$, and so correlators of the classical problem will scale differently along the x and τ directions. Furthermore, as we note below, there is no guarantee that the Gibbs weights are positive, and they could even be complex valued.

Given this simple, and ubiquitous, quantum–classical mapping $\mathcal{QC}$, one can now legitimately raise the question: why does one need a separate theory of quantum phase transitions? Is it not possible to simply lift results from the corresponding classical theory and obtain all needed properties of the quantum system? The answer to the second question is an emphatic "no," and a direct treatment of the quantum problems is certainly needed. The reasons for this should become clearer to the reader on proceeding through the book, but we note some important points here:

- Note that the quantum–classical mapping $\mathcal{QC}$ yields quantum correlation functions that are in *imaginary* time. The most interesting properties of the quantum critical point are often related to their *real*-time dynamics (e.g. their energy spectra, inelastic neutron scattering cross-sections, or relaxation rates as measured in NMR experiments). To obtain these, one needs to analytically continue the imaginary time results to real time. The crucial point is that this analytic continuation is an ill-posed problem; that is, it is possible to continue exact imaginary time results to real time, but anything short of an exact result leads to unreliable, and usually unphysical, results. In particular, existing analytic results in the theory of classical critical phenomena (with the exception of a single exact result in two spatial dimensions that we shall consider in Chapter 10) are totally inadequate for obtaining $T > 0$ dynamic properties of the corresponding quantum critical points; approximation schemes which work in imaginary time usually fail after analytic continuation to real time, i.e. the operations of expanding in a control parameter, and analytic continuation, do not commute. The problem is particularly severe for the long time limit $t \gg \hbar / k_B T$, which is usually of the greatest practical interest. These correlations are essentially impossible to reconstruct from the equivalent classical problem, which only yields imaginary time correlations in the domain $0 \leq \tau \leq \hbar / k_B T$. It is therefore of crucial importance that the theory be constructed using the physical concepts of the quantum critical point and that it formulate the dynamic analysis directly in real time at all stages.
- We will see in the following chapters that a fundamental new time scale characterizing the dynamic properties of systems near a quantum critical point is the *phase coherence*

time, τ_φ. Loosely speaking, τ_φ is the time over which the wavefunction of the many-body system retains memory of its phase. Local measurements separated by times shorter than τ_φ will display quantum interference effects. Precise definitions of τ_φ have to be tailored to the physical situation at hand, and these will be presented later for the models and regimes considered. In most cases τ_φ is closely related to the thermal equilibration time, τ_{eq}, discussed in Section 1.2. The phase coherence time has no analog near the corresponding classical critical point in D dimensions. Note from (2.5) that an infinite D-dimensional classical system maps onto a d-dimensional quantum system at $T = 0$; in all the models we shall consider in this book, the latter will have either a unique ground state or one with a degeneracy small enough that the entropy is not thermodynamically significant: under these circumstances we can expect that it is always possible to define a τ_φ that is infinite at $T = 0$, and therefore the quantum system has perfect phase coherence at sufficiently low temperatures. From the infinite D-dimensional classical point of view, however, this result may seem extremely peculiar. Most such systems have a high-"temperature" disordered phase in which there is no long-range order and all correlations decay exponentially over very short scales. Yet we are claiming that such a disordered state maps onto a corresponding "quantum-disordered" state, which is characterized by correlations that have an *infinite* correlation time (there is also a long length scale, the distance excitations can travel in a time τ_φ – for related remarks from experimentalists' perspectives, the reader should see the articles by Mason *et al.* [320] and Aeppli *et al.* [4]); for this reason we shall eschew the commonly used "quantum-disordered" appellation and refer to this state, as noted earlier, as a *quantum paramagnet*. This peculiarity is closely related to the ill-posed nature of the analytic continuation noted above. Quantum systems at $T = 0$ really do have a genuinely different long-range phase correlation in time that is almost completely hidden once the mapping to imaginary time and the corresponding classical system has been performed. Only for $T > 0$ does the τ_φ of the quantum system become finite. An important purpose of this book is to show how to introduce a characterization of quantum states that demonstrates the perfect coherence at $T = 0$, to show how to compute τ_φ for $T > 0$, and to highlight the crucial role played by τ_φ in the structure of the dynamic correlations. The manner in which $\tau_\varphi \to \infty$ as $T \to 0$ is an important diagnostic in characterizing the different $T > 0$ regions in the vicinity of the quantum critical point. We shall find that, in all of the models we study, the time L_τ in (2.5) appears as a lower bound on the rate of divergence of τ_φ as $T \to 0$, that is,

$$\tau_\varphi \geq \mathcal{C}\frac{\hbar}{k_B T} \text{ as } T \to 0, \tag{2.13}$$

where $\mathcal{C}$ is a number of order unity. Our estimates of τ_{eq} in Section 1.2 are clearly consistent with (2.13). In the quantum critical region of Fig. 1.3, the inequality in (2.13) is saturated; this region will be of particular interest to us. Its dynamical properties have not been studied until recently, and we will find that they have many remarkably universal characteristics even though their saturating the lower bound on τ_φ implies that their physics is maximally incoherent. Because of this shortest possible τ_φ, the quantum critical region realizes a "nearly perfect" fluid, as we will discuss briefly in Section 15.5.

- For a large class of interesting, physically relevant quantum critical points, the corresponding classical critical points are rather artificial and not of a class that have been studied earlier. In random systems, the classical problems have disorder that is infinitely correlated along the imaginary time direction. Moreover, even in nonrandom systems, the classical problems often have complex-valued Boltzmann weights. These complex weights are clearly a consequence of the underlying quantum mechanics and are often best understood as "Berry phase" factors (see [464] for an elementary introduction to Berry phases; the Berry phases are complex even in imaginary time). We will study quantum critical points of these types in Part IV of this book. We will see that this leads to a whole new class of phenomena, which have no analogs in the classical theory.

- Even for those quantum critical points that do have well-studied classical analogs, note that we need the classical correlation functions in a rather curious slab geometry: one which is infinite in $D - 1$ dimensions and of finite length L_τ in one direction. There are very few existing results in such a geometry, and one often has to reconstruct the needed correlators from scratch.

Despite these caveats, it should be evident that it will pay to push the quantum–classical mapping $\mathcal{QC}$ as far as possible, for this will allow us to get maximum mileage from the sophisticated and profound developments in the theory of classical critical phenomena. This is the strategy of this book. We begin in Parts II and III by thoroughly examining a class of quantum phase transitions that *do* have simple and well-studied classical analogs. In this manner, we will introduce many of the central concepts needed in a somewhat more familiar environment. Then, as noted above, we will proceed in Part IV to many other physically important quantum phase transitions that involve Berry phases in a crucial way, but which do not have useful classical analogs.

There have also been discussions of the dynamical properties of quantum field theories at finite temperature in particle physics literature [50, 247, 259, 381]. However, these are exclusively concerned with physics in $D = 4$ in models that do not satisfy "hyperscaling" [59] properties, and this leads to significant differences from the systems we shall examine here. Some of these studies [50, 247] have examined the model $\mathcal{S}_\phi$ in (2.11) for the case $D = 4$, $N = 1$, which turns out to be essentially a free field theory at low energies. As a result, inelastic decoherence effects are rather weak and nonuniversal. This will be discussed further in Chapter 14. There is also interest in the high-temperature dynamical properties of non-Abelian gauge theories [247, 381]. These are asymptotically free at high energies (i.e. scattering between the elementary excitations is negligibly small at high energies), and as a result the high-temperature behavior is controlled by a Gaussian and classical fixed point. We will see an analogous phenomenon here in a much simpler context in Section 12.3; the simplicity will allow us to make greater progress than has so far been possible for the gauge theories. The models of primary interest in this book satisfy hyperscaling and are *not* asymptotically free at high energies; such models have not been studied in particle physics literature.

PART II

A FIRST COURSE

3 Classical phase transitions

Given the motivation outlined in Chapter 2, we begin by discussing phase transitions in the context of classical statistical mechanics. This is a vast subject, and the reader can find many other books which explore many subtle issues. Here, our purpose is to summarize the main ideas which transfer easily to our subsequent discussions of quantum phase transitions.

We will consider the most important models of classical phase transitions: ferromagnets with N component spins residing on the sites, i, of a hypercubic lattice. For $N = 1$, this is the familiar Ising model with a partition function already met in (2.2):

$$\mathcal{Z} = \sum_{\{\sigma_i^z = \pm 1\}} \exp(-H),$$

$$H = -K \sum_{\langle i,j \rangle} \sigma_i^z \sigma_j^z. \tag{3.1}$$

Note that we have refrained from inserting an explicit factor of temperature above, as the symbol T will be reserved for the temperature of the quantum systems we consider later. As is discussed in introductory statistical mechanics texts, this Ising model describes the vicinity of the liquid–gas critical point, with the average value $\langle \sigma_i^z \rangle$ measuring the density in the vicinity of site i. It can also model loss of ferromagnetism with increasing temperature in magnets in which the electronic spins preferentially align along a particular crystalline axis: this "easy-axis" behavior can be induced by the spin–orbit interaction. For $N > f1$, we generalize (3.1) to the model of (2.3)

$$\mathcal{Z} = \prod_i \int D\mathbf{n}_i \delta \left(\mathbf{n}_i^2 - 1 \right) \exp \left(K \sum_{\langle i,j \rangle} \mathbf{n}_i \cdot \mathbf{n}_j \right). \tag{3.2}$$

The $N = 3$ case (known as the "Heisenberg" model) describes ferromagnetism in materials with sufficiently weak, small spin–orbit couplings, so that the spins can freely orient along any direction. The $N = 2$ case (known as the "XY" model) describes the superfluid–normal transition in liquid helium and other superfluids, as we will see in Section 8.3.

The observables, $\mathcal{O}$, of the classical models will be arbitrary functions of the $\mathbf{n}_i$, and we are interested in their expectation values defined by

$$\langle \mathcal{O} \rangle \equiv \frac{1}{\mathcal{Z}} \sum_{\{\sigma_i^z = \pm 1\}} \exp(-H)\mathcal{O} \tag{3.3}$$

for $N = 1$, and similarly for $N \geq 1$.

As discussed in Chapter 2, the models in (3.1) and (3.2) undergo phase transitions at some critical $K = K_c$. We are interested in describing the nature of the spin correlations in the vicinity of this critical point, and especially their universal aspects. We begin in Section 3.1 by describing this transition using a variational method which leads to a "mean-field" theory. A more general formulation of the mean-field results appears in the framework of Landau theory in Section 3.2, which allows easy treatment of spatial variations. Finally, corrections to Landau theory are considered in Section 3.3.

3.1 Mean-field theory

First, we give a heuristic derivation of mean-field theory. Here, and below, for notational simplicity we will focus on the $N = 1$ Ising case, although the generalization to $N > 1$ is not difficult.

We focus on the fluctuations of a particular spin σ_i^z. This spin feels the local Hamiltonian

$$-K\sigma_i^z \left(\sum_{j \text{ neighbor of } i} \sigma_j^z \right) \approx -K\sigma_i^z \left(\sum_{j \text{ neighbor of } i} \left\langle \sigma_j^z \right\rangle \right) = -2DKN_0\sigma_i^z. \tag{3.4}$$

The mean-field approximation is in the center, where we replace all the neighboring spins by their average value. Here N_0 is the ferromagnetic moment, defined by

$$N_0 \equiv \langle \sigma_i^z \rangle \tag{3.5}$$

in the full theory, where the translational invariance of H guarantees the independence of N_0 on the site i. Given the simple effective Hamiltonian for site i in Eq. (3.4), we can now evaluate

$$\langle \sigma_i^z \rangle = \frac{\sum_{\sigma_i^z = \pm 1} \sigma_i^z \exp\left(2DKN_0\sigma_i^z\right)}{\sum_{\sigma_i^z = \pm 1} \exp\left(2DKN_0\sigma_i^z\right)} = \tanh(2DKN_0). \tag{3.6}$$

Combining (3.5) and (3.6), we have our central mean-field equation

$$N_0 = \tanh(2DKN_0) \tag{3.7}$$

for the value of N_0. We will discuss the nature of its solutions shortly.

Let us now give a more formal derivation of (3.7) using the variational method. This method relies on the choice of an arbitrary mean-field Hamiltonian, H_{MF}. Naturally, we want to choose H_{MF} so that we are able to easily evaluate its partition function $\mathcal{Z}_{MF}$, and the expectation values of all the observables, which we denote $\langle \mathcal{O} \rangle_{MF}$ after evaluation as in (3.3) but with H replaced by H_{MF}. We now want to optimize the choice of H_{MF} by a variational principle which bounds the exact free energy $\mathcal{F} = -\ln \mathcal{Z}$. Of course, the best possible choice is $H_{MF} = H$, but this does not allow easy evaluation of correlations. The variational principle descends from the theorem

$$\mathcal{F} \leq \mathcal{F}_{MF} + \langle H - H_{MF} \rangle_{MF}. \tag{3.8}$$

The proof of the theorem proceeds as follows (here, and below, we use the symbol "Tr" to denote the sum over all the σ_i^z):

$$
\begin{aligned}
e^{-\mathcal{F}} &= \mathrm{Tr}\, e^{-H} \\
&= \mathrm{Tr}\, e^{-(H-H_{MF})-H_{MF}} \\
&= e^{-\mathcal{F}_{MF}} \left\langle e^{-(H-H_{MF})} \right\rangle_{MF}.
\end{aligned}
\tag{3.9}
$$

We now use the statement of the convexity of the exponential function, which is

$$
\left\langle e^{-\mathcal{O}} \right\rangle \geq e^{-\langle \mathcal{O} \rangle}.
\tag{3.10}
$$

Taking logarithms of both sides of (3.9), we finally obtain (3.8).

Returning to the Ising model, we choose the simplest H_{MF} consisting of a set of decoupled spins in a "mean field" h_{MF}:

$$
H_{MF} = -h_{MF} \sum_i \sigma_i^z.
\tag{3.11}
$$

Then, we see immediately that

$$
\mathcal{F}_{MF} = -M \ln(2 \cosh h_{MF}),
\tag{3.12}
$$

where M is the total number of sites on the lattice, and

$$
N_0 = \langle \sigma_i^z \rangle_{MF} = -\frac{1}{M} \frac{\partial \mathcal{F}_{MF}}{\partial h_{MF}} = \tanh(h_{MF}).
\tag{3.13}
$$

Using (3.8) to bound the free energy, we have

$$
\begin{aligned}
\mathcal{F} &\leq \mathcal{F}_{MF} - K \sum_{\langle ij \rangle} \left\langle \sigma_i^z \sigma_j^z \right\rangle + h_{MF} \sum_i \langle \sigma_i^z \rangle \\
&\leq \mathcal{F}_{MF} - MKDN_0^2 + Mh_{MF}N_0.
\end{aligned}
\tag{3.14}
$$

We now have upper bounds for the free energy for every value of the, so far, undetermined parameter h_{MF}. Clearly we want to choose h_{MF} to minimize the right hand side of (3.14); we will declare the resulting upper bound as our approximate result for $\mathcal{F}$ – this is the mean-field approximation. Actually, it is helpful to trade the variational parameter h_{MF} with the value of the ferromagnetic moment N_0, as they are related to each other by (3.13). So our variational parameter is now N_0, and our mean-field free energy is a function of N_0 given by

$$
\mathcal{F}(N_0)/M = \mathcal{F}_{MF}(N_0)/M - KDN_0^2 + N_0 h_{MF}(N_0),
\tag{3.15}
$$

where the functions $\mathcal{F}_{MF}(N_0)$ and $h_{MF}(N_0)$ are defined by (3.12) and (3.13). Using these expressions, we obtain the explicit expression

$$
\frac{\mathcal{F}(N_0)}{M} = -KDN_0^2 + \frac{1+N_0}{2} \ln \frac{1+N_0}{2} + \frac{1-N_0}{2} \ln \frac{1-N_0}{2},
\tag{3.16}
$$

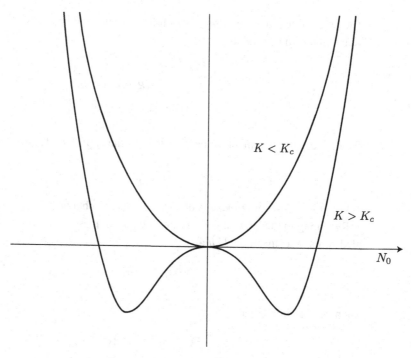

Fig. 3.1   Plot of (3.16) as a function of N_0 for different K.

which can be interpreted as the sum of estimates of the internal energy and entropy of the Ising spins (see Exercise 3.1). Our task is now to minimize (3.16) over values of N_0 for each K. Before examining the results of this, let us examine the nature of the stationarity condition by taking the derivative of (3.15):

$$\frac{1}{M}\frac{\partial \mathcal{F}}{\partial N_0} = \frac{1}{M}\frac{\partial \mathcal{F}}{\partial h_{MF}}\frac{\partial h_{MF}}{\partial N_0} - 2KDN_0 + h_{MF} + N_0\frac{\partial h_{MF}}{\partial N_0}. \tag{3.17}$$

Using (3.13) we observe that the first and last terms cancel, and so the stationarity condition is simply $h_{MF} = 2KDN_0$, which is finally equivalent to our earlier heuristic result in (3.7).

Rather than solving (3.7), it is more instructive to examine the solution by plotting (3.16) as a function of N_0 for different K. This is shown in Fig. 3.1. We notice a qualitative change in the nature of the minimization at $K = K_c = 1/(2D)$. For $K < K_c$ (high temperatures) the free energy is minimized by $N_0 = 0$: this corresponds to the high-temperature "paramagnetic" phase. However, for $K > K_c$ we have two degenerate minima at nonzero values of N_0 which have the same magnitude but opposite signs. The system will "spontaneously" choose one of these equivalent minima, leading to ferromagnetic order. Note that this choice is not invariant under the spin–flip symmetry of the underlying H, and so this is a simple illustration of the phenomenon of "spontaneous symmetry breaking." The critical point $K = K_c$ is the position of the phase transition between the paramagnetic and ferromagnetic phases.

3.2 Landau theory

The main idea of mean-field theory has been to represent the 2^M degrees of freedom in the Ising model by a single mean magnetization, N_0. The free energy is determined as a function of N_0, and then minimized to obtain the optimal equilibrium state.

Landau theory retains the idea of free energy optimization, but generalizes N_0 to a continuum field $\phi_\alpha(x)$. Here $\alpha = 1 \ldots N$, and x is a D-dimensional coordinate associated with a hypercubic lattice of spacing a. We have now returned to a consideration of the theory for general N. The central actor in Landau theory will be a free energy *functional*, $\mathcal{F}[\phi_\alpha(x)]$, which has to be minimized with respect to variations in $\phi_\alpha(x)$.

For now, we will keep the definition of $\phi_\alpha(x)$ somewhat imprecise. Physically, $\phi_\alpha(x)$ represents a coarse-grained average of the local magnetization $n_{i\alpha}$ in the vicinity of $x = x_i$, as we discussed in (2.10) for the corresponding quantum model

$$\phi_\alpha(x) \sim \sum_{i \in \mathcal{N}(x)} n_{i\alpha}. \tag{3.18}$$

As discussed in the text following (2.10), we view ϕ_α as a "soft" spin whose magnitude can vary freely over all positive values.

The Landau free energy functional $\mathcal{F}[\phi_\alpha(x)]$ is now derived from a few basic principles:

- The Hamiltonian is invariant under a common O(N) rotation, $n_{i\alpha} \rightarrow R_{\alpha\beta} n_{i\beta}$, where $R_{\alpha\beta}$ is a rotation matrix applied to all sites i. So the free energy should also be invariant under global rotations of the ϕ_α. We saw an example of this for $N = 1$: then the only symmetry is $\sigma_i^z \rightarrow -\sigma_i^z$, and consequently, (3.16) is an even function of N_0.

- Near the critical point at $K = K_c$, the average value N_0 was smaller than the natural value $|\sigma_i^z| = 1$. We expect this to hold also for $N > 1$. After the coarse-graining in (3.18), we expect that ϕ_α is small in a similar sense. Our main interest is in the vicinity of K_c, and therefore the Landau functional will be expanded in powers of ϕ_α.

- A key step by which Landau theory improves mean-field theory is that it allows for spatial variations in the local magnetic order. We will assume here that the important spatial variations occur on a scale which is much larger than a lattice spacing. This assumption will be seen to be valid later, provided we are close to the critical point $K = K_c$. With this assumption, we will be able to expand the free energy functional in gradients of ϕ_α.

We are now prepared to write down the important terms in $\mathcal{F}[\phi_\alpha(x)]$. Expanding in powers and gradients of $\phi_\alpha(x)$ we have

$$\mathcal{F} = \int d^D x \left\{ \frac{1}{2} \left[\mathcal{K} (\nabla_x \phi_\alpha)^2 + r \phi_\alpha^2(x) \right] + \frac{u}{4!} \left(\phi_\alpha^2(x) \right)^2 \right\}, \tag{3.19}$$

expressed in terms of the parameters $\mathcal{K}, r, u$. We may regard these as unknown phenomenological parameters that have to be determined by fitting to experimental or numerical data. However, we can obtain initial estimates by matching to the results on mean-field theory in

Section 3.1. First, we set the overall normalization of ϕ_α by setting its average value equal to that of the $n_{i\alpha}$:

$$\langle \phi_\alpha(x_i) \rangle = \langle n_{i\alpha} \rangle . \tag{3.20}$$

Then comparing (3.19) with the expansion of (3.16) to quartic order, we obtain for $N = 1$

$$r = a^{-D}(1 - 2DK), \quad u = 2a^{-D}. \tag{3.21}$$

Of course, this method does not yield the value of $\mathcal{K}$, because mean-field theory does not have any spatial variations. To estimate $\mathcal{K}$ we may examine the energy of a domain wall at low temperatures between two oppositely oriented ferromagnetic domains with $\sigma_i^z = \pm 1$. Such a domain wall has energy $2K$ per unit length; computing the domain wall energy using (3.19), we obtain the estimate $\mathcal{K} \approx 2Ka^{2-D}$.

An interesting feature of (3.21) is that $r = 0$ at $K = K_c$. In fact, we expect quite generally that $r \sim K_c - K$, as a simple argument now shows. The optimum value of $\phi_\alpha(x)$ under (3.19) is clearly given by a space-independent solution (provided $\mathcal{K} > 0$). For $r > 0$ (and assuming that $u > 0$ generally), the minimum of $\mathcal{F}$ is obtained by $\phi_\alpha(x) = 0$. This is clearly the paramagnetic phase. In contrast, for $r < 0$, it will pay to choose a space-independent but nonzero ϕ_α. The O(N) invariance of $\mathcal{F}$ guarantees that there is a degenerate set of minima that map onto each other under O(N) rotations. So let us orient ϕ_α along the $\alpha = 1$ axis, and write

$$\phi_\alpha = \delta_{\alpha,1} N_0. \tag{3.22}$$

Inserting this into (3.19) and minimizing for $r < 0$ we obtain

$$N_0 = \sqrt{\frac{-6r}{u}}. \tag{3.23}$$

This shows that N_0 vanishes as $r \nearrow 0$, and the approach to the critical point allows us to introduce the *critical exponent* β defined by

$$N_0 \sim (-r)^\beta. \tag{3.24}$$

Both mean-field and Landau theory predict that $\beta = 1/2$, as can also be verified by a minimization of the full expression in (3.16). Related analyses of other observables allow us to obtain other critical exponents, as we explore in Exercise 3.4.

3.3 Fluctuations and perturbation theory

We now want to proceed beyond the mean-field treatment of the phase transition at $K = K_c$. We expect that the value of K_c will have corrections to its mean-field value of $1/(2D)$: we will not focus on these here because they are *nonuniversal*, i.e. dependent upon specific details of the microscopic Hamiltonian. Rather, our focus will be on universal quantities like the critical exponent β. The structure of Landau theory already suggests reasons why β may be universal: notice that the value of β depended only on the quartic polynomial structure of the free energy, which in turn followed from symmetry considerations. Modifying the form of H, e.g. by adding second-neighbor ferromagnetic couplings, would not change the arguments leading to the Landau free energy functional, and we would still obtain $\beta = 1/2$ although K_c would change.

Indeed, the universality suggested by Landau theory is too strong. It indicates that any ferromagnet with O(N) symmetry always has $\beta = 1/2$. We now see that there are fluctuation corrections to Landau theory, and that the universal quantities depend not only on symmetry, but also on the dimensionality, D. The Landau theory predictions are correct for $D > 4$, while there are corrections for $D \leq 4$. We will need the renormalization group approach to compute universal quantities for $D < 4$, and this is described in Chapter 4.

One way to address fluctuation corrections is to return to the underlying partition function in (3.1), and expand it as a power series in K or in $1/K$. Such series expansions have been carried out to very high orders, and they are efficient and accurate methods for describing the behavior at high and low temperatures. However, they are not directly suited for addressing the vicinity of the critical point $K = K_c$. Instead, we would like to use a method which builds on the success of Landau theory, and yields its results at leading order. The coarse-graining arguments associated with (3.18) suggest a route to achieving this: rather than summing over all the individual spins in (3.1), we should integrate over all values of the collective field variable $\phi_\alpha(x)$. In other words, we should regard the expression in (3.19) not as the free energy functional, but as the Hamiltonian (or "action") of a classical statistical mechanics problem in which the degrees of freedom are represented by the field $\phi_\alpha(x)$. The partition function is therefore represented by the functional integral

$$\mathcal{Z} = \int \mathcal{D}\phi_\alpha(x) \exp(-\mathcal{S}_\phi),$$

$$\mathcal{S}_\phi = \int d^D x \left\{ \frac{1}{2} \left[(\nabla_x \phi_\alpha)^2 + r\phi_\alpha^2(x) \right] + \frac{u}{4!} \left(\phi_\alpha^2(x) \right)^2 \right\}. \tag{3.25}$$

Here the symbol $\int \mathcal{D}\phi_\alpha(x)$ represents an infinite dimensional integral over the values of the field $\phi_\alpha(x)$ at every spatial point x. Whenever in doubt, we will interpret this somewhat vague mathematical definition by discretizing x to a set of lattice points of small spacing $\sim 1/\Lambda$. Equivalently, we will Fourier transform $\phi_\alpha(x)$ to $\phi_\alpha(k)$, and impose a cutoff $|k| < \Lambda$ in the set of allowed wavevectors.

We have set the coefficient of the gradient term $\mathcal{K}$ equal to unity in (3.25). This is to avoid clutter of notation, and is easily accomplished by an appropriate rescaling of the field ϕ_α and the spatial coordinates.

An immediate advantage of the representation in (3.25) is that Landau theory is obtained simply by making the saddle-point approximation to the functional integral. We can also see that, as described in more detail below, systematic corrections to Landau theory appear in an expansion in powers of the quartic coupling u. The remainder of this chapter is devoted to explaining how to compute the terms in the u expansion. Each term has an efficient representation in terms of "Feynman diagrams," from which an analytic expression can also be obtained.

3.3.1 Gaussian integrals

We introduce the technology of Feynman diagrams in the simplest possible setting. Let us discretize space, and write the $\phi_\alpha(x_i)$ variables as y_i; we drop the α label to avoid clutter of indices. Then consider the multidimensional integral

$$\mathcal{Z}(u) = \int \mathcal{D}y \exp\left(-\frac{1}{2}\sum_{ij} y_i A_{ij} y_j - \frac{u}{24}\sum_i y_i^4\right), \tag{3.26}$$

where the off-diagonal terms in the matrix A arise from the spatial gradient terms in $\mathcal{S}_\phi$. In this section, we consider A to be an arbitrary positive definite, symmetric matrix. The positive definiteness requires that $r > 0$, i.e. $K < K_c$. Also, we have defined

$$\int \mathcal{D}y = \prod_i \int_{-\infty}^{\infty} \frac{dy_i}{\sqrt{2\pi}}, \tag{3.27}$$

and are interested in the expansion of $\mathcal{Z}(u)$ in powers of u. Thinking of (3.26) as a statistical mechanics ensemble, we will also be interested in the power series expansion of correlators like

$$C_{ij}(u) \equiv \langle y_i y_j \rangle \equiv \frac{1}{\mathcal{Z}(u)} \int \mathcal{D}y\, y_i y_j \exp\left(-\frac{1}{2}\sum_{k\ell} y_k A_{k\ell} y_\ell - \frac{u}{24}\sum_k y_k^4\right). \tag{3.28}$$

First, we note the exact expressions for these quantities at $u = 0$. The partition function is

$$\mathcal{Z}(0) = (\det A)^{-1/2}. \tag{3.29}$$

This result is most easily obtained by performing an orthogonal rotation of the y_i to a basis which diagonalizes the matrix A_{ij} before performing the integral (Exercise 3.5). Also useful for the u expansion is the identity

$$\int \mathcal{D}y \exp\left(-\frac{1}{2}\sum_{ij} y_i A_{ij} y_j - \sum_i J_i y_i\right)$$

$$= (\det A)^{-1/2} \exp\left(\frac{1}{2}\sum_{ij} J_i A_{ij}^{-1} J_j\right), \tag{3.30}$$

which is obtained by shifting the y_i variables to complete the square in the argument of the exponential. By taking derivatives of this identity with respect to the J_i, and then setting $J_i = 0$, we can generate expressions of all the correlators at $u = 0$. In particular, the two-point correlator is

$$C_{ij}(0) = A_{ij}^{-1}. \tag{3.31}$$

For the $2n$-point correlator, we have an expression known as Wick's theorem:

$$\langle y_1 y_2 \ldots y_{2n} \rangle = \sum_P \langle y_{P1} y_{P2} \rangle \ldots \langle y_{P(2n-1)} y_{P2n} \rangle, \tag{3.32}$$

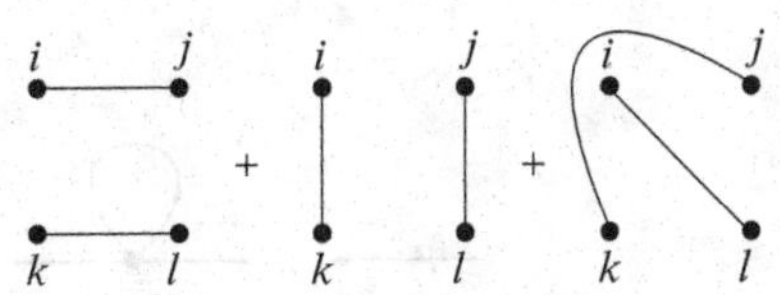

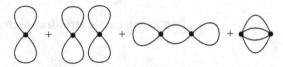

Fig. 3.2

Representation of the 4-point correlator in (3.33). Each line is a factor of the propagator in (3.31).

Fig. 3.3

Diagrams for the partition function to order u^2.

where the summation over P represents the sum over all possible products of pairs, and we reiterate that both sides of the equation are evaluated at $u = 0$. Thus for the 4-point correlator, we have

$$\langle y_i y_j y_k y_\ell \rangle = \langle y_i y_j \rangle \langle y_k y_\ell \rangle + \langle y_i y_k \rangle \langle y_j y_\ell \rangle + \langle y_i y_\ell \rangle \langle y_j y_k \rangle. \tag{3.33}$$

There is a natural diagrammatic representation of the right-hand side of (3.33): we represent each distinct field y_i by a dot, and then draw a line between dots i and j to represent each factor of $C_{ij}(0)$ see Fig. 3.2.

We can now generate the needed expansions of $\mathcal{Z}(u)$ and $C_{ij}(u)$ simply by expanding the integrands in powers of u, and evaluating the resulting series term by term using Wick's theorem. What follows is simply a set of very useful diagrammatic rules for efficiently obtaining the answer at each order. However, whenever in doubt about the value of a diagram, it is often easiest to go back to this primary definition.

For $\mathcal{Z}(u)$, expanding to order u^2, we obtain the diagrams shown in Fig. 3.3 which evaluate to the expression

$$\frac{\mathcal{Z}(u)}{\mathcal{Z}(0)} = 1 - \frac{u}{8} \sum_i \left(A_{ii}^{-1} \right)^2 + \frac{1}{2} \left(\frac{u}{8} \sum_i \left(A_{ii}^{-1} \right)^2 \right)^2 + \frac{u^2}{16} \sum_{i,j} A_{ii}^{-1} A_{jj}^{-1} \left(A_{ij}^{-1} \right)^2$$

$$+ \frac{u^2}{48} \sum_{i,j} \left(A_{ij}^{-1} \right)^4 + \mathcal{O}(u^3). \tag{3.34}$$

We are usually interested in the free energy, which is obtained by taking the logarithm of the above expression, yielding

$$\ln \frac{\mathcal{Z}(u)}{\mathcal{Z}(0)} = -\frac{u}{8} \sum_i \left(A_{ii}^{-1} \right)^2 + \frac{u^2}{16} \sum_{i,j} A_{ii}^{-1} A_{jj}^{-1} \left(A_{ij}^{-1} \right)^2 + \frac{u^2}{48} \sum_{i,j} \left(A_{ij}^{-1} \right)^4 + \mathcal{O}(u^3). \tag{3.35}$$

Now, note an important feature of (3.35): the terms here correspond precisely to the subset of the terms in Fig. 3.3 associated with the *connected* diagrams. These are diagrams in which all points are connected to each other by at least one line, and this result is an

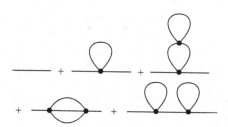

Diagrams for the two-point correlation function to order u^2.

example of the "linked cluster theorem." We will not prove this very useful result here: at all orders in u, we can obtain the perturbation theory for the free energy by keeping only the connected diagrams in the expansion of the partition function.

Now let us consider the u expansion of the two-point correlator, $C_{ij}(u)$, in (3.28). Here, we have to expand the numerator and denominator in (3.28) in powers of u, evaluate each term using Wick's theorem, and then divide the result series. Fortunately, the linked cluster theorem simplifies things a great deal here too. The result of the division is simply to cancel all the disconnected diagrams. Thus, we need only expand the numerator, and keep only connected diagrams. The diagrams are shown in Fig. 3.4 to order u^2, and they evaluate to

$$C_{ij}(u) = A_{ij}^{-1} - \frac{u}{2} \sum_k A_{ik}^{-1} A_{kk}^{-1} A_{kj}^{-1} + \frac{u^2}{4} \sum_{k,\ell} A_{ik}^{-1} A_{kk}^{-1} A_{k\ell}^{-1} A_{\ell\ell}^{-1} A_{\ell j}^{-1}$$

$$+ \frac{u^2}{4} \sum_{k,\ell} A_{ik}^{-1} \left(A_{k\ell}^{-1} \right)^2 A_{\ell\ell}^{-1} A_{kj}^{-1} + \frac{u^2}{6} \sum_{k,\ell} A_{ik}^{-1} \left(A_{k\ell}^{-1} \right)^3 A_{\ell j}^{-1}. \qquad (3.36)$$

We now state the useful *Dyson's theorem*. For this, it is helpful to consider the expansion of the inverse of the C_{ij} matrix, and write it as

$$C_{ij}^{-1} = A_{ij} - \Sigma_{ij}, \qquad (3.37)$$

where the matrix Σ_{ij} is called the "self-energy," for historical reasons not appropriate here. Using (3.36), some algebra shows that to order u^2

$$\Sigma_{ij}(u) = -\delta_{ij} \frac{u}{2} A_{ii}^{-1} + \delta_{ij} \frac{u^2}{4} \sum_k \left(A_{ik}^{-1} \right)^2 A_{kk}^{-1} + \frac{u^2}{6} \left(A_{ij}^{-1} \right)^3, \qquad (3.38)$$

and these are shown graphically in Fig. 3.5. Dyson's theorem states that we can obtain the expression for the Σ_{ij} directly from the graphs for C_{ij} in Fig. 3.4 by two modifications: (*i*) drop the factors of A^{-1} associated with external lines, and (*ii*) keep only the graphs which are one-particle irreducible (1PI). The latter are graphs which do not break into disconnected pieces when one internal line is cut; the last graph in Fig. 3.4 is one-particle reducible, and so does not appear in (3.38) and Fig. 3.5.

The expression in (3.38) will be the basis for much of the analysis in Part II.

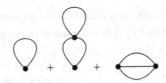

Fig. 3.5

Diagrams for the self-energy to order u^2.

3.3.2 Expansion for susceptibility

We now apply the results of Section 3.3.1 to our functional integral representation in (3.25) for the vicinity of the phase transition.

The problem defined by (3.25) has an important simplifying feature not shared by our general analysis of (3.26): translational invariance. This means that correlators depend only upon the differences of spatial coordinates, and that the analog of the matrix A can be diagonalized by a Fourier tranformation. So now we define the correlator

$$C_{\alpha\beta}(x - y) = \langle \phi_\alpha(x)\phi_\beta(y)\rangle - \langle \phi_\alpha(x)\rangle \langle \phi_\beta(y)\rangle, \tag{3.39}$$

where the subtraction allows generalization to the ferromagnetic phase; we will consider only the paramagnetic phase here.

The subtraction in (3.39) is also needed for the fluctuation–dissipation theorem. We will discuss the full version of this theorem in Section 7.1, but note a simpler version. We consider the susceptibility, $\chi_{\alpha\beta}$, the response of the system to an applied "magnetic" field h_α, under which the action changes as

$$S_\phi \to S_\phi - \int d^D x \, h_\alpha(x)\phi_\alpha(x). \tag{3.40}$$

Then

$$\chi_{\alpha\beta}(x - y) = \frac{\delta \langle \phi_\alpha(x)\rangle}{\delta h_\beta(y)} = C_{\alpha\beta}(x - y), \tag{3.41}$$

where the last equality follows from taking the derivative with respect to the field. Below we set $h_\alpha = 0$ after taking the derivative. The Fourier transform of the susceptibility $\chi_{\alpha\beta}$ is

$$\chi_{\alpha\beta}(k) = \int d^D x \, e^{-ikx} \chi_{\alpha\beta}(x). \tag{3.42}$$

In the paramagnetic phase, $\chi_{\alpha\beta}(k) \equiv \delta_{\alpha\beta}\chi(k)$, and the susceptibility $\chi(k)$ will play a central role in our analysis.

We can also Fourier transform the field $\phi_\alpha(x)$ to $\phi_\alpha(k)$, and so obtain the following representation of the action from (3.25):

$$S_\phi = \frac{1}{2} \int \frac{d^D k}{(2\pi)^D} |\phi_\alpha(k)|^2 (k^2 + r)$$
$$+ \frac{u}{4!} \int \frac{d^D k}{(2\pi)^D} \frac{d^D q}{(2\pi)^D} \frac{d^D p}{(2\pi)^D} \phi_\alpha(k)\phi_\alpha(q)\phi_\alpha(p)\phi_\alpha(-k - p - q). \tag{3.43}$$

In this representation it is clear that the quadratic term in the action is diagonal, and so the inversion of the matrix A is immediate. In particular, from (3.31) we have the susceptibility at $u = 0$

$$\chi_0(k) = \frac{1}{k^2 + r},$$
(3.44)

where we have defined $\chi_0(k)$ to be the value of $\chi(k)$ at $u = 0$. Dyson's theorem in (3.37) becomes a simple algebraic relation

$$\chi(k) = \frac{1}{1/\chi_0(k) - \Sigma(k)} = \frac{1}{k^2 + r - \Sigma(k)}.$$
(3.45)

We will shortly obtain an explicit expression for $\Sigma(k)$.

Let us now explore some of the consequences of the $u = 0$ result in (3.44), which describes Gaussian fluctuations about mean-field theory in the paramagnetic phase, $r > 0$. The zero momentum susceptibility, which we denote simply as $\chi \equiv \chi(k=0) = 1/r$, diverges as we approach the phase transition at $K = K_c$ from the high temperature paramagnetic phase. This divergence is a key feature of the phase transition, and its nature is encoded in the critical exponent γ defined by

$$\chi \sim (K_c - K)^{-\gamma}.$$
(3.46)

At this leading order in u we have $\gamma = 1$.

We can also examine the spatial correlations in the $u = 0$ theory. Performing the inverse Fourier transform to $C_{\alpha\beta}(x) = \delta_{\alpha\beta} C(x)$ we find

$$C(x) = \int \frac{d^D k}{(2\pi)^D} \frac{e^{ikx}}{(k^2 + r)} = \frac{(2\pi)^{-D/2}}{(x\xi)^{(D-2)/2}} K_{(D-2)/2}(x/\xi),$$
(3.47)

where here K is the modified Bessel function, and we have introduced a characteristic length scale, ξ, defined by

$$\xi = 1/\sqrt{r}.$$
(3.48)

This is the correlation length, and is a measure of the distance over which fluctuations of ϕ_α (or the underlying spins σ_i^z) are correlated. This is evident from the limiting forms of (3.47) in various asymptotic regimes:

$$C(x) \sim \begin{cases} \dfrac{1}{x^{D-2}}, & x \ll \xi \\[2mm] \dfrac{e^{-x/\xi}}{x^{(D-1)/2}\xi^{(D-3)/2}}. & x \gg \xi \end{cases}$$
(3.49)

As could be expected of a correlation length, the correlations decay exponentially to zero at distances larger than ξ.

An important property of our expression in (3.48) for the correlation length is that it diverges upon the approach to the critical point. This divergence is also associated with a critical exponent, ν, defined by

$$\xi \sim (K_c - K)^{-\nu},$$
(3.50)

and our present theory yields $\nu = 1/2$. In the vicinity of the phase transition, this large value of ξ provides an a-posteriori justification of our taking a continuum perspective on the fluctuations. In other words, it supports our mapping from the lattice models in (3.1) and (3.2) to the classical field theory in (3.25), where we replaced the lattice spin variables by the collective field ϕ_α using (3.18).

Let us now move beyond the $u = 0$ theory, and consider the corrections at order u. After mapping to Fourier space, the result in (3.38) for the self-energy yields

$$\Sigma(k) = -u \frac{(N+2)}{6} \int \frac{d^D p}{(2\pi)^D} \frac{1}{p^2 + r}. \tag{3.51}$$

Here, and below, there is an implicit upper bound of $k < \Lambda$ needed to obtain finite answers for the wavevector integrals. The N dependence comes from keeping track of the spin index α along each line of the Feynman diagram, and allowing for the different possible contractions of such indices at each u interaction point. We then have from (3.45) our main result for the correction in the susceptibility:

$$\frac{1}{\chi(k)} = k^2 + r + u \frac{(N+2)}{6} \int \frac{d^D p}{(2\pi)^D} \frac{1}{p^2 + r} + \mathcal{O}(u^2). \tag{3.52}$$

The first consequence of (3.52) is a shift in the position of the critical point. From (3.46), a natural way to define the position of the phase transition is by the zero of $1/\chi$. The order u correction in (3.52) shows that the critical point is no longer at $r = r_c = 0$, but at

$$r_c = -u \frac{(N+2)}{6} \int \frac{d^D p}{(2\pi)^D} \frac{1}{p^2} + \mathcal{O}(u^2). \tag{3.53}$$

Now, let us combine (3.52) and (3.53) to determine the behavior of χ as $r \searrow r_c$. We introduce the coupling s defined by

$$s \equiv r - r_c, \tag{3.54}$$

which measures the deviation of the system from the critical point. Rewriting (3.52) in terms of s rather than r (we will always use s in favor of r in all subsequent analysis), we have

$$\frac{1}{\chi} = s + u \left(\frac{N+2}{6} \right) \int^\Lambda \frac{d^D p}{(2\pi)^D} \left(\frac{1}{p^2 + s} - \frac{1}{p^2} \right). \tag{3.55}$$

We are interested in the vicinity of the critical point, at which $s \to 0$.

A crucial point is that the nature of this limit depends sensitively on whether D is greater than or less than four. For $D > 4$, we can simply expand the integrand in (3.55) in powers of s and obtain

$$\frac{1}{\chi} = s(1 - c_1 u \Lambda^{D-4}), \tag{3.56}$$

where c_1 is a nonuniversal constant dependent upon the nature of the cutoff. Thus the effects of interactions appear to be relatively innocuous: the static susceptibility still diverges with the mean-field form $\chi(0) \sim 1/s$ as $s \to 0$, with the critical exponent $\gamma = 1$. This is in fact the generic behavior to all orders in u, and all the mean-field critical exponents apply for $D > 4$.

For $D < 4$, we notice that the integrand in (3.55) is convergent at high momenta, and so it is permissible to send $\Lambda \to \infty$. We then find that the correction to first order in u has a universal form

$$\frac{1}{\chi} = s\left[1 - \left(\frac{N+2}{6}\right)\frac{2\Gamma((4-D)/2)}{(D-2)(4\pi)^{D/2}}\frac{u}{s^{(4-D)/2}}\right]. \tag{3.57}$$

Note that no matter how small u is, the correction term eventually becomes important for a sufficiently small s, and indeed it diverges as $s \to 0$. So for sufficiently large ξ, the mean-field behavior cannot be correct, and a resummation of the perturbation expansion in u is necessary.

The situtation becomes worse at higher orders in u. As suggested by (3.57), the perturbation series for $1/(s\chi)$ is actually in powers of $u/s^{(4-D)/2}$, and so each successive term diverges more strongly as $s \to 0$. Thus the present perturbative analysis is unable to describe the vicinity of the critical point for $D < 4$. We will show that this problem is cured by a renormalization group treatment in the following chapter.

Exercises

3.1 Consider an Ising model on a system of N sites. Let $N_\uparrow$ be the number of up spins. Calculate $\Gamma(N_\uparrow, N)$, the total number of ways these $N_\uparrow$ spins could have been placed among the N sites. Obtain the entropy $S = k_B \ln \Gamma$ as a function of the magnetization $m = (N_\uparrow - N_\downarrow)/N$, where $N_\downarrow = N - N_\uparrow$. Combine this computation of the entropy with a mean-field estimate of the average internal energy to obtain (3.16).

3.2 *Ising antiferromagnet:* We consider the Ising antiferromagnet on a square lattice. The Hamiltonian is

$$H_I = J \sum_{\langle ij \rangle} \sigma_i \sigma_j, \tag{3.58}$$

where i, j extend over the sites of a square lattice, $\langle ij \rangle$ refers to nearest neighbors, and $\sigma_i = \pm 1$. Note that there is no minus sign in front of J. We take $J > 0$, so the ferromagnetic state, with all σ_i parallel, is the *highest* energy state. The ground states have the pattern of a chess board: $\sigma_i = 1$ on one sublattice (A) and $\sigma_i = -1$ on the other sublattice (B), and vice versa. Use mean-field theory to describe the phase diagram of this model. Argue that the mean-field Hamiltonian should have two fields, h_A and h_B, on the two sublattices, and correspondingly, two magnetizations m_A and m_B. Obtain equations for m_A and m_B and determine the value of T_c.

3.3 *XY model:* We generalize the Ising model (with binary spin variables σ_i) to a model of *vector spins*, $\vec{S}_i$, of unit length ($\vec{S}_i^2 = 1$ at each i). This is a model of ferromagnetism in materials, e.g. iron, in which the electron spin is free to rotate in all directions (rather than being restricted to be parallel or anti-parallel to a given direction, as in the Ising model). For simplicity, let us assume that the spin is only free to rotate within the x–y plane, i.e. $\vec{S}_i = (\cos\theta_i, \sin\theta_i)$. So the degree of freedom is an angle $0 \leq \theta_i < 2\pi$ on each site i. The Hamiltonian of these XY spins is

$$\mathcal{H} = -J \sum_{\langle ij \rangle} \vec{S}_i \cdot \vec{S}_j = -J \sum_{\langle ij \rangle} \cos(\theta_i - \theta_j), \tag{3.59}$$

and the partition function is

$$Z = \prod_i \int_0^{2\pi} d\theta_i \, e^{-\beta\mathcal{H}}. \tag{3.60}$$

Use the variational approach to obtain a mean-field theory for the mean magnetization $\vec{m} = \langle \vec{S}_i \rangle$ at a temperature T. Use a trial Hamiltonian $\mathcal{H}_0 = -\sum_i \vec{h} \cdot \vec{S}_i$, and bound the free energy by $F \leq \langle \mathcal{H} - \mathcal{H}_0 \rangle_0 + F_0$. Argue that because of spin rotational invariance (about the z axis) we can choose $\vec{h}$ and $\vec{m}$ to point along the x-axis, and hence $\mathcal{H}_0 = -\sum_i h \cos\theta_i$. Also, in the variational approach we define

$$m \equiv \langle \cos\theta \rangle_0 = -\frac{1}{N} \frac{\partial F_0(h)}{\partial h}. \tag{3.61}$$

We use this equation for m to solve for h as a function of m, and so now we can consider F to be a function of the variational parameter m. Show that F equals

$$F = -\frac{NqJm^2}{2} + Nhm + F_0(h). \tag{3.62}$$

Use these equations to evaluate $\partial F / \partial m$ and so obtain the mean-field equation for m (note that you do not need an explicit form for F_0 to obtain this) and determine the critical temperature on a lattice with coordination number q.

3.4 *Critical exponents:* In mean-field theory, the free energy density of the Ising model near its critical point can be obtained by minimizing the functional of the magnetization density m:

$$F = \frac{a}{2}m^2 + \frac{b}{4}m^4 - hm + F_0(T).$$

Here F represents the Helmholtz free energy density; $F_0(T)$ is a smooth background function of T, all of whose derivatives are finite and nonzero at $T = T_c$. Assume b is independent of T, while a is approximated by a linear T dependence $a = a_0(T - T_c)$.

(a) The critical exponent β is defined by the manner in which m vanishes upon approaching T_c at $h = 0$:

$$m \sim (T_c - T)^\beta.$$

What is the mean-field value of β?

(b) The critical exponent δ is determined by the h dependence of m at $T = T_c$:

$$m \sim h^{1/\delta}.$$

What is the mean-field value of δ?

(c) The critical exponents γ and γ' are defined by the manner in which the magnetic susceptibility $\chi = \partial m / \partial h|_{h \to 0}$ behaves above and below T_c:

$$\chi \sim A(T - T_c)^{-\gamma}, \quad T > T_c$$
$$\chi \sim A'(T_c - T)^{-\gamma'}. \quad T < T_c$$

Determine the values of γ, γ', and A/A'.

(d) Similarly, the behavior of the specific heat, $C_V = -T(\partial^2 F / \partial T^2)$ at $h = 0$ is specified by

$$C_V \sim B(T - T_c)^{-\alpha}, \quad T > T_c$$
$$C_V \sim B'(T_c - T)^{-\alpha'}. \quad T < T_c$$

Determine the values of α, α', B, and B'.

3.5   Establish (3.29) by changing variables of integration in (3.26) so that the matrix A is diagonal in the new basis. This will involve working with the eigenvectors and eigenvalues of A.

3.6   Redo the computations in Section 3.3.1 for an N component field $y_{i\alpha}$.

3.7   *Landau theory for the XY model:* This is expressed in terms of a vector field $\vec{m}(r) = (m_1(r), m_2(r))$. In zero applied field, the Landau free energy has the form

$$\mathcal{F} = \int d^3 r \left[\frac{K}{2} \sum_{i=x,y,z} \sum_{a=1,2} (\partial_i m_a)^2 + \frac{\alpha}{2} \left(\sum_{a=1,2} m_a^2 \right) + \frac{\beta}{4} \left(\sum_{a=1,2} m_a^2 \right)^2 \right]$$

$$(3.63)$$

and the critical point is at $\alpha = 0$ (assume, $K, \beta > 0$). Determine the correlation functions $G_{11}(r) = \langle m_1(r) m_1(0) \rangle - \langle m_1(r) \rangle \langle m_1(0) \rangle$ and $G_{22}(r) = \langle m_2(r) m_2(0) \rangle - \langle m_2(r) \rangle \langle m_2(0) \rangle$ for both signs of α. You can compute these correlation functions by applying an external field $\vec{h}(r)$ to the system, under which

$$\mathcal{F} \to \mathcal{F} - \int d^3 r \sum_{a=1,2} h_a(r) m_a(r),$$

$$(3.64)$$

and then computing the change in $\langle m_a \rangle$ due to the presence of the field. As for the Ising model, we have, to linear order in $\vec{h}$

$$\langle m_a(r) \rangle|_{\vec{h}} = \langle m_a(r) \rangle|_{\vec{h}=0} + \frac{1}{k_B T} \int d^3 r' G_{aa}(r - r') h_a(r') + \dots$$

$$(3.65)$$

By writing $\vec{m}$ in the above form, you can read off the values of G_{aa}. Above the critical temperature ($\alpha > 0$), you should find $G_{11} = G_{22}$, which is a simple consequence of rotational invariance. Below the critical temperature, ($\alpha < 0$), choose the state with $\langle m_1(r) \rangle = \sqrt{|\alpha|/\beta}$ and $\langle m_2(r) \rangle = 0$ in zero field. You should find $G_{11} \neq G_{22}$, and determine both functions in 3 dimensions.

The renormalization group

In Chapter 3 we developed the basic tools to describe the phase transition in the classical Ising model, and its cousins with N-component spins. We argued that the vicinity of the critical point at $K = K_c$ could be described by the classical field theory in (3.25). However, we observed that an expansion of the observable properties in powers of the quartic coupling u broke down near the critical point for dimensions $D < 4$. We will now show how the renormalization group circumvents this breakdown.

In its full generality, the renormalization group (RG) is a powerful tool with applications in many fields of physics, and covered at great lengths in other texts. Our treatment here will be relatively brief, and will be directed towards addressing the critical properties of the classical field theory in (3.25).

The key to the success of the RG is that it exposes a new symmetry of the critical point at $K = K_c$, which is not present in the underlying Hamiltonian. This symmetry can be understood to be a consequence of the divergence of the correlation length, ξ, at $K = K_c$, as indicated in (3.50). With the characteristic length scale equal to infinity, we may guess that the structure of the correlations is the same at all length scales, i.e. the physics is invariant under a *scaling transformation* under which the coordinates change as

$$x \rightarrow x' = x/b, \tag{4.1}$$

where b is the rescaling factor. In other words, the basic structure of all correlations should be invariant under a transformation from the x to the x' coordinates. An example of this invariance appears in our result for the two-point correlation function $C(x) \sim x^{2-D}$ for $x \ll \xi$ in (3.49); this is only rescaled by an overall prefactor under the transformation (4.1). We should note here that the lattice statistical mechanics model does have a characteristic length scale, which is a the lattice spacing; the invariance under (4.1) holds only at lengths much larger than a, and only at the critical point.

We will see that the role of the scale invariance is similar to that of other, more familiar, symmetries. As an example, symmetry under rotations is due to an invariance of the Hamiltonian under a change in angular coordinates $\theta' = \theta + b$; (4.1) is an analogous coordinate transformation. Further, we know that rotational invariance classifies observables according to how they respond to the coordinate change: scalars, vectors, tensors, . . . , which are labeled by different values of the angular momentum. We will similarly find that scale invariance labels observables by their *scaling dimension*.

4.1 Gaussian theory

It is useful to begin with a simplified model for which the scaling transformations can be exactly computed. This is the free field theory obtained from (3.25) at $u = 0$: let us write it down here explicitly for completeness:

$$\mathcal{Z} = \int \mathcal{D}\phi_\alpha(x) \exp\left(-\frac{1}{2}\int d^D x \left[(\nabla_x \phi_\alpha)^2 + r\phi_\alpha^2(x)\right]\right). \tag{4.2}$$

It is now easy to see that all correlations associated with this ensemble will maintain their form if we combine the rescaling transformation (4.1) with the definitions

$$\phi'_\alpha(x') = b^{(D-2)/2}\phi_\alpha(x),$$
$$r' = b^2 r. \tag{4.3}$$

The powers of b appearing in (4.3) are the scaling dimensions of the respective variables, which we denote as

$$\dim[\phi] = (D-2)/2,$$
$$\dim[r] = 2. \tag{4.4}$$

Also, by definition, we always have from (4.1) that $\dim[x] = -1$.

These scaling transformations now place strong restrictions on the form of the correlation functions. Thus for the two-point correlation of ϕ_α, we have from (4.3) that

$$\left\langle \phi'_\alpha(x')\phi'_\beta(0) \right\rangle = b^{(D-2)} \left\langle \phi_\alpha(x)\phi_\beta(0) \right\rangle. \tag{4.5}$$

However, both correlators are evaluated in the same ensemble, and ϕ_α and ϕ'_α are merely dummy variables of integration which can be relabeled at will. Therefore the correlators must have the same functional dependence on the spatial coordinates and the couplings constants, and so

$$C(x/b; b^2 r) = b^{D-2}C(x; r), \tag{4.6}$$

where we have indicated the dependence of the correlator on the coupling r, which was previously left implicit. This is the payoff equation from the RG transformation, and places a nontrivial constraint on the form of the correlations; it can be checked that the results (3.47) and (3.48) do obey (4.6).

Although it is strictly not necessary here, it will be useful to build the rescaling transformation by the factor b via a series of infinitesimal transformations. This is analogous to the use of angular momentum to generate infinitesimal rotation in quantum mechanics. Here, we set $b = 1 + d\ell$, where $d\ell \ll 1$, and build up to a finite rescaling $b = e^\ell$ by a repeated action of infinitesimal rescalings. These rescalings are most conveniently represented in terms of differential equations representing the RG flow of the coupling constants. In the present case, we have only the coupling r, and by (4.3) or (4.4), its flow is represented by

$$\frac{dr}{d\ell} = 2r. \tag{4.7}$$

The constraint on the correlator in (4.6) can be written as

$$C(e^{-\ell}x, r(\ell)) = e^{(D-2)\ell}C(x, r).$$ (4.8)

The result of the RG flow (4.7) is very simple. For an initial value $r > 0$, we have $r \to \infty$ as ℓ increases: this represents the physics of the paramagnetic phase. Similarly for $r < 0$, we have $r \to -\infty$ as ℓ increases, representing the ferromagnetic phase. In between these two divergent flows, we have the *fixed point* $r = r^* = 0$, where r is ℓ-independent. This fixed point represents the phase transition between the two phases. Note that at an RG fixed point, we can set $r = r^*$ on both sides of a rescaling equation like (4.8), and so then the correlations are invariant under a rescaling of coordinates alone. Here, we have $C(e^{-\ell}x) = e^{(D-2)\ell}C(x)$, whose solution is

$$C(x) \sim x^{2-D}, \quad r = r^*,$$ (4.9)

which agrees with (3.49).

The above connection between RG fixed points and scale-invariance at phase transitions is very general, and we will meet numerous other examples. The "homogeneity" relationship (4.8) for the correlation function is also in a form that applies in many more complex situations.

Our main result so far has been the identification of the "Gaussian fixed point," $r^* = 0$, of the RG transformation. Let us now look at the stability of this fixed point to other perturbations to the action. The simplest, and most important, is the quartic coupling u already contained in (3.25). Applying the transformation (4.3) we now see that u transforms as

$$u' = b^{4-D}u.$$ (4.10)

Equivalently, this can be written as the RG flow equation

$$\frac{du}{d\ell} = (4 - D)u.$$ (4.11)

Thus in the more complete space of the two couplings r and u, the Gaussian fixed point is identified by $r^* = 0$ and $u^* = 0$.

We also notice a crucial dichotomy in the flow of u away from this Gaussian fixed point. For $D > 4$, the flow of u is back towards $u^* = 0$ as ℓ increases. In RG parlance, the Gaussian fixed point is *stable* towards u perturbations for $D > 4$. This result is entirely equivalent to our more pedestrian observation in (3.56), where we found that perturbation theory in u did not change the leading critical singularity for $D > 4$. Conversely, for $D < 4$, the Gaussian fixed point is *unstable* to u perturbations. This means that we have to flow away from $u^* = 0$. As we will see below, the flow is towards a new fixed point with $u^* \neq 0$, known as the Wilson–Fisher fixed point. It is the Wilson–Fisher fixed point which will cure the problem with perturbation theory in $D < 4$ identified in (3.57).

Before embarking on our search for the stable fixed point for $D < 4$, let us also consider other possible perturbations to the Gaussian fixed point. A simple example is the six-order coupling v defined by

$$\mathcal{S}_\phi \to \mathcal{S}_\phi + v \int d^D x \left(\phi_\alpha^2(x) \right)^3.$$ (4.12)

Application of (4.3) easily yields

$$\frac{dv}{d\ell} = (6 - 2D)v, \tag{4.13}$$

and so $v^* = 0$ is stable for $D > 3$. In alternative common terminology, v is "irrelevant" for $D > 3$. A similar analysis can be applied to all possible local couplings which are invariant under O(N) symmetry, and it is found that they are all irrelevant for $D \geq 3$. This was the underlying reason for our focus on the field theory in (3.25), with only a single quartic nonlinearity. The coupling v is "marginal" at the Gaussian fixed point in $D = 3$, but we will see that it does not play an important role at the Wilson–Fisher fixed point.

4.2 Momentum shell RG

Our RG analysis so far has not been much more than glorified dimensional analysis. Nevertheless, the simple setting of the Gaussian field theory has been a useful place to establish notation and introduce key concepts. We are now ready to face the full problem of classical field theory (3.25) in $D \leq 4$ in the presence of the quartic nonlinearity.

The structure of the perturbation expansion in Section 3.3.2 shows that well-defined expressions for the fluctuation corrections are obtained only in the presence of a wavevector cutoff Λ. So in our scaling transformations we also have to keep track of the length scale Λ^{-1}. True scale invariance at the critical point appears only at length scales much larger than Λ^{-1}, and we need a more general RG procedure which allows such an asymptotic scale invariance to develop.

The needed new idea is that of decimation of the degrees of freedom. The first step in the RG will be a partial integration (or "decimation") of some of the short-distance degrees of freedom. This results in a new problem with a smaller number of degrees of freedom. In our formulation with a momentum cutoff Λ, the new problem has a smaller cutoff $\widetilde{\Lambda} < \Lambda$. We choose $\widetilde{\Lambda} = \Lambda/b$, and integrate degrees of freedom in the shell in momentum space between these momenta.

The second step of the RG is the rescaling transformation already discussed. Given the mapping of length scale $x' = x/b$, the mapping of the cutoff Λ is

$$\Lambda' = b\widetilde{\Lambda} = \Lambda. \tag{4.14}$$

Note that after the complete RG transformation, the initial and final cutoffs are equal. Thus our RG will be defined at a *fixed* cutoff Λ. This is very useful, because we need no longer keep track of factors of cutoff in any of the scaling relations, and can directly compare theories simply by comparing the values of coupling constants like $r, u, \ldots$

Now we only need an implementation of the first decimation step. The second rescaling step will proceed just as in Section 4.1.

The key to the decimation procedure is the decomposition

$$\phi_\alpha(x) = \phi_\alpha^<(x) + \phi_\alpha^>(x), \tag{4.15}$$

where the two components lie in different regions of momentum space

$$\phi_\alpha^<(x) = \int_0^{\Lambda/b} \frac{d^D k}{(2\pi)^D} \phi_\alpha^<(k) e^{ikx}, \quad \phi_\alpha^>(x) = \int_{\Lambda/b}^{\Lambda} \frac{d^D k}{(2\pi)^D} \phi_\alpha^>(k) e^{ikx}. \tag{4.16}$$

Then we evaluate the partition function as usual, but integrate only over the fields at large momenta $\phi^>$:

$$\exp\left(-\mathcal{S}_{\phi^<}\right) = \int \mathcal{D}\phi_\alpha^>(x) \exp\left(-\mathcal{S}_\phi\right). \tag{4.17}$$

The resulting functional integral defines a functional $\mathcal{S}_{\phi^<}$ of the low momentum fields alone, which we will now compute to the needed order. An important property which facilitates the present analysis is that the $\phi^<$ and $\phi^>$ decouple at the Gaussian fixed point, i.e. the Gaussian action is a sum of terms which involve only $\phi^<$ or $\phi^>$, but not both. This decoupling is a consequence of their disjoint support in the momentum space, and consequently we can write

$$\mathcal{S}_\phi = \int d^D x \left\{ \frac{1}{2} \left[(\nabla_x \phi_\alpha^<)^2 + r\phi_\alpha^{<2}(x) + (\nabla_x \phi_\alpha^>)^2 + r\phi_\alpha^{>2}(x) \right] \right.$$
$$\left. + \frac{u}{4!} \left(\left(\phi_\alpha^<(x) + \phi_\alpha^>(x) \right)^2 \right)^2 \right\}. \tag{4.18}$$

Inserting (4.18) into (4.17) we obtain

$$\mathcal{S}_{\phi^<} = \frac{1}{2} \int d^D x \left[(\nabla_x \phi_\alpha^<)^2 + r\phi_\alpha^{<2}(x) \right] - \ln \mathcal{Z}^>$$
$$- \ln \left\langle \exp\left(-\frac{u}{4!} \int d^D x \left(\left(\phi_\alpha^<(x) + \phi_\alpha^>(x) \right)^2 \right)^2 \right) \right\rangle_{\mathcal{Z}^>}. \tag{4.19}$$

Here $\mathcal{Z}^>$ is the *free* Gaussian ensemble defined by

$$\mathcal{Z}^> = \int \mathcal{D}\phi_\alpha^>(x) \exp\left(-\frac{1}{2} \int d^D x \left[(\nabla_x \phi_\alpha^>)^2 + r\phi_\alpha^{>2}(x) \right] \right), \tag{4.20}$$

and the expectation value in the last term in (4.19) is taken under this Gaussian ensemble (as indicated by the subscript). The second term in (4.19) is an additive constant, and is important in computations of the free energy; we do not include it below because it plays no role in renormalization of the coupling constants.

It now remains to evaluate the expectation value in the last term in (4.19). This is easily done in powers of u using the methods described in Section 3.3.1. Notice that $\phi^<$ appears as a source term, whose powers will multiply various correlation functions of the $\phi^>$; the latter can be evaluated by Wick's theorem, or equivalently by Feynman diagrams. Thus all internal lines in the Feynman diagram represent $\phi^>$, while external lines are $\phi^<$. We show the Feynman diagrams that are important to us in Fig. 4.1 The spatial dependence of $\phi_\alpha^<(x)$ determines the momenta that are injected into the Feynman diagrams by the external vertices. We are ultimately interested in a spatial gradient expansion of the resulting action functional of $\phi^<$, and this means that we can expand the Feynman diagrams in powers of the external momenta.

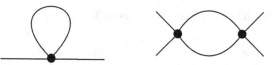

Fig. 4.1 Feynman graphs important for the Wilson–Fisher RG equations. The external lines represent $\phi_<$, while the internal lines are $\phi^>$ propagators. The first graph renormalizes r, while the second renormalizes u.

In the action functional for $\phi^<$ so obtained, we obtain terms which have the same form as those in the original $\mathcal{S}_\phi$ in (3.25). We also obtain a number of other terms which involve higher powers of $\phi^<$ or additional gradients: we have just argued at the end of Section 4.1 that these other terms are not important, and so we can safely drop them here. The first graph in Fig. 4.1 renormalizes terms that are quadratic in $\phi^<$, and this turns out to be independent of momentum: this is related to the k independence of the self-energy $\Sigma(k)$ in (3.51). The second graph in Fig. 4.1 is quartic in $\phi^<$, and this can be evaluated at zero external momenta because we are not interested in gradients of the quartic term. Collecting terms in this manner, the final results of (4.19) can be written as

$$\mathcal{S}_{\phi^<} = \int d^D x \left\{ \frac{1}{2} \left[(\nabla_x \phi_\alpha^<)^2 + \tilde{r} \phi_\alpha^{<2}(x) \right] + \frac{\tilde{u}}{4!} \left(\phi_\alpha^{<2}(x) \right)^2 \right\}. \tag{4.21}$$

Note that the coefficient of the gradient term does not renormalize: this is an artifact of the low-order expansion, and will be repaired below. The other terms do renormalize, and have the modified values

$$\tilde{r} = r + u \frac{(N+2)}{6} \int_{\Lambda/b}^{\Lambda} \frac{d^D p}{(2\pi)^D} \frac{1}{(p^2 + r)},$$

$$\tilde{u} = u - u^2 \frac{(N+8)}{6} \int_{\Lambda/b}^{\Lambda} \frac{d^D p}{(2\pi)^D} \frac{1}{(p^2 + r)^2}. \tag{4.22}$$

Finally, it should be reiterated that the theory $\mathcal{S}_{\phi^<}$ in (4.21) has an implicit momentum cutoff of $\tilde{\Lambda} = \Lambda/b$.

We can now immediately implement the second rescaling step of the RG and obtain the new couplings

$$r' = b^2 \tilde{r}, \quad u' = b^{4-D} \tilde{u}. \tag{4.23}$$

Finally, we specialize to an infinitesimal rescaling $b = 1 + d\ell$, expand everything to first order in $d\ell$, and obtain the RG flow equations

$$\frac{dr}{d\ell} = 2r + u \frac{(N+2)}{6} \frac{S_D}{(1+r)}.$$

$$\frac{du}{d\ell} = (4-D)u - u^2 \frac{(N+8)}{6} \frac{S_D}{(1+r)^2}. \tag{4.24}$$

Because the RG flow is computed at fixed Λ, we have conveniently set $\Lambda = 1$, and will henceforth measure all lengths in units of Λ^{-1}. The phase space factor $S_D = 2/(\Gamma(D/2) (4\pi)^{D/2})$ arises from the surface area of a sphere in D dimensions. The equations (4.24)

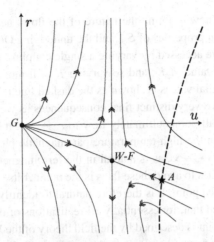

Fig. 4.2 Plot of the flow equations in (4.24). There are two fixed points, the Gaussian (G) and Wilson–Fisher (W–F). The dashed
line represents a possible line of initial values of r and u, as the coupling K is changed in the underlying lattice model.
The critical point of such a lattice model is the point A, which flows into the Wilson–Fisher fixed point.

update (4.7) and (4.11) to the next order in u, and are the celebrated Wilson–Fisher RG
equations.

Let us momentarily ignore concerns about the range of validity of RG flow equations,
and examine the consequences of integrating (4.24). The results are shown in Fig. 4.2.
In addition to the $r^* = u^* = 0$ Gaussian fixed point already found, we observe there is a
second fixed point at nonzero values r^* and u^*: this is the Wilson–Fisher fixed point, and
it will be the focus of our attention.

However, before embarking upon a study of the Wilson–Fisher fixed point, we need a
systematic method of assessing the reliability of our results. Wilson and Fisher pointed out
that a very useful expansion is provided by

$$\epsilon \equiv 4 - D. \tag{4.25}$$

Even though the physical values of D are integer, all our expressions for the perturbative
expansions, Feynman diagrams, and flow equations have been analytic functions of D. So
we can consider an analytic continuation to the complex D plane, and hence to small ϵ.
The expansion in powers of ϵ has since established itself as an invaluable tool in describing
a variety of classical and quantum critical points.

We can now systematically determine the values of r^* and u^* by an ϵ expansion of
(4.24). We find

$$u^* = \frac{6\epsilon}{(N+8)S_4} + \mathcal{O}(\epsilon^2), \quad r^* = -\frac{\epsilon(N+2)}{2(N+8)} + \mathcal{O}(\epsilon^2). \tag{4.26}$$

All the omitted higher order terms in (4.24), and all other terms higher order in ϕ_α, or with
additional gradients that could have been added to $\mathcal{S}_\phi$ will not modify the results in (4.26).
This includes the effect of the coupling v in (4.12), which is actually irrelevant for small ϵ.

Let us now examine the nature of the flows near r^* and u^*, and their implication for the critical properties of $\mathcal{S}_\phi$, and the underlying O(N) spin models. The phases of the spin models are accessed by varying a single coupling K. The initial values of r and u depend upon the value of K, and so varying K will move us along a line in the r–u plane: this line of initial values is shown as the dashed line in Fig. 4.2. Now notice that the RG flows predict two very distinct final consequences as we vary K. On the right of the point A in Fig. 4.2, all points ultimately flow to $r \to \infty$; it is natural to identify all such points as residing in the high-temperature paramagnetic phase. In contrast, points to the left of A flow to $r \to -\infty$, as is natural in the ferromagnetic phase. The only point on the line of initial values to avoid these fates is the point A itself, and it flows directly into the Wilson–Fisher fixed point. It is therefore natural to identify A as the point $K = K_c$. Thus we have established that, for essentially all realizations of the O(N) spin model, the physics of the critical point is described by the field theory of the Wilson–Fisher fixed point. This is a key step in the "proof" of the hypothesis of universality: independent of the set of microscopic couplings in H, the critical point is described by the same universal field theory.

Let us examine the structure of the flows near the Wilson–Fisher fixed point. Defining, $r = r^* + \delta r$ and $u = u^* + \delta u/S_D$, (4.24) yields the linearized equations

$$\frac{d}{d\ell}\begin{pmatrix} \delta r \\ \delta u \end{pmatrix} = \begin{pmatrix} 2 - \dfrac{(N+2)\epsilon}{(N+8)} & \dfrac{(N+2)}{6} + \dfrac{(N+2)^2\epsilon}{(N+8)} \\ 0 & -\epsilon \end{pmatrix}\begin{pmatrix} \delta r \\ \delta u \end{pmatrix}. \qquad (4.27)$$

These equations are most easily integrated by diagonalizing the coefficient matrix: this is done by defining two new eigen-couplings $w_{1,2}$ which are different, linearly-independent combinations of δr and δu. The flow of the eigen-couplings is simply

$$\frac{dw_i}{d\ell} = \lambda_i w_i, \quad i = 1, 2 \qquad (4.28)$$

with the eigenvalues

$$\lambda_1 = 2 - \frac{(N+2)\epsilon}{(N+8)} + \mathcal{O}(\epsilon^2), \quad \lambda_2 = -\epsilon + \mathcal{O}(\epsilon^2). \qquad (4.29)$$

The flows in (4.29) are now finally as simple as those for the Gaussian theory in Section 4.1. Regardless of its initial value, the coupling w_2 is attracted to $w_2^* = 0$ and so can be regarded as irrelevant. Setting $w_2^* = 0$ puts us on the track to identifying the universal properties of the critical point. Notice however, that the eigenvalue λ_2 has a small magnitude, and so the flow towards the universal theory will be slow: we expect this flow of w_2 to provide the leading corrections to the universal critical behavior.

Our entire RG analysis has therefore been reduced to the flow of a *single* relevant coupling w_1, with the simple flow equation (4.28). Just as was the case with the coupling r in Section 4.1, the critical point is at $w_1 = w_1^* = 0$, and so may identify this coupling

$$w_1 \sim K_c - K \qquad (4.30)$$

as a measure of the deviation from criticality. The flow of w_1 when combined with the rescaling of the field ϕ_α leads to a number of interesting physical consequences which we explore below.

4.3 Field renormalization

The entire analysis in Section 4.2 did not modify the original rescaling of the field ϕ_α defined in (4.3). This is actually an artifact of working to first order in ϵ. At higher orders in ϵ we do find a momentum in the self-energy, and consequently a renormalization of the gradient term in $\mathcal{S}_\phi$, leading to change in the RG rescaling of ϕ_α. A complete treatment of these effects is far more efficiently carried out using a more formal field-theoretical RG which we will not describe now. Rather we will be satisfied with a shortcut which yields the leading nonvanishing renormalization after some reasonable assumptions.

Anticipating the field scale renormalization, we define the RG rescaling of ϕ_α by

$$\phi'_\alpha(x') = b^{(D-2+\eta)/2}\phi_\alpha(x), \tag{4.31}$$

where η is known as the anomalous dimension of the field ϕ_α. This is of course equivalent to

$$\dim[\phi_\alpha] = (D - 2 + \eta)/2. \tag{4.32}$$

Assuming (4.32), the generalization of the relation (4.8) for the correlator $C(x)$ at the critical point $K = K_c$ or $w_1 = 0$ is

$$C(e^{-\ell}x) = e^{(D-2+\eta)\ell}C(x). \tag{4.33}$$

This implies $C(x) \sim x^{-(D-2+\eta)}$. Taking the Fourier transform, we have for the susceptibility

$$\chi(k) \sim \frac{1}{k^{2-\eta}}. \tag{4.34}$$

Let us see how (4.34) could emerge from an analysis of the Wilson–Fisher fixed point. We set $r = r^*$ and $u = u^*$, and compute $\chi(k)$ using (3.45) and a perturbative expansion for the self-energy $\Sigma(k)$. Because we are at the critical point, $\chi^{-1}(0)$ should vanish, and hence we should have $r - \Sigma(0) = 0$; thus $\chi^{-1}(k) = k^2 - \Sigma(k) + \Sigma(0)$. At first order in u, $\Sigma(k)$ is k-independent, and so we have no correction to the free field behavior. However, at second order in u, we obtain a momentum-dependent Σ given by the Feynman diagram in Fig. 4.3 (which corresponds to the last term in (3.38)), which yields

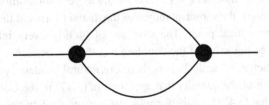

Graph contributing at order ϵ^2 to the anomalous dimension η.

$$\Sigma(k) = \ldots + u^2 \frac{(N+2)}{18} \int \frac{d^D p}{(2\pi)^D} \frac{d^D q}{(2\pi)^D} \frac{1}{(p^2 + r)(q^2 + r)((p + q + k)^2 + r)}.$$

$$(4.35)$$

To leading nonvanishing order in ϵ, we can set $u = u^*$, $r = 0$, and $D = 4$ above. The resulting two-loop integral requires some technical results from the mathematical theory of Feynman graphs to evaluate; the reader can find a concise and useful discussion, with a valuable table of integrals, in the book by Ramond [392]. Evaluating the integral in this manner we find

$$\Sigma(k) = \ldots - \left(u^* S_4\right)^2 \frac{(N+2)}{72} k^2 \ln \frac{\Lambda}{k}.$$

$$(4.36)$$

Inserting this into $\chi^{-1}(k) = k^2 - \Sigma(k) + \Sigma(0)$, and assuming the logarithm is the first term in a series which exponentiates, we obtain the form in (4.34). Using the value of u^* in (4.26), we obtain our needed result for the anomalous dimension of ϕ_α:

$$\eta = \frac{(N+2)}{2(N+8)^2} \epsilon^2 + \mathcal{O}(\epsilon^3).$$

$$(4.37)$$

4.4 Correlation functions

The results of Sections 4.2 and 4.3 will now be collected and applied to determine the form of correlations near the critical point.

Given the form of the field scaling relation (4.32), and the flow of the relevant coupling w_1 in (4.28), the most complete form of the homogeneity relation (4.8) for the two-point ϕ_α correlator is

$$C(e^{-\ell} x; w_1 e^{\lambda_1 \ell}) = e^{(D-2+\eta)\ell} C(x; w_1).$$

$$(4.38)$$

Recall that we have already set the leading irrelevant coupling $w_2 = w_2^* = 0$, and so we are dealing exclusively with the "universal theory." The relation (4.38) holds for any ℓ, and so let us evaluate it at $\ell = \ell^*$ where $w_1 e^{\lambda_1 \ell^*} = \pm 1$; the choice in sign will depend upon the sign of the initial value of w_1, i.e. whether we are above or below the critical point. Then it takes the form

$$C(x; w_1) = \xi^{-(D-2+\eta)} F_\pm(x/\xi),$$

$$(4.39)$$

where we have set $\xi = e^{\ell^*}$ and $F_\pm$ are as yet undetermined functions, known as the scaling functions; the subscript indicates the distinct forms of the scaling function on the two sides of the critical point. The structure of (4.39) is very informative: it indicates that as we change the value of the coupling w_1, the x dependence of the correlations changes at the characteristic scale ξ. It is therefore natural to identify ξ with the correlation length, the analog of the quantity that appeared in (3.47) in the Gaussian theory; indeed, it is easy to see that (3.47) is of the form (4.39), with an explicit result for the scaling function F_+, and an exponent $\eta = 0$ as expected at the Gaussian fixed point. We will mainly be interested in

(4.39) at the Wilson–Fisher fixed point, where η is given by (4.37), and we can make an analogous ϵ expansion for the scaling functions $F_{\pm}$.

The above results also specify the divergence of ξ at the critical point. Using the expressions just before and after (4.39), we have

$$\xi \sim |w_1|^{-1/\lambda_1} \sim |K_c - K|^{-\nu}. \tag{4.40}$$

We meet again the correlation exponent ν, whose value is given quite generally by

$$\nu = \frac{1}{\lambda_1}. \tag{4.41}$$

In other words, the exponent ν is the inverse of the relevant RG eigenvalue. Implicit in our discussion is the assumption that there is only *one* such relevant eigenvalue, and that *all* other perturbations are irrelevant at the Wilson–Fisher fixed point. The reader will recognize that we have in fact established this strong and powerful result to leading order in ϵ. There is overwhelming numerical evidence that such a result is also true at the physically important value of $\epsilon = 1$, and this is behind the success of the ϵ expansion.

We will also have occasion to meet RG fixed points with either zero or more than one relevant perturbation. Those with zero relevant perturbations describe *critical phases* rather than critical points; for all couplings not too far from the fixed point, the flow is towards the fixed point, and hence the long-distance correlations have characteristics independent of all microscopic parameters. The extent of such critical phases is determined by the domain of attraction of the RG fixed point. RG fixed points with more than one relevant perturbation describe "multicritical points." Reaching such multicritical points requires that we tune more than one linearly independent coupling in the underlying Hamiltonian: the number of tuning parameters is equal to the number of relevant perturbations.

Finally, let us show that (4.39) also allows us to determine all other critical exponents we have defined so far. Integrating (4.39) over all x, we obtain using (3.42) the result for the wavevector-dependent susceptibility

$$\chi(k) = \xi^{2-\eta}\widetilde{F}_{\pm}(k\xi), \tag{4.42}$$

where the scaling functions $\widetilde{F}_{\pm}$ are Fourier transforms of $F_{\pm}$. This result clearly generalizes (4.34) away from the critical point. As long as we are not at the critical point, we expect the susceptibility to be nonsingular as $k \to 0$, and so the $\widetilde{F}_{\pm}$ will approach nonzero constants in the limit of zero argument. Consequently, the uniform static susceptibility behaves as

$$\chi \sim \xi^{2-\eta}, \tag{4.43}$$

or the critical exponent γ in (3.46) is given by the exact "scaling relation"

$$\gamma = (2 - \eta)\nu. \tag{4.44}$$

It remains to determine the critical exponent β for the ferromagnetic moment $\langle \phi_\alpha \rangle = N_0 \hat{e}_\alpha$, where $\hat{e}_\alpha$ is an arbitrary unit N-component vector. As discussed regarding (3.24), this moment vanishes as

$$N_0 \sim (K - K_c)^{\beta}, \tag{4.45}$$

as we approach the critical point from low temperatures. Rather than deducing directly from (4.39), let us reapply the RG transformation from scratch. Application of the fundamental scaling relation (4.31) tells us that

$$N_0(w_1 e^{\lambda_1 \ell}) = e^{(D-2+\eta)\ell/2} N_0(w_1),\tag{4.46}$$

where again we neglect the influence of irrelevant couplings like w_2. Evaluating (4.46) at $\ell = \ell^*$ as before, we obtain (4.45) with the scaling relation

$$\beta = (D - 2 + \eta)\nu/2.\tag{4.47}$$

More generally, we can regard (4.47) as the most important application of a scaling relation determining the singular contribution to the average value of any observable $\mathcal{O}$

$$[\langle \mathcal{O} \rangle]_{\text{sing}} \sim \xi^{-\dim[\mathcal{O}]}.\tag{4.48}$$

An important case of the above result is the scaling of the free energy density $\widetilde{\mathcal{F}} = -(1/V) \ln \mathcal{Z}$, where V is the D-dimensional volume of the system. The partition function $\mathcal{Z}$ is a RG invariant, and so the scaling dimension of $\ln \mathcal{Z}$ must be zero. Consequently

$$\dim[\widetilde{\mathcal{F}}] = D,\tag{4.49}$$

and the singular part of the free energy density scales as ξ^{-D}. This result is often stated in terms of the exponent of the specific heat (see Exercise 3.4) $-\partial^2 \widetilde{\mathcal{F}}/\partial r^2 \sim |K - K_c|^{-\alpha}$ for which we have the "hyperscaling relation"

$$\alpha = 2 - D\nu.\tag{4.50}$$

Exercises

4.1 Consider the RG flow of the sixth-order coupling v in (4.12). Compute the one-loop correction to the RG flow in (4.13) by determining the coefficient of the term of order uv on the right-hand side. Hence show that the Wilson–Fisher fixed point has $v^* = 0$, and the fixed-point eigenvalue of the six-order operator is

$$\lambda_v = 6 - 2D - \frac{(n+14)}{2} S_4 u^* + \mathcal{O}(\epsilon^2)$$
$$= -2 - \frac{(n+26)}{(n+8)}\epsilon + \mathcal{O}(\epsilon^2).$$

4.2 This exercise is adapted from [382]. We consider the consequences of anisotropy in the $O(N)$ symmetry of the Wilson–Fisher fixed point. In some applications to classical ferromagnets, spin–orbit interactions may introduce a weak anisotropy in which the $r\phi_\alpha^2$ term is replaced by

$$r_s \sum_{\alpha < N} \phi_\alpha^2 + r_N \phi_N^2,\tag{4.51}$$

while the quartic term is replaced by

$$\frac{u_1}{24} \sum_{\alpha,\beta<N} \phi_\alpha^2 \phi_\beta^2 + \frac{u_2}{12} \sum_{\alpha<N} \phi_\alpha^2 \phi_N^2 + \frac{u_3}{24} \phi_N^4. \tag{4.52}$$

Clearly, the original problem with full $O(N)$ symmetry is the case $r_s = r_n$ and $u_1 = u_2 = u_3$. The model with $r_s = \infty$, $u_1 = u_2 = 0$ is the field theory of the Ising model, while the model with $O(N-1)$ symmetry is $r_n = \infty$, $u_2 = u_3 = 0$.

(a) Show that the one-loop RG flow equations for this model are:

$$\frac{dr_s}{d\ell} = 2r_s + \frac{(N+1)}{6(1+r_s)} S_D u_1 + \frac{1}{6(1+r_n)} S_D u_2,$$

$$\frac{dr_n}{d\ell} = 2r_n + \frac{(N-1)}{6(1+r_s)} S_D u_2 + \frac{1}{2(1+r_n)} S_D u_3,$$

$$\frac{du_1}{d\ell} = \epsilon u_1 - \frac{(n+7)}{6(1+r_s)^2} S_D u_1^2 - \frac{1}{6(1+r_n)^2} S_D u_2^2,$$

$$\frac{du_2}{d\ell} = \epsilon u_2 - \frac{2}{3(1+r_s)(1+r_n)} S_D u_2^2$$

$$\qquad - \frac{(N+1)}{6(1+r_s)^2} S_D u_1 u_2 - \frac{1}{2(1+r_n)^2} S_D u_2 u_3,$$

$$\frac{du_3}{d\ell} = \epsilon u_3 - \frac{3}{2(1+r_n)^2} S_D u_3^2 - \frac{(N-1)}{6(1+r_s)^2} S_D u_2^2. \tag{4.53}$$

(b) Show that these equations reduce to the expected equations in the limits corresponding to the models with $O(N)$, Ising, and $O(N-1)$ symmetry just noted.

(c) Consider the fixed point of the flow equations with $O(N)$ symmetry: $r_s = r_n = r^*$, and $u_1 = u_2 = u_3 = u^*$. Show that, to leading order in ϵ, and for $n \le 4$, this fixed point has *two* relevant eigenvalues $2 - (n+2)\epsilon/(n+8)$ and $2 - 2\epsilon/(n+8)$.

(d) Assume the experimental conditions are such that the u couplings are close to the $O(N)$ fixed point. Describe, qualitatively, the behavior of the susceptibility for $T > T_c$ for the two cases $r_s > r_n$ and $r_n > r_s$.

5 The quantum Ising model

This chapter returns to the quantum Ising model (1.7) which was introduced in Section 1.4.1:

$$H_I = -Jg \sum_i \hat{\sigma}_i^x - J \sum_{\langle ij \rangle} \hat{\sigma}_i^z \hat{\sigma}_j^z. \tag{5.1}$$

We will examine the eigenstates of H_I in more detail in the large and small-g limits. These limits were also studied briefly in Section 1.4.1, where we argued that there was a quantum phase transition at some intermediate value $g = g_c$. We will also study the quantum to classical mapping introduced in Chapter 2, and carefully establish [127, 378] the mapping to the classical Ising model of (2.2). This mapping will eventually lead us to a field-theoretical analysis of the vicinity of the quantum critical point at g_c.

Before embarking on the large and small-g expansions of H_I, we introduce the effective Hamiltonian method in Section 5.1. This is an indispensable tool for characterizing the spectrum of this and other models. This method is described in textbooks on quantum mechanics, and we only recall a basic result which we will put to extensive use.

We then use the effective Hamiltonian method to examine the spectrum of H_I under strong-($g \gg 1$) and weak-($g \ll 1$) coupling limits, which were discussed briefly in Section 1.4.1. The analysis is relatively straightforward in these limits, and two very different physical pictures emerge. In $d = 1$, the model H_I is exactly solvable, and this is described later in Chapter 10: this exact solution shows that there is a critical point exactly at $g = g_c = 1$, but that the qualitative properties of the ground states for $g > g_c$ ($g < g_c$) are very similar to those for $g \gg 1$ ($g \ll 1$). We argue below, and in the following chapters, that these features also hold for $d > 1$, where no exact solution is possible. One of the two limiting descriptions is therefore always appropriate, and only the critical point $g = g_c$ has genuinely different properties at $T = 0$.

5.1 Effective Hamiltonian method

We consider a Hamiltonian of the form $H = H_0 + H_1$, where the eigenstates of H_0 are easily determined, and we are interested in describing the influence of H_1 in perturbation theory. Further, we assume that the eigenvalues of H_0 are separated into distinct groups of closely-spaced levels, such that the energy separation between two levels within the same group is always much smaller than the separation between two levels in distinct groups.

We use the symbols α, β, ... to denote the groups, and i, j, ... to denote levels within each group. Thus the eigenstates of H_0 are $|i, \alpha\rangle$, with eigenenergies $E_{i\alpha}$, and so

$$|E_{i\alpha} - E_{j\alpha}| \ll |E_{i\alpha} - E_{j\beta}| \quad \text{for } \alpha \neq \beta. \tag{5.2}$$

We are often interested in the structure of the levels within a given group, and would like to understand their behavior without reference to levels in other groups. However, in general, H_1 will have nonzero matrix elements between states belonging to different groups: consequently a conventional perturbative analysis will require repeated reference to states lying outside the group of interest. The idea of the effective Hamiltonian method is to perform a unitary transformation which eliminates these inter-group matrix elements. After the unitary transformation, we obtain a new Hamiltonian H_{eff} which has nonzero matrix elements only within each group.

We skip the straightforward, but tedious, analysis needed to obtain H_{eff} order-by-order in H_1. We quote here only the final result to second order in H_1. The new Hamiltonian H_{eff} is defined by the following nonzero matrix elements between any two levels, $|i, \alpha\rangle$ and $|j, \alpha\rangle$ belonging to the same group α:

$$\langle i, \alpha | H_{\text{eff}} | j, \alpha \rangle = E_{i\alpha} \delta_{ij} + \langle i, \alpha | H_1 | j, \alpha \rangle$$
$$+ \sum_{k, \beta \neq \alpha} \frac{\langle i, \alpha | H_1 | k, \beta \rangle \langle k, \beta | H_1 | j, \alpha \rangle}{2} \left(\frac{1}{E_{i\alpha} - E_{k\beta}} + \frac{1}{E_{j,\alpha} - E_{k\beta}} \right) \tag{5.3}$$

Naturally, we have $\langle i, \alpha | H_{\text{eff}} | j, \beta \rangle = 0$ for all $\alpha \neq \beta$, ensuring that H_{eff} is block diagonal, and we can work independently within each group α.

5.2 Large-*g* expansion

In the interests of simplicity, we will restrict our discussion of the large-g expansion to the case $d = 1$. The generalization to $d > 1$ involves only minor differences, and these will be noted explicitly where needed. The situation is very different for the small-g expansion discussed in the following section: there the cases $d = 1$ and $d > 1$ require a separate analysis.

The $g = \infty$ ground state was presented in (1.9), where we also discussed the nature of the $1/g$ corrections. We found a quantum paramagnetic ground state, invariant under the Z_2 symmetry (1.13), with exponentially decaying $\hat{\sigma}^z$ correlations as in (1.11). Conventional perturbation theory can be used to obtain the ground state wavefunction and energy in powers of $1/g$. On a system of M sites, with periodic boundary conditions, the ground state energy is

$$E_0 = -MJg \left(1 + 1/(4g^2) + \mathcal{O}(1/g^3) \right), \tag{5.4}$$

where the leading corrections arise from virtual states with two left-pointing spins created by the exchange term in H_I.

What about the excited states? For $g = \infty$ these can also be listed exactly, and the levels appear in groups as needed for the effective Hamiltonian method. The lowest excited states are

$$|i\rangle = |\leftarrow\rangle_i \prod_{j \neq i} |\rightarrow\rangle_j, \qquad (5.5)$$

obtained by flipping the state on site i to the other eigenstate of $\hat{\sigma}^x$ (the eigenstates of $\hat{\sigma}^x$ were defined in (1.10)). There are M such states, and they are degenerate with energy $E_0 + 2Jg$. We will refer to them as the "single-particle" states. Similarly, the next degenerate manifold of states is the two-particle states $|i, j\rangle$, obtained by flipping the spins at sites i and j; clearly, there are $M(M-1)/2$ such states, and they have energy $E_0 + 4Jg$. Generalizing, we can construct the p particle states: there are $M!/((M-p)!p!)$ states with energy $E_0 + 2pJg$. Identifying the ground state as the $p = 0$ case, the values $p = 0, 1 \ldots M$ clearly span the entire Hilbert space of 2^M states. Of course, when p is of order M, this particle labelling will not be particularly useful.

We now consider the nature of the effective Hamiltonian in the different particle number subspaces in turn.

5.2.1 One-particle states

For the one-particle states, the exchange term $\hat{\sigma}_i^z \hat{\sigma}_{i+1}^z$ in H_I is not diagonal in the basis of the $|\rightarrow\rangle$, $|\leftarrow\rangle$ states and leads only to the off-diagonal matrix element

$$\langle i | H_I | i+1 \rangle = -J, \qquad (5.6)$$

which hops the "particle" between nearest-neighbor sites. As in the tight-binding models of solid state physics [28], the Hamiltonian is therefore diagonalized by going to the momentum space basis

$$|k\rangle = \frac{1}{\sqrt{M}} \sum_j e^{ikx_j} |j\rangle. \qquad (5.7)$$

This eigenstate has energy

$$\varepsilon_k = Jg[2 - (2/g)\cos(ka) + \mathcal{O}(1/g^2)], \qquad (5.8)$$

where a is the lattice spacing, and we have dropped an additive term of the ground state energy E_0. Henceforth, all particle excitation energies will be measured relative to E_0. The lowest energy one-particle state is therefore at $\varepsilon_0 = 2gJ - 2J$.

At next order in $1/g$, H_I mixes the one-particle states with the zero- and two-particle states. Their influence on the one-particle subspace can be described by direct application of (5.3): we find terms by which the particle can hop to second neighbors. (Actually, initially it appears that there are terms by which the particle can hop arbitrary distances, but these cancel between the contributions of the zero- and two-particle subspaces.) The wavefunction (5.7) still diagonalizes the effective Hamiltonian in the one-particle subspace, and then (5.8) improves to (Exercise 5.1)

$$\varepsilon_k = Jg[2 - (2/g)\cos(ka) + (1 - \cos(2ka))/(2g^2) + \mathcal{O}(1/g^3)]. \qquad (5.9)$$

This is a convenient point to define another useful concept: the *quasiparticle residue*, $\mathcal{A}$. The operator $\hat{\sigma}_i^z$ flips the *i*th spin between $|\rightarrow\rangle$ and $|\leftarrow\rangle$, and so moves states up or down one step between the *p*-particle subspaces. So we can regard $\hat{\sigma}^z$ as the sum of a particle creation and annihilation operator. It is useful to consider the operator at a definite momentum *k* by defining, as in (5.7),

$$\hat{\sigma}^z(k) = \frac{1}{\sqrt{M}} \sum_j e^{ikx_j} \hat{\sigma}_j^z. \tag{5.10}$$

Then, the quasiparticle residue is defined by the overlap between the actual one-particle state at momentum $k = 0$, and that obtained by creating a particle in the ground state by the particle creation operator

$$\mathcal{A} \equiv |\langle k = 0|\hat{\sigma}^z(k)|0\rangle|^2. \tag{5.11}$$

The computation of $\mathcal{A}$ in the $1/g$ expansion is discussed in Exercise 5.1. We see later that $\mathcal{A}$ appears naturally in correlation functions associated with neutron scattering experiments. Also, $\mathcal{A}$ is nonzero in the entire paramagnetic phase, but vanishes at the quantum critical point for $d \leq 3$.

5.2.2 Two-particle states

Now consider the two-particle states. At $g = \infty$, the subspace of two-particle states is spanned by the states (generalizing (5.5))

$$|i, j\rangle = |\leftarrow\rangle_i |\leftarrow\rangle_j \prod_{h \neq i, j} |\rightarrow\rangle_h, \tag{5.12}$$

where $i \neq j$. Also note that $|i, j\rangle = |j, i\rangle$, and so we may restrict our attention to $i > j$. Alternatively, we can say that the states are symmetric under interchange of the particle positions i, j, and so we treat the particles as bosons. At first order in $1/g$, these states will be mixed by the matrix element (5.6); this will couple $|i, j\rangle$ to $|i \pm 1, j\rangle$ and $|i, j \pm 1\rangle$ for all $i > j + 1$, while $|i, i - 1\rangle$ will couple only to $|i + 1, i - 1\rangle$ and $|i, i - 2\rangle$. For i and j well separated, we can ignore this last case, and the two particles will be independent of each other, with the matrix elements for each particle identical to those considered above for single particles. So the particles will acquire momenta k_1, k_2 (say), and the total energy of the two-particle state will be $E_k = \varepsilon_{k_1} + \varepsilon_{k_2}$, with a total momentum $k = k_1 + k_2$. However, when i and j approach each other, we have to consider mixing between these momentum states arising from the restrictions in the matrix elements noted above. This is a problem in ordinary scattering theory, treated in many elementary quantum mechanics texts.

Thus we have an important general result: among the two-particle states is a set of scattering eigenstates. These states are labelled by a pair of momenta k_1, k_2 (their ordering is unimportant), and their eigenenergy is *exactly* the sum of the single-particle energies determined in Section 5.2.1, $\varepsilon_{k_1} + \varepsilon_{k_2}$. Clearly, this result is only true in the thermodynamic limit $M \to \infty$ where it is always possible to separate the single-particle states in well separated regions of the sample. It is also clear that this result holds in all d (integrability in $d = 1$ plays no role in this conclusion), and is also true for the *p*-particle states (when we will need p unordered momenta to label the states). The result fails only when p becomes of order M, when it is no longer possible to find well separated regions. In other words, we should take the limit $M \to \infty$ at fixed p.

In some cases, there will also be states other than scattering states among the multiparticle states. These are *bound states* in which two or more particles move together as in a "molecule." Each bound state will be labelled by a single center-of-mass momentum, independent of the number of particles in the bound state. Consideration of the multiparticle Schrödinger equation in the $1/g$ expansion (discussed below) shows that there are no bound states in the present situation. However, we will meet bound states shortly in the discussion on the $g \ll 1$ expansion.

Let us now consider the two-particle scattering states in $d = 1$ more carefully. The scattering of the two incoming particles with momenta k_1, k_2 will conserve total energy and, up to a reciprocal lattice vector $2\pi/a$, total momentum. For small k_1, k_2 (which is our primary interest here), these conservation laws allow only one solution in $d = 1$: the momenta of the particles in the final state are also k_1 and k_2. The existence of a single final state is a special feature of $d = 1$, while a sum over an infinite number of momenta in the final state is required for the problems in $d > 1$ we consider later. By this reasoning, we can conclude that the wavefunction of the two-particle state will have the following form for $i \gg j$:

$$\left(e^{i(k_1 x_i + k_2 x_j)} + S_{k_1 k_2} e^{i(k_2 x_i + k_1 x_j)}\right)|i, j\rangle. \tag{5.13}$$

The quantity $S_{k_1 k_2}$ is of central importance and is the S matrix for two-particle scattering. Upon interpreting the stationary scattering state in (5.13) from the perspective of a time-dependent scattering problem, in which particles scatter from an incoming wave corresponding to the first term in (5.13), to an outgoing wave corresponding to the second term, the S matrix can be related (just as in familiar scattering theory) to the time-evolution operator of H_I from the infinite past to the infinite future; it must therefore be a unitary matrix. In the present situation with a single final state, the S matrix is a complex number of unit modulus. The computation of the S matrix from the Schrödinger equation at order $1/g$ is discussed in Exercise 5.2. The result turns out to be remarkably simple; we find

$$S_{k_1 k_2} = -1, \tag{5.14}$$

for all momenta k_1, k_2. We will not give an explicit derivation of this result here (a detailed discussion of the computation of such S matrices in general spin models may be found in [107]). Instead, we present a simple argument in the next paragraph that shows that a result such as (5.14) holds in the limit of small k_1, k_2 for a generic Ising chain with additional further neighbor exchange couplings; the validity of (5.14) at all momenta is a special feature of the nearest neighbor exchange model (5.1) (also considered later in (10.1)). Our argument will also show that (5.14) continues to hold at higher orders in $1/g$ for small k_1, k_2.

Transform to the center-of-mass frame of the two particles, and consider the Schrödinger equation for their relative coordinates $x = x_i - x_j$. Taking, for simplicity, a repulsive delta function potential $u\delta(x)$ between them (the result does not require this special form), we can write down the schematic Schrödinger equation

$$\left(-\frac{d^2}{dx^2} + u\delta(x)\right)\psi(x) = E\psi(x), \tag{5.15}$$

where x is their relative coordinate and $\psi(x)$ is the wavefunction in the center of mass frame. We make a simple argument based upon dimensional analysis. Note from (5.15) that u has the dimensions of inverse length. The S matrix is a dimensionless quantity, and it can be a function only of u and the relative momentum $k = k_1 - k_2$. Dimensionally, this can only be of the form $S = f(u/k)$, where f is some unknown function. We are interested in the limit $k \to 0$, which is given by the value of $f(\infty)$. However, conceptually, it is much simpler to obtain $f(\infty)$ by taking $u \to \infty$ at fixed k. Thus, to slowly moving particles, the potential appears effectively impenetrable. This means that $\psi(x)$ should vanish at $x = 0$, and its bosonic symmetry under particle exchange implies that it has the form $\psi \sim \sin(k|x|/2)$ for small x. Comparing with (5.13), we conclude that $f(\infty) = -1$, and so (5.14) holds universally in the limit of small momenta.

Similar considerations can be applied to scattering states in dimensions $d > 1$. In general, the coupling u in (5.15) has dimensions of (length)$^{(d-2)}$. So the S matrix is a dimensionless function of uk^{d-2}. For $d > 2$, this means that the $k \to 0$ limit is equivalent to the $u \to 0$ limit. Consequently, there is no scattering at low momenta, and the S matrix equals $+1$. Note the striking contrast from $d = 1$, where the low momentum S matrix is generically -1. The $d = 2$ case is marginal, and is discussed further in Chapter 16: there S matrix reaches $+1$ at low momenta, but only logarithmically slowly.

We have now described the manner in which $1/g$ perturbations lift the degeneracy of the $g = \infty$ two-particle eigenstates (5.12). The energy of a two-particle state with total momentum k is given by $E_k = \varepsilon_{k_1} + \varepsilon_{k_2}$, where $k = k_1 + k_2$. Note that, for a fixed k, there is still an arbitrariness in the single-particle momenta $k_{1,2}$ and so the total energy E_k can take a range of values. There is thus no definite energy–momentum relation, and we have instead a "two-particle continuum." It should be clear, however, that the lowest energy two-particle state in the infinite system (its "threshold") is at $2\varepsilon_0$.

Most of the above analysis can be generalized to the $p > 2$ particle states: there are no bound states, and the scattering states have thresholds at $p\varepsilon_0$.

After accounting for the finite bandwidth of the p-particle states, it is possible that as g becomes smaller, the eigenenergies of states with different numbers of particles can start to overlap. At this point, our effective Hamiltonian method must break down, because the energy differences between states in two different groups can vanish. The most important new phenomenon that appears now is the possibility of the decay of particles into multiparticle final states: the rate of this decay can be determined by Fermi's golden rule. The simplest example of this is the decay of a single particle into a three-particle final state (the decay to two particles is forbidden by the Z_2 symmetry), and this becomes possible for a particle with sufficiently large momentum not too far from the quantum critical point. However, the one-particle states with momenta over a finite range near $k = 0$ always remain rigorously stable in the paramagnetic phase. This is because their decay is forbidden by energy–momentum conservation as long as the energy gap, ε_0, is nonzero: there is not enough kinetic energy to overcome the cost of creating two additional particles, each costing energy at least ε_0, at these low momenta.

We present a more complete discussion of the stability of single particle states, and of decay processes, after we have developed the field-theoretic methods for the vicinity of the quantum critical point.

Upon explicitly carrying out these higher order computations for scattering between multiparticle states for the particular nearest-neighbor model H_I, some rather miraculous features emerge for this special Hamiltonian in $d = 1$. As already noted, the result (5.14) holds not only at small k_1, k_2, but at all momenta and at all orders in $1/g$. There are also no processes involving the decay of particles, even though this might be energetically permitted. This remarkable fact appears quite mysterious at this stage, but is explained rather simply in Section 10.1 using a mapping of H_I to fermionic variables.

5.3 Small-g expansion

The $g = 0$ ground states were given in (1.12). They are twofold degenerate and possess long-range correlations in the magnetic order parameter $\hat{\sigma}^z$. The spontaneous magnetization N_0 equals $\pm \langle \hat{\sigma}^z \rangle$ in the two ground states, corresponding to spontaneous breaking of the Z_2 symmetry (1.13). All of the statements made in this paragraph clearly hold for $g = 0$, and they will hold for some $g > 0$ provided that the perturbation theory in g has a nonzero radius of convergence. The exact solution of the $d = 1$ model discussed in Chapter 10 verifies that this is indeed the case.

Much of the analysis in the small-g limit parallels that of the large-g expansion in the previous section. The analog of the expansion in the ground state energy in (5.4) now yields

$$E_0 = -MJ \left(d + g^2/(4d) + \mathcal{O}(g^3) \right), \tag{5.16}$$

in d dimensions, where the second term is the contribution of fluctuations to states with one spin flipped from the majority direction. The spontaneous magnetization can be computed by expanding (1.15), and yields (Exercise 5.3)

$$N_0 = 1 - g^2/(8d^2) + \mathcal{O}(g^4). \tag{5.17}$$

We find that the states are labelled by particle number, but the physical interpretations of the particles are now different from the large-g case. Indeed, the nature of the particle states is very different for $d \geq 2$ and $d = 1$, and so we consider these cases separately. The $d > 2$ cases are, however, quite similar to $d = 2$, and so we only describe $d = 2$ and $d = 1$ in the subsections below.

5.3.1 $d = 2$

Here, and below, we only describe the excited states of the ferromagnetic state with moment pointing up, $| \uparrow \rangle$ in (1.12). Those of the $| \downarrow \rangle$ state have the corresponding structure after a global spin flip.

Throughout, our $d = 2$ analysis is carried out on the square lattice, although the generalization to other lattices is straightforward.

The single-particle states are quite similar to those of Section 5.2.1, after a ninety degree rotation in spin space. At $g = 0$, they consist of a single down spin in a background of up spins. Following (5.5), we can write these as

$$|i\rangle = |\downarrow\rangle_i \prod_{j \neq i} |\uparrow\rangle_j. \qquad (5.18)$$

There are M such one-particle states on a lattice of M sites, and at $g = 0$ they all have energy $E_0 + 8J$.

At order g, we find that the transverse field term has vanishing matrix elements between the single-particle states. This is different from the $1/g$ expansion, where (5.6) already induced a nearest-neighbor hopping among the single-particle states. We therefore have to use (5.3) to find a nearest-neighbor hopping at order g^2. This involves using intermediate states with zero or two particles; we find that their contributions to (5.3) cancel among each other, except when the initial and final sites are nearest neighbors. (see Exercise 5.3). Then, forming momentum eigenstates just as in (5.7), we find that the dispersion of the one-particle states is (we subtract E_0 from all energies from now on)

$$\varepsilon_k = J \left[8 - \frac{g^2}{4}(1 + \cos k_x + \cos k_y) + \mathcal{O}(g^3) \right]. \qquad (5.19)$$

Next we turn to the two-particle states. When the two flipped spins are separated from each other, the energy of such states at $g = 0$ is $16J$. However, when the flipped spins are nearest neighbors, their energy takes the smaller value of $12J$. These nearest neighbor states therefore form bound states below the two-particle continuum. There are $2M$ such nearest neighbor states: after moving to momentum eigenstates, this means that there will be two bound states at each momentum k. Again, we need the effective Hamiltonian method to compute the dispersion of these bound states at order g^2: this is discussed in Exercise 5.3. We find that the bound states are symmetric and antisymmetric combinations of the horizontal and vertical pairs of spins. In other words, the bound states have an internal angular momentum, and are s and d wave pairs of the single-particle states.

Apart from the $2M$ bound states, the remaining $M(M - 1)/2 - 2M$ two-particle states form scattering states with energy $16J + \mathcal{O}(g)$. We will not compute their scattering S matrix, but it again involves application of (5.3).

Similar considerations apply to $p > 2$ particle states. Their scattering states are at $8pJ$, but there are always bound states at lower energies. As an example, we show three- and four-particle bound states with energy $14J$ and $16J$ in Fig. 5.1.

The three-particle bound state with energy $14J$ is therefore below the two-particle continuum. In the present magnetically ordered state, there is no symmetry prohibiting a nonzero matrix element between different particle number states at some high order in the effective Hamiltonian. Nevertheless, for a finite range of small-g, energy conservation will prevent decay of the three-particle bound state with energy $14J$ into the two-particle continuum, and this bound state is therefore stable. However, the bound states with energy

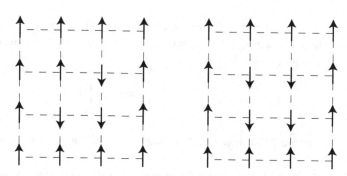

Fig. 5.1
Three- and four-particle bound states in the magnetically ordered phase.

$16J$ will decay into the two-particle continuum. For p-particle bound states with larger p, there are many more channels for decay into scattering states, and so these are much less likely to be stable.

5.3.2 $d = 1$

We can initially try to describe the low-lying excited states of the $d = 1$ Ising chain also by the M states in (5.18), which have energy $4J$. However, they are degenerate with $\sim M^2$ states of the following type:

$$|\cdots \uparrow\uparrow\uparrow\downarrow\downarrow\downarrow\downarrow\uparrow\uparrow\uparrow \cdots\rangle. \tag{5.20}$$

These states can be interpreted as pairs of *domain walls* between the $|\uparrow\rangle$ and $|\downarrow\rangle$ ferromagnetic states. There is no additional energy cost to moving the domain walls apart from each other (i.e. in changing the number of down spins in (5.20)), and we can also view the state in (5.18) as a pair of domain walls on nearest neighbor links. It is not possible to give a one-particle interpretation to these degenerate domain wall-pair states. Particle-like states are labelled by a single momentum, and this can only take $\sim M$ values.

However, it is clear that a *single* domain wall (or "kink") can serve as the needed elementary particle excitation. The one-particle state located between sites i and $i + 1$ is

$$\cdots |\uparrow\rangle_{i-2} |\uparrow\rangle_{i-1} |\uparrow\rangle_i |\downarrow\rangle_{i+1} |\downarrow\rangle_{i+2} |\downarrow\rangle_{i+3} |\downarrow\rangle_{i+4} \cdots$$

At $g = 0$ the energy of such a state is $2J$, and there are $2M$ such states. At nonzero g, the transverse field term induces nearest neighbor hops between such domain wall states, already at first order in g. Consequently, we can form momentum eigenstates, and these have the dispersion

$$\varepsilon_k = J(2 - 2g\cos(ka) + \mathcal{O}(g^2)). \tag{5.21}$$

A novel feature of these one-particle kink states is that they require a large deformation away from the ground state wavefunction: we needed to flip all spins to the right (say) of the particle location. Therefore, they have a nonlocal "topological" character. As a consequence, these particles do not show up as single particle-like excitations in experiments with local probes: this will become clear in our subsequent discussion of correlation functions in Chapter 10. Nevertheless, they are necessary for a proper and complete understanding of the spectrum of the quantum Ising chain in its ferromagnetic phase.

At energies higher than $2J$, the spectrum can be completely interpreted in terms of p-particle scattering states of the kinks; this includes the state in (5.18) which is now part of the two-particle continuum. The S matrix for the collision of two domain walls can be computed in a perturbation theory in g, and we find results very similar to those in the strong-coupling $1/g$ expansion: For the generic Ising chain we find $S_{kk'} = -1$ in the low momentum limit, but for the particular nearest-neighbor chain (10.1) we find that there is no particle production, and $S_{kk'} = -1$ at all momenta to all orders in g. These special features are consequences of the integrability of the quantum Ising chain, which is discussed in Chapter 10. There we will see that the kinks are best interpreted as free fermions.

5.4 Review

Our methods have so far given a satisfactory description of the two phases of the quantum Ising model in all d. For $g \gg 1$, we have the paramagnetic phase, above which there are particle excitations consisting of "left"-pointing spins. For $g \ll 1$, we have the ferromagnetic phase with a broken Z_2 symmetry and two degenerate ground states with long-range magnetic order. Above these ground states we have excitations consisting either of "bubbles" of spins oriented opposite to the ferromagnetic moment (in $d \geq 2$) or of kinks (in $d = 1$).

We would now like to see how these descriptions meet at the quantum critical point at $g = g_c$ between them. We will see that the two descriptions remain qualitatively valid at the lowest energies all the way up to, but not including, the critical point. At higher energies, and also exactly at the critical point, we need a new description.

We reach the needed description by a detour. We now establish the promised equivalence between the d-dimensional quantum Ising model (5.1) and the $D = d+1$ dimensional classical Ising model (3.1). Then, we transfer the insights developed for the classical critical point in Chapters 3 and 4 to the quantum case.

We begin in Section 5.5 by first considering the $D = 1$ case.

5.5 The classical Ising chain

Here we consider the $D = 1$, $N = 1$ classical spin ferromagnet, more commonly known as the ferromagnetic Ising chain [238]. This chain has the partition function

$$\mathcal{Z} = \sum_{\{\sigma_\ell^z = \pm 1\}} \exp(-H), \tag{5.22}$$

where σ_ℓ^z are Ising spins on sites ℓ of a chain, which take the values ± 1, and H is given by

$$H = -K \sum_{\ell=1}^{M_\tau} \sigma_\ell^z \sigma_{\ell+1}^z - h \sum_{\ell=1}^{M_\tau} \sigma_\ell^z. \tag{5.23}$$

In all of our discussion of classical statistical mechanics models we absorb "temperature" into the definition of the coupling constants, as we have done above for K and h; in contrast, the temperature of quantum-mechanical models will always be explicitly indicated, and we will reserve the symbol T for it (as we see below, the total length of the classical model will determine T). There are a total of M_τ Ising spins (M_τ large), and for convenience we have also added a uniform magnetic field h acting on all the spins. We assume periodic boundary conditions, and therefore $\sigma^z_{M_\tau+1} \equiv \sigma^z_1$.

We evaluate the partition function exactly following the original solution of Ising [238]. The trick is to write $\mathcal{Z}$ as a trace over a matrix product, with one matrix for every site on the chain. Note that the partition functions involve the exponential of a sum of terms on the sites of the chain. Rewrite this as the product of exponentials of each term, and we easily obtain

$$\mathcal{Z} = \sum_{\{\sigma^z_\ell\}} \prod_{\ell=1}^{M_\tau} T_1(\sigma^z_\ell, \sigma^z_{\ell+1}) T_2(\sigma^z_\ell), \tag{5.24}$$

where $T_1(\sigma^z_1, \sigma^z_2) = \exp(K\sigma^z_1\sigma^z_2)$ and $T_2(\sigma^z) = \exp(h\sigma^z)$. Now note that (5.24) has precisely the structure of a matrix product, if we interpret the two possible values of σ^z_ℓ as the index labeling the rows and columns of a 2×2 matrix T_1; T_2 has only one index and so should be interpreted as a diagonal matrix. Thus we have

$$\mathcal{Z} = \mathrm{Tr}\,(T_1 T_2 T_1 T_2 \cdots M_\tau \text{ times} \cdots), \tag{5.25}$$

where the summation over the $\{\sigma^z_\ell\}$ has been converted to a matrix trace because of the periodic boundary conditions, and

$$T_1 = \begin{pmatrix} e^K & e^{-K} \\ e^{-K} & e^K \end{pmatrix}, \qquad T_2 = \begin{pmatrix} e^h & 0 \\ 0 & e^{-h} \end{pmatrix}. \tag{5.26}$$

The matrix $T_1 T_2$ is identified as the "transfer matrix" of the Ising chain, H (5.23), the nomenclature suggesting that it transfers the trace over spins from each site to its neighbor. We can manipulate (5.25) into

$$\begin{aligned} \mathcal{Z} &= \mathrm{Tr}\,(T_1 T_2)^{M_\tau} \\ &= \mathrm{Tr}\,\left(T_2^{1/2} T_1 T_2^{1/2}\right)^{M_\tau} \\ &= \epsilon_1^{M_\tau} + \epsilon_2^{M_\tau}, \end{aligned} \tag{5.27}$$

where $\epsilon_{1,2}$ are the eigenvalues of the symmetric matrix

$$T_2^{1/2} T_1 T_2^{1/2} = \begin{pmatrix} e^{K+h} & e^{-K} \\ e^{-K} & e^{K-h} \end{pmatrix}, \tag{5.28}$$

given by

$$\epsilon_{1,2} = e^K \cosh(h) \pm (e^{2K} \sinh^2(h) + e^{-2K})^{1/2}. \tag{5.29}$$

With these eigenvalues, (5.27) leads to an exact result for the free energy $F = -\ln \mathcal{Z}$. We will return to interpreting this result for F shortly.

Now we show how the above approach can also lead to exact information on correlation functions. For simplicity, we consider only the case $h = 0$ (the generalization to nonzero h is not difficult) and describe the two-point spin correlator, like that in (3.39)

$$C(\ell - \ell') \equiv \langle \sigma_\ell^z \sigma_{\ell'}^z \rangle = \frac{1}{\mathcal{Z}} \sum_{\{\sigma_\ell^z\}} \exp(-H) \sigma_\ell^z \sigma_{\ell'}^z. \qquad (5.30)$$

Going through exactly the same steps as those in the derivation of (5.27) we see that

$$\langle \sigma_\ell^z \sigma_{\ell'}^z \rangle = \frac{1}{\mathcal{Z}} \mathrm{Tr} \left(T_1^{M_\tau - \ell'} \hat{\sigma}^z T_1^{\ell' - \ell} \hat{\sigma}^z T_1^\ell \right), \qquad (5.31)$$

where we have assumed that $\ell' \geq \ell$, and σ^z (without a site index) is also interpreted as a 2×2 diagonal Pauli matrix $\hat{\sigma}^z$ in (1.8). The trace in (5.31) can be evaluated in closed form in the basis in which T_1 is diagonal. The eigenvectors of T_1 are the states in (1.10) and the corresponding eigenvalues are $\epsilon_1 = 2 \cosh(K)$ and $\epsilon_2 = 2 \sinh(K)$. Using the matrix elements $\langle \rightarrow | \sigma^z | \rightarrow \rangle = \langle \leftarrow | \sigma^z | \leftarrow \rangle = 0$ and $\langle \rightarrow | \sigma^z | \leftarrow \rangle = \langle \rightarrow | \sigma^z | \leftarrow \rangle = 1$ we obtain from (5.27) and (5.31)

$$\langle \sigma_\ell^z \sigma_{\ell'}^z \rangle = \frac{\epsilon_1^{M_\tau - \ell' + \ell} \epsilon_2^{\ell' - \ell} + \epsilon_2^{M_\tau - \ell' + \ell} \epsilon_1^{\ell' - \ell}}{\epsilon_1^{M_\tau} + \epsilon_2^{M_\tau}}. \qquad (5.32)$$

Equations (5.31) and (5.32) are our main results on the Ising chain with an arbitrary number of sites, M_τ. While simple, they contain a great deal of useful information, as we will now show; much of the structure we extract below generalizes to more complex Ising models with nonnearest-neighbor interactions.

Let us examine the form of the correlations in (5.32) in the limit of an infinite chain ($M_\tau \rightarrow \infty$); then we have

$$\langle \sigma_\ell^z \sigma_{\ell'}^z \rangle = (\tanh(K))^{\ell' - \ell}. \qquad (5.33)$$

It is useful for the following discussion to label the spins not by the site index i, but by a physical length coordinate τ; we have chosen the symbol τ, rather than the more conventional x, because we will shortly interpret this "length" as the imaginary time direction of a quantum problem. So if we imagine that the spins are placed on a lattice of spacing a, then $\sigma^z(\tau) \equiv \sigma_\ell^z$ where

$$\tau = \ell a. \qquad (5.34)$$

With this notation, we can write (5.33) as

$$C(\tau) \equiv \langle \sigma^z(\tau) \sigma^z(0) \rangle = e^{-|\tau|/\xi}, \qquad (5.35)$$

where the correlation length, ξ, is given by

$$\frac{1}{\xi} = \frac{1}{a} \ln \coth(K). \qquad (5.36)$$

We emphasize that the symbol ξ always represents the actual correlation length at $h = 0$; the actual correlation length for $h \neq 0$ will, of course, be different. In the large-K limit, the correlation length becomes much larger than the lattice spacing, a:

$$\frac{\xi}{a} \approx \frac{1}{2} e^{2K} \gg 1, \quad K \gg 1. \tag{5.37}$$

In the sequel, we shall primarily be interested in physics on the scale of order ξ, in the regime where ξ is much greater than a. It is precisely in this situation that the concepts of the scaling limit and universality become useful, and they are introduced in the following subsections.

5.5.1 The scaling limit

The simplest way to think of the scaling limit is to first divide all lengths into "large" and "small" lengths. For the Ising chain, we take the correlation length ξ, the observation scale τ, and the system size

$$L_\tau \equiv M_\tau a \tag{5.38}$$

as our large lengths, and the lattice spacing, a, as the only small length. The scaling limit of an observable is then defined as its value when all corrections involving the ratio of small to large lengths are neglected.

There are two conceptually rather different, but equivalent, ways of thinking about the scaling limit. We can either send the small length a to zero while keeping the large lengths fixed (as particle physicists are inclined to do) or send all the large lengths to infinity while keeping a fixed (as is more common among condensed matter physicists). Because the physics can only depend upon the ratio of lengths, it is clear that the two methods are equivalent. We shall choose among these points of view at our convenience and show that it is often very useful to straddle this cultural divide and use the insights of both perspectives.

To complete the definition of the scaling limit, we also have to discuss the manner in which the parameters K and h must be treated. From (5.36), we see that K can be expressed in terms of the ratio of lengths ξ/a; we can use this to eliminate explicit dependence upon K, and then the scaling limit is specified by the already specified $\xi/a \to \infty$ limit. It remains to discuss the behavior of h. In general, there is no a-priori way of determining this and one has to examine the structure of the correlation functions to determine the appropriate limit. Let us guess the answer here by a physical argument. The scaling limit involves the study of large K, when the spin correlation length becomes large. Under these conditions, spins a few lattice spacings apart invariably point in the same direction, and they should therefore be sensitive to the mean magnetic field h per unit length. This is measured by $\tilde{h}$, defined by

$$\tilde{h} \equiv \frac{h}{a}. \tag{5.39}$$

So we take the scaling limit $a \to 0$ while keeping $\tilde{h}$ fixed; any other choice would result in a limiting theory with spins under the influence of a field with either infinite or vanishing strength. Alternatively stated, we have chosen $1/\tilde{h}$, a quantity with the dimensions of length, as one of our large length scales.

We have assembled all the necessary steps for the scaling limit. Express any observable in terms of the physical length τ, replace the number of sites M_τ by L_τ/a, solve (5.36) to express K in terms of ξ/a, and use (5.39) to replace h by $\tilde{h}$. Then take the limit $a \to 0$ at fixed τ, L_τ, ξ, and $\tilde{h}$.

We first describe the results for the free energy. The quantity with the finite scaling limit should clearly be the free energy density, $\mathcal{F}$:

$$\mathcal{F} = -(\ln \mathcal{Z})/(M_\tau a). \tag{5.40}$$

First, from (5.29) we obtain in the scaling limit

$$\epsilon_{1,2} \approx \left(\frac{2\xi}{a}\right)^{1/2}\left[1 \pm \frac{a}{2\xi}(1 + 4\tilde{h}^2\xi^2)^{1/2}\right]. \tag{5.41}$$

Inserting this into (5.29), and using the identity $\lim_{y\to\infty}(1 + c/y)^y = e^c$, we obtain

$$\mathcal{F} = E_0 - \frac{1}{L_\tau}\ln\left[2\cosh\left(L_\tau\sqrt{1/(4\xi^2) + \tilde{h}^2}\right)\right], \tag{5.42}$$

where $E_0 = -K/a$ is the ground state energy per unit length of the chain in zero external field.

In a similar manner, we can take the scaling limit of the correlation function in (5.32), which we recall was in zero external field $\tilde{h} = 0$. We obtain

$$C(\tau) = \langle \sigma^z(\tau)\sigma^z(0) \rangle = \frac{e^{-|\tau|/\xi} + e^{-(L_\tau - |\tau|)/\xi}}{1 + e^{-L_\tau/\xi}}. \tag{5.43}$$

The results (5.42) and (5.43) are the main conclusions of this subsection.

5.5.2 Universality

The assertion of *universality* is that the results of the scaling limit are not sensitive to the precise microscopic model being used. This can be seen as the formal consequence of the physically reasonable requirement that correlations at the scale of large ξ should not depend upon the details of the interactions on the scale of the lattice spacing, a.

Let us describe this by an explicit example. Suppose, instead of using the model H in (5.23), we worked with a Hamiltonian H_1 with both first (K_1) and second (K_2) neighbor exchanges between the Ising spins σ^z. This model can also be solved by the transfer matrix methods (one needs a basis of four sites corresponding to the four states of two near-neighbor spins, and the transfer matrix is 4×4), but we will not present the explicit solution here. From the solution we can determine the correlation length, ξ of H_1, which will be a function of both K_1 and K_2. Now, as in Section 5.5.1, express the free energy density in terms of ξ, and take the limit $a \to 0$ at fixed ξ, L_τ, and $\tilde{h}$. The implication of universality is that the result will be precisely identical to (5.42), with E_0 given by the ground state energy density of H_1 in zero field: $E_0 = -(K_1 + K_2)/a$. The reader is invited to check this assertion for this simple example.

We can make the above assertion more precise by introducing the concept of a *universal scaling function*. We write (5.42) in the form

$$\mathcal{F} = E_0 + \frac{1}{L_\tau} \Phi_{\mathcal{F}} \left(\frac{L_\tau}{\xi}, \tilde{h} L_\tau \right), \tag{5.44}$$

where $\Phi_{\mathcal{F}}$ is the universal scaling function, whose explicit value can be easily deduced by comparing with (5.42). Notice that the arguments of $\Phi_{\mathcal{F}}$ are simply the two dimensionless ratios that can be made out of the three large lengths at our disposal: L_τ, ξ, and $1/\tilde{h}$. The prefactor, $1/L_\tau$, in front of $\Phi_{\mathcal{F}}$ is necessary because the free energy density has dimensions of inverse length.

As its name implies, the scaling function $\Phi_{\mathcal{F}}$ is independent of microscopic details. In contrast, E_0, the ground state energy of the Ising chain, clearly depends sensitively on the values of the microscopic exchange constants; it is therefore identified as a *nonuniversal* additive contribution to $\mathcal{F}$.

In a similar manner, we can introduce a universal scaling function of the two-point correlation function of (3.39). We have

$$C(\tau) = \langle \sigma^z(\tau) \sigma^z(0) \rangle = \Phi_\sigma \left(\frac{\tau}{L_\tau}, \frac{L_\tau}{\xi}, \tilde{h} L_\tau \right), \tag{5.45}$$

where Φ_σ is another universal scaling function, and there is now no nonuniversal additive constant. Again Φ_σ is a function of all the independent dimensionless combinations of large lengths; there is no prefactor because the correlator is clearly dimensionless. We can read off the value of $\Phi_\sigma(y_1, y_2, 0)$ by comparing (5.45) with (5.43), but determining the full function $\Phi(y_1, y_2, y_3)$ requires knowledge of the lattice correlator in the presence of a nonzero h, which is somewhat tedious to obtain. A simpler method becomes apparent in the following subsection.

5.5.3 Mapping to a quantum model: Ising spin in a transverse field

We show that the statistical mechanics of the Ising chain can be mapped onto the quantum mechanics of a *single* Ising spin [152, 496]. Further, as stated in the introduction to this chapter, correlators of the quantum spin precisely reproduce the scaling limit of the classical Ising chain.

Let us return to the expressions (5.25) and (5.26) and write the transfer matrices T_1, T_2 in terms of ratios of "large" to small length scales. We have

$$T_1 = e^K (1 + e^{-2K} \hat{\sigma}^x)$$

$$\approx e^K (1 + (a/2\xi) \hat{\sigma}^x)$$

$$\approx \exp(a(-E_0 + (1/2\xi) \hat{\sigma}^x)),$$

$$T_2 = \exp(a\tilde{h}\hat{\sigma}^z), \tag{5.46}$$

where $\hat{\sigma}^{x,z}$ are the Pauli matrices in (1.8). Note that both $T_{1,2}$ have the form e^{aO}, where O is some operator, acting on the $|\uparrow, \downarrow\rangle$ states, that is independent of a. Using the fact that $e^{aO_1}e^{aO_2} = e^{a(O_1+O_2)}(1 + \mathcal{O}(a^2))$, we can write (5.25) in the limit $a \to 0$ as

$$T_1 T_2 \approx \exp(-a H_Q),$$

$$\mathcal{Z} = (T_1 T_2)^M \approx \mathrm{Tr}\exp(-H_Q/T), \tag{5.47}$$

where

$$H_Q = E_0 - \frac{\Delta}{2}\hat{\sigma}^x - \tilde{h}\hat{\sigma}^z, \tag{5.48}$$

with

$$T \equiv \frac{1}{L_\tau}, \quad \Delta \equiv \frac{1}{\xi}. \tag{5.49}$$

We have introduced the fundamental *quantum* Hamiltonian H_Q. It describes the dynamics of a single Ising quantum spin, whose Hilbert space consists of the two states $|\uparrow, \downarrow\rangle$, and which is under the influence of a longitudinal field $\tilde{h}$ and a *transverse* field Δ; it is the single-site version of (1.7) with an additional longitudinal field. Note, from the first relation in (5.47), that the transfer matrix of the classical chain H is the quantum evolution operator $e^{-H_Q \tau}$ over an imaginary time $\tau = a$, the lattice spacing. Thus the transfer from one site to the next is similar to evolution in imaginary time, and length coordinates for the classical chain translate into imaginary time coordinates for the quantum model H_Q. The energy Δ is also the *gap* between the ground and excited states of H_Q in zero (longitudinal) field, and it is precisely equal to the inverse of the correlation length of the classical Ising chain, as expected from the length to time mapping. Further, the partition function of the quantum spin is taken at a temperature T that precisely equals the inverse of the total length of the classical chain. These correspondences between a gap of a quantum system and a correlation length of the corresponding classical model along the "time" direction, and between the temperature of the quantum system and the total length of the classical model, are extremely general and apply to essentially all of the models we consider in this book.

We can use (5.47) and (5.48) to quickly evaluate the free energy of the quantum spin, $\mathcal{F} = -T \ln \mathcal{Z}$. The eigenenergies of H_Q are $E_0 \pm [(\Delta/2)^2 + \tilde{h}^2]^{1/2}$, and we have

$$\mathcal{F} = E_0 - T \ln\left[2\cosh\left(\sqrt{(\Delta/2)^2 + \tilde{h}^2}/T\right)\right], \tag{5.50}$$

which agrees precisely with the scaling limit of the classical Ising chain (5.42). Indeed, the single spin quantum Hamiltonian H_Q is precisely *the* theory describing the universal scaling properties of the entire class of classical Ising chains with short-range interactions. Statements of this type are often shortened to "H_Q is the scaling theory of H."

The correspondence between H_Q and H also extends to correlation functions. Let us define the *time-ordered* correlator, C, of H_Q in imaginary time by

$$C(\tau_1, \tau_2) = \begin{cases} \frac{1}{\mathcal{Z}}\mathrm{Tr}\left(e^{-H_Q/T}\hat{\sigma}^z(\tau_1)\hat{\sigma}^z(\tau_2)\right) & \text{for } \tau_1 > \tau_2, \\ \frac{1}{\mathcal{Z}}\mathrm{Tr}\left(e^{-H_Q/T}\hat{\sigma}^z(\tau_2)\hat{\sigma}^z(\tau_1)\right) & \text{for } \tau_1 < \tau_2, \end{cases} \tag{5.51}$$

where $\hat{\sigma}^z(\tau)$ is defined by the imaginary time evolution under the H_Q:

$$\hat{\sigma}^z(\tau) \equiv e^{H_Q \tau} \hat{\sigma}^z e^{-H_Q \tau}. \tag{5.52}$$

Now, upon carrying through the mapping described above for the free energy for the case of the correlation function, we find that the correlator C above is indeed the same as the classical model correlation function in (3.39), (5.30), and (5.35)

$$C(\tau_1, \tau_2) = \lim_{a \to 0} \langle \sigma^z(\tau_1) \sigma^z(\tau_2) \rangle_H, \tag{5.53}$$

where we have emphasized by the subscript that the average on the right-hand side is for the classical model with Hamiltonian H. The time-ordered functions appear in the quantum problem for the same reason we had to assume $\ell' \geq \ell$ in (5.31): as the transfer matrix evolves the system from "earlier" sites to "later" sites, the earlier $\hat{\sigma}^z$ operators appear first in the trace.

The representation (5.51) also makes the origin of the mapping between the quantum gap, Δ, and the classical correlation length, ξ, in (5.49) quite clear. We can evaluate (5.51) at $T = 0$ by inserting a complete set of H_Q eigenstates and obtain the general representation

$$C(\tau_1, \tau_2) = \sum_n |\langle 0|\sigma^z|n\rangle|^2 e^{-(E_n - E_0)|\tau_1 - \tau_2|}, \tag{5.54}$$

where $|n\rangle$ are all the eigenstates of H_Q with eigenvalues E_n, and $|0\rangle$ is the ground state. For sufficiently large $|\tau_1 - \tau_2|$, the sum over n will be dominated by the lowest energy state for which the matrix element is nonzero, and this gives an exponential decay of the correlation function over a "length" $\xi = 1/(E_1 - E_0) = 1/\Delta$. Of course, in the present simple system there are only a total of two states, but this result is clearly more general.

It is quite easy to evaluate (5.51) for H_Q, and the direct quantum computation is much simpler than the use of classical mapping in (5.53). We find

$$C(\tau_1, \tau_2) = \Phi_\sigma \left(T(\tau_1 - \tau_2), \frac{\Delta}{T}, \frac{\tilde{h}}{T} \right), \tag{5.55}$$

where Φ_σ is precisely the same scaling function that appeared in (5.45) and can be computed from (5.51) to be

$$\Phi_\sigma(y_1, y_2, y_3) = \frac{4y_3^2}{y_2^2 + 4y_3^2} + \frac{y_2^2}{y_2^2 + 4y_3^2} \frac{\cosh\left(\sqrt{y_2^2 + 4y_3^2}(1 - 2|y_1|)/2\right)}{\cosh\left(\sqrt{y_2^2 + 4y_3^2}/2\right)}. \tag{5.56}$$

It can be checked that the $y_3 = 0$ case of this result agrees with the combination of (5.43) and (5.45).

5.6 Mapping of the quantum Ising chain to a classical Ising model

We now move beyond the single-site quantum case, and apply the methods of Section 5.5 to the original quantum Ising model in (5.1). We only consider the case $d = 1$, $D = 2$ here;

the generalization to higher d is immediate, and does not involve any new subtleties. Our presentation here is the inverse of that in Section 5.5: we begin with the quantum model and derive an equivalent classical model.

As in (5.47), we consider the transfer matrix associated with imaginary time evolution over a short time a. Thus we can write the partition function as

$$\mathcal{Z} = \text{Tr} \exp\left(-H_I/T\right) = \text{Tr}\left(\exp\left(-aH_I\right)\right)^{M_\tau},   \tag{5.57}$$

where, as in (5.38) and (5.49), $L_\tau = M_\tau a = 1/T$. Referring to Fig. 2.1, we can view this as the statistical mechanics of a system of M_τ rows in τ direction, with $\exp\left(-aH_Q\right)$ being the transfer matrix from one row to the next. We have M sites in the quantum model along the x direction, and hence this transfer matrix is $2^M \times 2^M$ dimensional, much more complex than the simple 2×2 transfer matrices we met in Section 5.5.

As in Section 5.5, we are interested here in the limit of $a \to 0$ and $M_\tau \to \infty$ at fixed $M_\tau a = 1/T$. The exponential of the original Hamiltonian H_I in (5.1) is difficult to evaluate, so let us use the same trick as in (5.47) to write

$$\exp\left(-aH_I\right) = T_1 T_2 + \mathcal{O}(a^2),   \tag{5.58}$$

where

$$T_1 = \exp\left(Jga \sum_i \hat{\sigma}_i^x\right), \quad T_2 = \exp\left(Ja \sum_i \hat{\sigma}_i^z \hat{\sigma}_{i+1}^z\right).   \tag{5.59}$$

The matrix elements of the operators T_1 and T_2 can be evaluated exactly, and we will do so shortly.

The last remaining step is to insert a complete set of states between each $T_1 T_2$ term in (5.57). It is convenient to choose these states to be eigenstates of all the $\hat{\sigma}_i^z$. Let us denote these states as $|\{m_i\}\rangle$, where $m_i = \pm 1$ are the eigenvalues of $\hat{\sigma}_i^z$, and there are a total of 2^M such states. It is immediately evident that T_2 is diagonal in this basis, whereas T_1 is not; in particular

$$T_2 |\{m_i\}\rangle = \exp\left(Ja \sum_i m_i m_{i+1}\right) |\{m_i\}\rangle.   \tag{5.60}$$

As in Section 5.5, we label the M_τ time steps in (5.57) by the index ℓ, and so we write the corresponding states as $|\{m_i(\ell)\}\rangle$. Then (5.57) can be written as

$$\mathcal{Z} = \sum_{\{m_i(\ell)\}} \prod_{\ell=1}^{M_\tau} \langle\{m_i(\ell)\}| T_1 T_2 |\{m_i(\ell+1)\}\rangle.   \tag{5.61}$$

Note that the summation above is over the 2^{MM_τ} possible values of $m_i(\ell) = \pm 1$. So the above expression is starting to look like that for a two-dimensional classical Ising model as in (2.2), with i and ℓ the coordinates along the x and τ directions, respectively. The expression for the weights in the partition function does not yet look like that in (2.2). After evaluating the matrix element of T_2 using (5.60), we need the following matrix element of T_1:

$$\langle\{m_i(\ell)\}|\, T_1\,|\{m_i(\ell+1)\}\rangle = \prod_i \langle\{m_i(\ell)\}|\exp\left(Jga\hat{\sigma}_i^x\right)|\{m_i(\ell+1)\}\rangle. \qquad (5.62)$$

Note that the right-hand side is a product of terms, each of which only involves the 2×2 Ising subspace at site i. Consequently, we can use the analog of the manipulations below (5.26) and (5.46) to derive a useful identity valid for a single Ising spin:

$$\langle m|\exp\left(gJa\,\hat{\sigma}^x\right)|m'\rangle = A\exp\left(Bmm'\right), \qquad (5.63)$$

where $m, m' = \pm 1$ and

$$A = \frac{1}{2}\cosh(2Jga), \quad \exp(-2B) = \tanh(Jga). \qquad (5.64)$$

The identity (5.63) is easily verified by evaluating both sides at the four possible values of m and m'.

Collecting the results in (5.60–5.63), we have our main expression for the partition function of the classical Ising model

$$\mathcal{Z} = \sum_{\{m_i(\ell)\}} \exp\left(\sum_{i,\ell}\left[Ja\,m_i(\ell)m_{i+1}(\ell) + B\,m_i(\ell)m_i(\ell+1)\right]\right), \qquad (5.65)$$

where we have dropped an overall normalization associated with powers of A. This has precisely the same structure as the Ising model in (2.2), with the only difference that the "exchange" couplings along the x and τ directions are not equal. This was, however, to be expected, because there is no fundamental equivalence between the space and imaginary time directions for the quantum Ising models.

All of our discussion of classical statistical mechanics in Chapters 3 and 4 can now be directly applied to (5.65): the anisotropic couplings do not cause important modifications to any of the arguments. Thus we obtain a description of the critical point in terms of the D-dimensional field theory in (3.25) with an $N=1$ component for the field ϕ_α. The only change appearing from the anisotropic coupling is that the coefficients of the gradient terms along the space and time directions can be different. This finally maps (3.25) to the d-dimensional quantum field theory described by the partition function in (2.11) at $N=1$: the anisotropy determines the value of the "velocity" c. It is clear now that c can be scaled away by a change of the time or space coordinates, and so does not modify any properties of the theory.

To complete our goal of describing the quantum critical properties of the quantum Ising model (2.2), we now need to analyze the spectrum and correlations of the quantum field theory (2.11) for $N=1$. We defer this discussion to the end of Chapter 6, where we treat the case of general N. The $N>1$ cases will apply to the quantum rotor model, which we also discuss in Chapter 6.

Exercises

5.1   (a) Perform the standard Rayleigh–Schrödinger perturbation theory on the ground state in (1.9) and the first excited state (5.7) to obtain the $1/g^2$ correction to the quasiparticle energy in (5.9). In this perturbation theory, the unperturbed Hamiltonian H_0 is the first term in (5.1), and the perturbation H_1 is the second term in (5.1). Note that because of momentum conservation $\langle k|H_1|k'\rangle \propto \delta_{kk'}$. Thus the perturbation theory splits into distinct sectors for k. Within each sector, the first excited state is nondegenerate, and so this computation can use nondegenerate perturbation theory, and not worry about the issues discussed in Section 5.1. The latter issues do becomes important at higher order in $1/g$.

       (b) Compute the value of $\mathcal{A}$ in (5.11) to order $1/g$. First note that $\hat{\sigma}^z(k)|0\rangle$ has terms of order 1 and of order $1/g$ in the one-particle sector, and only terms of order $1/g$ in the three-particle sector. Similarly for $H_1|k\rangle$. Consequently, show that all contributions to (5.11) from the three-particle sector are of order $1/g^2$. By working within the one-particle sector show that

$$\langle k|\hat{\sigma}^z(k)|0\rangle = 1 + \frac{\cos k}{2g}. \tag{5.66}$$

A curious feature of the above result is that quasiparticle residue at $k = 0$ actually increases initially as g is reduced from large values. We expect it to vanish at the quantum critical point (see Chapter 10, and so $\mathcal{A}$ has a nonmonotonic approach to the quantum critical point.

5.2   Consider the two-particle subspace described by the $N(N - 1)/2$ states (5.12) on a d-dimensional lattice. At leading order in $1/g$, we may consider the projection of H_I onto the two-particle subspace, and ignore all matrix elements to other states. Thus H_I only hops the particles between nearest-neighbor sites, with an on-site hard core repulsion. Determine eigenstates of this projected Hamiltonian. This is done by transforming to center-of-mass and relative coordinates, and solving the scattering problem in the relative coordinates; this determines the S matrix. Verify (5.14) in $d = 1$.

5.3   The questions below refer to the ferromagnetic phase of H_I in (5.1) on the square lattice, where $g \ll 1$, and we can perform perturbation theory in g. The lattice has N sites, and you can assume periodic boundary conditions where needed.

       (a) Compute the ground state energy per site to order g^2.

       (b) Compute the ferromagnetic moment $N_0 = \langle \sigma_i^z \rangle$ to order g^2, where i is any site. You only need the ground state wavefunction to order g to do this, but you need the normalization to order g^2. More precisely, write the ground state wavefunction, $|G \uparrow\rangle$ as

$$|G \uparrow\rangle = (1 + \alpha g^2)|\text{all spins up}\rangle + g \sum \beta|\text{states with one down spin}\rangle$$
$$+ g^2 \sum \gamma|\text{states with two down spins}\rangle + \cdots \tag{5.67}$$

Argue that to determine N_0 to order g^2, you need the values of α and β, but do *not* need to know the value of γ. Determine β by first-order perturbation theory, and α by the requirement that the wavefunction is normalized.

(c) Find and diagonalize the effective Hamiltonian in the one-particle subspace to order g^2. This is the subspace with one flipped spin, and has N states with energy $8J$ above the ground state at leading order. You should find that the effective Hamiltonian involves only local moves of the flipped spin. The effective Hamiltonian has contributions from intermediate states with two or zero flipped spins, and those leading to nonlocal hops of the flipped spin exactly cancel against each other.

(d) Find and diagonalize the effective Hamiltonian in the two-particle-bound-state subspace to order g^2. This is the subspace with two neighboring flipped spins, and has $2N$ states with energy $12J$ above the ground state at leading order.

The quantum rotor model

This chapter analyzes the spectrum of the quantum rotor model (1.25), whose Hamiltonian we reproduce here

$$H_R = \frac{J\tilde{g}}{2} \sum_i \hat{\mathbf{L}}_i^2 - J \sum_{\langle ij \rangle} \hat{\mathbf{n}}_i \cdot \hat{\mathbf{n}}_j. \tag{6.1}$$

Our analysis parallels that of the quantum Ising model in Chapter 10. We begin by a perturbative analysis of both phases: the paramagnetic phase at $\tilde{g} \gg 1$, and the magnetically ordered phase at $\tilde{g} \ll 1$. We then describe the mapping of the partition function of the quantum rotor model in d dimensions to the classical O(N) spin model in (2.3) and (3.2). This mapping allows us to address the vicinity of the critical point using the methods of Chapters 3 and 4.

We begin with the perturbative analyses, which are expected to hold on either side of a quantum critical point at $\tilde{g} = \tilde{g}_c$, which separates the ordered and the quantum paramagnetic phases. We will see later that $\tilde{g}_c = 0$ in $d = 1$, but $\tilde{g}_c > 0$ for $d > 1$.

6.1 Large-$\tilde{g}$ expansion

The strong-coupling expansion was discussed in [197] and briefly noted in Section 1.4.2. At $\tilde{g} = \infty$, the exchange term in H_R can be neglected, and the Hamiltonian decouples into independent sites and can be diagonalized exactly. The eigenstates on each site are the eigenstates of $\mathbf{L}^2$; for $N = 3$ these are the states of (1.24):

$$|\ell, m\rangle_i \quad \ell = 0, 1, 2, \ldots, \quad -\ell \leq m \leq \ell, \tag{6.2}$$

and have eigenenergy $J\tilde{g}\ell(\ell + 1)/2$. The ground state of H_R in the large $\tilde{g}$ limit consists of the quantum paramagnetic state with $\ell = 0$ on every site:

$$|0\rangle = \prod_i |\ell = 0, m = 0\rangle_i. \tag{6.3}$$

Compare this with the strong-coupling ground state (1.9) of the Ising model. Indeed, the remainder of the large g analysis of the quantum Ising model in Section 5.2 can be borrowed here for the rotor model, and we can therefore be quite brief. The lowest excited state is a "particle" in which a single site has $\ell = 1$, and this excitation hops from site to site. An important difference from the Ising model is that this particle is three-fold degenerate, corresponding to the three allowed values $m = -1, 0, 1$. The single-particle states are labeled by a momentum k and an azimuthal angular momentum m and have energy

$$\varepsilon_{\vec{k},m} = J\tilde{g}\left(1 - (2/3\tilde{g})\sum_{\mu}\cos(k_\mu a) + \mathcal{O}(1/\tilde{g}^2)\right), \tag{6.4}$$

where the sum over μ extends over the d spatial directions. This result is the analog of (5.8). Multiparticle states can be analyzed as in Section 5.2, the only change being that the states and the S-matrices now carry O(N) indices.

6.2 Small-$\tilde{g}$ expansion

For small $\tilde{g}$, the ground state breaks O(N) symmetry, and all the $\hat{\mathbf{n}}_i$ vectors orient themselves in a common, but arbitrary direction. This is similar to the broken Z_2 symmetry of the quantum Ising model in Section 5.3.

Excitations above this state consist of "spin waves," which can have an arbitrarily low energy (i.e. they are "gapless"). This is a crucial difference from the Ising model in Section 5.3, in which there was an energy gap above the ground state. The presence of gapless spin excitations is a direct consequence of the continuous O(N) symmetry of H_R: we can make very slow deformations in the orientation of $\langle\hat{\mathbf{n}}\rangle$, obtaining an orthogonal state whose energy is arbitrarily close to that of the ground state. Explicitly, for $N = 3$ and a ground state polarized along $(0, 0, 1)$, we parameterize

$$\hat{\mathbf{n}}(x, t) = \left(u_1(x, t), u_2(x, t), \left(1 - u_1^2 - u_2^2\right)^{1/2}\right), \tag{6.5}$$

where $|u_1|, |u_2| \ll 1$. In this limit, the commutation relations (1.21) become

$$[\hat{L}_1, u_2] = i, \quad [\hat{L}_2, u_1] = -i, \tag{6.6}$$

i.e. $u_1, \hat{L}_2$ and $u_2, -\hat{L}_1$ are canonically conjugate pairs. Also, in the limit where $u_{1,2}$ are small, the rotor momenta are also in the 1, 2 plane, and hence the third component of the rotor angular momentum is negligibly small, $\hat{L}_3 \approx 0$; so by (1.21), $\hat{L}_1$ and $\hat{L}_2$ are commuting variables. We now insert (6.5) into (6.1), and focus on the long-wavelength excitations by taking the continuum limit: this yields the Hamiltonian

$$H_R = \int \frac{d^dx}{a^d}\left[\frac{J\tilde{g}}{2}\left(\hat{L}_1^2 + \hat{L}_2^2\right) + \frac{Ja^2}{2}\left((\nabla u_1)^2 + (\nabla u_2)^2\right)\right], \tag{6.7}$$

where a is the lattice spacing. The reader will now recognize that (6.7) and the commutation relations (6.6) define the dynamics of a set of harmonic oscillators, two for each wavevector k. Explicitly, let us make the following normal mode expansion in terms of the harmonic oscillator creation and annihilation operators:

$$u_\lambda(x) = \int \frac{d^d k}{(2\pi)^d} \frac{J\tilde{g}}{\sqrt{2a^d \varepsilon_k}} \left(a_\lambda(k) e^{i\vec{k}\cdot\vec{x}} + a_\lambda^\dagger(\vec{k}) e^{-i\vec{k}\cdot\vec{x}} \right),$$

$$\epsilon_{\lambda\lambda'} L_{\lambda'}(x) = -i \int \frac{d^d k}{(2\pi)^d} \sqrt{\frac{a^d \varepsilon_k}{2J\tilde{g}}} \left(a_\lambda(\vec{k}) e^{i\vec{k}\cdot\vec{x}} - a_\lambda^\dagger(\vec{k}) e^{-i\vec{k}\cdot\vec{x}} \right), \qquad (6.8)$$

where $\lambda = 1, 2$ is a polarization index, $\epsilon_{\lambda\lambda'}$ is the unit antisymmetric tensor. Then it can be verified that if the $a(\vec{k}, t)$ operators satisfy the familiar harmonic oscillator equal-time commutation relations

$$\left[a_\lambda(\vec{k}), a_{\lambda'}^\dagger(\vec{k}') \right] = \delta_{\lambda\lambda'} (2\pi)^d \delta^d(\vec{k} - \vec{k}'),$$
$$\left[a_\lambda(\vec{k}), a_{\lambda'}(\vec{k}') \right] = 0, \qquad (6.9)$$

the commutation relations (6.6) are obeyed. Further, the Hamiltonian explicitly displays the simple sum over independent harmonic oscillators

$$H_R = \sum_\lambda \int \frac{d^d k}{(2\pi)^d} \varepsilon_k \left[a_\lambda^\dagger(\vec{k}) a_\lambda(\vec{k}) + 1/2 \right]. \qquad (6.10)$$

Here the oscillation frequency is

$$\varepsilon_k = ck, \quad c = Ja\sqrt{\tilde{g}}, \qquad (6.11)$$

where c is the spin-wave velocity.

Thus the excitation spectrum of the magnetically ordered phase consists of two polarizations of quantized spin waves with dispersion $\varepsilon_k = ck$; for general N, there are $N - 1$ spin waves. It is useful to recall how quantization of electromagnetic waves led to the concept of a particle-like excitation called the photon: the particle is just a wavepacket. Similarly, here we can interpret the quantized spin waves as a set of $N - 1$ quantized particles.

The reader should note the distinction between the $N - 1$ particles in the ordered phase with the N particles obtained in the quantum paramagnet in the strong coupling expansion above. In the ordered phase, rotations about the axis of $\langle \hat{\mathbf{n}} \rangle$ do not produce a new state, and so there are only $N - 1$ independent rotations about axes orthogonal to $\langle \hat{\mathbf{n}} \rangle$ that lead to gapless spin-wave modes.

The ground state wavefunction of the magnetically ordered state includes quantum zero-point motion of the spin waves about the fully polarized state. One consequence of the zero-point motion is that the ordered moment on each site is reduced at order $\tilde{g}$:

$$\langle \hat{n}_3 \rangle = \langle (1 - u_1^2 - u_2^2)^{1/2} \rangle$$
$$\approx 1 - (1/2) \langle u_1^2 + u_2^2 \rangle$$
$$= 1 - \frac{\sqrt{\tilde{g}} a^{d-1}}{2} \int \frac{d^d k}{(2\pi)^d} \frac{1}{k}. \qquad (6.12)$$

In the last step we have evaluated the expectation value in the quantized harmonic oscillator ground state after using the normal mode expansion (6.8). The integral over momenta k is cut off at large k by the inverse lattice spacing, but there is no cutoff at small k. We therefore notice a small k divergence in $d = 1$, indicating an instability in the small-$\tilde{g}$ expansion. We will see that the small $\tilde{g}$ prediction of a state with magnetic long-range order is never valid in $d = 1$, and the physical picture of the quantum paramagnet introduced by the large-$\tilde{g}$ expansion holds for all $\tilde{g}$. In contrast, the small-$\tilde{g}$ expansion appears stable for $d > 1$, and we do expect magnetically ordered states to exist. In this case, comparison of the small and large-$\tilde{g}$ expansions correctly suggests the existence of a quantum phase transition at intermediate $\tilde{g}$.

The above was an analysis in the linearized, harmonic limit. The nonlinearities neglected above lead to nonzero spin-wave scattering amplitudes, which we show later are quite innocuous at low enough energies in dimensions $d > 1$. Precisely in $d = 1$, spin-wave interactions are very important and destroy the long-range order of the ground state, as was already apparent from (6.12). For the classical ferromagnet (3.2), to which the present model maps, this corresponds to the absence of long-range order in $D = 2$ and is known as the Hohenberg–Mermin–Wagner theorem.

6.3 The classical *XY* chain and an O(2) quantum rotor

We will consider the $D = 1$, $N = 2$ classical ferromagnet; this is also referred to as the *XY* ferromagnet. We generalize (5.22) and (5.23) to $N = 2$ by replacing σ_ℓ^z by a two-component unit-length variable $\mathbf{n}_\ell$. This modifies (5.22) to

$$\mathcal{Z} = \prod_\ell \int D\mathbf{n}_i \delta\left(\mathbf{n}_\ell^2 - 1\right) \exp\left(-H\right). \qquad (6.13)$$

For H we modify (5.23) to

$$H = -K \sum_{\ell=1}^{M_\tau} \mathbf{n}_\ell \cdot \mathbf{n}_{\ell+1} - \sum_{\ell=1}^{M_\tau} \mathbf{h} \cdot \mathbf{n}_\ell, \qquad (6.14)$$

where, as in the Ising case, we have added a uniform field $\mathbf{h} = (h, 0)$. It is convenient to parameterize the unit-length classical spins, $\mathbf{n}_\ell$, by

$$\mathbf{n}_\ell = (\cos\theta_\ell, \sin\theta_\ell), \qquad (6.15)$$

where the continuous angular variables, θ_ℓ, run from 0 to 2π. In these variables, H takes the form

$$H = -K \sum_{\ell=1}^{M_\tau} \cos(\theta_\ell - \theta_{\ell+1}) - h \sum_{\ell=1}^{M_\tau} \cos\theta_\ell, \qquad (6.16)$$

and the partition function is

$$\mathcal{Z} = \int_0^{2\pi} \prod_{\ell=1}^{M_\tau} \frac{d\theta_\ell}{2\pi} \exp(-H). \qquad (6.17)$$

We again assume periodic boundary conditions with $\theta_{M_\tau+1} \equiv \theta_1$. Note that in zero field, H remains invariant if all the spins are rotated by the same angle ϕ, $\theta_\ell \to \theta_\ell + \phi$, and so our results will not depend upon the particular orientation chosen for $\mathbf{h}$. The partition function can be evaluated by transfer matrix methods [144, 253] quite similar to those used for the Ising chain. Although we will not use such a method to obtain our results, we nevertheless describe the main steps for completeness. First write $\mathcal{Z}$ in the form

$$\mathcal{Z} = \int_0^{2\pi} \prod_{i=1}^{M_\tau} \frac{d\theta_\ell}{2\pi} \langle \theta_1 | \hat{T} | \theta_2 \rangle \langle \theta_2 | \hat{T} | \theta_3 \rangle \cdots \langle \theta_{M_\tau} | \hat{T} | \theta_1 \rangle$$
$$= \text{Tr}\,\hat{T}_\tau^M, \tag{6.18}$$

where the symmetric transfer matrix operator $\hat{T}$ is defined by

$$\langle \theta | \hat{T} | \theta' \rangle = \exp\left(K \cos(\theta - \theta') + \frac{h}{2}(\cos\theta + \cos\theta') \right), \tag{6.19}$$

and the trace is clearly over continuous angular variable θ. As in the Ising case, we have to diagonalize the transfer matrix $\hat{T}$ by solving the eigenvalue equation,

$$\int_0^{2\pi} \frac{d\theta'}{2\pi} \langle \theta | \hat{T} | \theta' \rangle \Psi_\mu(\theta') = \lambda_\mu \Psi_\mu(\theta), \tag{6.20}$$

for the eigenfunctions $\Psi_\mu(\theta)$ (with $\Psi_\mu(\theta + 2\pi) = \Psi_\mu(\theta)$) and corresponding eigenvalues λ_μ. Then the partition function $\mathcal{Z}$ is simply

$$\mathcal{Z} = \sum_\mu \lambda_\mu^{M_\tau}, \tag{6.21}$$

where the sum extends over the infinite number of eigenvalues λ_μ. The solution of (6.20) is quite involved, and the present approach is a rather convoluted method of obtaining the universal properties of H.

Instead, it is useful to approach the problem with a little physical insight and take the scaling limit at the earliest possible stage. We anticipate, from our experience with the Ising model, that the universal scaling behavior will emerge at large values of K. For this case, θ_ℓ is not expected to vary much from one site to the next, suggesting that it should be useful to expand in terms of gradients of θ_ℓ. So we define a continuous coordinate $\tau = \ell a$, where a is the lattice spacing, and the label τ anticipates its eventual interpretation as the imaginary time coordinate of a quantum problem. Then, to lowest order in the gradients of the function $\theta(\tau = \ell a) \equiv \theta_\ell$, the Hamiltonian H takes the continuum form H_c:

$$H_c[\theta(\tau)] = \int_0^{L_\tau} d\tau \left[\frac{\xi}{4} \left(\frac{d\theta(\tau)}{d\tau} \right)^2 - \tilde{h} \cos\theta(\tau) \right], \tag{6.22}$$

where

$$\xi = 2Ka, \quad \tilde{h} = \frac{h}{a}, \tag{6.23}$$

and as before $L_\tau = M_\tau a$. The coefficient of the gradient squared term is clearly a length (along the time direction) and we have written this length in terms of the symbol ξ: the parameterization anticipates some of our subsequent results where we see that ξ is the $h = 0$ correlation length of an infinite XY chain. With this new form of H, the partition function becomes a functional integral

$$Z_c = \sum_{p=-\infty}^{\infty} \int_{\theta(L_\tau)=\theta(0)+2\pi p} \mathcal{D}\theta(\tau) \exp\left(-H_c[\theta(\tau)]\right). \tag{6.24}$$

The integral is taken over all functions $\theta(\tau)$ that satisfy the specified boundary conditions. As we can continuously follow the value of θ from $\tau = 0$ to $\tau = L_\tau$, its actual value, and not just the angle modulo 2π, becomes significant; so we allow for an overall phase winding by $2\pi p$ in the boundary conditions. This boundary condition is the only remnant of the periodicity of the original lattice problem as $\theta(\tau)$ is allowed to assume all real values. We have also absorbed an overall normalization factor into the definition of the functional integral, and we will therefore not keep track of additive nonuniversal constant contributions to the free energy, such as E_0 of Section 5.5.

We now assert that Z_c and H_c are the universal scaling theories of H and Z in (6.16) and (6.17). Hence if we started with a different microscopic model, its universal properties would also be described by Z_c, with the only change being in the values of ξ and $\tilde{h}$. For instance, if we had a Hamiltonian like (6.16), but with jth neighbor interactions K_j, its continuum limit would also be H_c, with the same value for $\tilde{h}$, but ξ modified to

$$\xi = 2a \sum_{j=1}^{\infty} K_j j^2. \tag{6.25}$$

This continuum limit is valid for all models in which the summation over j in (6.25) converges. The universality of H_c also applies to models in which the constraint $\mathbf{n}_i^2 = 1$ is not imposed rigidly and fluctuations in the amplitude of $\mathbf{n}_i$ are allowed about their mean value. The prescription for determining the input value of ξ is, however, still very simple: set the magnitude of $\mathbf{n}_i$ to its optimum value and measure the energy change of a uniform twist. Corrections due to the fluctuations in the magnitude of $\mathbf{n}_i$ about this optimum value will not modify the universal scaling theory (6.23).

Before turning to an evaluation of Z_c and its associated correlators, let us describe the scaling forms expected in the universal theory. These can be deduced by simple dimensional analysis. In the present case ξ, L_τ, and $\tilde{h}$ are the large lengths of the theory, and we simply make the appropriate dimensionless combinations. We thus have for the free energy $\mathcal{F} = -(\ln Z_c)/L_\tau$ and the two-point correlator:

$$\mathcal{F} = \frac{1}{L_\tau} \Phi_{\mathcal{F}}\left(\frac{L_\tau}{\xi}, \tilde{h}L_\tau\right),$$

$$\langle \mathbf{n}(\tau) \cdot \mathbf{n}(0)\rangle = \Phi_n\left(\frac{\tau}{L_\tau}, \frac{L_\tau}{\xi}, \tilde{h}L_\tau\right), \tag{6.26}$$

where $\Phi_{\mathcal{F}}$ and Φ_n are universal functions, portions of which are determined explicitly below.

Let us evaluate $\mathcal{Z}_c$ in zero field ($\tilde{h} = 0$). To satisfy the boundary conditions let us decompose

$$\theta(\tau) = \frac{2\pi p \tau}{L_\tau} + \theta'(\tau), \tag{6.27}$$

where $\theta'(\tau)$ satisfies periodic boundary conditions $\theta'(L_\tau) = \theta'(0)$. Inserting this into (6.22) we find that the cross term between the two pieces of $\theta(\tau)$ vanishes because of the periodic boundary conditions on θ', and (6.24) becomes

$$\mathcal{Z}_c(\tilde{h} = 0) = \left(\sum_{p=-\infty}^{\infty} \exp\left(-\frac{\pi^2 p^2 \xi}{L_\tau} \right) \right)$$
$$\times \int_{\theta'(L_\tau)=\theta'(0)} \mathcal{D}\theta'(\tau) \exp\left(-\frac{\xi}{4} \int_0^{L_\tau} d\tau \left(\frac{d\theta'}{d\tau} \right)^2 \right). \tag{6.28}$$

Now note that the last functional integral is simply the familiar Feynman path integral for the amplitude of a single quantum mechanical free particle, of mass $\xi/2$ with coordinate θ', to return to its starting position after imaginary time L_τ. Using the standard expression for this we find finally

$$\mathcal{Z}_c(\tilde{h} = 0) = (2\pi) \left(\frac{\xi}{4\pi L_\tau} \right)^{1/2} A(\pi \xi / L_\tau), \tag{6.29}$$

where the factor of 2π comes from the integral over $\theta'(0)$, and $A(y)$ is the elliptic theta function defined by

$$A(y) = \sum_{p=-\infty}^{\infty} e^{-\pi p^2 y}. \tag{6.30}$$

This result is clearly consistent with the scaling form for the free energy density $\mathcal{F} = -(\ln \mathcal{Z}_c)/L_\tau$ in (6.26).

Let us push the analogy with the quantum mechanics of a particle a bit further and complete the quantum–classical mapping by obtaining an explicit expression for the quantum Hamiltonian, H_Q, which describes the scaling limit. Note that $\mathcal{Z}_c$ in (6.24), with the summation over p included, can be interpreted as the Feynman path integral of a particle constrained to move on a *circle* of unit radius; the angular coordinate of the particle is θ, and p represents the number of times the particle winds around the circle in its motion from imaginary time $\tau = 0$ to $\tau = L_\tau$. The term proportional to $\tilde{h}$ is then a potential energy term that preferentially locates the particle at $\theta = 0$. The Hamiltonian of this quantum particle is then

$$H_Q = -\Delta \frac{\partial^2}{\partial \theta^2} - \tilde{h} \cos\theta, \tag{6.31}$$

where, as we will see shortly, Δ is defined as in the Ising case to be the gap of H_Q in zero external field. As the mass of the quantum particle is $1/(2\Delta)$, we have by comparing with (6.22)

$$\Delta = \frac{1}{\xi}. \tag{6.32}$$

This is precisely of the form (5.49), and it is another realization of the fact that the gap of the quantum model is equal to the correlation "length" of the classical model along the imaginary time direction. For some of our subsequent discussion it is useful to express H_Q solely in terms of quantum operators. Let $\hat{\mathbf{n}}$ be the Heisenberg operator corresponding to $\mathbf{n}$. Let us also define $\hat{L}$ as the angular momentum operator of the rotor:

$$\hat{L} = \frac{1}{i} \frac{\partial}{\partial \theta}. \tag{6.33}$$

Then we have the commutation relation

$$[\hat{L}, \hat{n}_\alpha] = i\epsilon_{\alpha\beta}\hat{n}_\beta, \tag{6.34}$$

where α, β extend over the two coordinate axes, x, y in the spin plane, and $\epsilon_{xy} = -\epsilon_{yx} = 1$, with other components zero. These are precisely the $N = 2$ case of the commutation relations following from (1.17) and (1.20). The Hamiltonian H_Q is clearly

$$H_Q = \Delta \hat{L}^2 - \tilde{\mathbf{h}} \cdot \hat{\mathbf{n}}, \tag{6.35}$$

which is simply the quantum rotor model (1.22) in the presence of field $\tilde{\mathbf{h}}$, $1/(2\Delta)$ is the moment of inertia of the rotor, and commutation relation (6.34) is the $N = 2$ analog of (1.21). We have established the needed result: the scaling limit of the $D = 1$ classical XY ferromagnet is given exactly by the Hamiltonian of a single O(2) quantum rotor.

The Hamiltonian H_Q is related to the transfer matrix $\hat{T}$ (in (6.19)) of the lattice XY model by a relationship identical to that found in (5.47). By a gradient expansion of (6.19) the reader can verify that

$$\hat{T} \approx \exp(-aH_Q) \tag{6.36}$$

to leading order in the lattice spacing a. So again, the transfer matrix "evolves" the system by an imaginary time a.

We can use the quantum–classical mapping and obtain explicit expressions for the universal scaling functions of the classical problem in (6.26). First, using the mapping (2.5) $T = 1/L_\tau$, let us write down the scaling forms (6.26) in the quantum language:

$$\mathcal{F} = T\Phi_{\mathcal{F}}\left(\frac{\Delta}{T}, \frac{\tilde{h}}{T}\right),$$

$$\langle \mathbf{n}(\tau) \cdot \mathbf{n}(0) \rangle = \Phi_n\left(T\tau, \frac{\Delta}{T}, \frac{\tilde{h}}{T}\right). \tag{6.37}$$

We see here a structure that was used in (5.55), and which is used throughout the book. We characterize the universal properties by the "small" energy scales Δ, $\tilde{h}$ (these are the analogs of the "large" length scales of the corresponding classical problem, while the nonuniversal behavior at "small" length scales in the classical system maps onto high-energy physics in the quantum system, which is not of interest here). These "small" energy scales then appear in universal scaling functions of dimensionless ratio of these energies with the physical temperature, T.

Let us turn to evaluation of the scaling functions. The eigenstates, $\psi_\mu(\theta)$, and eigenvalues, ϵ_μ, of H_Q are determined by solving the Schrödinger equation

$$H_Q \psi_\mu(\theta) = \epsilon_\mu \psi_\mu(\theta), \tag{6.38}$$

subject to the boundary condition $\psi_\mu(0) = \psi_\mu(2\pi)$. The equation (6.38) can be considered as the continuum scaling limit of the eigenvalue equation (6.20), with the correspondence in (6.36) and $\lambda_\mu = \exp(-a\epsilon_\mu)$. The continuum limit partition function $\mathcal{Z}_c$ can be expressed directly in terms of H_Q:

$$\mathcal{Z}_c = \text{Tr}\exp(-H_Q/T)$$
$$= \sum_\mu \exp(-\epsilon_\mu/T), \tag{6.39}$$

where $T = 1/L_\tau$. The two-point correlator of $\hat{n}$ can also be expressed in the quantum language

$$\langle \mathbf{n}(\tau) \cdot \mathbf{n}(0) \rangle = \frac{1}{\mathcal{Z}_c} \text{Tr}\left(e^{-H_Q/T} e^{H_Q\tau} \hat{\mathbf{n}} e^{-H_Q\tau} \cdot \hat{\mathbf{n}} \right)$$
$$= \frac{1}{\mathcal{Z}_c} \sum_{\mu,\nu} |\langle \mu|\hat{\mathbf{n}}|\nu\rangle|^2 e^{-\epsilon_\mu/T} e^{-(\epsilon_\nu - \epsilon_\mu)\tau}, \tag{6.40}$$

where the summation over μ, ν extends over all the eigenstates of H_Q, and we have assumed $\tau > 0$.

The solution of (6.38), combined with (6.39) and (6.40), provides the complete solution of the universal scaling properties of the classical *XY* chain. An elementary solution of the eigenvalue equation (6.38) is only possible at $\tilde{h} = 0$, to which we restrict our attention from now on. In zero field, the eigenstates are $\psi_m(\theta) \propto e^{im\theta}$, where m is an arbitrary integer, and the corresponding eigenvalues are Δm^2 (these are the states of (1.23)). The ground state has zero energy ($m = 0$), and, as promised, the gap to the lowest excited states ($m = \pm 1$) is Δ. We can therefore evaluate the partition function

$$\mathcal{Z}_c(\tilde{h} = 0) = \sum_{m=-\infty}^{\infty} \exp\left(-\frac{\Delta m^2}{T}\right)$$
$$= A(\Delta/\pi T), \tag{6.41}$$

a result that satisfies (6.37); the function $A(y)$ was defined in (6.30). If we compare this with (6.29), and use (6.32) and (5.49), the equivalence of the two expressions for $\mathcal{Z}_c$ is not immediately obvious. However, equality can be established by use of the following inversion identity, which the reader is invited to establish as a simple application of the Poisson summation formula:

$$A(y) = \frac{A(1/y)}{\sqrt{y}}. \tag{6.42}$$

In terms of the original classical model, the expression (6.29) for $\mathcal{Z}_c$ is useful for large ξ (or large values of K, corresponding to a low classical "temperature," which has been absorbed into the definition of K) when its series converges rapidly; conversely the dual expression (6.41) is most useful for small ξ (or small K and high classical "temperatures").

Let us also discuss the form of the correlation functions at $\tilde{h} = 0$. Recalling (6.15), and the wavefunction $\psi_m(\theta) \propto e^{im\theta}$, we have the very simple matrix element

$$|\langle m|\hat{\mathbf{n}}|m+1\rangle|^2 = 1, \tag{6.43}$$

and all others vanish; the correlation function follows simply from (6.40), and it is clear that the result agrees with (6.37). In particular, at $T = 0$ or $L_\tau = \infty$ we have

$$\langle \mathbf{n}(\tau) \cdot \mathbf{n}(0) \rangle = e^{-\Delta|\tau|}, \tag{6.44}$$

which establishes, as in the Ising chain, the inverse of the gap Δ as the correlation length of the classical chain.

6.4 The classical Heisenberg chain and an O(3) quantum rotor

We now generalize the results of the previous section to the $D = 1$, $N = 3$ case. The $N = 3$ classical ferromagnet is also known as the classical Heisenberg chain. The partition function is still given by (6.13) and the classical Hamiltonian by (6.14), with the only change being that $\mathbf{n}$ is now a three-component unit vector. Taking its continuum limit as for $N = 2$, we replace (6.22) and (6.24) by the partition function

$$\mathcal{Z}_c = \int \mathcal{D}\mathbf{n}(\tau)\delta(\mathbf{n}^2 - 1)\exp\left(-H_c[\mathbf{n}(\tau)]\right),$$

$$H_c[\mathbf{n}(\tau)] = \int_0^{L_\tau} d\tau \left[\frac{(N-1)\xi}{4}\left(\frac{d\mathbf{n}(\tau)}{d\tau}\right)^2 - \tilde{\mathbf{h}} \cdot \mathbf{n}\right], \tag{6.45}$$

with $\mathbf{n}(0) = \mathbf{n}(L_\tau)$; now $\xi = 2Ka/(N-1)$ and $\tilde{\mathbf{h}} = \mathbf{h}/a$ as in (6.23). We have chosen the definition of ξ by anticipating a later computation in which ξ will be seen to be the correlation length. We only consider the case $N = 3$ in this subsection and have quoted, without proof, the form for general N; note that (6.45) agrees with (6.22) for $N = 2$. Unlike (6.22), it is not possible to evaluate the partition function (6.45) in this form. Recall that for the $N = 2$ case of (6.22) we had a simple angular parameterization in which H became purely quadratic in the angular variable. One could parameterize the three-component $\mathbf{n}$ using spherical coordinates, but the resulting H is not simply quadratic.

Further progress toward the evaluation of $\mathcal{Z}_c$ can, however, be made after the quantum–classical mapping. To do this, note, as in (6.28), that the functional integral in (6.45) can be interpreted as the imaginary-time Feynman path integral for a particle moving in a three-dimensional space with coordinate $\mathbf{n}$. Then the term with $(\partial\mathbf{n}/\partial\tau)^2$ is its kinetic energy and its mass is $1/\xi$, and the term proportional to $\tilde{\mathbf{h}}$ is like a "gravitational potential energy." The constraint that $\mathbf{n}^2 = 1$ may be viewed as a very strong potential that prefers that the particle move on the surface of a unit sphere. We can therefore perform the quantum–classical mapping simply by writing down the Schrödinger Hamiltonian, H_Q, for this particle. The restriction that the motion take place on the surface of a sphere simply means that the radial kinetic energy term of the particle should be dropped. The resulting H_Q generalizes (6.35) to $N = 3$:

$$H_Q = \frac{\Delta}{2}\hat{\mathbf{L}}^2 - \tilde{\mathbf{h}} \cdot \mathbf{n}, \tag{6.46}$$

where the angular momentum operator $\hat{\mathbf{L}}$ has three components (in general it has $N(N - 1)/2$ components); again this is simply the $N = 3$ single rotor model H_K in (1.22) in the presence of a field $\tilde{\mathbf{h}}$. The operators $\hat{\mathbf{L}}$ and $\hat{\mathbf{n}}$ obey the commutation relations in (1.21). The parameter Δ is again the energy gap at $\tilde{\mathbf{h}} = 0$, as we will see below, and is given by $\Delta = 1/\xi$, as in (6.32). If we determine all the eigenvalues ϵ_μ of H_Q, then the explicit expression for $\mathcal{Z}_c$ is given by (6.39). Determination of the eigenvalues of H_Q can, for instance, be done by solving the Schrödinger differential equation for a wavefunction $\psi_\mu(\mathbf{n})$ on the surface of a unit sphere. The Hamiltonian in Schrödinger's equation is given by H_Q, with $\hat{\mathbf{L}}$ a differential operator:

$$L_\alpha = -i\epsilon_{\alpha\beta\gamma} n_\beta \frac{\partial}{\partial n_\gamma}. \tag{6.47}$$

In summary, the complete solution of the classical partition function $\mathcal{Z}_c$ is given by mapping the problem to the dynamics of an O(3) quantum rotor with Hamiltonian H_Q defined by Equations (2.69), (1.19), and (2.70), where the value of $\mathcal{Z}_c$ is given by (6.39).

We conclude this section by explicitly determining the eigenvalues for $\tilde{\mathbf{h}} = 0$. In this case, it is evident that the eigenfunctions ψ_μ are simply the spherical harmonics, and the eigenvalues are

$$(\Delta/2)\ell(\ell + 1), \quad \ell = 0, 1, 2 \ldots \infty \tag{6.48}$$

with degeneracy $2\ell + 1$ (as in (1.24)), so that

$$\mathcal{Z}_c(\tilde{\mathbf{h}} = 0) = \mathrm{Tr}\, e^{-H_Q/T}$$
$$= \sum_{\ell=0}^{\infty} (2\ell + 1) \exp\left(-\frac{\Delta}{2T}\ell(\ell + 1)\right), \tag{6.49}$$

replacing (6.41), and as before $T = 1/L_\tau$. The ground state is the nondegenerate $\ell = 0$ state, and it can be checked that the energy gap is Δ. The correlations continue to obey (6.44), and so there is no long-range order in the classical Heisenberg chain, and the correlation length $= 1/\Delta$.

6.5 Mapping to classical field theories

Continuing our analysis parallel to that of the Ising model in Section 5.6, here we apply the result of Sections 6.3 and 6.4 to obtain a representation of the original d-dimensional quantum rotor model (6.1) as a classical statistical mechanics model in D dimensions.

Here our analysis is actually simpler than that in Section 5.6 because we can keep the imaginary time coordinate τ continuous at all stages. For the Ising model, we were forced to discretize time into M_τ steps of size a: this is because the Ising spins could evolve only in discrete steps. Here, we are dealing with continuous spin variables which can evolve in continuous time, and so the limit $a \to 0$ can be taken at the outset.

As in Section 5.6, we begin with the d-dimensional quantum rotor model, and then apply the results of Sections 6.3 and 6.4 independently to the quantum rotor on each site i. In particular, using the equivalence between (6.22) and (6.31) for $N = 2$, and that between (6.45) and (6.46) for general N, we can now write

$$\mathcal{Z} = \text{Tr} \exp\left(-H_R/T\right)$$

$$= \prod_i \int \mathcal{D}\mathbf{n}_i(\tau)\delta(\mathbf{n}_i^2(\tau) - 1) \exp(-\mathcal{S}_n),$$

$$\mathcal{S}_n = \int_0^{1/T} d\tau \left[\frac{1}{2J\tilde{g}} \sum_i \left(\frac{d\mathbf{n}_i(\tau)}{d\tau} \right)^2 - J \sum_{\langle ij \rangle} \mathbf{n}_i \cdot \mathbf{n}_j \right], \qquad (6.50)$$

along with the periodic boundary conditions $\mathbf{n}_i(1/T) = \mathbf{n}_i(0)$. This expression is of the needed form: it involves a summation over the orientation of a "classical" O(N) spin located on a $D + 1$-dimensional spacetime, with the temporal direction having the form of a circle of circumference $1/T$, i.e. the spacetime has the topology of a cylinder. After discretizing the time direction, it takes the form of the model (3.2) considered in Chapter 3, although we will not need to take that step here. A complementary mapping is obtained by taking the spatial continuum limit of (6.50): then we obtain the D-dimensional O(N) nonlinear sigma model in (2.12), with c given in (6.11) and the coupling g given by

$$g = N\sqrt{\tilde{g}}a^{d-1}. \qquad (6.51)$$

Finally, we can apply the arguments of Chapter 3 to motivate the D-dimensional model of the field ϕ_α ($\alpha = 1 \ldots N$) given by (2.11) or (3.25). Just as was the case in Chapters 3 and 4, the quantum field theory in (2.11) will be the most convenient formulation to understand the behavior of the quantum Ising and rotor models across the quantum critical point.

6.6 Spectrum of quantum field theory

We have finally assembled all ingredients to describe the quantum phase transition in the quantum Ising and rotor models. We have argued that the quantum field theory (2.11) completely describes all low energy properties in the vicinity of the quantum critical point: in particular it captures the excitations of both phases and of the critical point.

Having used the classical analysis of Chapters 3 and 4 to motivate the "soft-spin" ϕ_α continuum formulation in (2.11), let us now work backwards from this classical theory to a Hamiltonian description in terms of a continuum quantum model. This will be a convenient way of describing the ground state and its excitations. Just as was the case for the classical model, we can analyze the continuum quantum model perturbatively in powers of the coupling u. As argued in Chapter 4, we can expect this perturbative computation to fail at the quantum critical point for $D < 4$, and we will then have to invoke the renormalization group analysis for a complete picture. However, in this section we will be satisfied by the mean-field description at lowest order in u: as we saw in Chapter 3, this gives an adequate description of both phases and of the critical point, failing mainly in the values of the critical exponents for $D < 4$.

Applying the arguments used to obtain (6.50) in reverse, we conclude that (2.11) is equivalent to the continuum quantum model with the Hamiltonian

$$\mathcal{H} = \int d^d x \left\{ \frac{1}{2} \left[\pi_\alpha^2 + c^2 (\nabla_x \phi_\alpha)^2 + r\phi_\alpha^2(x) \right] + \frac{u}{4!} \left(\phi_\alpha^2(x) \right)^2 \right\}. \qquad (6.52)$$

Here $\pi_\alpha(x, t)$ is the canonical momentum to the field ϕ_α, and they therefore satisfy the equal-time commutation relations

$$[\phi_\alpha(x), \pi_\beta(x')] = i\delta_{\alpha\beta}\delta(x - x'). \qquad (6.53)$$

The remainder of this section analyzes the theory defined by (6.52) and (6.53), to leading order in u. Our analysis is the quantum analog of the classical considerations in Section 3.2.

6.6.1 Paramagnet

First, let us consider the paramagnetic phase, $r > 0$. The effective potential has a minimum near $\phi_\alpha = 0$, and so the low-lying excitations are small fluctuations of ϕ_α about this minimum. For these, we can ignore the quartic u term. Then (6.52) becomes a harmonic theory, which we can diagonalize into normal modes just as we did earlier for (6.7). Now the normal mode expansion in terms of harmonic oscillator operators is

$$\phi_\alpha(x) = \int \frac{d^d k}{(2\pi)^d} \frac{1}{\sqrt{2\varepsilon_k}} \left(a_\alpha(k)e^{i\vec{k}\cdot\vec{x}} + a_\alpha^\dagger(\vec{k})e^{-i\vec{k}\cdot\vec{x}} \right),$$

$$\pi_\alpha(x) = -i \int \frac{d^d k}{(2\pi)^d} \sqrt{\frac{\varepsilon_k}{2}} \left(a_\alpha(\vec{k})e^{i\vec{k}\cdot\vec{x}} - a_\alpha^\dagger(\vec{k})e^{-i\vec{k}\cdot\vec{x}} \right), \qquad (6.54)$$

where the creation and annihilation operators satisfy the analog of the commutation relations in (6.9):

$$\left[a_\alpha(\vec{k}), a_\beta^\dagger(\vec{k}') \right] = \delta_{\alpha\beta}(2\pi)^d \delta^d(\vec{k} - \vec{k}'),$$

$$\left[a_\alpha(\vec{k}), a_\beta(\vec{k}') \right] = 0. \qquad (6.55)$$

Again, these commutation relations ensure that (6.53) is obeyed, and the Hamiltonian is just the sum of harmonic oscillators as in (6.10):

$$\mathcal{H} = \int \frac{d^d k}{(2\pi)^d} \varepsilon_k \left[a_\alpha^\dagger(\vec{k})a_\alpha(\vec{k}) + 1/2 \right]. \qquad (6.56)$$

The energy of these normal modes is

$$\varepsilon_k = (c^2 k^2 + r)^{1/2}. \tag{6.57}$$

So our main result is that the low-lying excitations of the paramagnetic phase consist of N particles, which transform under the fundamental representation of $O(N)$. This spectrum is seen to be in perfect correspondence with earlier results from the $g \gg 1$ expansions. For the $N = 1$ case, the particle in Section 5.2.1 is equivalent to the present excitations: both are created by the action of the order parameter $\phi_\alpha \sim \hat{\sigma}^z$ on the ground state. For $N > 1$, we have the N-fold degenerate particles discussed in Section 6.1.

The energy gap above the paramagnetic state, from (6.57), is $\Delta = \sqrt{r}$. Unlike our previous $g \gg 1$ analysis, we can now follow the evolution of this gap all the way up to the quantum critical point at $r = 0$. This gap vanishes at the critical point as in (1.1), thus identifying the mean-field exponent $z\nu = 1/2$. Recall that we noted earlier in Section 2.1 that the quantum Ising and rotor models have dynamic exponent $z = 1$.

6.6.2 Quantum critical point

Right at the critical point $r = 0$, we have our first result for the nature of the excitation spectrum: there are N particles, all dispersing as

$$\varepsilon_k = ck. \tag{6.58}$$

The linear dispersion is consistent with dynamic exponent $z = 1$. It should be contrasted with the small momentum $\sim k^2$ dispersion of particles in the paramagnetic phase in (6.57). We see in Chapter 7 that the particles are not stable excitations of the critical point for $D < 4$: the strong interactions from the quartic coupling u make them susceptible to decay into multiple lower energy excitations, and the quasiparticle residue Z is equal to zero at the quantum critical point. In contrast, the particles in (6.57) are stable in the paramagnetic phase, with a nonzero Z.

6.6.3 Magnetic order

For $r < 0$, just as in Section 3.2, the potential in (6.52) is minimized at $\phi_\alpha = N_0 \delta_{\alpha,1}$, where N_0 was given in (3.23)

$$N_0 = \sqrt{\frac{-6r}{u}}. \tag{6.59}$$

We have arbitrarily chosen the magnetic order oriented along the $\alpha = 1$ direction, without loss of generality. Now let us write

$$\phi_\alpha(x) = N_0 \delta_{\alpha,1} + \tilde{\phi}_\alpha(x), \tag{6.60}$$

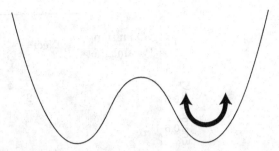

Fig. 6.1   The longitudinal mode corresponding to the oscillations of ϕ about the nonzero minimum value. There are also $N-1$ gapless transverse modes which move orthogonally to the plane of the page which are not shown.

and expand the Hamiltonian in (6.52) to quadratic order in the $\tilde{\phi}_\alpha$. A straightforward computation yields

$$\mathcal{H} = \frac{1}{2} \int d^d x \left\{ \sum_{\alpha=1}^{N} \left(\pi_\alpha^2 + c^2 (\nabla_x \tilde{\phi}_\alpha)^2 \right) + 2|r| \tilde{\phi}_1^2 \right\}. \tag{6.61}$$

We can now quantize just as in Section 6.6.1, and find two types of excitation:

$$\varepsilon_k = ck, \quad N-1 \text{ particles,}$$
$$\varepsilon_k = (c^2 k^2 + 2|r|)^{1/2}, \quad 1 \text{ particle.} \tag{6.62}$$

The gapless $N-1$ particles are easy to identify: they are clearly the spin waves we met in Section 6.2 in (6.11).

For $N > 1$, the single particle with energy gap $\sqrt{2|r|}$ is not one we have met before. It corresponds to small longitudinal oscillations of the ϕ_1 field about the minimum at $\phi_1 = N_0$, see Fig. 6.1. It is the analog of what is known in particle theory literature as the Higgs particle. In general, this Higgs particle can decay into multiple lower-energy spin waves. It has been argued that such decay processes dominate for $d < 3$, and the Higgs particle is therefore not a stable excitation. However, in $d = 3$, the Higgs is stable; indeed neutron scattering experiments [414] on TlCuCl$_3$ have observed the Higgs excitation [430], see Fig. 6.2.

For $N = 1$, we have only the particle excitation with the energy gap $\sqrt{2|r|}$. For $d > 1$, we claim this is the same as the low-lying particle excitation found in Section 5.3.1 in the small-g expansion. In the latter approach, the excitation was a "bubble" of a down spin moving in a ferromagnetic background of up spins. In the present field-theoretical analysis we have fluctuations of ϕ_1 about N_0, which also contribute to decrease in the local ferromagnetic moment.

Finally, what about the case $N = 1$, $d = 1$? We found in the small-g expansion in Section 5.3.2 that the stable excitations were domain walls or "kinks" that interpolated between the ground states with magnetization $\pm N_0$. Here, such a domain wall involves interpolating ϕ_1 between the minima in the effective potential at $\pm N_0$ over the maximum at $\phi_1 = 0$, see Fig. 6.3. This domain wall is also a local minimum of the potential in the

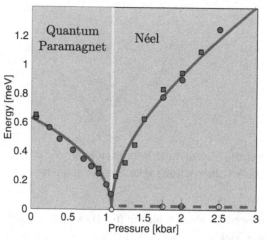

From Ref. [414]. Measurements of the excitation energies in TlCuCl$_3$ across the quantum critical point induced by applied pressure (horizontal axis). This transition is described by the O(3) quantum rotor model in $d = 3$. Below the critical pressure, we are in the paramagnetic phase, and the three quasiparticle modes are shown by the symbols to the left of the critical pressure. Above the critical pressure, in the magnetically ordered phase, the two spin wave modes are shown by the symbols at zero energy. There is also an additional Higgs mode shown by the nonzero energy symbols to the right of the critical pressure. The ratio of the energy of the Higgs mode to the triplet mode on the paramagnetic side is close to $\sqrt{2}$, as predicted [430] by the ratio of (6.57) and (6.62).

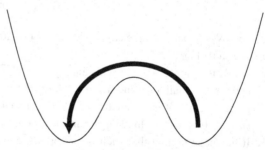

A domain wall between two Ising ferromagnetic states interpolates from one minimum of the effective potential to the other.

Hamiltonian but with a space-dependent $\phi_1(x)$. Indeed, taking the variational derivative of (6.52) with respect to variations in $\phi_1(x)$, we obtain the saddle-point equation

$$-\partial_x^2\phi_1 - |r|\phi_1 + u\phi_1^3/6 = 0. \tag{6.63}$$

This equation has the usual solutions $\phi_1 = \pm N_0$, but also the space-dependent solution

$$\phi_1(x) = \pm N_0 \tanh\left((x - x_0)\sqrt{|r|}/2\right), \tag{6.64}$$

where x_0 is abitrary. This is the domain wall centered at $x = x_0$. Quantization of the motion of x_0 leads to the domain wall particle. The Higgs particle of small ϕ_1 oscillations about

$\pm N_0$ is evidently unstable to decay into a pair of domain wall particles: we discussed this phenomenon in Section 5.3.2. We study these domain walls more completely in Chapter 10, where we will find that they are fermions, and provide a complete description of the transition into the paramagnetic state.

Exercises

6.1 Establish the connection between the quantum rotor model of Section 1.4.2, and the spin ladder model described in Section 1.4.3. Note that the matrix elements of $\mathbf{n}_i$ in the angular momentum eigenstates are equal to those of the position operator between the spherical harmonics. Determine the spectrum of its angular momentum $\ell = 1$ excitations to leading order in g. By mapping this spectrum to that of the spin ladder, determine suitable values of K and g.

6.2 *O(2) rotors with long-range interactions:* An array of charged superconducting dots at the sites i (with positions r_i) of a d-dimensional cubic lattice is described by the Hamiltonian

$$H = \sum_i \frac{\hat{n}_i^2}{2C} + \sum_{i<j} \hat{n}_i \hat{n}_j \frac{e^{*2}}{|r_i - r_j|} - J \sum_{\langle \alpha\alpha' \rangle} \cos\left(\hat{\phi}_i - \hat{\phi}_j\right), \qquad (6.65)$$

where $\hat{n}_i$ is the number operator for Cooper pairs on dot i (each Cooper pair has charge $e^* = 2e$), $\hat{\phi}_i$ is the conjugate phase operators, and the only nonvanishing commutation relation is

$$\left[\hat{\phi}_i, \hat{n}_j\right] = i\delta_{\alpha\alpha'}. \qquad (6.66)$$

Assume we are in the large J superconducting phase where the phases are all aligned at $\hat{\phi}_i = 0$ (say). Obtain the Heisenberg equations of motion for $\hat{\phi}_i$ and $\hat{n}_i$, and linearize them for small fluctuations about the ground state. By determining the normal mode spectrum of these equations, obtain the long-wavelength form of the "plasmon" oscillations in $d = 1, 2, 3$.

6.3 Consider the two-particle sector of the quasiparticle excitations of the paramagnetic state described in Section 6.6.1. The particles have a two-body interaction proportional to u. Compute the matrix elements of this interaction $\langle k_1, \alpha_1 | u | k_2, \alpha_2 \rangle$. Set up the two-body scattering problem, and discuss qualitative features of the low momentum scattering amplitude in $d = 1, 2, 3$. Compare your results with those obtained in the $1/g$ expansion of the Ising model in Exercise 5.5.2.

7 Correlations, susceptibilities, and the quantum critical point

We have so far described our quantum phases and critical points in terms of the wave-functions and energies of the eigenstates of the Hamiltonian. However, as we saw in our treatment of D-dimensional classical statistical mechanics in Chapters 3 and 4, a more subtle and complete characterization is obtained by considering correlation functions of various observable operators. These correlation functions are also amenable to a Feynman graph expansion and the renormalization group transformation, which was crucial in our full treatment of the classical critical point. This chapter considers correlation functions of the d-dimensional quantum model, and applies them to obtain an improved understanding of the quantum phases and the quantum critical point.

Section 5.5.3 has already presented a detailed description of the connection between the correlation functions of the $D = 1$ classical Ising chain and the single-site (i.e. $d = 0$) quantum Ising model. This mapping is immediately extended to the general D case, following the reasoning in Sections 5.6 and 6.5. From this we obtain the fundamental result that the two-point correlation function, C, of ϕ_α in (3.39) of the D-dimensional classical field theory (2.11) is precisely the same as the time-ordered correlation function of the operator ϕ_α under the Hamiltonian $\mathcal{H}$ in (6.52). Specifically, the latter correlation function is

$$C_{\alpha\beta}(x, \tau_1; y, \tau_2) = \begin{cases} \frac{1}{Z}\text{Tr}\left(e^{-\mathcal{H}/T}\hat{\phi}_\alpha(x, \tau_1)\hat{\phi}_\beta(y, \tau_2)\right) & \text{for } \tau_1 > \tau_2, \\ \frac{1}{Z}\text{Tr}\left(e^{-\mathcal{H}/T}\hat{\phi}_\beta(y, \tau_2)\hat{\phi}_\alpha(x, \tau_1)\right) & \text{for } \tau_1 < \tau_2, \end{cases} \quad (7.1)$$

where $\hat{\phi}_\alpha(x, \tau)$ is defined by imaginary time evolution under the $\mathcal{H}$:

$$\hat{\phi}_\alpha(x, \tau) \equiv e^{\mathcal{H}\tau}\phi_\alpha(x)e^{-\mathcal{H}\tau}. \quad (7.2)$$

This quantum correlator is then precisely the classical correlator C in (3.39) after the coordinates (x, τ) are mapped to the D-dimensional coordinate x in (3.39).

The quantum correlation function in (7.1) is thus a technically useful quantity. All the methods developed in Chapters 3 and 4 can be immediately applied towards its computation: this leads to a very efficient method for determining key characteristics of the phase diagram of the quantum model. However, C is not directly measurable in experiments on the quantum system. The purpose of the first two sections below is to establish a connection between the imaginary-time correlation C and real-time correlation functions naturally related to experimental probes. We then use these connections to better characterize the phases and the critical point of the quantum model.

7.1 Spectral representation

The first step in our analysis is to express C in (7.1) in the so-called spectral representation. Actually, we have already used spectral representations at $T = 0$ in (5.54) and (6.40), and present a more complete discussion here.

To clean up the notation, we drop the spin components in (7.1) because they play no essential role, and consider the case $N = 1$. Because the correlator in (7.1) is periodic in time with period $1/T$, it is useful to define its Fourier transform at the "Matsubara" frequency, ω_n, which must be an integer multiple of $2\pi T$, $\omega_n = 2\pi n T$, by

$$
\begin{aligned}
\chi(x, \omega_n) &\equiv \int_0^{1/T} d\tau\, e^{i\omega_n \tau} C(x, \tau; 0, 0) \\
&= \frac{1}{Z} \int_0^{1/T} d\tau\, e^{i\omega_n \tau} \mathrm{Tr}\left(e^{-\mathcal{H}/T} \hat{\phi}(x, \tau)\hat{\phi}(0, 0)\right),
\end{aligned} \tag{7.3}
$$

where we have used spatial and temporal translation invariance to set the arguments of the second $\hat{\phi}$ at the origin of spacetime.

Now imagine we know all the eigenstates and eigenenergies of the continuum Hamiltonian $\mathcal{H}$ in (6.52). In general, these states will occupy a continuum of energies, but by placing the field theory in a d-dimensional cubic box of size L (we will eventually take $L \to \infty$) we can obtain a discrete spectrum in which the exact eigenstates are labeled by the index m. Thus a complete set of orthonormal eigenstates is $|m\rangle$, and their eigenenergies are m. These eigenstates satisfy the completeness identity:

$$
\sum_m |m\rangle\langle m| = \hat{1}, \tag{7.4}
$$

where $\hat{1}$ is the identity operator. We now insert this identity before and after the first $\hat{\phi}$ operator to obtain

$$
\begin{aligned}
\chi(x, \omega_n) &= \sum_{m,m'} \frac{\langle m'|\phi(x)|m\rangle\langle m|\phi(0)|m'\rangle}{Z} \int_0^{1/T} d\tau\, e^{(i\omega_n - E_m + E'_m)\tau - E_{m'}/T} \\
&= \frac{1}{Z} \sum_{m,m'} \langle m'|\phi(x)|m\rangle\langle m|\phi(0)|m'\rangle \frac{\left(e^{-E_m/T} - e^{-E_{m'}/T}\right)}{(i\omega_n - E_m + E_{m'})};
\end{aligned} \tag{7.5}
$$

in the last step we used the fact that $e^{i\omega_n/T} = 1$ at all Matsubara frequencies. We can now write this in its final form, known as the spectral representation

$$
\chi(x, \omega_n) = \int_{-\infty}^{\infty} \frac{d\Omega}{\pi} \frac{\rho(x, \Omega)}{\Omega - i\omega_n}, \tag{7.6}
$$

where the spectral density $\rho(x, \Omega)$ is given by

$$
\begin{aligned}
\rho(x, \Omega) &\equiv \frac{\pi}{Z} \sum_{m,m'} \langle m'|\phi(x)|m\rangle\langle m|\phi(0)|m'\rangle \\
&\quad \times (e^{-E_{m'}/T} - e^{-E_m/T})\delta(\Omega - E_m + E_{m'}).
\end{aligned} \tag{7.7}
$$

This spectral density is the key quantity connecting various correlation functions in both real and imaginary time. Indeed, once we know the spectral density, we can easily obtain all needed correlation functions. In particular, from (7.6) we immediately obtain the correlation function at the Matsubara frequencies of imaginary time. The inverse problem is much more difficult: from a knowledge of $\chi(x, \omega_n)$ at all ω_n, it is not easy to find $\rho(x, \Omega)$. Indeed, this problem is ill-posed: very small errors in the values of $\chi(x, \omega_n)$ lead to large errors in $\rho(x, \Omega)$. However, when exact analytic expressions for $\chi(x, \omega_n)$ are available, it is possible to determine $\rho(x, \Omega)$; we will use this method on a number of occasions.

7.1.1 Structure factor

Let us now turn to correlation functions in *real* time, t, which are directly observable in the laboratory. We define time evolution of operators in the Heisenberg picture by (compare (7.2))

$$\hat{\phi}_\alpha(x, t) \equiv e^{i\mathcal{H}t}\phi_\alpha(x)e^{-i\mathcal{H}t}. \tag{7.8}$$

Then the real-time analog of (7.1) is the correlation function

$$\widetilde{C}_{\alpha\beta}(x, t; x', t') = \frac{1}{\mathcal{Z}}\text{Tr}\left(e^{-\mathcal{H}/T}\hat{\phi}_\alpha(x, t)\hat{\phi}_\beta(x', t')\right). \tag{7.9}$$

As above, we drop the indices α, β below, and deal only with the case $N = 1$.

The dynamic structure factor, $S(k, \omega)$ is defined by a Fourier transform of the real-time correlation (compare (7.3)):

$$S(k, \omega) = \int d^d x \int_{-\infty}^{\infty} dt\,\widetilde{C}(x, t; x', t')e^{-i\vec{k}\cdot(\vec{x}-\vec{x}')+i\omega(t-t')}. \tag{7.10}$$

Note that the time integration extends over all real values of t, unlike the limited domain between 0 and $1/T$ for imaginary time.

The dynamic structure factor is the quantity naturally measured in scattering experiments, such as neutron, X-ray, or light scattering of solid-state systems. This becomes clear from the spectral representation: proceeding as in (7.5) by repeated insertions of the identity (7.4), it is easy to show that

$$S(k, \omega) = \frac{2\pi}{\mathcal{Z}V}\sum_{m,m'}e^{-E_{m'}/T}|\langle m'|\phi(\vec{k})|m\rangle|^2\delta(\omega - E_m + E_{m'}), \tag{7.11}$$

where V is the volume of the system and $\phi(\vec{k})$ is the spatial Fourier transform of the operator $\phi(x)$. The expression (7.11) has the structure of a transition rate computed using Fermi's golden rule. The system is initially in the state $|m'\rangle$ with the thermal probability $e^{-E_{m'}/T}/\mathcal{Z}$; an external perturbation (the incoming photon or neutron) couples linearly to the operator $\phi(\vec{k})$, and (7.11) computes the transition probability per unit time to the final state $|m\rangle$. The result is clearly proportional to the Born scattering cross-section of the photon or neutron with momentum transfer $\vec{k}$ and energy transfer ω. Note that we are making the Born approximation only on the coupling between the probe and the system: in principle, (7.11) treats all interactions within the system exactly.

Comparing the expression (7.11) with the spectral density in (7.7), we obtain the exact identity

$$S(k, \omega) = \frac{2}{1 - e^{-\omega/T}} \rho(k, \omega), \tag{7.12}$$

where $\rho(k, \omega)$ is the spatial Fourier transform of $\rho(x, \omega)$. This is the first of our needed connections between real- and imaginary-time correlations, relating the dynamic structure factor to the spectral density, which in turn determines the correlator at the imaginary Matsubara frequencies by (7.6). The identity (7.12) is one statement of the "fluctuation–dissipation" theorem, and the reason for this terminology will become clearer in the following subsection.

7.1.2 Linear response

Now we consider another experimentally useful quantity: the time-dependent response to an external perturbation. For simplicity, we consider an external time and space-dependent "field" $h_\alpha(x, t)$ which couples linearly to the field operator $\phi(x)$, and so changes the Hamiltonian by

$$\mathcal{H} \to \mathcal{H} - \int d^d x \phi_\alpha(x) h_\alpha(x, t). \tag{7.13}$$

This perturbation is the analog of (3.40) in the classical model.

Because of the presence of $h_\alpha(x, t)$, all observables now have space and time dependence, and the system is no longer in thermal equilibrium. We would like to compute the change in the observables from equilibrium to linear order in $h_\alpha(x, t)$. This is given by a very general expression known as the Kubo formula. Without any specific knowledge of $\mathcal{H}$, we can write the shift away from equilibrium for an arbitrary observable $\mathcal{O}(x)$ in the following form

$$\delta\langle \mathcal{O}(x)\rangle(t) = \int d^d x' \int_{-\infty}^{\infty} dt' \chi_{\mathcal{O}\alpha}(x - x', t - t') h_\alpha(x', t'), \tag{7.14}$$

where the initial δ indicates "change due to external field," and the expectation value on the left-hand side is evaluated in the density matrix describing the state of the system in the presence of h. The coefficient on the right-hand side is the dynamic susceptibility χ: it is a characteristic of $\mathcal{H}$ in the absence of h, and so it is invariant under time and space translations. Finally, the expression (7.14) must obey the important constraint of causality:

$$\chi(x, t) = 0 \text{ for } t < 0, \tag{7.15}$$

because the response can only depend upon the values of h at earlier times. This identifies χ as the so-called "retarded" response function.

The Kubo formula is a general result for the susceptibility χ. Its derivation involves a simple exercise in first-order time-dependent perturbation theory: we start from an initial thermal state described by a density matrix $\exp(-\mathcal{H}/T)/\mathcal{Z}$, and compute its evolution under the change (7.13) by integrating the equations of motion to first order in h. The computation is discussed further in Exercise 7.1, and leads to the main result

$$\chi_{\mathcal{O}\alpha}(x - x', t - t') = i\theta(t - t')\frac{1}{\mathcal{Z}}\mathrm{Tr}\left(e^{-\mathcal{H}/T}[\hat{\mathcal{O}}(x, t), \hat{\phi}_\alpha(x', t')]\right), \tag{7.16}$$

where $\theta(t)$ is the unit step function, $\mathcal{H}$ is the Hamiltonian in the absence of h, and the time evolution of the operators is specified as in (7.8).

For our subsequent analysis we focus on the observable $\mathcal{O} = \phi$, and drop the α index by considering $N = 1$. Then the susceptibility of interest is

$$\chi(x - x', t - t') = i\theta(t - t')\frac{1}{\mathcal{Z}}\mathrm{Tr}\left(e^{-\mathcal{H}/T}[\hat{\phi}(x, t), \hat{\phi}(x', t')]\right). \tag{7.17}$$

It is useful to consider this susceptibility in momentum and frequency space by defining

$$\chi(k, \omega) = \int d^d x \int_0^\infty dt\, \chi(x, t)e^{-i\vec{k}\cdot\vec{x}+i\omega t}. \tag{7.18}$$

Note the limits on the time integration, which are a consequence of (7.16). Because of these limits, if we consider ω as a complex number, the integral in (7.18) is well-defined for ω in the upper half-plane: the oscillatory factor $e^{i\omega t}$ becomes a decaying exponential for ω in the upper half-plane, and so the integral (7.18) converges. The function $\chi(k, \omega)$ is therefore an analytic function of ω in the upper half-plane, and we define its value on the real ω axis by analytic continuation from the upper half-plane. Alternatively stated, we map $\omega \to \omega + i\eta$, where η is a small positive number, at intermediate stages of the computation, and take the limit $\eta \to 0$ at the end; this procedure leads to convergent results at all stages.

Let us now obtain a spectral representation of (7.17) and (7.18) as before. We insert (7.4) around the ϕ operators, and perform the Fourier transform to obtain

$$\chi(k, \omega) = \frac{1}{\mathcal{Z}V}\sum_{m,m'}|\langle m'|\phi(k)|m\rangle|^2\frac{\left(e^{-E_{m'}/T} - e^{-E_m/T}\right)}{\omega + i\eta - E_m + E_{m'}} \tag{7.19}$$

in the limit $\eta \to 0^+$. Now comparing (7.19) with (7.7), we obtain our main result

$$\chi(k, \omega) = \int_{-\infty}^\infty \frac{d\Omega}{\pi}\frac{\rho(k, \Omega)}{\Omega - \omega - i\eta} \tag{7.20}$$

connecting the retarded response function to the spectral density. The relations (7.6), (7.12), and (7.20) are the key results of this section, connecting the spectral density to the imaginary-time correlations, the real-time dynamic structure factor, and retarded susceptibility. Also note that $\chi(k, \omega = 0) \equiv \chi(k)$ is the *static* susceptibility.

A key feature of our results is the close similarity between (7.6) and (7.20). They show that the imaginary-time susceptibility $\chi(k, \omega_n)$ and the retarded response function $\chi(k, \omega)$ are part of the same analytic function $\chi(k, z)$ defined by

$$\chi(k, z) = \int_{-\infty}^\infty \frac{d\Omega}{\pi}\frac{\rho(k, \Omega)}{\Omega - z} \tag{7.21}$$

for a general complex frequency z. For $z = i\omega_n$ on the imaginary axis, $\chi(k, z)$ is the imaginary-time correlation at the Matsubara frequencies. And for $z = \omega + i\eta$, just above the real axis, we obtain the retarded response functions of the Kubo formula. Thus we can map the imaginary-time correlation to the retarded response function by analytic continuation. Also, our notation for the frequency argument of χ, ω_n vs. ω, implicitly determines whether we are considering response functions on the imaginary or real axis.

For the case where the Kubo formula (7.17) involves the commutator of a field with its Hermitian conjugate, the associated spectral density $\rho(k, \Omega)$ in (7.7) is real. Then we can write (7.20) as

$$\rho(k, \omega) = \mathrm{Im}\chi(k, \omega) = \frac{(1 - e^{-\omega/T})}{2} S(k, \omega), \tag{7.22}$$

where we used (7.12). The structure factor on the right-hand side is a measure of fluctuations of the field ϕ, while $\mathrm{Im}\chi(k, \omega)$ measures the response of ϕ which is out of phase with the applied field (from (7.14)). As in the damped harmonic oscillator, it is the out-of-phase component which measures the energy absorbed by the system from the external field, thus justifying the name "fluctuation–dissipation" theorem.

We have now finally assembled all the tools necessary to make full use of the connection between classical and quantum critical points. The following sections contain the payoff in our understanding of quantum phase transitions. We are now able to take the imaginary-time "classical" correlations in D dimensions discussed in Chapters 3 and 4 and use them to understand the properties of the quantum Ising and rotor models of Chapters 5 and 6 at a deeper level.

7.2 Correlations across the quantum critical point

This section addresses the same problem as Section 6.6: a description of the spectrum of the quantum field theory (6.52) across the quantum critical point. Rather than using the perturbative arguments of Section 6.6, here we are able to go further by analytically continuing the RG-improved results of Chapter 4. As in Section 6.6, we consider the paramagnetic phase, the magnetically ordered phase, and the quantum critical point in turn.

7.2.1 Paramagnet

We begin with the Gaussian result (3.44) for the D-dimensional classical theory. Analytically continuing this to the quantum theory in d dimensions, we map $k^2 \to c^2 k^2 - \omega^2$, and so obtain the retarded response function

$$\chi(k, \omega) = \frac{1}{c^2 k^2 + r - (\omega + i\eta)^2}. \tag{7.23}$$

Taking its imaginary part, we have the spectral density

$$\rho(k, \omega) = \frac{\mathcal{A}}{2\varepsilon_k} [\delta(\omega - \varepsilon_k) - \delta(\omega + \varepsilon_k)], \tag{7.24}$$

where $\varepsilon_k = (c^2 k^2 + r)^{1/2}$ is the same dispersion relation as in (6.57), and we have introduced a "quasiparticle residue" $\mathcal{A} = 1$. Thus the spectral density has delta functions at precisely the energy of the N-fold degenerate quasiparticles of Section 6.6.

Now let us move beyond the Gaussian theory, and look at perturbative corrections in u. Then the D-dimensional susceptibility is given by (3.45), which includes a self-energy $\Sigma(k)$. The perturbative expression for the self-energy to order u^2 was given in (4.35). The one-loop terms for the self-energy only contribute a renormalization of r, and the effects we are interested in arise from the analytical continuation of (4.35) under $k^2 \rightarrow c^2 k^2 - \omega^2$, so that

$$\chi(k, \omega) = \frac{1}{c^2 k^2 + r - (\omega + i\eta)^2 - \Sigma(k, \omega)}. \tag{7.25}$$

We continue to identify the position of the pole of $\chi(k, \omega)$ (if present) as a function of ω as a determinant of the spectrum of the quasiparticle, and the residue of the pole as the quasiparticle residue $\mathcal{A}$. The real part of the self-energy $\Sigma(k, \omega)$ will serve to modify the quasiparticle dispersion relation, and the value of $\mathcal{A}$, but will not remove the pole from the real ω axis. To understand possible decay of the quasiparticle, we need to consider the imaginary part of the self-energy.

We made general arguments for the absolute stability of the quasiparticle, provided that k is not too large, in Chapters 5 and 6, and these continue to apply here. Their consequence here is the relation

$$\mathrm{Im}\,\Sigma(k, \omega = \varepsilon_k) = 0 \tag{7.26}$$

at $T = 0$. This can be explicitly verified by a somewhat lengthy evaluation of (4.35), and an analytic continuation of the result: see Exercise 7.2. An immediate consequence is that the dynamic susceptibility has a delta function contribution which is given exactly by (7.24). All the higher order corrections only serve to renormalize r, and reduce the quasiparticle residue $\mathcal{A}$ from unity; the dispersion relation continues to retain the form in (6.57) by relativistic invariance. The stability of the delta function reflects the stability of the single quasiparticle excitations: a quasiparticle with momentum k not too large cannot decay into any other quasiparticle states and still conserve energy and momentum.

However, $\Sigma(k, \omega)$ does have some more interesting consequences at higher ω. We discussed in Sections 5.2.2 and 6.1 the existence of multiparticle continua. Here ω is the energy inserted by ϕ into the ground state, and so for $\omega > pr$, with p integer, we expect the creation of p particle states. The global O(N) or Z_2 symmetry actually restricts p to be odd, and so the lowest energy multiparticle states that will appear in χ are at $\omega = 3r$. Consonant with this, we find that the self-energy acquires a nonzero imaginary part at zero momentum only for $\omega > 3r$, i.e. there is a threshold for three-particle creation at $\omega = 3r$. The form of $\mathrm{Im}\,\Sigma(0, \omega)$ at the threshold can be obtained by analytically continuing (4.35): this is explored in Exercise 7.2 and leads to

$$\mathrm{Im}\,\Sigma(0, \omega) \propto \mathrm{sgn}(\omega)\theta(|\omega| - 3r)(|\omega| - 3r)^{(d-1)} \tag{7.27}$$

for ω around $3r$.

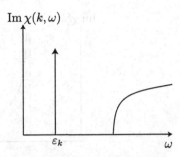

Fig. 7.1 The spectral density in the paramagnetic phase at $T = 0$ and a small k. Shown are a quasiparticle delta function at $\omega = \varepsilon_k$ and a three-particle continuum at higher frequencies. There are additional n-particle continua ($n \geq 5$ and odd) at higher energies, which are not shown.

Taking the imaginary part of (7.25), we obtain the generic form of the spectral density shown in Fig. 7.1.

We now present a simple physical argument for the nature of the threshold singularity in (7.27), to supplement the more formal computation in Exercise 7.2 from (4.35). Just above threshold, we have a particle with energy $3r + \delta\omega$ which decays into three particles with energies just above r. The particles in the final state will also have a small momentum, and so we can make a nonrelativistic approximation for their dispersion: $r + c^2 k^2/(2r)$. Because the rest mass contributions, r, add up to the energy of the initial state, we can neglect them from now. The decay rate, by Fermi's golden rule is proportional to the density of final states, which yields

$$\text{Im}\Sigma(0, 3r + \delta\omega) \propto \int_0^{\delta\omega} d\Omega_1 d\Omega_2 \int \frac{d^d p}{(2\pi)^d} \frac{d^d q}{(2\pi)^d} \delta\left(\Omega_1 - \frac{c^2 p^2}{2r}\right)$$

$$\times \delta\left(\Omega_2 - \frac{c^2 q^2}{2r}\right) \delta\left(\delta\omega - \Omega_1 - \Omega_2 - \frac{c^2(p+q)^2}{2r}\right)$$

$$\sim (\delta\omega)^{(d-1)}, \tag{7.28}$$

in agreement with (7.27).

We expect this perturbative estimate of the threshold singularity to be exact in all $d \geq 2$. In $d = 1$, there are strong final state corrections from the interactions between the quasiparticles, and these are better explored using the methods of Chapter 10.

7.2.2 Quantum critical point

Here our present methods yield a qualitatively new result, beyond the reach of the perturbative arguments of Chapters 5 and 6.

We analytically continue the classical critical point result in (4.34) to obtain the dynamic susceptibility at the quantum critical point at $T = 0$:

$$\chi(k, \omega) \sim \frac{1}{(c^2 k^2 - \omega^2)^{1-\eta/2}}. \tag{7.29}$$

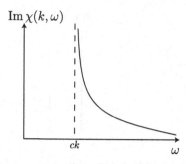

Fig. 7.2   The spectral density at the quantum critical point. Note the absence of a quasiparticle pole, like that in Fig. 7.1

The key feature differentiating this result from (7.23) is that this susceptibility does *not* have poles on the real frequency axis. Rather, there are branch cuts going out from $\omega = \pm ck$ to infinity. Taking the imaginary part, we obtain a continuous spectral weight at $|\omega| > ck$

$$\text{Im}\chi(k,\omega) \sim \frac{\text{sgn}(\omega)\theta(|\omega| - ck)}{(\omega^2 - c^2k^2)^{1-\eta/2}}, \tag{7.30}$$

see Fig. 7.2. The absence of a pole indicates that there are no well-defined quasiparticle excitations. Instead we have a dissipative continuum of critical excitations at all $|\omega| > ck$: any perturbation will not create a particle-like pulse, but decay into a broad continuum. This is a generic property of a strongly-coupled quantum critical point.

More generally, we can use the scaling from (4.42) to describe the evolution of the spectrum as r approaches the critical point at $r = r_c$ from the paramagnetic phase at $T = 0$. Because of the relativistic invariance, the energy gap $\Delta \sim \xi^{-z}$ with $z = 1$, where the correlation length ξ diverges as in the classical model $\xi \sim (r - r_c)^{-\nu}$. In terms of Δ, analytic continuation of (4.42) yields

$$\chi(k,\omega) = \frac{1}{\Delta^{2-\eta}} \widetilde{F}\left(\frac{ck}{\Delta}, \frac{\omega}{\Delta}\right). \tag{7.31}$$

In the paramagnetic phase, the N quasiparticles have dispersion $\varepsilon_k = (c^2k^2 + \Delta^2)^{1/2}$ (the momentum dependence follows from relativistic invariance). Comparing (7.31) with (7.24), we see that the two expressions are compatible if the quasiparticle residue scales as

$$\mathcal{A} \sim \Delta^\eta, \tag{7.32}$$

so the quasiparticle residue vanishes as we approach the quantum critical point. Above the quasiparticle pole, the susceptibility of the paramagnetic phase also has p particle continua having thresholds at $\omega = (c^2k^2 + p^2\Delta^2)^{1/2}$, with $p \geq 3$ and p odd. As $\Delta \to 0$ upon approaching the quantum critical point, these multiparticle continua merge to a common threshold at $\omega = ck$ to yield the quantum critical spectrum in (7.30).

7.2.3 Magnetic order

Now $r < r_c$, and we have to expand about the magnetically ordered saddle point (6.60). The first important consequence of the magnetic order is that the nonzero limit in (1.14) combines with (7.10) to yield a delta function in the dynamic structure factor

$$S(k, \omega) = N_0^2 (2\pi)^{d+1} \delta(\omega) \delta^d(k) + \ldots, \tag{7.33}$$

where the *ellipses* represent contributions at nonzero ω. This delta function is easily detectable in elastic neutron scattering, and is a clear signature of the presence of magnetic long-range order.

We now discuss the finite ω contributions to (7.33). We assume the ordered moment is oriented along the $\alpha = 1$ direction. From Gaussian fluctuations about the saddle point in (6.60) we obtain, as in (6.61), susceptibilities which are diagonal in the spin index, with the longitudinal susceptibility

$$\chi_{11}(k, \omega) = \frac{1}{c^2 k^2 - (\omega + i\eta)^2 + 2|r|}, \tag{7.34}$$

and the transverse susceptibility

$$\chi_{\alpha\alpha}(k, \omega) = \frac{1}{c^2 k^2 - (\omega + i\eta)^2}. \qquad \alpha > 1 \tag{7.35}$$

The poles in these expressions correspond to the $N - 1$ spin waves and the Higgs particle of Section 6.6.3.

It now remains to study the perturbative corrections to (7.34) and (7.35). The $N - 1$ spin waves are expected to be generically stable and well-defined, as they owe their existence to the broken symmetry. We study the general structure of the corrections to (7.35) in Chapter 8. In contrast, as we noted in Section 6.6.3, there is nothing protecting the stability of the longitudinal Higgs particle, and we now examine its decay into multiple spin-wave excitations.

Let us return to the classical perspective, and expand the action in (3.25) in powers of $\tilde{\phi}_\alpha$ defined in (6.60); then the corrections to (6.61) are

$$\mathcal{S}_\phi = \int d^D x \left\{ \frac{1}{2} \left[(\nabla_x \tilde{\phi}_\alpha)^2 + 2|r| \tilde{\phi}_1^2 \right] + \sqrt{\frac{|r|u}{6}} \tilde{\phi}_1 \tilde{\phi}_\alpha^2 + \frac{u}{4!} (\tilde{\phi}_\alpha^2)^2 \right\}. \tag{7.36}$$

Note the new cubic term, which plays a key role in our considerations below. We would like to understand the influence of the nonlinearities in (7.36) on the pole at $\omega = (c^2 k^2 + 2|r|)^{1/2}$ in (7.35). For this purpose, let us introduce the self-energy of the longitudinal mode

$$\chi_{11}(k) = \frac{1}{k^2 + 2|r| - \Sigma_{11}(k)}. \tag{7.37}$$

To lowest order in u, the leading term in the self-energy comes from two cubic interactions, as shown in Fig 7.3, and evaluates to

$$\Sigma_{11}(k) = \frac{(N-1)|r|u}{3} \Pi(k),$$

$$\Pi(k) \equiv \int \frac{d^D p}{(2\pi)^D} \frac{1}{p^2 (p+k)^2} = \frac{C_D}{k^{4-D}}, \tag{7.38}$$

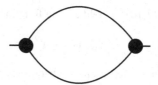

Fig. 7.3 Leading diagram for the self-energy of the longitudinal response. All internal lines involve propagators of $\tilde{\phi}_\alpha$ with $\alpha > 1$.

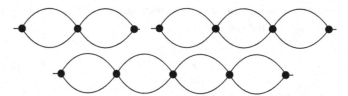

Fig. 7.4 More singular terms in the self-energy of the longitudinal response. Again all internal lines involve propagators of $\tilde{\phi}_\alpha$ with $\alpha > 1$.

where in the last step we have given the answer for $D < 4$, with C_D a phase space factor dependent upon the spacetime dimension ($C_3 = 1/8$). For $D > 4$, the answer depends upon the spacetime dimension, and at small k we can simply take $\Pi(k)$ to be a constant. For $D < 4$, this contribution to the self-energy is divergent in the limit of low momentum. This divergence is a consequence of the gapless spin-wave modes, which apparently have a singular effect on the longitudinal response. This divergence should raise the concern that higher order terms in the self-energy will be even more singular. This is indeed the case, and we have shown some of the more singular graphs in Fig. 7.4. Fortunately, it is possible to re-sum the most singular terms at each order in u because they form a geometric series. We find

$$\Sigma_{11}(k) = \frac{(N-1)|r|u}{3}\Pi(k)\left[1 - \frac{(N+1)u}{6}\Pi(k) + \frac{(N+1)^2u^2}{36}\Pi^2(k) + \ldots\right]$$

$$= \frac{(N-1)|r|u\Pi(k)/3}{1 + (N+1)u\Pi(k)/6}. \tag{7.39}$$

Now note that for $d < 3$, the divergence of $\Pi(k \to 0)$ and (7.39) imply that $\Sigma_{11}(k \to 0)$ is a finite constant. However, there are many other contributions to Σ_{11} which yield a constant, and so this raises a natural concern on the extent to which we can trust (7.39). We need a parameter other than the nonlinearity u to systematically control the computation. A convenient choice is to take N, the number of components of ϕ_α, and to take the large N limit. We study this expansion in much greater detail in Part III, starting in Chapter 11. Here the $N \to \infty$ limit has to be taken after defining $u = \bar{u}/N$, and keeping r and $\bar{u}$ fixed. It can then be shown that (7.39) is the leading contribution to Σ_{11} in this limit, and that all other corrections are suppressed by a factor of $1/N$ (see Exercise 7.3). Taking the $N \to \infty$ limit in (7.39), we now write

$$\Sigma_{11}(k) = \frac{|r|\,\overline{u}\Pi(k)/3}{1 + \overline{u}\Pi(k)/6}. \quad N \to \infty \qquad (7.40)$$

Let us now analytically continue (7.37) and (7.40) to real frequencies to obtain the retarded dynamic susceptibility. For $d > 3$, where we can take $\Pi(k)$ to be a constant, there is little effect apart from a renormalization of r, and the "Higgs" pole in (7.17) survives. However, there is a much stronger effect in $d < 3$ for $N > 1$. Here, after mapping $k^2 \to c^2 k^2 - \omega^2$, we have at low momenta and frequency

$$\chi_{11}(k, \omega) = \frac{1}{c^2 k^2 - \omega^2 + \dfrac{12|r|}{C_{d+1}\overline{u}}(c^2 k^2 - \omega^2)^{(3-d)/2}}. \quad N \to \infty \qquad (7.41)$$

The pole at the position of the Higgs energy $\omega = \pm (c^2 k^2 + 2|r|)^{1/2}$ has disappeared, and we only have a branch cut having its onset at the spin-wave energy $\omega = \pm ck$. Thus, for $d < 3$ and $N > 1$, there is no Higgs particle, and only a broad continuum of multiple spin-wave excitations in the longitudinal response. Note also that, unlike the Gaussian result in (7.34), the static longitudinal susceptibility $\chi(k \to 0, \omega = 0) \sim k^{d-3}$ is divergent; this is also a generic consequence [372, 426, 559] of the breaking of the continuous O(N) symmetry in $D < 4$. The Higgs particle does survive at $N = 1$ for all $d > 1$, corresponding to the excitation studied in Section 5.3.1.

The $d = 3$ case is marginal, and we will not evaluate the detailed form for $\Pi(k)$ in this case. The Higgs particle is at the boundary of stability, and this accounts for its observation in TlCuCl$_3$, as discussed in Section 6.6.3.

Exercises

7.1 Write down the equation for motion of the density matrix, and integrate it to first order in the external perturbation. Hence obtain the Kubo formula (7.16).

7.2 Evaluate the self-energy of the quasiparticle in the paramagnetic phase by computing the Feynman graph in (4.35). Analytically continue the result to real frequencies to obtain (7.26) and (7.27).

7.3 Write down the complete expression for $\Sigma_{11}(k)$ to order u^2. Show that the result agrees with (7.40) in the limit $N \to \infty$ at fixed $\overline{N} = Nu$.

Broken symmetries

This chapter continues our study of d-dimensional quantum field theory (2.11) by a closer examination of the structure of the magnetically ordered phase. The considerations below apply also to the D-dimensional classical model (3.25), and given our improved understanding of the close connections between the two models in Chapter 7, we will freely move back-and-forth between the classical and quantum cases.

We have so far characterized the magnetically ordered phase by the presence of long-range correlations in the two-point correlations of the order parameter field ϕ_α. After defining the two-point correlation function (without subtraction of the disconnected pieces as in (3.39)) by

$$\overline{C}_{\alpha\beta}(x - x') = \langle \phi_\alpha(x)\phi_\beta(x')\rangle, \tag{8.1}$$

the magnetically ordered phase was identified by

$$\lim_{|x|\to\infty} C_{\alpha\beta}(x) \neq 0. \tag{8.2}$$

The right-hand side is the square of the spontaneous magnetization N_0. The existence of a nonzero N_0 signals a spontaneous breakdown of the $O(N)$ symmetry of the underlying degrees of freedom.

While this definition is formally adequate, we now explore another definition which is closer to our physical intuition on the structure of a magnetically ordered state. We often think of ordered phases as having an intrinsic "rigidity," i.e. they respond little to external perturbations, and prefer to revert to their original configuration. Thus, a solid, with broken translational symmetry, is resistant to shear deformations, in contrast to a liquid which preserves translational symmetry and shears freely. This chapter provides a characterization of the rigidity of the magnetically ordered phase of the quantum field theory (2.11).

8.1 Discrete symmetry and surface tension

Let us consider the Ising case first, as described by the $N = 1$ case of the d-dimensional quantum Hamiltonian in (6.52). The key idea is to consider changes in its total free energy under changes of the boundary conditions. We have so far considered infinite systems, but now let us place the field theory on a d-dimensional cubic box of side L. We single out a preferred direction, say x_1, and take periodic boundary conditions of ϕ along all the $(d-1)$

transverse directions $x_\perp$. Let $0 \leq x_1 \leq L$, and we allow ϕ to vary freely as a function of x_1 but modify the Hamiltonian by the following boundary terms

$$\mathcal{H} \to \mathcal{H} - h_\ell \int d^{d-1}x_\perp \phi(x_1 = 0, x_\perp) - h_r \int d^{d-1}x_\perp \phi(x_1 = L, x_\perp). \qquad (8.3)$$

In other words, we apply a field h_ℓ on the left edge, and a field h_r on the right edge.

Now follow the distinct behaviors of the system in the limit of large $h_\ell = h_r$, from that for $h_\ell = -h_r$.

For $h_\ell = h_r$, the fields at the edges will pin ϕ along a common direction. In the paramagnetic phase, the memory of this boundary pinning decays in a length of order ξ as we move into the bulk of the system. However, in the magnetically ordered phase, the "rigidity" ensures that preferred edge orientations of ϕ propagate all the way across the system. Because the two edges are oriented in the same way, we will clearly have a smooth polarization of the spontaneous magnetization along the common direction.

In contrast, for $h_\ell = -h_r$, we get new physics only on the magnetically ordered side. Now the magnetic polarizations along the edges are oriented in opposite directions, and they will have to reconcile this incompatibility somewhere in the middle of the system. A domain wall, like that in (6.64), is forced in by the boundary conditions, and this will clearly increase the free energy of the system.

It is now clear that the free energy *difference*, $\Delta\mathcal{F}$, between the two boundary conditions contains useful information. It is dominated by the presence of the domain wall for the $h_\ell = -h_r$ boundary conditions, and includes information on all the quantum and thermal fluctuations of the domain wall around the mean-field configuration of (6.64). From the considerations above, we can conclude that the free energy difference behaves in the following manner in the limit of large L

$$\Delta\mathcal{F} = \begin{cases} \sim e^{-L/\xi} & \text{for } r > r_c \\ \Sigma L^{d-1} & \text{for } r < r_c \end{cases}. \qquad (8.4)$$

The exponential decay in the paramagnetic phase characterizes its insensitivity to changes in the boundary conditions. In contrast, in the magnetically ordered phase, the rigidity forces in a domain wall whose free energy is proportional to L^{d-1}, the volume of its transverse dimensions. The proportionality constant is the *surface tension*, Σ. The above definition makes it clear that Σ has dimensions of energy/(length)$^{(d-1)}$.

We can now use the scaling arguments of Chapter 4 to deduce the behavior of Σ as we approach the critical point. The surface tension is directly related to the free energy $\mathcal{F} = -T \ln \mathcal{Z}$, and so its transformations under the renormalization group follow directly from that of the partition function. As noted in (4.49), the RG is defined to leave $\mathcal{Z}$ invariant, and so $\mathcal{F}$ scales, as expected, like an energy; in other words

$$\text{dim}[\mathcal{F}] = z, \qquad (8.5)$$

where z is called the dynamic critical exponent. We define z more carefully in Section 10.2, and all the models considered so far have $z = 1$. Hence, from (8.4), we deduce that

$$\text{dim}[\Sigma] = d + z - 1. \qquad (8.6)$$

From this it follows that Σ vanishes as we approach the critical point $r \nearrow r_c$ as

$$\Sigma \sim \xi^{-(d+z-1)}. \tag{8.7}$$

We can actually say more for the special case of the quantum Ising model under consideration. Here $z = 1$, and energy scales are connected to length scales by the velocity c. After use of this energy scale, there are no arbitrary scales left in the definition of Σ, and so we can conclude that

$$\Sigma = \Upsilon_1 \hbar c \xi^{-d}, \tag{8.8}$$

where Υ_1 is a universal dimensionless number, i.e. it is the same for all quantum Ising models at their quantum critical point.

The above arguments also allow us to deduce the behavior of $\Delta\mathcal{F}$ precisely at the quantum critical point $r = r_c$. Here the only available length scale is L, and so $\Delta\mathcal{F} \sim L^{-z}$. For the $z = 1$ quantum Ising model

$$\Delta\mathcal{F} = \Upsilon_2 \hbar c / L, \tag{8.9}$$

where, again, Υ_2 is a universal constant.

8.2 Continuous symmetry and the helicity modulus

We now generalize the considerations of Section 8.1 from $N = 1$ to $N \geq 2$. The presence of a continuous O(N) symmetry dramatically modifies the response to changes in boundary conditions. Rather than having a sharp domain wall between oppositely oriented states, the optimal state can greatly lower its free energy by spreading out the difference between the two edges over a gradual change in orientation of the magnetic order. Such a gradual change was clearly not an option for the Ising case.

We present our remaining discussion for $N = 2$, although all the results below have an immediate generalization to $N > 2$. We take the same geometry as in Section 8.1, and write the boundary fields as

$$\mathcal{H} \rightarrow \mathcal{H} - h_{\ell\alpha} \int d^{d-1}x_\perp \phi_\alpha(x_1 = 0, x_\perp) - h_{r\alpha} \int d^{d-1}x_\perp \phi_\alpha(x_1 = L, x_\perp). \tag{8.10}$$

Let us assume the fields have a common large magnitude h, and differ only in their orientations:

$$h_{\ell\alpha} = h_0(\cos(\theta_\ell), \sin(\theta_\ell)), \quad h_{r\alpha} = h_0(\cos(\theta_r), \sin(\theta_r)). \tag{8.11}$$

Thus we have imposed a *twist* in the phase of the *XY* order parameter. The twisted boundary conditions are clearly characterized by the angular difference $\Delta\theta = \theta_r - \theta_\ell$. Clearly, we want $\Delta\theta$ to have the smallest possible absolute value, and so we define the phases so that $-\pi \leq \Delta\theta \leq \pi$. Now we can expect that the lowest free energy state will have the smallest possible local phase gradient. So it pays to spread the twist across the entire sample, yielding a local phase gradient of $\Delta\theta/L$. This phase gradient costs free energy: for large L

the phase gradient is small, and so we can expand the local free energy density in powers of $\Delta\theta/L$. By time-reversal symmetry, the free energy can only involve even powers of the phase gradient, and so we conclude that the cost to the free energy density is proportional to $(\Delta\theta/L)^2$. Integrating the free energy density over the entire sample, we conclude that for $r < r_c$

$$\Delta\mathcal{F} = \frac{\rho_s}{2}\left(\frac{\Delta\theta}{L}\right)^2 L^d. \tag{8.12}$$

This equation replaces the result (8.4) for $N \geq 2$. There is no change in the results of Section 8.1 between $N = 1$ and $N > 1$ for $r \geq r_c$.

The coefficient in (8.12) defines the *helicity modulus*, ρ_s. Depending upon the physical context, it is also referred to as the spin stiffness or (as we will see below), the superfluid density. The helicity modulus is nonzero only in the magnetically ordered phase, and is a measure of the rigidity of the response to twists in the phase of the order parameter.

The behavior of ρ_s as we approach the critical point can be deduced as in Section 8.1 We have

$$\dim[\rho_s] = d + z - 2, \tag{8.13}$$

and hence the helicity modulus vanishes as $\rho_s \sim \xi^{-(d+z-2)}$. For $z = 1$, we have

$$\rho_s = \Upsilon_3 \hbar c \xi^{-(d-1)}, \tag{8.14}$$

where Υ_3 is a universal number.

Our considerations so far have defined ρ_s in terms of the full free energy, which is computed by integrating out all degrees of freedom. However, the structure of (8.12) suggests that we can use the RG to obtain a local definition of ρ_s after a partial integration of the degrees of freedom. We have seen above that the dominant low-energy excitations in the magnetically ordered phase are slow twists in the orientation of the local magnetic order. In RG terms, this means that after we integrate out to length scales longer than the spin correlation length ξ, the state of the system can be characterized by a slowly varying field $\theta(r)$, representing the local twists in the order parameter. Thus, at long scales, we have an effective Hamiltonian for an emergent field $\theta(x)$, which is given by

$$\mathcal{H}_{\text{eff}} = \frac{\rho_s}{2}\int d^d x \, (\nabla_x \theta)^2. \tag{8.15}$$

For the case of the quantum rotor model, we can use the relativistic invariance of (2.11) to write down an effective Lagrangian for $\theta(x, \tau)$:

$$\mathcal{L}_{\text{eff}} = \frac{\rho_s}{2}\int d\tau \int d^d x \left\{(\nabla_x \theta)^2 + \frac{1}{c^2}(\partial_\tau \theta)^2\right\}. \tag{8.16}$$

So we have a simple and powerful result. At length scales larger than ξ, the quantum statistical mechanics of the magnetically ordered phase is described by quantum and thermal fluctuations of the field $\theta(x, \tau)$, which is controlled by the Gaussian quantum field theory in (8.16). The value of ρ_s in (8.16) is exactly the same as that in (8.12), and so its determination requires a full computation in the underlying field theory. The following section provides a specific method by which ρ_s can be computed perturbatively from (2.11).

8.2.1 Order parameter correlations

The structure of (8.15) also allows us to make a general and exact statement about the two-point correlation functions of the field ϕ_α. As before, we derive the results for $N = 2$, but the final result holds for all N.

We assume the magnetic order is in the $\alpha = 1$ direction, with $\langle \phi_1 \rangle = N_0$. We now compute the transverse susceptibility, $\chi_{22}(k, 0)$ exactly in the limit $k \to 0$; this susceptibility will also be denoted as $\chi_\perp$. It is the response of the system to a very slowly varying static field, $h(x)$, which couples linearly to the $\alpha = 2$ component of ϕ_α. The system will respond to such an external field by a slowly varying shift in the angular orientation of the order parameter, and the net energy cost will change from (8.15) to

$$\mathcal{H}_{\text{eff}} = \int d^d x \left[\frac{\rho_s}{2} (\nabla_x \theta)^2 - h N_0 \sin \theta \right]. \tag{8.17}$$

Minimizing the energy cost with respect to variations in θ to linear order in h, we obtain in Fourier space

$$\langle \phi_2(k) \rangle \approx N_0 \theta(k)$$
$$= \frac{N_0^2}{\rho_s k^2} h(k). \tag{8.18}$$

This gives us the exact result

$$\lim_{k \to 0} \chi_\perp(k, 0) = \frac{N_0^2}{\rho_s k^2}. \tag{8.19}$$

8.3 The London equation and the superfluid density

Here we use $\mathcal{L}_{\text{eff}}$ in (8.16) as an effective description of the model of $N = 2$ quantum rotors discussed in Section 1.4.3. There, we met the Hamiltonian (1.35) as a description of bosons hopping between potential minima at the site i. These bosons could represent either Cooper pairs of electrons in a superconducting array of Josephson junctions, or ultracold bosonic atoms in an optical lattice. It is useful to imagine that each boson carries an electrical charge e^*. For the Cooper pair, $e^* = 2e$; the ultracold atoms are neutral, but it is useful to endow them with a nonzero charge e^* as a technical device for characterizing the physical properties of the ordered state.

We now place the rotor model (1.35) in an external magnetic field associated with the vector potential $\vec{A}(x)$. The coupling of the underlying quantum particles to this vector potential implies that the boson hopping term $\exp(i(\theta_j - \theta_i))$ in (1.35) will be modified by the Aharanov–Bohm phase factor to

$$\exp\left(i\theta_j - i\theta_i - i\frac{e^*}{\hbar c_\ell} \int_{x_i}^{x_j} d\vec{x} \cdot \vec{A}(x) \right). \tag{8.20}$$

Here c_ℓ is the actual velocity of light, not to be confused with the velocity c appearing in the field theory (2.11). Naturally, the structure of (8.20) was fixed by requiring that the theory be invariant under electromagnetic gauge transformations under which

$$\vec{A} \to \vec{A} + \nabla_x \vartheta, \quad \theta_i \to \theta_i + \frac{e^*}{\hbar c_\ell} \vartheta, \tag{8.21}$$

where ϑ is an arbitrary gauge transformation. All our subsequent analyses of the rotor model (1.35) should respect this gauge invariance: its mapping to the field theory (2.11), and the subsequent RG procedure by which (8.16) is derived as the low-energy effective field theory of the ordered phase. This gauge invariance immediately allows us to deduce the form of the effective theory $\mathcal{L}_{\text{eff}}$ in the presence of the externally imposed $\vec{A}(x)$: we simply demand that the effective theory also be gauge invariant, and this leads to the unique result

$$\mathcal{L}_{\text{eff}} = \frac{\rho_s}{2} \int d\tau \int d^d x \left\{ \left(\nabla_x \theta - (e^*/(\hbar c_\ell))\vec{A} \right)^2 + \frac{1}{c^2} (\partial_\tau \theta)^2 \right\}. \tag{8.22}$$

With (8.22) in our hands, we are now in a position to compute a number of interesting response functions of the quantum rotor model. In particular, we can easily obtain the electrical current $\vec{J}$ that is induced by the external vector potential $\vec{A}$. As in standard electromagnetic theory, the current operator $\vec{J}$ is defined by the functional derivative of the action with respect to $\vec{A}$:

$$\vec{J} = -c_\ell \frac{\delta \mathcal{S}}{\delta \vec{A}}, \tag{8.23}$$

which for (8.22) yields

$$\vec{J} = \rho_s \frac{e^*}{\hbar} \left(\nabla_x \theta - \frac{e^*}{\hbar c_\ell} \vec{A} \right). \tag{8.24}$$

Just as in Section 7.1.2, we are now ready to define the expectation value of the current using linear response in the perturbation imposed by $\vec{A}$, and to compute the associated response function using a Kubo formula. Thus a Fourier transform of (7.14) yields (after generalizing to both a spatial and imaginary time dependence in both $\vec{J}$ and $\vec{A}$):

$$\langle J_a(k, \omega_n) \rangle = K_{ab}(k, \omega_n) A_b(k, \omega_n), \tag{8.25}$$

where $a, b = x, y, \ldots$ are indices representing spatial directions, and K_{ab} is the linear response function for the electrical current, which is often referred to as the current–current correlation function. This function plays an important role in characterizing the properties of the ordered state.

It is now a simple matter to evaluate the expectation value of (8.24) under the Gaussian theory (8.22) and so obtain the low-energy structure of K_{ab} in the ordered state; we find

$$K_{ab}(k, \omega_n) = -\frac{\rho_s e^{*2}}{\hbar^2 c_\ell} \left(\delta_{ab} - \frac{k_a k_b}{k^2 + \omega_n^2/c^2} \right). \tag{8.26}$$

This result has a number of important physical implications, which we will now describe. Before turning to these, let us note that (8.26) can be used to obtain another formal definition of ρ_s:

$$\rho_s = \frac{\hbar^2 c_\ell}{(d-1)e^{*2}} \lim_{k \to 0} \lim_{\omega \to 0} \left(\delta_{ab} - \frac{k_a k_b}{k^2} \right) K_{ab}(k, \omega). \tag{8.27}$$

The order of limits is significant, and this relationship identifies the helicity modulus as the zero momentum limit of the static current–current correlation function. We use (8.27) to provide an explicit computation of ρ_s in the quantum field theory of the rotor model.

Let us evaluate (8.26) at zero frequency, and insert the result into (8.9). In the Coulomb gauge $\nabla_x \cdot \vec{A} = 0$, we obtain

$$\vec{J} = -\frac{\rho_s e^{*2}}{\hbar^2 c_\ell} \vec{A}. \tag{8.28}$$

This is the important London equation, characterizing the electromagnetic response of a superconductor. It combines with the familiar Maxwell equation for the magnetic field $\vec{B} = \nabla_x \times \vec{A}$

$$\nabla_x \times \vec{B} = \frac{4\pi}{c_\ell} \vec{J}, \tag{8.29}$$

to yield the equation which predicts the Meissner effect:

$$\nabla_x^2 \vec{B} = \frac{1}{\lambda_L^2} \vec{B}. \tag{8.30}$$

The solution of this equation near the boundary between the system and free space shows that the $\vec{B}$ field decays exponentially to zero at a length-scale λ_L. This is the London penetration depth, given here by

$$\frac{1}{\lambda_L^2} = \frac{4\pi \rho_s e^{*2}}{\hbar^2 c_\ell^2}. \tag{8.31}$$

This is an exact relationship between the penetration depth and helicity modulus of the superconductor. Note that it involves only fundamental constants of nature, and ρ_s is the only material-dependent quantity. Thus, ρ_s provides a complete description of the response to external magnetic fields, and can be directly measured by observations of the London penetration depth.

Finally, (8.25) and (8.26) also contain information on the electrical conductivity of the ordered phase. Recall that we can also induce an electrical field using the vector potential by making it time-dependent: $\vec{E} = -(1/c_\ell)(\partial \vec{A}/\partial t)$. So from (8.26) we see that the electrical current is given by $\vec{J} = \sigma \vec{E}$, where the electrical conductivity is given by (for real frequencies ω, and for a spatially isotropic system)

$$\sigma(\omega) = \frac{c_\ell}{i\omega} K_{xx}(k \to 0, \omega). \tag{8.32}$$

From (8.26), we have

$$\text{Re}[\sigma(\omega)] = \frac{\rho_s e^{*2}}{\hbar^2} \delta(\omega). \tag{8.33}$$

The zero frequency delta function shows why the "magnetically" ordered phase of the O(2) rotor model is a superconductor. For neutral particles, we can use similar arguments to deduce the response to an external force (rather than an electrical field), and so identify this phase as a superfluid. The relationship (8.33) also shows why ρ_s is identified as the superfluid density.

8.3.1 The rotor model

Let us now evaluate the current response function for the quantum field theory (2.11). We proceed using the perturbative expansion described in Sections 6.6.3 and 7.2.3, and obtain the result to leading order in u. Corrections at higher order in u are discussed later in more detail in Chapter 15.

For $N = 2$, we can introduce a complex field $\psi = \phi_1 + i\phi_2$. Then coupling to the external vector potential is obtained by requirements of gauge invariance, under which the spatial gradient term in the action is modified by

$$|\nabla_x \psi|^2 \rightarrow |(\nabla_x - i(e^*/(\hbar c_\ell))\vec{A})\psi|^2. \tag{8.34}$$

Evaluating the current from (8.23), we obtain

$$\vec{J} = \frac{e^*}{i\hbar}\left(\psi^*\nabla_x\psi - \nabla_x\psi^*\psi\right) - \frac{e^{*2}}{\hbar^2 c_\ell}2|\psi|^2\vec{A}. \tag{8.35}$$

We can now evaluate $\langle \vec{J} \rangle$ using the Kubo formula, using the perturbation theory of Section 6.6.3. We parameterize $\psi = N_0(1 + \rho)e^{i\theta}$, and expand the resulting action in powers of ρ. At leading order, we find that the effective action for θ has exactly the form of (8.22), while the current (8.35) reduces to (8.24), provided that we identify

$$\rho_s = 2N_0^2. \tag{8.36}$$

The fluctuations in ρ lead to corrections to this expression at higher order in u, as described in Chapter 15.

Exercises

8.1 *Conductivity across the superconductor–insulator transition.* This exercise anticipates results that are explored in more detail in Chapter 15. In the vicinity of the superconductor–insulator transition (with no long-range Coulomb interactions), the low-energy states are described by the $N = 2$ ϕ_α^4 field theory. We write this in terms of a complex field $\Psi = (\phi_1 + i\phi_2)/\sqrt{2}$ with action

$$S = \int d^d x d\tau \left[|\nabla_x\Psi|^2 + |\partial_\tau\Psi|^2 + r|\Psi|^2 + \frac{u}{2}|\Psi|^4\right]. \tag{8.37}$$

Consider the response of this theory to an external vector potential $\vec{A}(x, \tau)$ under which

$$\nabla_x \Psi \rightarrow (\nabla_x - ie^* \vec{A})\Psi. \tag{8.38}$$

The imaginary-time current–current correlation function is then given by

$$K_{\alpha\beta}(x - x', \tau - \tau') = \langle J_\alpha(x, \tau) J_\beta(x', \tau') \rangle = \left. \frac{\delta^2 \ln \mathcal{Z}}{\delta A_\alpha(x, \tau) \delta A_\beta(x', \tau')} \right|_{\vec{A}=0}, \tag{8.39}$$

where α, β are components of the spatial coordinate x. From the zero-momentum, finite frequency component of K we can obtain the conductivity, σ, at imaginary frequencies by (see e.g. the text by Mahan)

$$\sigma(\omega_n) = \frac{1}{\omega_n} K_{11}(q = 0, \omega_n). \tag{8.40}$$

After analytic continuation to real frequencies, we can obtain the physical conductivity $\sigma(\omega)$. Obtain expressions for $\text{Re}[\sigma(\omega)]$ at $T = 0$ to the leading nonvanishing order in an expansion in u for the $r > 0$ (insulating) and $r < 0$ (superconducting) phases, as described below:

(a) In the superconducting phase, $r < 0$, the leading contribution is of order $1/u$, and you are only required to obtain this. This contribution arises from the condensate, and yields a $\text{Re}[\sigma(\omega)] \sim \delta(\omega)$. Obtain only the coefficient of the delta function.

(b) In the insulating phase, $r > 0$, the leading contribution is of order u^0, and you are required only to obtain this. Your expression will have two terms, the so-called "paramagnetic" (with two Ψ propagators) and "diamagnetic" (with one Ψ propagator) contributions. First evaluate the integral over the internal imaginary frequency. The paramagnetic term depends upon ω_n, while the diamagnetic term is independent of ω_n. *Before* evaluating the momentum integral, analytically continue $i\omega_n \rightarrow \omega$ and determine $\text{Re}[\sigma(\omega)]$ – only the paramagnetic term should contribute to this. Finally, examine the momentum integral in this last expression – you should find that the integrand is nonzero only below a maximum momentum, and so the integral is always convergent.

9 Boson Hubbard model

This chapter finally moves beyond the quantum rotor models which have been the complete focus of our attention so far in Part II. Our motivation is two-fold: to introduce the coherent state path integral, which plays an important role in developing the field theory for many interesting quantum phase transitions; and to provide a deeper and more complete explanation of our claimed connection between the $N = 2$ rotor model and the experiments on ultracold bosonic atoms in an optical lattice which was claimed in Sections 1.3 and 1.4.3. We do this by studying the boson Hubbard model, which has a direct connection to the microscopic Hamiltonian of the ultracold atoms.

The Hubbard model was originally introduced as a description of the motion of electrons in transition metals, with the motivation of understanding their magnetic properties. This original model remains a very active subject of research today, and important progress has been made in recent years by examining its properties in the limit of large spatial dimensionality [160, 165].

In this chapter, we examine only the much simpler "boson Hubbard model," following the analysis in an important paper by Fisher *et al.* [148]. As the name implies, the elementary degrees of freedom in this model are spinless bosons, which take the place of the spin-1/2 fermionic electrons in the original Hubbard model. These bosons could represent Cooper pairs of electrons undergoing Josephson tunneling between superconducting islands, helium atoms moving on a substrate, or ultracold atoms in an optical lattice. Processes in which the Cooper pair boson decays into a pair of electrons are neglected in this simple model, and this caveat must be kept in mind while discussing applications to superconductors.

Many of the results discussed in this chapter were also obtained in early literature on quantum transitions in anisotropic magnets in the presence of an applied magnetic field. These are reviewed by Kaganov and Chubukov [256], who also gave an extensive discussion of experimental applications. We will, however, not use their formulation here.

Apart from its direct physical applications, the importance of the boson Hubbard model lies in providing one of the simplest realizations of a quantum phase transition that does not map onto a previously studied classical phase transition in one higher dimension. The continuum theory describing this transition includes complex Berry phase terms, which, in the simplest formulation of the theory, do not become real even after analytic continuation to imaginary time. We shall meet some genuinely new physical phenomena associated with quantum critical points in a relatively simple context, and the insight will be generally applicable to more complicated models in subsequent chapters.

However, as noted above, there are also quantum phase transitions in the boson Hubbard model which map precisely onto those of the $N = 2$ quantum rotor model. We establish this here, and also show how the quantum rotor universality is destroyed as we vary parameters.

Let us define the degrees of freedom of the model of interest. We introduce the boson operator $\hat{b}_i$, which annihilates bosons on the sites, i, of a regular lattice in d dimensions. These Bose operators and their Hermitian conjugate creation operators obey the commutation relation

$$[\hat{b}_i, \hat{b}_j^\dagger] = \delta_{ij}, \tag{9.1}$$

while two creation or annihilation operators always commute. It is also useful to introduce the boson number operator

$$\hat{n}_{bi} = \hat{b}_i^\dagger \hat{b}_i, \tag{9.2}$$

which counts the number of bosons on each site. We allow an arbitrary number of bosons on each site. Thus the Hilbert space consists of states $|\{m_j\}\rangle$, that are eigenstates of the number operators

$$\hat{n}_{bi}|\{m_j\}\rangle = m_i|\{m_j\}\rangle, \tag{9.3}$$

and every m_j in the set $\{m_j\}$ is allowed to run over all nonnegative integers. This includes the "vacuum" state with no bosons at all $|\{m_j = 0\}\rangle$.

The Hamiltonian of the boson Hubbard model is

$$H_B = -w \sum_{\langle ij \rangle} \left(\hat{b}_i^\dagger \hat{b}_j + \hat{b}_j^\dagger \hat{b}_i\right) - \mu \sum_i \hat{n}_{bi} + (U/2) \sum_i \hat{n}_{bi}(\hat{n}_{bi} - 1). \tag{9.4}$$

The first term, proportional to w, allows hopping of bosons from site to site ($\langle ij \rangle$ represents nearest neighbor pairs); if each site represents a superconducting grain, then w is the Josephson tunneling that allows Cooper pairs to move between grains. The second term, μ, represents the chemical potential of the bosons: changing the value of μ changes the total number of bosons. Depending upon the physical conditions, a given system can either be constrained to be at a fixed chemical potential (the grand canonical ensemble) or have a fixed total number of bosons (the canonical ensemble). Theoretically it is much simpler to consider the fixed chemical potential case, and results at fixed density can always be obtained from them after a Legendre transformation. Finally, the last term, $U > 0$, represents the simplest possible repulsive interaction between the bosons. We have taken only an on-site repulsion. This can be considered to be the charging energy of each superconducting grain. Off-site and longer-range repulsion are undoubtedly important in realistic systems, but these are neglected in this simplest model.

There is a basic similarity between the boson Hubbard model and the $O(N)$ rotor Hamiltonian H_R in (1.25) that is useful in understanding their respective physical properties. First, let us consider the issue of symmetries. The rotor Hamiltonian H_R was invariant under global $O(N)$ rotation of the rotor fields $\hat{\mathbf{n}}_i$ and $\hat{\mathbf{L}}_i$; the present H_B is invariant under a global $U(1) \equiv O(2)$ phase transformation under which

$$\hat{b}_i \to \hat{b}_i e^{i\phi}. \tag{9.5}$$

Now note that the w term in H_B is quite similar to the J term in H_R: both couple neighboring sites in a manner that prefers a state that breaks the global symmetry. However, these terms compete with the Jg term in H_R, or the U term in H_B, both of which are completely local and prefer states that are invariant under their respective symmetry transformations. So, by analogy with H_R, we may expect a quantum phase transition in H_B as a function of w/U between a state in which the U(1) symmetry (9.5) is unbroken to one in which it is broken.

There is, however, a crucial difference between H_R and H_B that requires a more careful discussion of the symmetries in the two models. Recall that a consequence of the O(N) symmetry of H_R was the conservation of total angular momentum in H_R; similarly we have the conservation of the total number of bosons

$$\hat{N}_b = \sum_i \hat{n}_{bi}; \tag{9.6}$$

it is easily verified that $\hat{N}_b$ commutes with $\hat{H}$. Note that to H_R we can add an external field $\mathbf{H}$ coupled to the conserved total angular momentum, as we do in (11.1); the term analogous to this is the chemical potential μ in H_B, which couples to $\hat{N}_b$. This correspondence also brings out the difference. Recall that all of our analysis of H_R was carried out in zero field, $\mathbf{H} = \mathbf{0}$, and we examined only the linear response to an infinitesimal external field $\mathbf{H}$. However, the choice $\mathbf{H} = \mathbf{0}$ was a natural one, as it was *only* for this value that the remainder of H_R was O(N) invariant (at least for $N \geq 3$). In contrast, note that the μ term in H_B does not break any symmetries, and H_B remains invariant under (9.5) for any value of μ. Hence there is no natural symmetry criterion by which we can prefer a specific value of μ, and we have no choice but to examine H_B for all μ. (Even for the case $N = 2$, the choice $\mathbf{H} = \mathbf{0}$ for H_R can be made from the requirement of a "particle–hole" symmetry under which $\mathbf{n}_i \rightarrow -\mathbf{n}_i$, while $\mathbf{L}_i$ remains invariant; there is no such corresponding symmetry for H_B.) It turns out that the results for H_B for general μ also allow us to understand H_R for finite nonzero $\mathbf{H}$.

We begin our study of H_B by introducing a simple mean-field theory in Section 9.1. The coherent state path integral representation of the boson Hamiltonian is then developed in Section 9.2. The continuum quantum theories describing fluctuations near the quantum critical points are introduced in Section 9.3.

9.1 Mean-field theory

The strategy, as in any mean-field theory, is to model the properties of H_B by the best possible sum, H_{MF}, of single-site Hamiltonians:

$$H_{\text{MF}} = \sum_i \left(-\mu \hat{n}_{bi} + (U/2)\,\hat{n}_{bi}(\hat{n}_{bi} - 1) - \Psi_B^* \hat{b}_i - \Psi_B \hat{b}_i^\dagger \right), \tag{9.7}$$

where the complex number Ψ_B is a variational parameter. We have chosen a mean-field Hamiltonian with the same on-site terms as H_B and have added an additional term with a "field" Ψ_B to represent the influence of the neighboring sites; this field has to be self-consistently determined. Note that this term breaks the U(1) symmetry and does not conserve the total number of particles. This is to allow for the possibility of broken-symmetric phases, whereas symmetric phases appear at the special value $\Psi_B = 0$. As we saw in the analysis of H_R, the state that breaks the U(1) symmetry will have a nonzero stiffness to rotations of the order parameter; in the present case this stiffness is the super-fluid density characterizing a superfluid ground state of the bosons.

Another important assumption underlying (9.7) is that the ground state does not sponta-neously break a translational symmetry of the lattice, as the mean-field Hamiltonian is the same on every site. Such a symmetry breaking is certainly a reasonable possibility, but we ignore this complication here for simplicity.

We will determine the optimum value of the mean-field parameter Ψ_B by a standard procedure. First, determine the ground state wavefunction of H_{MF} for an arbitrary Ψ_B; because H_{MF} is a sum of single-site Hamiltonians, this wavefunction is simply a product of single-site wavefunctions. Next, evaluate the expectation value of H_B in this wavefunction. By adding and subtracting H_{MF} from H_B, we can write the mean-field value of the ground state energy of H_B in the form

$$\frac{E_0}{M} = \frac{E_{MF}(\Psi_B)}{M} - Zw\langle\hat{b}^\dagger\rangle\langle\hat{b}\rangle + \langle\hat{b}\rangle\Psi_B^* + \langle\hat{b}^\dagger\rangle\Psi_B, \qquad (9.8)$$

where $E_{MF}(\Psi_B)$ is the ground state energy of H_{MF}, M is the number of sites of the lattice, Z is the number of nearest neighbors around each lattice point (the "coordination num-ber"), and the expectation values are evaluated in the ground state of H_{MF}. The final step is to minimize (9.8) over variations in Ψ_B. We have carried out this step numerically and the results are shown in Fig. 9.1.

Note that even on a single site, H_{MF} has an infinite number of states, corresponding to the allowed values $m \geq 0$ of the integer number of bosons on each site. The numerical procedure necessarily truncates these states at some large occupation number, but the errors are not difficult to control. In any case, we will show that all the essential properties of the

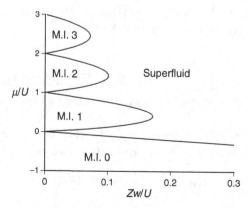

Fig. 9.1 Mean-field phase diagram of the ground state of the boson Hubbard model H_B in (9.4). The notation M.I. n refers to a Mott insulator with $n_0(\mu/U) = n$.

phase diagram can be obtained analytically. Also, by taking the derivative of (9.8) with respect to Ψ_B, it is easy to show that at the optimum value of Ψ_B

$$\Psi_B = Zw\langle \hat{b} \rangle; \tag{9.9}$$

this relation, however, does not hold at a general point in parameter space.

First, let us consider the limit $w = 0$. In this case the sites are decoupled, and the mean-field theory is exact. It is also evident that $\Psi_B = 0$, and we simply have to minimize the on-site interaction energy. The on-site Hamiltonian contains only the operator $\hat{n}$, and the solution involves finding the boson occupation numbers (which are the integer-valued eigenvalues of $\hat{n}$) that minimize H_B. This is simple to carry out, and we get the ground state wavefunction

$$|m_i = n_0(\mu/U)\rangle, \tag{9.10}$$

where the integer-valued function $n_0(\mu/U)$ is given by

$$n_0(\mu/U) = \begin{cases} 0, & \text{for } \mu/U < 0, \\ 1, & \text{for } 0 < \mu/U < 1, \\ 2, & \text{for } 1 < \mu/U < 2, \\ \vdots & \quad\quad \vdots \\ n, & \text{for } n - 1 < \mu/U < n. \end{cases} \tag{9.11}$$

Thus each site has exactly the same integer number of bosons, which jumps discontinuously whenever μ/U goes through a positive integer. When μ/U is exactly equal to a positive integer, there are two degenerate states on each site (with boson numbers differing by 1) and so the entire system has a degeneracy of 2^M. This large degeneracy implies a macroscopic entropy; it will be lifted once we turn on a nonzero w.

We now consider the effects of a small nonzero w. As is shown in Fig. 9.1, the regions with $\Psi_B = 0$ survive in lobes around each $w = 0$ state (9.10) characterized by a given integer value of $n_0(\mu/U)$. Only at the degenerate point with $\mu/U =$ integer does a nonzero w immediately lead to a state with $\Psi_B \neq 0$. We consider the properties of this $\Psi_B \neq 0$ later, but now we discuss the properties of the lobes with $\Psi_B = 0$ in some more detail. In mean-field theory, these states have wavefunctions still given exactly by (9.10). However, it is possible to go beyond mean-field theory and make an important exact statement about each of the lobes: the expectation value of the number of bosons in each site is given by

$$\langle \hat{b}_i^\dagger \hat{b}_i \rangle = n_0(\mu/U), \tag{9.12}$$

which is the same result one would obtain from the product state (9.10) (which, we emphasize, is not the exact wavefunction for $w \neq 0$). There are two important ingredients behind the result (9.12): the existence of an energy gap and the fact that $\hat{N}_b$ commutes with H_B. First, recall that at $w = 0$, provided that μ/U was not exactly equal to a positive integer, there was a unique ground state, and there was a nonzero energy separating this state from all other states (this is the energy gap). As a result, when we turn on a small nonzero w, the ground state moves adiabatically without undergoing any level crossings with any other state. Now the $w = 0$ state is an exact eigenstate of $\hat{N}_b$ with eigenvalue $Mn_0(\mu/U)$, and the perturbation arising from a nonzero w commutes with $\hat{N}_b$. Consequently, the ground state will remain an eigenstate of $\hat{N}_b$ with precisely the same eigenvalue, $Mn_0(\mu/U)$, even for small nonzero w. Assuming translational invariance, we then immediately have the exact result (9.12). Note that this argument also shows that the energy gap above the ground state will survive everywhere within the lobe. These regions with a quantized value of the density and an energy gap to all excitations are known as "Mott insulators." Their ground states are very similar to, but not exactly equal to, the simple state (9.10): they involve, in addition, terms with bosons undergoing virtual fluctuations between pairs of sites, creating particle–hole pairs. The Mott insulators are also known as "incompressible" because their density does not change under changes of the chemical potential μ or other parameters in H_B:

$$\frac{\partial \langle \hat{N}_b \rangle}{\partial \mu} = 0. \tag{9.13}$$

It is worth re-emphasizing here the remarkable nature of the exact result (9.12). From the perspective of classical critical phenomena, it is most unusual to find the expectation value of any observable to be pinned at a quantized value over a finite region of the phase diagram. However, as we will see, quantum field theories of a certain structure allow such a phenomenon, and we meet different realizations of it in subsequent chapters. The existence of observables such as $\hat{N}_b$ that commute with the Hamiltonian is clearly a crucial ingredient.

The numerical analysis shows that the boundary of the Mott insulating phases is a second-order quantum phase transition (i.e. a nonzero Ψ_B turns on continuously). With the benefit of this knowledge, we can determine the positions of the phase boundaries. By the usual Landau theory argument, we simply need to expand E_0 in (9.8) in powers of Ψ_B,

$$E_0 = E_{00} + r|\Psi_B|^2 + \mathcal{O}(|\Psi_B|^4), \tag{9.14}$$

and the phase boundary appears when r changes sign. The value of r can be computed from (9.8) and (9.7) by second-order perturbation theory, and we find

$$r = \chi_0(\mu/U)\left[1 - Zw\chi_0(\mu/U)\right], \tag{9.15}$$

where

$$\chi_0(\mu/U) = \frac{n_0(\mu/U) + 1}{Un_0(\mu/U) - \mu} + \frac{n_0(\mu/U)}{\mu - U(n_0(\mu/U) - 1)}. \tag{9.16}$$

The function $n_0(\mu/U)$ in (9.11) is such that the denominators in (9.16) are positive, except at the points at which the boson occupation number jumps at $w = 0$. The solution of the simple equation $r = 0$ leads to the phase boundaries shown in Fig. 9.1.

Finally, we turn to the phase with $\Psi_B \neq 0$. The mean-field parameter Ψ_B varies continuously as the parameters are varied. As a result all thermodynamic variables also change, and the density does not take a quantized value; by a suitable choice of parameters, the average density can be varied smoothly across any real positive value. So this is a compressible state in which

$$\frac{\partial \langle \hat{N}_b \rangle}{\partial \mu} \neq 0. \tag{9.17}$$

As we noted earlier, the presence of a $\Psi_B \neq 0$ implies that the U(1) symmetry is broken, and there is a nonzero stiffness to twists in the orientation of the order parameter. The fluctuation analysis discussed in Section 9.3 can be combined with the methods of Chapters 8 and 15 to show that this state is a superfluid and that the stiffness is just the superfluid density.

Corrections to the mean-field phase diagram of Fig. 9.1 have been considered by Freericks and Monien [153, 154]. They find singularities in the shape of the Mott lobes at the positions of the $z = 1$ transitions. Monte Carlo simulations of the phase diagram have also been carried out [39, 364], and confirm the physical picture discussed above.

We also note that extensions of the boson Hubbard model with interactions beyond nearest neighbor can spontaneously break translational symmetry at certain densities. If this symmetry breaking coexists with the superfluid order, one can obtain a "supersolid" phase. These issues have been discussed in [21, 40, 155, 170, 352, 483, 520, 521]. See also Exercise 9.1.

9.2 Coherent state path integral

This section applies to the boson Hubbard model the analog of the mapping described in Section 6.5 for the quantum rotor model. We apply this using a general method called the coherent state path integral. While the path integral in Section 6.5 is over all possible quantum trajectories in the configuration space, $\mathbf{n}_i(\tau)$, of the quantum rotor model, the coherent state path integral is over phase space; we choose to work in phase space because there is often no choice of configuration space which does not break some important symmetry of the Hamiltonian. Thus the integral involves integrands which do not commute with each other in the conventional Hamiltonian formulation, and this is reflected by the presence of *Berry phase* terms in the action for the path integral. Consequently, the action is not real, even in imaginary time. One subtle consequence of the presence of Berry phases is that the path integral is not a well-defined mathematical quantity, unlike the integrals in Section 6.5 which can be rigorously defined by their connection to equivalent statistical models. So formally, the coherent state path integral requires additional temporal derivative terms for a proper regularization: we generally ignore such issues here, because the naive analysis based upon assuming a well defined time continuum limit is invariably correct: the reader is referred to [357] for a more complete discussion.

To avoid inessential indices, we present the derivation of the coherent state path integral by focusing on a single site, and drop the site index. We first derive the result in a general notation, to allow subsequent application to quantum spin systems in Section 19.1. So we consider a general Hamiltonian $H(\hat{\mathbf{S}})$, dependent upon operators $\hat{\mathbf{S}}$ which need not commute with each other. So for the boson Hubbard model, $\hat{\mathbf{S}}$ is a two-dimensional vector of operators $\hat{b}$ and $\hat{b}^\dagger$ which obey (9.1). When we apply the results below to quantum spin systems, $\hat{\mathbf{S}}$ represents the usual spin operators $\hat{S}_{x,y,z}$, which obey (1.29).

Our first step is to introduce the coherent states. These are an infinite set of states $|\mathbf{N}\rangle$, labeled by the continuous vector $\mathbf{N}$ (in two or three dimensions for the two cases above). They are normalized to unity,

$$\langle \mathbf{N}|\mathbf{N}\rangle = 1, \tag{9.18}$$

but are not orthogonal $\langle \mathbf{N}|\mathbf{N}'\rangle \neq 0$ for $\mathbf{N} \neq \mathbf{N}'$. They do, however, satisfy a completeness relation

$$\mathcal{C}_N \int d\mathbf{N}\,|\mathbf{N}\rangle\langle\mathbf{N}| = 1, \tag{9.19}$$

where $\mathcal{C}_N$ is a normalization constant. Because of their nonorthogonality, these states are called "over-complete." Finally, they are chosen with a useful property: the diagonal expectation values of the operators $\hat{\mathbf{S}}$ are very simple:

$$\langle \mathbf{N}|\hat{\mathbf{S}}|\mathbf{N}\rangle = \mathbf{N}. \tag{9.20}$$

This property implies that the vector $\mathbf{N}$ is a classical approximation to the operators $\hat{\mathbf{S}}$. The relations (9.18), (9.19), and (9.20) define the coherent states, and are all we need here to set up the coherent state path integral.

We also need the diagonal matrix elements of the Hamiltonian in the coherent state basis. Usually, it is possible to arrange the operators such that

$$\langle \mathbf{N}|H(\hat{\mathbf{S}})|\mathbf{N}\rangle = H(\mathbf{N}); \tag{9.21}$$

i.e. $H(\mathbf{N})$ has the same functional dependence upon $\mathbf{N}$ as the original Hamiltonian has on $\mathbf{S}$. For the boson Hubbard model, this corresponds, as we will see, to normal-ordering the creation and annihilation operators. In any case, the right-hand side could have a distinct functional dependence on $\mathbf{N}$, but we will just refer to the diagonal matrix element as above.

We proceed to the derivation of the coherent state path integral for the partition function

$$\mathcal{Z} = \mathrm{Tr}\exp(-H(\hat{\mathbf{S}})/T). \tag{9.22}$$

The transformation of $\mathcal{Z}$ into a path integral proceeds along the same lines as in Section 5.6. We break up the exponential into a large number of exponentials of infinitesimal time evolution operators

$$\mathcal{Z} = \lim_{M\to\infty} \prod_{i=1}^{M} \exp(-\Delta\tau_i H(\hat{\mathbf{S}})), \tag{9.23}$$

where $\Delta\tau_i = 1/MT$, and insert a set of coherent states between each exponential by using the identity (9.19); we label the state inserted at a "time" τ by $|\mathbf{N}(\tau)\rangle$. We can then evaluate the expectation value of each exponential by use of the identity (9.20)

$$
\begin{aligned}
\langle \mathbf{N}(\tau)| & \exp(-\Delta\tau H(\hat{\mathbf{S}}))|\mathbf{N}(\tau - \Delta\tau)\rangle \\
&\approx \langle \mathbf{N}(\tau)|(1 - \Delta\tau H(\hat{\mathbf{S}}))|\mathbf{N}(\tau - \Delta\tau)\rangle \\
&\approx 1 - \Delta\tau \langle \mathbf{N}(\tau)|\frac{d}{d\tau}|\mathbf{N}(\tau)\rangle - \Delta\tau H(\mathbf{N}) \\
&\approx \exp\left(-\Delta\tau \langle \mathbf{N}(\tau)|\frac{d}{d\tau}|\mathbf{N}(\tau)\rangle - \Delta\tau H(\mathbf{N})\right).
\end{aligned}
\tag{9.24}
$$

In each step we have retained expressions correct to order $\Delta\tau$. Because the coherent states at time τ and $\tau + \Delta\tau$ can in principle have completely different orientations, a priori, it is not clear that expanding these states in derivatives of time is a valid procedure. This is a subtlety that afflicts all coherent state path integrals and has been discussed more carefully by Negele and Orland [357]. The conclusion of their analysis is that except for the single "tadpole" diagram where a point-splitting of time becomes necessary, this expansion in derivatives of time always leads to correct results. In any case, the resulting coherent state path integral is a formal expression that cannot be directly evaluated, and in case of any doubt one should always return to the original discrete time product in (9.23).

Keeping in mind the above caution, we insert (9.24) into (9.23), take the limit of small $\Delta\tau$, and obtain the following functional integral for $\mathcal{Z}$:

$$
\mathcal{Z} = \int_{\mathbf{N}(0)=\mathbf{N}(1/T)} \mathcal{D}\mathbf{N}(\tau) \exp\left\{-\mathcal{S}_B - \int_0^{1/T} d\tau\, H(\mathbf{N}(\tau))\right\},
\tag{9.25}
$$

where

$$
\mathcal{S}_B = \int_0^{1/T} d\tau\, \langle \mathbf{N}(\tau)|\frac{d}{d\tau}|\mathbf{N}(\tau)\rangle,
\tag{9.26}
$$

and $H(S\mathbf{N})$ is obtained by replacing every occurrence of $\hat{\mathbf{S}}$ in the Hamiltonian by $S\mathbf{N}$. The promised Berry phase term is $\mathcal{S}_B$, and it represents the overlap between the coherent states at two infinitesimally separated times. It can be shown straightforwardly from the normalization condition, $\langle \mathbf{N}|\mathbf{N}\rangle = 1$, that $\mathcal{S}_B$ is pure imaginary.

9.2.1 Boson coherent states

We now apply the general formalism above to the boson Hubbard model. As before, we drop the site index i.

For the state label, we replace the two-dimensional vector $\mathbf{N}$ by a complex number ψ, and so the coherent states are $|\psi\rangle$, with one state for every complex number. A state with the properties (9.18), (9.19), and (9.20) turns out to be

$$
|\psi\rangle = e^{-|\psi|^2/2} \exp\left(\psi \hat{b}^\dagger\right)|0\rangle,
\tag{9.27}
$$

where $|0\rangle$ is the boson vacuum state (one of the states in (9.3)). This state is normalized as required by (9.18), and we can now obtain its diagonal matrix element

$$\langle \psi | \hat{b} | \psi \rangle = e^{-|\psi|^2} \frac{\partial}{\partial \psi^*} \langle 0 | e^{\psi^* \hat{b}} e^{\psi \hat{b}^\dagger} | 0 \rangle$$

$$= e^{-|\psi|^2} \frac{\partial}{\partial \psi^*} e^{|\psi|^2} = \psi, \tag{9.28}$$

which satisfies the requirement (9.20). For the complete relation, we evaluate

$$\int d\psi d\psi^* |\psi\rangle \langle \psi| = \sum_{n=0}^{\infty} \frac{|n\rangle\langle n|}{n!} \int d\psi d\psi^* |\psi|^{2n} e^{-|\psi|^2}$$

$$= \pi \sum_{n=0}^{\infty} |n\rangle\langle n|, \tag{9.29}$$

where $|n\rangle$ are the number states in (9.3), $d\psi d\psi^* \equiv d\mathrm{Re}(\psi) d\mathrm{Im}(\psi)$, and we have picked only the diagonal terms in the double sum over number states because the off-diagonal terms vanish after the angular ψ integration. This result identifies $\mathcal{C}_N = 1/\pi$. So we have satisfied the properties (9.18), (9.19), and (9.20) required of all coherent states.

For the path integral, we need the Berry phase term in (9.26). This is a path integral over trajectories in the complex plane, $\psi(\tau)$, and we have

$$\langle \psi(\tau) | \frac{d}{d\tau} | \psi(\tau) \rangle = e^{-|\psi(\tau)|^2} \langle 0 | e^{\psi^*(\tau) \hat{b}} \frac{d}{d\tau} | e^{\psi(\tau) \hat{b}^\dagger} | 0 \rangle = \psi^* \frac{d\psi}{d\tau}. \tag{9.30}$$

We are now ready to combine (9.30) and (9.25) to obtain the coherent state path integral of the boson Hubbard model.

9.3 Continuum quantum field theories

Returning to our discussion of the boson Hubbard model, here we describe the low-energy properties of the quantum phase transitions between the Mott insulators and the superfluid found in Section 9.1. We find that it is crucial to distinguish between two different cases, each characterized by its own universality class and continuum quantum field theory. The important diagnostic distinguishing the two possibilities is the behavior of the boson density across the transition. In the Mott insulator, this density is of course always pinned at some integer value. As one undergoes the transition to the superfluid, depending upon the precise location of the system in the phase diagram of Fig. 9.1, there are two possible behaviors of the density: either (A) the density remains pinned at its quantized value in the superfluid in the vicinity of the quantum critical point, or (B) the transition is accompanied by a change in the density. We show below that case (A) is described by the $N = 2$ case of the quantum rotor field theory (3.25) which we have already studied, and study in greater detail in Part III. Case (B) leads to a different field theory whose properties are examined further in Chapter 16.

We begin by writing the partition function of H_B, $\mathcal{Z}_B = \mathrm{Tr}\,e^{-H_B/T}$ in the coherent state path integral representation derived in Section 9.2:

$$
\mathcal{Z}_B = \int \mathcal{D}b_i(\tau)\mathcal{D}b_i^\dagger(\tau) \exp\left(-\int_0^{1/T} d\tau \mathcal{L}_b\right),
$$
$$
\mathcal{L}_b = \sum_i \left(b_i^\dagger \frac{db_i}{d\tau} - \mu b_i^\dagger b_i + (U/2)\, b_i^\dagger b_i^\dagger b_i b_i\right) - w\sum_{\langle ij\rangle}\left(b_i^\dagger b_j + b_j^\dagger b_i\right). \qquad (9.31)
$$

Here we have changed notation $\psi(\tau) \to b(\tau)$, as is conventional; we are dealing exclusively with path integrals from now on, and so there is no possibility of confusion with the operators $\hat{b}$ in the Hamiltonian language. Also note that the repulsion proportional to U in (9.4) becomes the product of four boson operators above after normal ordering, and we can then use (9.21).

It is clear that the critical field theory of the superfluid–insulator transition should be expressed in terms of a spacetime-dependent field $\Psi_B(x,\tau)$, which is analogous to the mean-field parameter Ψ_B appearing in Section 9.1. Such a field is most conveniently introduced by the well-known Hubbard–Stratanovich transformation on the coherent state path integral. We decouple the hopping term proportional to w by introducing an auxiliary field $\Psi_{Bi}(\tau)$ and transforming $\mathcal{Z}_B$ to

$$
\mathcal{Z}_B = \int \mathcal{D}b_i(\tau)\mathcal{D}b_i^\dagger(\tau)\mathcal{D}\Psi_{Bi}(\tau)\mathcal{D}\Psi_{Bi}^\dagger(\tau) \exp\left(-\int_0^{1/T} d\tau \mathcal{L}_b'\right),
$$
$$
\mathcal{L}_b' = \sum_i \left(b_i^\dagger \frac{db_i}{d\tau} - \mu b_i^\dagger b_i + (U/2)\, b_i^\dagger b_i^\dagger b_i b_i - \Psi_{Bi} b_i^\dagger - \Psi_{Bi}^* b_i\right) + \sum_{i,j}\Psi_{Bi}^* w_{ij}^{-1}\Psi_{Bj}.
$$
$$
\qquad (9.32)
$$

We have introduced the symmetric matrix w_{ij} whose elements equal w if i and j are nearest neighbors and vanish otherwise. The equivalence between (9.32) and (9.31) can be established easily by simply carrying out the Gaussian integral over Ψ_B; this also generates some overall normalization factors, but these have been absorbed into a definition of the measure $\mathcal{D}\Psi_B$. Let us also note a subtlety we have glossed over: strictly speaking, the transformation between (9.32) and (9.31) requires that all the eigenvalues of w_{ij} be positive, for only then are the Gaussian integrals over Ψ_B well defined. This is not the case for, say, the hypercubic lattice, which has negative eigenvalues for w_{ij}. This can be repaired by adding a positive constant to all the diagonal elements of w_{ij} and subtracting the same constant from the on-site b part of the Hamiltonian. We will not explicitly do this here as our interest is only in the long-wavelength modes of the Ψ_B field, and the corresponding eigenvalues of w_{ij} are positive.

For our future purposes, it is useful to describe an important symmetry property of (9.32). Note that the functional integrand is invariant under the following time-dependent U(1) gauge transformation:

$$b_i \rightarrow b_i e^{i\phi(\tau)},$$
$$\Psi_{Bi} \rightarrow \Psi_{Bi} e^{i\phi(\tau)},$$
$$\mu \rightarrow \mu + i\frac{\partial\phi}{\partial\tau}. \tag{9.33}$$

The chemical potential μ becomes time dependent above, and so this transformation takes one out of the physical parameter regime; nevertheless (9.33) is very useful, as it places important restrictions on subsequent manipulations of $\mathcal{Z}_B$.

The next step is to integrate out the b_i, $b_i^\dagger$ fields from (9.32). This can be done exactly in powers of Ψ_B and Ψ_B^*: the coefficients are simply products of Green's functions of the b_i. The latter can be determined in closed form because the Ψ_B-independent part of $\mathcal{L}_b'$ is simply a sum of single-site Hamiltonians for the b_i: these were exactly diagonalized in (9.10), and all single-site Green's functions can also be easily determined. We re-exponentiate the resulting series in powers of Ψ_B, Ψ_B^* and expand the terms in spatial and temporal gradients of Ψ_B. The expression for $\mathcal{Z}_B$ can now be written as [148]

$$\mathcal{Z}_B = \int \mathcal{D}\Psi_B(x,\tau)\mathcal{D}\Psi_B^*(x,\tau) \exp\left(-\frac{V\mathcal{F}_0}{T} - \int_0^{1/T} d\tau \int d^dx \mathcal{L}_B\right),$$

$$\mathcal{L}_B = K_1 \Psi_B^* \frac{\partial\Psi_B}{\partial\tau} + K_2\left|\frac{\partial\Psi_B}{\partial\tau}\right|^2 + K_3 |\nabla\Psi_B|^2 + \tilde{r}|\Psi_B|^2 + \frac{u}{2}|\Psi_B|^4 + \cdots \tag{9.34}$$

Here $V = Ma^d$ is the total volume of the lattice, and a^d is the volume per site. The quantity $\mathcal{F}_0$ is the free energy density of a system of decoupled sites; its derivative with respect to the chemical potential gives the density of the Mott insulating state, and so

$$-\frac{\partial\mathcal{F}_0}{\partial\mu} = \frac{n_0(\mu/U)}{a^d}. \tag{9.35}$$

The other parameters in (9.34) can also be expressed in terms of μ, U, and w but we will not display explicit expressions for all of them. Most important is the parameter $\tilde{r}$, which can be seen to be

$$\tilde{r}a^d = \frac{1}{Zw} - \chi_0(\mu/U), \tag{9.36}$$

where χ_0 was defined in (9.16). Notice that $\tilde{r}$ is proportional to the mean-field r in (9.15); in particular, $\tilde{r}$ vanishes when r vanishes, and the two quantities have the same sign. The mean-field critical point between the Mott insulator and the superfluid appeared at $r = 0$, and it is not surprising that the mean-field critical point of the continuum theory (9.34) is given by the same condition.

Of the other couplings in (9.34), K_1, the coefficient of the first-order time derivative also plays a crucial role. It can be computed explicitly, but it is simpler to note that the value of

K_1 can be fixed by demanding that (9.34) be invariant under (9.33) for small ϕ: a simple calculation shows that we must have

$$K_1 = -\frac{\partial \tilde{r}}{\partial \mu}. \tag{9.37}$$

This relationship has a very interesting consequence. Note that K_1 vanishes when $\tilde{r}$ is μ-independent; however, this is precisely the condition that the Mott insulator–superfluid phase boundary in Fig. 9.1 have a vertical tangent (i.e. at the tips of the Mott insulating lobes). This is significant because at the value $K_1 = 0$ it is evident that (9.34) is nothing but the $N = 2$ rotor model field theory action in (3.25), which is further studied in Part III. So the Mott insulator to superfluid transition is in the universality class of the O(2) quantum rotor model phase transition for $K_1 = 0$. In contrast, for $K_1 \neq 0$ we have a rather different field theory. We can now drop the K_2 term as it involves two time derivatives and so is irrelevant with respect to the single time derivative in the K_1 term. The resulting field theory is examined in some detail in the following chapter.

To conclude this discussion, we would like to correlate the above discussion on the distinction between the two universality classes with the behavior of the boson density across the transition. This can be evaluated by taking the derivative of the total free energy with respect to the chemical potential, as is clear from (9.4):

$$\langle \hat{b}_i^\dagger \hat{b}_i \rangle = -a^d \frac{\partial \mathcal{F}_0}{\partial \mu} - a^d \frac{\partial \mathcal{F}_B}{\partial \mu}$$
$$= n_0(\mu/U) - a^d \frac{\partial \mathcal{F}_B}{\partial \mu}, \tag{9.38}$$

where $\mathcal{F}_B$ is the free energy resulting from the functional integral over Ψ_B in (9.34). We examine the properties of (9.34) for general K_1, including fluctuations, in the following chapter. Here let us be satisfied by a simple mean-field treatment.

In mean-field theory, for $\tilde{r} > 0$, we have $\Psi_B = 0$, and therefore $\mathcal{F}_B = 0$, implying

$$\langle \hat{b}_i^\dagger \hat{b}_i \rangle = n_0(\mu/U), \quad \text{for } \tilde{r} > 0. \tag{9.39}$$

This clearly places us in a Mott insulator. As argued in Section 9.1, (9.39) is an exact result, and we obtain another verification of this in our analysis of the fluctuations of (9.34) in Chapter 16.

For $\tilde{r} < 0$, we have $\Psi_B = (-\tilde{r}/u)^{1/2}$, as follows from a simple minimization of $\mathcal{L}_B$; computing the resulting free energy we have

$$\langle \hat{b}_i^\dagger \hat{b}_i \rangle = n_0(\mu/U) + a^d \frac{\partial}{\partial \mu} \left(\frac{\tilde{r}^2}{2u} \right)$$
$$\approx n_0(\mu/U) + \frac{a^d \tilde{r}}{u} \frac{\partial \tilde{r}}{\partial \mu}. \tag{9.40}$$

In the second expression, we ignored the derivative of u as it is less singular as $\tilde{r}$ approaches 0; we will comment on the consequences of this shortly. Thus at the transition point at which $K_1 = 0$, by (9.37) we see that the leading correction to the density of the superfluid phase vanishes, and it remains pinned at the same value as in the Mott insulator. So, as claimed earlier, the transition with no density change is in the universality class of the O(2) quantum rotor model. Conversely, for the case $K_1 \neq 0$, the transition is always accompanied by a density change. This is a separate universality class, which is considered in the next chapter, and we will see there that we can also consider the density itself as an order parameter for the transition in this case.

We close by commenting on the consequences of the omitted higher order terms in (9.40) to the discussion above. Consider the trajectory of points in the superfluid with their density equal to some integer n. The implication of the above discussion is that this trajectory will meet the Mott insulator with $n_0(\mu/U) = n$ at its lobe. The O(2) quantum rotor model phase transition then describes the transition out of the Mott insulator into the superfluid along a direction tangent to the trajectory of density n. The approximations made above merely amounted to assuming that this trajectory was a straight line.

Exercises

9.1 *Supersolids.* Consider a double-layer boson Hubbard model of bosons $\hat{b}_{1i}$ and $\hat{b}_{2i}$ on two parallel layers 1,2:

$$H_{2B} = -w \sum_{\langle ij \rangle} \left(\hat{b}_{1i}^\dagger \hat{b}_{1j} + \hat{b}_{1j}^\dagger \hat{b}_{1i} + \hat{b}_{2i}^\dagger \hat{b}_{2j} + \hat{b}_{2j}^\dagger \hat{b}_{2i} \right) - w \sum_i \left(\hat{b}_{1i}^\dagger \hat{b}_{2i} + \hat{b}_{2i}^\dagger \hat{b}_{1i} \right)$$

$$+ \sum_i \left(-\mu \left[\hat{n}_{b1i} + \hat{n}_{b2i} \right] + \frac{U}{2} \left[\hat{n}_{b1i} (\hat{n}_{b1i} - 1) + \hat{n}_{b2i} (\hat{n}_{b2i} - 1) \right] \right)$$

$$+ V \sum_i \hat{n}_{b1i} \hat{n}_{b2i} + W \sum_{\langle ij \rangle} \left[\hat{n}_{b1i} \hat{n}_{b1j} + \hat{n}_{b2i} \hat{n}_{b2j} \right]. \qquad (9.41)$$

Thus bosons on the same layer have an on-site repulsion $U > 0$, bosons on opposite layers have a repulsion $V > 0$. Bosons on the same layer also have a nearest-neighbor interaction W, and we allow W to have either sign. We consider the case where the average boson density per site and per layer is exactly 1/2, and we take the limit $U \to \infty$: thus no site can have more than one boson. Use a variational approach to determine the ground state of H_{2B} as a function of V/w and W/w. The proposed mean-field variational wavefunction is

$$|G\rangle = \prod_i \left(\alpha_1 + \alpha_2 \hat{b}_{1i}^\dagger + \alpha_3 \hat{b}_{2i}^\dagger + \alpha_4 \hat{b}_{1i}^\dagger \hat{b}_{2i}^\dagger \right) |0\rangle, \qquad (9.42)$$

where $|0\rangle$ is the empty state, and α_1, α_2, α_3, and α_4 are variational parameters. Normalization of the wavefunction implies that

$$|\alpha_1|^2 + |\alpha_2|^2 + |\alpha_3|^2 + |\alpha_4|^2 = 1. \qquad (9.43)$$

(a) Show that the average density of 1/2 implies

$$|\alpha_2|^2 + |\alpha_3|^2 + 2|\alpha_4|^2 = 1. \qquad (9.44)$$

(b) Compute $\langle G|H_{2B}|G\rangle$ for a lattice with same-layer coordination number Z. Then minimize this as a function of the $\alpha_{1,2,3,4}$ subject to the constraints (9.43) and (9.44). The results yield a phase diagram, and the phases can be identified as discussed below.

(c) Argue that any phase with $\alpha_1 \neq 0$ must be a superfluid.

(d) Similarly, show that any phase with $\alpha_1 = \alpha_4 = 0$ is an insulator.

(e) The model H_{2B} has a layer interchange symmetry, and our mean-field allows this symmetry to be spontaneously broken. Show that this symmetry is broken in phases in which $|\alpha_2| \neq |\alpha_3|$. As such phases break a lattice symmetry, it is natural to refer to them as "solids."

(f) Are there any regimes which are both solids and superfluids? Such a phase would be a supersolid.

(g) Determine the order (first or second) of all quantum phase transitions in the phase diagram.

PART III

NONZERO TEMPERATURES

10

The Ising chain in a transverse field

Part II analyzed the properties of quantum Ising and rotor models in some detail at $T = 0$. We related the quantum phase transitions in these models to the N-component relativistic field theory (2.11), and used it to understand the critical properties.

The purpose of Part III is to extend this understanding to $T > 0$. We will demonstrate that the $T = 0$ quantum critical point controls a wide "quantum critical" region at $T > 0$, as illustrated in Fig. 1.2. We are especially interested in dynamic properties in this region: an interesting feature is that many "friction" coefficients are universal and depend only on fundamental constants of nature. We also explore the other regions of the phase diagrams in Fig. 1.2, including behavior in the vicinity of the phase transition at $T > 0$.

We begin this chapter by extending results of the $d = 1$ quantum Ising model of Chapter 5 to $T > 0$. This model does not have any phase transition at any $T > 0$, and so the crossover structure of the phase diagram is in the class in Fig. 1.2a. Phase transitions at $T > 0$ appear in models to be studied in the following chapter.

All of the results in this section are believed to be exact, but the physically oriented reader should not be turned off by this: we will keep technical details to a minimum and show how the exact results can be obtained by physical arguments that do much to illustrate the main underlying principles. Most of the important concepts of this book appear in the simple model under consideration; much of Part III is a description of similar phenomena in more complicated settings. This is thus one of the central chapters of Part III, and a careful reading is urged.

We will study the $d = 1$ case of (1.7), which is

$$H_I = -J \sum_i \left(g \hat{\sigma}_i^x + \hat{\sigma}_i^z \hat{\sigma}_{i+1}^z \right). \tag{10.1}$$

As we have discussed in Part I and Chapter 5, H_I exhibits a phase transition at $T = 0$ between an ordered state with the Z_2 symmetry broken and a quantum paramagnetic state where the symmetry remains unbroken. The quantum–classical mapping $\mathcal{QC}$ ensures that this transition is in the universality class of the $D = 2$ classical Ising model.

There has been a great deal of theoretical work on the ground state correlations of H_I [34, 299, 330, 378]. However, properties of the order parameter $\hat{\sigma}_z$ at $T > 0$, which are our primary interest here, have been studied much less. Methods relying upon knowledge of all the exact eigenstates and eigenfunctions of H_I do yield explicit results for equal-time correlators [36, 299, 327, 420], but results for unequal-time correlators have been restricted to $T = \infty$ [67, 376, 377] or to precisely the critical coupling [241, 331] (seen below to be $g = 1$). There is also an approach that relies upon deriving nonlinear partial differential equations satisfied by the $T > 0$ unequal-time correlators [278, 289, 332], but these have not yet been solved to yield the physical correlators. Our discussion of the low-T dynamics here follows the intuitive phenomenological approach developed recently in [442]. Despite its seeming inexactness, its results are believed to be asymptotically exact, and this will be supported by evidence from numerical computations.

In our discussion of the quantum critical point of H_I, we have occasion to meet again the scaling concepts encountered before in Chapter 4. These are extended here to define the *dynamic critical exponent z*. Another very useful, but much less familiar, concept is that of the *reduced scaling function*, and it is introduced as an essential tool toward understanding the mechanism of emergence of classical behavior in limiting regimes of the phase diagram.

We describe the properties of H_I using the analog of the correlation functions introduced in Chapter 7 for the quantum field theory. In the present context, this includes the dynamic two-point correlations of the order parameter $\hat{\sigma}^z$:

$$C(x_i, t) \equiv \langle \hat{\sigma}^z(x_i, t)\hat{\sigma}^z(0, 0) \rangle$$
$$= \text{Tr}\left(e^{-H_I/T} e^{iH_I t} \hat{\sigma}_i^z e^{-iH_I t} \hat{\sigma}_0^z\right)/\mathcal{Z}, \tag{10.2}$$

where $\mathcal{Z} = \text{Tr}(e^{-H_I/T})$ is the partition function, and $x_i = ia$ is the x-coordinate of the ith spins with a the lattice spacing. As before, we always use the symbol t to represent real physical time. Occasionally we also find it convenient to consider the correlation at an imaginary time τ; this is defined by the analytic continuation $it \to \tau$ from (10.2) with $\tau > 0$:

$$C(x_i, \tau) = \text{Tr}\left(e^{-H_I/T} e^{H_I \tau} \hat{\sigma}_i^z e^{-H_I \tau} \hat{\sigma}_0^z\right)/\mathcal{Z}. \tag{10.3}$$

Compare this definition with (5.51); from the discussion in Chapter 3 it should be clear that $C(x, \tau)$ is the correlator of the classical $D = 2$ Ising model (3.1) on an infinite strip of width $1/T$ and periodic boundary conditions along the "imaginary-time" direction. We also deal with the *dynamic structure factor*, $S(k, \omega)$; as in (7.10) this is the Fourier transform of $C(x, t)$ to wavevectors and frequencies:

$$S(k, \omega) = \int dx \int dt\, C(x, t) e^{-i(kx - \omega t)}. \tag{10.4}$$

This is a useful quantity because it is directly proportional to the cross-section in scattering experiments in which the probe (usually neutrons) couples to σ^z. If the energy of the scattered neutron is integrated over, then the cross-section is proportional to the *equal-time structure factor*, $S(k)$, defined by

$$S(k) \equiv \int \frac{d\omega}{2\pi} S(k, \omega), \tag{10.5}$$

which is clearly also the spatial Fourier transform of $C(x, 0)$. The number of arguments of S will specify whether we are referring to the dynamic or equal-time structure factor. The identity $(\hat{\sigma}_i^z)^2 = 1$ implies that $C(0, 0) = 1$ and leads to the following sum rule for the dynamic structure factor:

$$\int \frac{dk d\omega}{(2\pi)^2} S(k, \omega) = 1. \tag{10.6}$$

Finally, as in Chapter 7 we have the dynamic susceptibilities, $\chi(k, \omega)$ at real frequencies, and $\chi(k, \omega_n)$ at the Matsubara frequencies, which are related to the structure factor by the fluctuation–dissipation theorem. The spectral representation shows that $\chi(k, \omega)$ is an odd function of ω. The dynamic susceptibility measures the response of the magnetization σ^z to an external field that couples linearly to σ^z and is oscillating at a wavevector k and frequency ω. In the limit that the external field becomes time independent, the response is given by the *static susceptibility*, $\chi(k)$, defined by

$$\chi(k) \equiv \chi(k, \omega = 0). \tag{10.7}$$

Again, the number of arguments of χ will specify whether we are referring to the dynamic or static susceptibility.

The nature of the spectrum of H_I in the limits of small and large g has already been presented in Chapter 5. The distinct structures of these spectra implied that there had to be at least one quantum critical point at an intermediate g. The exact spectrum is determined here in Section 10.1 and this shows the existence of a unique quantum critical point at $g = 1$. The universal continuum quantum theory of the vicinity of $g = 1$ in $d = 1$ is obtained in Section 10.2. Equal-time correlators for $T > 0$ are discussed in Section 10.3, and the dynamical properties of the different $T > 0$ regimes are examined in Section 10.4.

We note that the reader may also wish to examine the recent book by Chakrabarti, Dutta, and Sen [74], which discusses aspects of quantum Ising models in one and higher dimensions.

10.1 Exact spectrum

The qualitative considerations of the previous section are quite useful in developing an intuitive physical picture. We now take a different route and set up a formalism that eventually leads to an exact determination of many physical correlators; these results vindicate the approximate methods for $g > 1$, $g < 1$ and also provide an understanding of the novel physics at $g = 1$.

The essential tool in the solution is the Jordan–Wigner transformation [251, 299]. This is a very powerful mapping between models with spin-1/2 degrees of freedom and spinless fermions. The central observation is that there is a simple mapping between the Hilbert space of a system with a spin-1/2 degree of freedom per site and that of spinless fermions hopping between sites with single orbitals. We may associate the spin-up state with an empty orbital on the site and a spin-down state with an occupied orbital. If the canonical fermion operator c_i annihilates a spinless fermion on site i, then this simple mapping immediately implies the operator relation

$$\hat{\sigma}_i^z = 1 - 2c_i^\dagger c_i. \tag{10.8}$$

It is also clear that the operation of c_i is equivalent to flipping the spin from down to up, or the operation of $\hat{\sigma}_i^+ = (\hat{\sigma}_i^x + i\hat{\sigma}_i^y)/2$; similarly, creating a fermion by $c_i^\dagger$ is equivalent to lowering the spin by $\hat{\sigma}_i^- = (\hat{\sigma}_i^x - i\hat{\sigma}_i^y)/2$. Although this equivalence works for a single site, we cannot yet equate the fermion operators with the corresponding spin operators for the many-site problem; this is because while two fermionic operators on different sites anticommute, two spin operators commute. The solution to this dilemma was found by Jordan and Wigner, who showed that the following representation satisfied both on-site and inter-site (anti)commutation relations:

$$\hat{\sigma}_i^+ = \prod_{j<i} \left(1 - 2c_j^\dagger c_j\right)c_i,$$

$$\hat{\sigma}_i^- = \prod_{j<i} \left(1 - 2c_j^\dagger c_j\right)c_i^\dagger. \tag{10.9}$$

The naive single-site correspondence has been modified by a "string" of operators, whose value is $+1$ (-1) if the total number of fermions on the sites to the left of site i are even (odd). Note that the spin operators have a highly nonlocal representation in terms of the fermion operators. This feature is also found in the inverse of (10.9):

$$c_i = \left(\prod_{j<i} \hat{\sigma}_j^z\right)\hat{\sigma}_i^+,$$

$$c_i^\dagger = \left(\prod_{j<i} \hat{\sigma}_j^z\right)\hat{\sigma}_i^-. \tag{10.10}$$

It can be verified that (10.8), (10.9), and (10.10) are consistent with the relations

$$\{c_i, c_j^\dagger\} = \delta_{ij}, \qquad \{c_i, c_j\} = \{c_i^\dagger, c_j^\dagger\} = 0,$$
$$[\hat{\sigma}_i^+, \hat{\sigma}_j^-] = \delta_{ij}\hat{\sigma}_i^z, \qquad [\hat{\sigma}_i^z, \hat{\sigma}_j^\pm] = \pm 2\delta_{ij}\hat{\sigma}_i^\pm, \tag{10.11}$$

where the curly brackets represent anticommutators and square brackets are commutators.

The above formulation of the Jordan–Wigner transformation is the conventional one, but in analysis of the Ising model it is convenient to rotate spin axes by 90 degrees about the y-axis so that

$$\hat{\sigma}^z \to \hat{\sigma}^x, \quad \hat{\sigma}^x \to -\hat{\sigma}^z. \tag{10.12}$$

The mapping becomes

$$\hat{\sigma}_i^x = 1 - 2c_i^\dagger c_i,$$
$$\hat{\sigma}_i^z = -\prod_{j<i}\left(1 - 2c_j^\dagger c_j\right)\left(c_i + c_i^\dagger\right). \tag{10.13}$$

Inserting (10.13) into H_I, we find that the resulting Hamiltonian is quadratic in the Fermi operators [325]:

$$H_I = -J\sum_i \left(c_i^\dagger c_{i+1} + c_{i+1}^\dagger c_i + c_i^\dagger c_{i+1}^\dagger + c_{i+1}c_i - 2gc_i^\dagger c_i + g\right). \tag{10.14}$$

This fermionic Hamiltonian has terms such as $c^\dagger c^\dagger$ that violate the fermion conservation number; from (10.13), this means that $\sum_i \hat{\sigma}_i^x$ is not conserved and $|\ \to\rangle$ spins can be flipped in pairs under time evolution, as we saw in the perturbation theory in Section 5.2. Hence the eigenstates of H_I will not have a definite fermion number. Nevertheless, the new terms are still quadratic in the fermion operators, and H_I can be diagonalized by elementary means. First, use the momentum eigenstates

$$c_k = \frac{1}{\sqrt{M}} \sum_j c_j e^{-ikr_j}. \tag{10.15}$$

where M is the number of sites, to get

$$H_I = J \sum_k \left(2\left[g - \cos(ka)\right] c_k^\dagger c_k + i \sin(ka)\left[c_{-k}^\dagger c_k^\dagger + c_{-k} c_k\right] - g \right). \tag{10.16}$$

Next, use the Bogoliubov transformation to map into a new set of fermionic operators (γ_k) whose number is conserved. These new operators are defined by a unitary transformation on the pair $c_k, c_{-k}^\dagger$:

$$\gamma_k = u_k c_k - i v_k c_{-k}^\dagger, \tag{10.17}$$

where u_k, v_k are real numbers satisfying $u_k^2 + v_k^2 = 1$, $u_{-k} = u_k$, and $v_{-k} = -v_k$. It can be checked that canonical fermion anticommutation relations for the c_k imply that the same relations are also satisfied by the γ_k, that is,

$$\{\gamma_k, \gamma_{k'}^\dagger\} = \delta_{k,k'}, \quad \{\gamma_k^\dagger, \gamma_{k'}^\dagger\} = \{\gamma_k, \gamma_{k'}\} = 0. \tag{10.18}$$

We also note the inverse of (10.17):

$$c_k = u_k \gamma_k + i v_k \gamma_{-k}^\dagger. \tag{10.19}$$

We insert (10.19) into (10.16) and demand that H_I not contain any terms like $\gamma^\dagger \gamma^\dagger$ that violate conservation of the γ fermions. The as yet undefined constants u_k, v_k can always be chosen to ensure this: we define $u_k = \cos(\theta_k/2)$, $v_k = \sin(\theta_k/2)$, and a simple calculation then shows that the choice

$$\tan \theta_k = \frac{\sin(ka)}{g - \cos(ka)} \tag{10.20}$$

satisfies our requirements. The final form of H_I is

$$H_I = \sum_k \varepsilon_k \left(\gamma_k^\dagger \gamma_k - 1/2 \right), \tag{10.21}$$

where

$$\varepsilon_k = 2J(1 + g^2 - 2g \cos k)^{1/2} \tag{10.22}$$

is the single-particle energy. As $\varepsilon_k \geq 0$, the ground state, $|0\rangle$, of H_I has no γ fermions and therefore satisfies $\gamma_k |0\rangle = 0$ for all k. The excited states are created by occupying the single-particle states; they can clearly be classified by the total number of occupied states, and an n-particle state has the form $\gamma_{k_1}^\dagger \gamma_{k_2}^\dagger \cdots \gamma_{k_n}^\dagger |0\rangle$, with all the k_i distinct.

The above structure of the spectrum confirms the approximate considerations of Sections 5.2 and 5.3. We have found that the particles are in fact free fermions, and two fermions will not scatter even when they are close to each other; alternatively they can be considered as hard-core bosons, which have an S matrix that does not allow particle production and that equals -1 at all momenta. We shall find it much more useful to take the latter point of view, as the bosonic particles have a simple, local interpretation in terms of the underlying spin excitations: for $g \gg 1$ the bosons are simply spins oriented in the $|\leftarrow\rangle$ direction, whereas for $g \ll 1$ they are domain walls between the two ground states. The fermionic representation is useful for certain technical manipulations, but the bosonic point of view is much more useful for making physical arguments, as we see below.

It is also reassuring to see that the exact single-particle excitation energy (10.22) agrees with (5.8) in the limit $g \gg 1$, and with (5.21) in the limit $g \ll 1$.

10.2　Continuum theory and scaling transformations

The excitation energy ε_k in (10.22) is nonzero and positive for all k provided $g \neq 1$. The energy gap, or the minimum excitation energy, is always at $k = 0$ and equals $2J|1 - g|$. This gap vanishes at $g = 1$, and it is natural to expect that $g = 1$ marks the phase boundary between the two qualitatively different phases discussed in Sections 5.2 and 5.3. Precisely at $g = 1$, fermions with low momenta can carry arbitrarily low energy, and they therefore must dominate the low-temperature properties. These properties suggest that the state at $g = 1$ is critical, and there is a universal continuum quantum field theory that describes the critical properties in its vicinity.

In Chapters 3 and 4 we mapped the classical Ising model to a field theory, which was then used to develop scaling and renormalization group ideas. A key to this analysis was the existence of the Gaussian field theory (4.2), which gave an analytically tractable fixed point of the RG, and then allowed a systematic analysis of corrections to the critical theory near dimension $D = 4$. We have obtained an exact solution of a quantum critical point above, for the quantum Ising model in $d = 1$. We therefore have the opportunity to subject it to an analysis which parallels that carried out earlier for the classical model. Upon taking its continuum limit, we find that the $d = 1$ quantum model is also described by a free (or "Gaussian") quantum field theory, but which is now expressed in terms of fermions. This free fermion theory is easily amenable to an RG analysis, just like that in Section 4.1 for the bosonic theory, and has a fixed point describing the quantum phase transition of the quantum Ising model. Unlike the situation in Section 4.1, we find here that the free fermion fixed point is in fact *stable* to all non-Gaussian interaction corrections. This allows us to obtain exact results for the critical properties of a generic quantum Ising model in one dimension.

A natural question then arises of the connection between the free fermion fixed point found below in $d = 1$, and the fixed points found in Chapter 4 for $d < 3$ (or $D < 4$). One of the central points of Parts I and II was that the D-dimensional classical Ising model, and $(d = D - 1)$-dimensional quantum Ising models are formally equivalent near their respective critical points. Should their RG fixed points then not be the same? The only reasonable conclusion from this equivalence, if correct, is that the $d = 1$ free fermion RG fixed point below is in fact the *same* as the Wilson–Fisher fixed point of the quantum field theory (2.11) for $N = 1$, when the latter is extended from d close to 3, all the way down to $d = 1$. There is a great deal of evidence that this remarkable claim is true: numerical analysis of the $(3 - d)$ expansion of the Wilson–Fisher fixed point at large orders yields results for critical properties which are strikingly close to the exact results presented below.

So let us turn to obtaining the critical theory of the $d = 1$ quantum Ising model. Because the important excitations are near $k = 0$, we expect that a naive gradient expansion will yield the required theory. We define the continuum Fermi field

$$\Psi(x_i) = \frac{1}{\sqrt{a}} c_i, \tag{10.23}$$

where the normalization has been chosen so that Ψ has units of inverse square root of length and

$$\{\Psi(x), \Psi^\dagger(x')\} = \delta(x - x'), \tag{10.24}$$

with the right-hand side a Dirac delta function in the continuum limit. We express H_I in terms of Ψ, and expand in spatial gradients, to obtain from (10.16) the continuum H_F:

$$H_F = E_0 + \int dx \left[\frac{c}{2} \left(\Psi^\dagger \frac{\partial \Psi^\dagger}{\partial x} - \Psi \frac{\partial \Psi}{\partial x} \right) + \Delta \Psi^\dagger \Psi \right] + \cdots, \tag{10.25}$$

where the ellipses represent terms with higher gradients, and E_0 is an uninteresting additive constant. The coupling constants in H_F are

$$\Delta = 2J(1 - g), \qquad c = 2Ja. \tag{10.26}$$

Note that at the critical point $g = 1$, we have $\Delta = 0$, and we have $\Delta > 0$ in the magnetically ordered phase and $\Delta < 0$ in the quantum paramagnet.

The continuum theory H_F in (10.25) can be viewed as having been obtained by replacing the dependence of the Hamiltonian on c_i, J, and g by Ψ, Δ, and c and then taking the limit $a \to 0$ at fixed Ψ, Δ, and c. Note from (10.26) that this limit requires $J \to \infty$ and $g \to 1$. Note also the similarity to the discussion in Section 5.5.1.

It is convenient to perform our subsequent scaling analysis in a path integral representation of the dynamics of H_F. The path integral is derived using the analog of the coherent state path integral discussed in Section 9.2, but applied to the Fock space of fermions. This procedure involves fermion coherent states, and leads to a path integral involving Grassman numbers. A full description of Grassman numbers, and of the fermion coherent states, would lead us on a long detour. Fortunately, there are many excellent discussions of the Grassman path integral in the literature, e.g. the book by Negele and Orland [357] or the text by Shankar [464]), and we refer the reader to them. In the end, the final form of the coherent state path integral is *identical* to that derived for bosons in Section 9.2: the only difference is that the variables of integration are noncommuting Grassman numbers rather than commuting complex numbers. In this manner, we obtain for the partition function $\mathcal{Z} = \text{Tr} e^{-H_F/T}$:

$$\mathcal{Z} = \int \mathcal{D}\Psi \mathcal{D}\Psi^\dagger \exp\left(-\int_0^{1/T} d\tau dx\, \mathcal{L}_I\right), \qquad (10.27)$$

where the functional integral is over complex Grassman fields Ψ, $\Psi^\dagger$ in space (x) and imaginary time (τ), and the Lagrangean density $\mathcal{L}_I$ is

$$\mathcal{L}_I = \Psi^\dagger \frac{\partial \Psi}{\partial \tau} + \frac{c}{2}\left(\Psi^\dagger \frac{\partial \Psi^\dagger}{\partial x} - \Psi \frac{\partial \Psi}{\partial x}\right) + \Delta \Psi^\dagger \Psi. \qquad (10.28)$$

The continuum theory $\mathcal{L}_I$ can be diagonalized much like the lattice model H_I, and the excitation energy now takes a "relativistic" form

$$\varepsilon_k = (\Delta^2 + c^2 k^2)^{1/2}, \qquad (10.29)$$

which shows that $|\Delta|$ is the $T = 0$ energy gap (we chose the sign of Δ to be different on the two sides of the critical value of g) and c is the velocity of the excitations, both measurable quantities. The form of ε_k correctly suggests that $\mathcal{L}_I$ is invariant under Lorentz transformations. This can be made explicit by writing the complex Grassman field Ψ in terms of two real Grassman fields, when the action becomes what is known as the field theory of Majorana fermions of mass Δ/c^2 [244]; we do not explicitly display this here.

The key to establishing that $\mathcal{L}_I$ is a *universal* critical theory is to examine its behavior under the *scaling transformations* of Chapter 4. Let us restate the physical meaning of this transformation in the present context. We think of $\mathcal{L}_I$ as an effective theory of an underlying lattice problem, applicable only at length scales larger than some lattice spacing a or with momenta smaller than $\Lambda = \pi/a$. We are ultimately interested in long-distance physics, and so it is useful to think of eliminating some short-distance degrees of freedom from $\mathcal{L}_I$: say all modes of the field Ψ with momenta between Λ and $\Lambda e^{-\ell}$, where $e^{-\ell}$ is a dimensionless rescaling factor. As (10.27) involves only a Gaussian functional integral, integrating these modes out will only add an overall additive constant to the free energy. We are left with a new theory having the same action as $\mathcal{L}_I$ but valid only at length scales larger than ae^ℓ. We complete the scaling transformation by rescaling lengths, times, and fields so that the resulting $\mathcal{L}_I$ has the same form and short-distance cutoff as the original $\mathcal{L}_I$. To this end we define (compare (4.1) and (4.3))

$$x' = xe^{-\ell},$$
$$\tau' = \tau e^{-z\ell},$$
$$\Psi' = \Psi e^{\ell/2}. \qquad (10.30)$$

The reader can easily check that the new $\mathcal{L}_I$ expressed in terms of x', τ', and Ψ', has the same form, and the same short-distance cutoff a, as the original $\mathcal{L}_I$ had in terms of x, τ, and Ψ at the position of the quantum critical point $\Delta = T = 0$. The parameter z is the *dynamic critical exponent* and determines the relative rescaling factors of space and time. In the present case, only the choice $z = 1$ leaves the velocity c invariant. Indeed, all the quantum rotor models will have $z = 1$ because they are related to classical problems that are fully isotropic in D spatial dimensions, as noted in Section 2.1. When viewed as a transformation on the continuum theory (i.e. for the case $a \to 0$), it is evident that (10.30) is an exact invariance of $\mathcal{L}_I$, and this is often a useful point of view to take.

Let us move away from the critical point $\Delta = 0$, $T = 0$, by changing Δ but keeping $T = 0$. Under the rescaling (10.30) the action $\mathcal{L}_I$ remains invariant only if we introduce a new Δ':

$$\Delta' = \Delta e^{\ell}. \qquad (10.31)$$

We see that any initially nonzero Δ grows indefinitely as one transforms to larger scales (larger ℓ); so it is a *relevant* perturbation, It destroys scale invariance at the largest scales.

Let us re-express the above results in terms of scaling dimensions. We have

$$\dim[\Delta] = 1. \qquad (10.32)$$

As in Chapter 4, we define the exponent ν as the inverse of the scaling dimension of the most relevant perturbation about a quantum critical point; in the present case, this will turn out to be Δ, and so

$$\nu = 1. \qquad (10.33)$$

For the fermion field operator we have

$$\dim[\Psi] = 1/2, \qquad (10.34)$$

while the spacetime dimensions have

$$\dim[x] = -1, \quad \dim[\tau] = -z. \qquad (10.35)$$

The temperature, T, is just an inverse time, and therefore

$$\dim[T] = z. \qquad (10.36)$$

This is positive, and so, not surprisingly, T is also a relevant perturbation at the quantum critical point. Let us also consider the scaling dimension of the free energy density $\widetilde{\mathcal{F}}$ of the system; we apply the tilde here, because earlier we had used $\mathcal{F}$ for the total free energy, not the free energy density (also, we always subtract out from $\widetilde{\mathcal{F}}$ the ground state energy at the quantum critical point $\Delta = 0$ and consider the singular behavior of the remainder). This is given by $\widetilde{\mathcal{F}} = -(T/V) \ln \mathcal{Z}$, where V is the volume and $\mathcal{Z}$ is the partition function. As the logarithm is dimensionless, and clearly $\dim[V] = d \dim[x]$, we have

$$\dim[\widetilde{\mathcal{F}}] = d + z, \qquad (10.37)$$

which is the generalization of the classical result in (4.49).

Finally, we also need the scaling dimension of the order parameter $\hat{\sigma}^z$. This is not a simple local function of the Fermi field Ψ and is therefore quite difficult to determine. We describe a relatively elaborate calculation in Section 10.3 that shows that

$$\dim[\hat{\sigma}^z] = 1/8. \qquad (10.38)$$

This is the first example here of an *anomalous dimension*. Indeed, the connection to the classical Ising model discussed in Chapter 5 suggests that the operator $\hat{\sigma}^z$ maps directly onto field ϕ of the field theory (2.11): both are the Ising order parameters which measure the broken symmetry across the phase transition. From (4.32), we have $\dim[\phi] = (D - 2 + \eta)/2 = \eta/2$. So we conclude from (10.38) that the $d = 1$ quantum Ising model has $\eta = 1/4$. We expect that $\eta = 1/4$ is also the exact result at the Wilson–Fisher fixed point in $D = 2$; i.e. the value of (4.37) for $\epsilon = 2$ and $N = 1$.

All previous scaling dimensions of the $d = 1$ Ising model coincided with their so-called *engineering dimension*, that is, that obtained by the familiar dimensional analysis of lengths and times in meters and seconds, with the additional freedom to use powers of the velocity c to convert all times into lengths; the anomalous dimension is defined by the *difference* of the scaling and engineering dimensions, and so all previous anomalous dimensions were 0. The engineering dimension of $\hat{\sigma}^z$, a dimensionless matrix, is clearly 0. Nevertheless, we see that it has a nonzero scaling dimension. This can happen without violating equality of engineering dimensions (which must always be preserved) because we have the additional freedom to use powers of the lattice energy scale J (or the lattice spacing, a) in defining the continuum limit of observables. Indeed, (10.38) implies that only the combination $J^{1/8}\hat{\sigma}^z$ has correlators that are finite in the continuum limit $a \to 0$ discussed below (10.26).

Armed with the knowledge of these scaling dimensions, we can put important general constraints on the structure of various universal scaling forms. We follow a simple, general convention in presenting these scaling forms. First pick the observable of interest and determine its scaling dimension. Then write down as a prefactor that power of T which has the same scaling dimension as the observable. This multiplies a dimensionless universal scaling function of a number of arguments; each argument should be a coupling or coordinate times a power of T so that the combination has net scaling dimension 0. Finally, powers of innocuous variables such as c with zero scaling dimension are inserted so that the engineering dimensions of the expressions are consistent.

As an example of such considerations, let us consider the scaling form satisfied by the two-point correlator $C(x, t)$ defined in (10.2):

$$C(x, t) = Z T^{1/4} \Phi_I \left(\frac{Tx}{c}, Tt, \frac{\Delta}{T} \right). \tag{10.39}$$

Here Z is an overall noncritical normalization constant, with zero scaling dimension, which depends on the details of the microscopic physics; its presence is related to the anomalous dimension of $\hat{\sigma}^z$ and consistency of (10.39) requires Z to have engineering dimension $-1/4$. We shortly relate Z to observable properties of the ground state. The scaling function Φ_I depends universally on its three arguments. The power of T in the front follows from (10.38) and (10.36). Note that the physics depends completely on the ratio of two energy scales, that of the $T = 0$ energy gap to temperature: Δ/T. The central purpose of this chapter is to present a fairly complete description of the physical properties of Φ_I as a function of Δ/T.

It is very important to note that the scaling form (10.39) will *not* satisfy the relationship $C(0, 0) = 1$, which is exactly obeyed by the lattice model; this short-distance property is lost once the continuum limit has been taken. Alternatively stated, the sum rule (10.6) will not be obeyed by the Fourier transform of (10.39).

We can also describe rather explicitly the sense in which Φ_I is universal; that is, what happens if we generalize H_I or $\mathcal{L}_I$ to include other short-range couplings? There are two different types of perturbation to $\mathcal{L}_I$ that are possible. The first type arises from higher spatial gradients in the mapping from the particular Hamiltonian H_I, and the simplest of these is

$$\lambda_1 \Psi^\dagger \frac{\partial^2 \Psi}{\partial x^2}. \tag{10.40}$$

The second type comes from additional terms we could add to H_I, such as $\hat{\sigma}_i^x \hat{\sigma}_{i+1}^x$, that respect the symmetry (1.13) and are therefore not expected to modify qualitative features of the transition; after the Jordan–Wigner transformation, and expansion in spatial gradients, such a term induces in the continuum limit a term

$$\lambda_2 \Psi^\dagger \frac{\partial \Psi^\dagger}{\partial x} \frac{\partial \Psi}{\partial x} \Psi; \tag{10.41}$$

note that two spatial gradients are required because the term with only one would vanish because of the Fermi statistics identity $\Psi^2 = 0$. A simple computation shows that

$$\dim[\lambda_1] = -1, \quad \dim[\lambda_2] = -2, \tag{10.42}$$

and so these couplings are *irrelevant*, they can be neglected in a discussion of the leading long-distance and low-T properties. This is in striking contrast to the bosonic Gaussian fixed point of Section 4.1, which was unstable to perturbations via the ϕ^4 interaction.

The absence of other relevant perturbations at $\Delta = 0$ implies that $\mathcal{L}_I$ is the universal continuum quantum field theory describing crossovers near the $\Delta = 0$, $T = 0$ quantum critical point. It is fortunate that this universal theory happens to be expressible as a free fermion model. Although our original motivation for examining H_I was its solvability, the arguments of this section have shown that this choice also happily coincides with that required for obtaining a universal critical theory.

There are two types of consequence of the irrelevant couplings. The first is that the values of the parameters Z, Δ, and c appearing in the scaling form (10.39) change; this change is quite difficult to compute, and therefore we should consider Z, Δ, and c to be defined by some experimental observable at $T = 0$: Δ is defined to be the energy gap at $T = 0$, c is the velocity of excitations at $\Delta = 0$, and Z is shown later to be related to certain amplitudes at $T = 0$. The second is that there are subleading corrections to the whole scaling form itself. The form of these corrections can be deduced from the general rules stated earlier, and we find that the result (10.39) has multiplicative corrections like

$$(1 + \lambda_1 T + \lambda_2 T^2 + \cdots). \tag{10.43}$$

These corrections are expected to be unimportant at low enough T.

Let us compute finite-temperature correlators of the free fermion field Ψ. These correlators are not related to any local observable of the Ising chain, and therefore they cannot be measured experimentally. Our main purpose in discussing them is to further illustrate the present scaling ideas in a simple context. The two-point Ψ correlators can be computed by performing the analog of the lattice Bogoliubov transformation on the continuum theory. We found for imaginary time $\tau > 0$

$$\left\langle \Psi(x,\tau)\Psi^\dagger(0,0)\right\rangle = \frac{1}{2}\int_{-\infty}^{\infty}\frac{dk}{2\pi}\frac{e^{ikx}}{e^{c|k|/T}+1}\left(e^{c|k|(1/T-\tau)}+e^{c|k|\tau}\right)$$

$$= \left(\frac{T}{4c}\right)\left(\frac{1}{\sin(\pi T(\tau-ix/c))}+\frac{1}{\sin(\pi T(\tau+ix/c))}\right). \qquad (10.44)$$

In a similar manner, we can find

$$\langle\Psi(x,\tau)\Psi(0,0)\rangle = \left(\frac{iT}{4c}\right)\left(\frac{1}{\sin(\pi T(\tau-ix/c))}-\frac{1}{\sin(\pi T(\tau+ix/c))}\right). \qquad (10.45)$$

The results (10.44) and (10.45) have precisely the scaling forms that would have been expected under the scaling dimensions in (10.30). At $T = 0$, (10.44) simplifies to

$$\left\langle\Psi(x,\tau)\Psi^\dagger(0,0)\right\rangle = \frac{1}{4\pi}\left(\frac{1}{c\tau-ix}+\frac{1}{c\tau+ix}\right). \qquad (10.46)$$

Now note that the transformation

$$c\tau\pm ix \rightarrow \frac{c}{\pi T}\sin\left(\frac{\pi T}{c}(c\tau\pm ix)\right) \qquad (10.47)$$

connects the $T = 0$ and $T > 0$ results. This mapping is actually an example of a very general connection between *all* $T = 0$ and $T > 0$ two-point correlators of the continuum theory $\mathcal{L}_I$. The existence of this mapping is due to a larger *conformal* symmetry of $\mathcal{L}_I$ [68]. (The reader is referred to [244] for further discussion on this point.) Here we defer discussion of this mapping to Chapter 20 where it arises as a simple consequence of the bosonization method.

10.3   Equal-time correlations of the order parameter

This section is of a technical nature. Its main purpose is to show how one may obtain the result (10.38) that $\dim[\hat{\sigma}^z] = 1/8$. We also obtain explicit expressions for certain crossover functions that cannot be obtained otherwise. The limiting forms of these crossover functions, and all of the interesting dynamical properties of the system, are obtained again later in Section 10.4 using simple physical arguments that rely on the bosonic picture of the excitations developed in Sections 5.2 and 5.3 using the large and small-g expansions. Most readers may therefore glance at the next paragraph where we outline the main results and omit the remainder of this section.

We begin by writing down the main result and then outline how it is obtained. The equal-time two-point correlation of the order parameter has the following long-distance limit at any $T > 0$ [420]:

$$\lim_{|x|\to\infty} C(x,0) = ZT^{1/4}G_I(\Delta/T)\exp\left(-\frac{T|x|}{c}F_I(\Delta/T)\right), \qquad (10.48)$$

where Z is the nonuniversal constant introduced earlier in (10.39), and $F_I(s)$ and $G_I(s)$ are universal scaling functions. Notice that (10.48) is completely consistent with the general scaling form (10.39). A crucial property of (10.48) is the prefactor of $T^{1/4}$, which establishes that $\dim[\hat{\sigma}^z] = 1/8$. A second important property is that the two-point correlations decay exponentially to zero at large enough x. Thus the $T = 0$ long-range order discussed in Section 5.3 disappears at any $T > 0$ (we will later give a simple physical explanation of this). The exponential decay defines a correlation length ξ that obeys

$$\xi^{-1} = \frac{T}{c} F_I\left(\frac{\Delta}{T}\right). \tag{10.49}$$

The exact, self-contained expression for the universal function F_I is [420]

$$F_I(s) = |s|\theta(-s) + \frac{1}{\pi} \int_0^\infty dy \ln \coth \frac{(y^2 + s^2)^{1/2}}{2}. \tag{10.50}$$

The $s > 0$ ($s < 0$) portion of F_I describes the magnetically ordered (paramagnetic) side. Despite its appearance, the function $F_I(s)$ is smooth as a function of s for all real s and is analytic at $s = 0$. The analyticity at $s = 0$ is required by the absence of any thermodynamic singularity at finite T for $\Delta = 0$. This is a key property, which was in fact used to obtain the answer in (10.50). The exact expression for the function $G_I(s)$ is also known [420]:

$$\ln G_I(s) = \int_s^1 \frac{dy}{y} \left[\left(\frac{dF_I(y)}{dy}\right)^2 - \frac{1}{4}\right] + \int_1^\infty \frac{dy}{y} \left(\frac{dF_I(y)}{dy}\right)^2, \tag{10.51}$$

and its analyticity at $s = 0$ follows from that of F_I. For the solvable model H_I, we chose the overall normalization of G_I such that $Z = J^{-1/4}$. In general, the value of Z is set by relating it to an observable, as we show below. Also note that Z has no dependence on Δ; it is therefore nonsingular at the quantum critical point.

We show a plot of the universal functions F_I and G_I in Fig. 10.1. Notice that they are perfectly smooth at $\Delta = 0$ ($s = 0$).

We now outline how to establish (10.48). We work with the lattice model H_I and consider the evaluation of $\langle \hat{\sigma}_i^z \hat{\sigma}_{i+n}^z \rangle$. The continuum limit for the correlators of $\mathcal{L}_I$ can only be taken at a relatively late stage. Using the fermionic representation (10.13) and the simple identity $1 - 2c_i^\dagger c_i = (c_i^\dagger + c_i)(c_i^\dagger - c_i)$, we obtain [299]

$$\langle \hat{\sigma}_i^z \hat{\sigma}_{i+n}^z \rangle = \left\langle (c_i^\dagger + c_i) \left[\prod_{j=i}^{i+n-1} (c_j^\dagger + c_j)(c_j^\dagger - c_j)\right] (c_{i+n}^\dagger + c_{i+n}) \right\rangle$$

$$= \left\langle (c_i^\dagger - c_i) \left[\prod_{j=i+1}^{i+n-1} (c_j^\dagger + c_j)(c_j^\dagger - c_j)\right] (c_{i+n}^\dagger + c_{i+n}) \right\rangle. \tag{10.52}$$

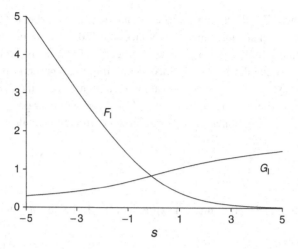

Fig. 10.1 The crossover functions for the correlation length (F_I) and the amplitude (G_I) as a function of $s = \Delta/T$.

Note that the string extends only between the sites i and $i + n$, with the operators on sites to the left of i having cancelled between the two strings. Now, using the notation

$$A_i = c_i^\dagger + c_i, \qquad B_i = c_i^\dagger - c_i, \tag{10.53}$$

we have

$$\langle \hat{\sigma}_i^z \hat{\sigma}_{i+n}^z \rangle = \langle B_i A_{i+1} B_{i+1} \cdots A_{i+n-1} B_{i+n-1} A_{i+n} \rangle. \tag{10.54}$$

Since the expectation values are with respect to a free Fermi theory, the expression on the right-hand side can be evaluated by the finite-temperature Wick's theorem [136], which relates it to a sum over products of expectation values of pairs of operators. The expectation value of any such pair is easily calculated:

$$\langle A_i A_j \rangle = \delta_{ij},$$
$$\langle B_i B_j \rangle = -\delta_{ij},$$
$$\langle B_i A_j \rangle = -\langle A_j B_i \rangle = D_{i-j+1}, \tag{10.55}$$

with

$$D_n \equiv \int_0^{2\pi} \frac{d\phi}{2\pi} e^{-in\phi} \tilde{D}(e^{i\phi}), \tag{10.56}$$

and

$$\tilde{D}(z = e^{i\phi}) \equiv \left(\frac{1 - gz}{1 - g/z} \right)^{1/2} \tanh\left[\frac{J}{T}((1 - gz)(1 - g/z))^{1/2} \right]; \tag{10.57}$$

note that the argument of the tanh (which arises from the thermal Fermi distribution function) is simply $\varepsilon_\phi/2T$. In determining $\langle B_i A_j \rangle$, we have used the representation (10.19) and evaluated expectation values of the γ_k under the free fermion Hamiltonian (10.22). Collecting the terms in the Wick expansion, we find

$$\langle \hat{\sigma}_i^z \hat{\sigma}_{i+n}^z \rangle = T_n \equiv \begin{vmatrix} D_0 & D_{-1} & \cdots & D_{-n+1} \\ D_1 & & & \\ & \cdot & & \\ & \cdot & & \\ & \cdot & D_0 & D_{-1} \\ D_{n-1} & & D_1 & D_0 \end{vmatrix}. \tag{10.58}$$

We are now faced with the mathematical problem of evaluating the determinant T_n: to obtain the universal scaling limit answer we need to take the limit $n \to \infty$ while keeping the system close to its critical point. The expression for T_n is in a special class of determinants known as Toeplitz determinants, and the limit $T_{n\to\infty}$ can indeed be evaluated in closed form using a fairly sophisticated mathematical theory. We do not present the details of this evaluation here, but refer the reader to the literature [36, 327, 332, 420]. The final, universal, result has already been quoted at the beginning of this section.

10.4 Finite temperature crossovers

The key result of the previous section is that equal-time correlations of the order parameter, $C(x, 0)$, decay exponentially to zero at any $T > 0$. The expression for the correlation length as a function of Δ/T is given in (10.49). From this result we can easily obtain the following important limiting forms, which are also rederived in this section using simpler physical arguments:

$$\xi = \begin{cases} c\sqrt{\dfrac{\pi}{2\Delta T}} e^{\Delta/T}, & \text{for } \Delta \gg T \\[2mm] \dfrac{4c}{\pi T}, & \text{for } |\Delta| \ll T \\[2mm] \dfrac{c}{|\Delta|}. & \text{for } \Delta \ll -T \end{cases} \tag{10.59}$$

Note that for $\Delta > 0$, the correlation length diverges exponentially as $T \to 0$. As we show explicitly in Section 10.4.1 below, this is a characteristic property of a state with long-range order only at $T = 0$. Precisely at $\Delta = 0$, the correlation length diverges as $\sim 1/T$, which agrees with the naive analysis of scaling dimensions at a quantum critical point $\sim T^{-1/z}$. Finally, for $\Delta < 0$, the correlation length reaches a finite value as $T \to 0$, suggesting a quantum paramagnetic ground state. These dependencies imply the important crossover phase diagram shown in Fig. 10.2. There are three distinct universal regimes, characterized by the limiting forms in (10.59), determined by the largest of the two characteristic energy scales, Δ or T, as we discussed near Fig. 1.3. A closely related phase diagram was discussed by Chakravarty, Halperin, and Nelson [75] in the context of a model we study

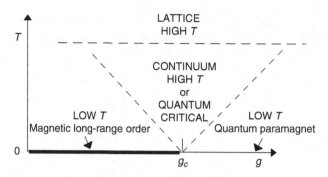

Fig. 10.2 Finite-T phase diagram of the $d = 1$ quantum Ising model, H_I, as a function of the coupling g and temperature T; compare Figs. 1.2a and 1.3. There is a quantum phase transition at $T = 0$, $g = g_c = 1$ with exponents $z = 1$, $\nu = 1$. Magnetic long-range order ($N_0 = \langle \hat{\sigma}_z \rangle \neq 0$) is present only for $T = 0$ and $g < g_c$. The ground state for $g > g_c$ is a quantum paramagnet. There is an energy gap above the ground state for all $g \neq g_c$. We use an energy scale $\Delta \sim g_c - g$ such that the energy gap is $|\Delta|$. The dashed lines are crossovers at $|\Delta| \sim T$. The low-T region on the magnetically ordered side ($\Delta > 0$, $g < g_c$) is studied in Section 10.4.1, and the low-T region on the quantum paramagnetic side ($\Delta < 0$, $g > g_c$) is studied in Section 10.4.2. The continuum high-T (or "quantum-critical") region is studied in Section 10.4.3; its properties are universal and determined by the continuum theory in (10.28). Finally there is also a "lattice high-T" region with $T \gg J$ where the properties are nonuniversal and determined by the lattice scale Hamiltonian.

in Chapter 13, with a different terminology for the various regimes. We find our choices more appropriate and convenient, although we briefly recall their notation in the following subsections.

We note that we finally have a first realization of the generic crossover phase diagram of the vicinity of the quantum critical point illustrated in Fig. 1.2a.

There are two low-T regimes with $T \ll |\Delta|$. The one for $\Delta > 0$, on the magnetically ordered side, has an exponentially diverging correlation length ξ as $T \to 0$; it is studied in Section 10.4.1. The other low-T regime with $\Delta < 0$ has a correlation length that saturates at a finite value as $T \to 0$; it is studied in Section 10.4.2. Then there is a novel continuum high-T regime, $T \gg |\Delta|$, where the physics is controlled primarily by the quantum critical point $\Delta = 0$ and its thermal excitations and is described by the associated continuum quantum field theory; its properties are discussed in Section 10.4.3. This is the analog of the "quantum-critical" regime of [75], but we prefer the term "high-T" as a more accurate description of the dynamical properties of this regime. It is implicit in our high-T limit here that we are not taking the temperature so high that the mapping to the universal continuum model breaks down, and we have to allow for corrections like those in (10.43): this implies that we should always satisfy $T \ll J$. There is therefore a second, nonuniversal high-T limit of the lattice model, also shown in Fig. 10.2, where $T \gg J$, but we have little to say about this regime here. The dynamic $T = \infty$ Ising model results of [67, 376, 377] fall into this last regime; more generally discussions of dynamics at $T = \infty$ may be found in [168] and references therein.

The three subsections below describe the universal dynamics of the Ising chain in the regions of Fig. 10.2. We pay particular attention to the central concept of the phase coherence time τ_φ, which was introduced in Section 2.2, where it was defined loosely as the time over which the wavefunction of the system retains phase memory, and so quantum interference is observable between local measurements separated by times up to τ_φ. We use more precise definitions here. We show that τ_φ obeys

$$\tau_\varphi \sim \hbar/k_B T \quad \text{in the "Continuum High-}T\text{" region, } T \gg \Delta,$$
$$\tau_\varphi \sim (\hbar/k_B T)e^{|\Delta|/k_B T} \quad \text{in the "Low-}T\text{" regions, } T \ll |\Delta|. \tag{10.60}$$

Note that τ_φ always diverges as $T \to 0$, for, as we argued in Section 2.2, the ground state of the system has perfect phase memory. On the magnetically ordered side ($\Delta > 0$, $g < 1$), the divergence of τ_φ is not surprising as it is also accompanied by divergence of the correlation length, as we saw in (10.59). However, on the quantum paramagnetic side ($\Delta < 0$, $g > 1$), the correlation length saturates as $T \to 0$; this clearly does not give a complete physical picture as the divergence of τ_φ indicates a certain temporal coherence. Therefore, as already noted in Section 2.2, the commonly used description of the $\Delta < -T$ region as "quantum disordered" is quite misleading: there are quite precise long-range correlations in time that characterize the perfect coherence of the paramagnetic ground state. Finally, in the continuum high-T region, we see that the lower bound on τ_φ in (2.13) is saturated – this is therefore the most incoherent region.

10.4.1 Low T on the magnetically ordered side, $\Delta > 0$, $T \ll \Delta$

In their study of the model of Chapter 13, Chakravarty, Halperin, and Nelson [75] called the analogous regime "renormalized classical" [75]. The reasons for this name become clear below; however, this is not the only regime that displays classical behavior, as we see in Section 10.4.2.

First, let us consider the results for the equal-time correlations. Assuming that it is valid to interchange the limits $T \to 0$ and $x \to \infty$ in (10.48), we can use the limiting values $F_I(\infty) = 0$, $G_I(s \to \infty) = s^{1/4}$ to deduce that (recall (1.14)):

$$N_0^2 \equiv \lim_{|x| \to \infty} C(x, 0) = Z\Delta^{1/4}, \quad \text{at } T = 0. \tag{10.61}$$

Thus, as claimed earlier, there is long-range order in the $g < 1$ ground state of H_I, with the order parameter $N_0 = \langle \hat{\sigma}^z \rangle = Z^{1/2}\Delta^{1/8}$. (Note that N_0 vanishes as g approaches g_c from below with the exponent $\beta = 1/8$.) We can also use the relationship (10.61) to relate the parameter Z to the observables N_0 and Δ. Turning next to nonzero T, for small $T \ll \Delta$, we obtain from the large s behavior of $F_I(s)$ (see (10.50)) that

$$C(x, 0) = N_0^2 e^{-|x|/\xi_c}, \quad \text{large } |x|, \tag{10.62}$$

where the correlation length

$$\xi_c^{-1} = \left(\frac{2|\Delta|T}{\pi c^2}\right)^{1/2} e^{-|\Delta|/T} \tag{10.63}$$

is finite at any nonzero T, showing that long-range order is present only precisely at $T = 0$. We have put a subscript c on the correlation length to emphasize that the system is expected to behave *classically* in this low-temperature region. This is a crucial characteristic of this region and the reason for classical behavior is quite simple and familiar. The excitations consist of particles (the kinks and antikinks of Section 5.3) whose mean separation ($\sim \xi_c \sim e^{\Delta/T}$) is much larger than their de Broglie wavelengths ($\sim (c^2/\Delta T)^{1/2}$), which is obtained by equating the kinetic energy $\varepsilon_k - \Delta \sim c^2 k^2/2\Delta$ to the thermal equipartition value $T/2$ as $T \to 0$, which is precisely the canonical condition for the applicability of classical physics. It is also reassuring to note that (10.62) has the form of equal-time correlations in the classical Ising chain at low T, which were discussed in Section 5.5. The prefactor N_0^2 is the true ground state magnetization including the effects of quantum fluctuations, and this is the reason for the adjective "renormalized" in the name for this region.

We show that it is possible to give a simple physical interpretation for the value of ξ_c in (10.63). The energy of a domain wall with a small momentum k is $\Delta + c^2 k^2/2\Delta$, and therefore classical Boltzmann statistics tells us that their density, ρ, is

$$\rho = \int \frac{dk}{2\pi} e^{-(\Delta + c^2 k^2/2\Delta)/T} = \left(\frac{T\Delta}{2\pi c^2}\right)^{1/2} e^{-\Delta/T}. \tag{10.64}$$

Comparing with (10.63), we see that $\xi_c = 1/2\rho$. This result follows if we assume that the domain walls are classical *point* particles, which are distributed independently with a density ρ. Consider a system of size $L \gg |x|$, and let it contain M thermally excited particles; then $\rho = M/L$. Let q be the probability that a given particle will be between 0 and x. Clearly,

$$q = \frac{|x|}{L}. \tag{10.65}$$

The probability that a given set of j particles are the only ones between 0 and x is then $q^j (1-q)^{M-j}$; as each particle reverses the orientation of the ground state, in this case $\hat{\sigma}^z(x, 0)\hat{\sigma}^z(0, 0) = N_0^2(-1)^j$. Summing over all possibilities we have

$$C(x, 0) = N_0^2 \sum_{j=0}^{M} (-1)^j q^j (1-q)^{M-j} \frac{M!}{j!(M-j)!}$$
$$= N_0^2(1-2q)^M \approx N_0^2 e^{-2qM} = N_0^2 e^{-2\rho|x|}, \tag{10.66}$$

thus establishing the desired result.

This semiclassical picture can also be extended to compute unequal-time correlations. In this computation it is essential to consider the collisions between the particles. Even though the particles are very dilute, they cannot really avoid each other in one dimension, and neighboring particles will always eventually collide (this is not true in higher dimensions where sufficiently dilute particles can be treated as noninteracting). During their collisions, the particles are certainly closer than their de Broglie wavelengths, and so the collisions must be treated quantum mechanically. Indeed, these collisions will be characterized by the two-particle S matrix, which was considered earlier in Section 5.3.2; however, the diluteness does allow us to consider the collisions of only pairs of particles.

To study dynamic correlations, let us re-examine the explicit expression for $C(x, t)$ in (10.2). We can show how it can be evaluated essentially exactly using some simple physical arguments. The key is to recall that classical mechanics emerges from quantum mechanics

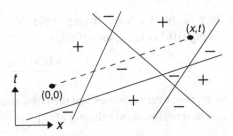

Fig. 10.3 A typical semiclassical contribution to the double time path integral for $\langle \hat{\sigma}^z(x, t)\hat{\sigma}^z(0, 0)\rangle$. Full lines are thermally excited particles that propagate forward and backward in time. The $\pm$ signs are significant only for $g < g_c$ and denote the orientation of the order parameter. For $g > g_c$, the dashed line is a particle propagating only forward in time from $(0, 0)$ to (x, t).

as a stationary phase evaluation of a first-quantized Feynman path integral. We therefore attempt to evaluate the expression in (10.2) by such a path integral. It is clear that the integral is over a set of trajectories moving forward in time, representing the operator $e^{-iH_I t}$, and a second set moving backwards in time, corresponding to the action of $e^{iH_I t}$. In the semiclassical limit, a stationary phase is achieved when the backward paths are simply the time reverse of the forward paths, and both sets are the classical trajectories. An example of a set of paths is shown in Fig. 10.3. Now observe that:

(i) The classical trajectories remain straight lines across collisions because the momenta before and after the collision are the same. This follows from the requirement of conservation of total momentum ($k_1 + k_2 = k_1' + k_2'$) and energy ($\varepsilon_{k_1} + \varepsilon_{k_2} = \varepsilon_{k_1'} + \varepsilon_{k_2'}$) in each two-particle collision, which has the unique solution $k_1 = k_1'$ and $k_2 = k_2'$ (or its equivalent permutation, which need not be considered separately because the particles are identical) in one dimension.

(ii) For each collision, the amplitude for the path acquires a phase $S_{k_1 k_2}$ along the forward path and its complex conjugate along the backward path. The net factor for the collision is therefore $|S_{k_1 k_2}|^2 = 1$.

These two facts imply that the trajectories are simply independently distributed straight lines, placed with a uniform density ρ along the x-axis, with an inverse slope

$$v_k \equiv \frac{d\varepsilon_k}{dk}, \qquad (10.67)$$

and with their momenta chosen with the Boltzmann probability density $e^{-\varepsilon_k/T}/\rho$ (Fig. 10.3).

Computing dynamic correlators is now an exercise in classical probabilities. As each particle trajectory is the boundary between domains with opposite orientations of spins, the value of $\hat{\sigma}^z(0, 0)\hat{\sigma}^z(x, t)$ is the square of the magnetization renormalized by quantum fluctuations (N_0^2) times $(-1)^j$, where j is the number of trajectories intersecting the dashed line in Fig. 10.3. Now it remains to average $N_0^2(-1)^j$ over the classical ensemble of trajectories defined above. This average can be carried out in a manner quite similar to that in the equal-time computation earlier. Again choosing a system size $L \gg |x|$ with M

particles, the probability q that a given particle with velocity v_k is between the points $(0, 0)$ and (x, t) in Fig. 10.3 is (compare (10.65))

$$q = \frac{|x - v_k t|}{L}. \tag{10.68}$$

We have to average over velocities and then evaluate the summation in (10.66). This gives one of the central results of this chapter [442]:

$$C(x, t) = N_0^2 R(x, t),$$
$$R(x, t) \equiv \exp\left(-\int \frac{dk}{\pi} e^{-\varepsilon_k/T} |x - v_k t|\right). \tag{10.69}$$

(This relaxation function also appeared in [115, 316] in a phenomenological analysis of related models by exponentiating a short-time expansion that ignored collisions.) The equal-time or equal-space form of the relaxation function $R(x, t)$ is quite simple:

$$R(x, 0) = e^{-|x|/\xi_c},$$
$$R(0, t) = e^{-|t|/\tau_\varphi}; \tag{10.70}$$

for general x, t the function R also decreases monotonically with increasing $|x|$ or $|t|$, but the decay is not simply an exponential. The spatial correlation length ξ_c is given in (10.63). We have identified the equal-space correlation time as the phase coherence time for obvious reasons: the long-range order in the ground state is clearly a manifestation of phase coherence, and its decay in time is a natural measure of τ_φ. We can determine τ_φ from (10.69), and remarkably we find that τ_φ is independent of the functional form of ε_k and depends only on the gap:

$$\begin{aligned}
\frac{1}{\tau_\varphi} &= \frac{2}{\pi} \int_0^\infty dk \frac{d\varepsilon_k}{dk} e^{-\varepsilon_k/T} \\
&= \frac{2}{\pi} \int_{|\Delta|}^\infty d\varepsilon_k e^{-\varepsilon_k/T} \\
&= \frac{2 k_B T}{\pi \hbar} e^{-|\Delta|/k_B T},
\end{aligned} \tag{10.71}$$

where we have momentarily inserted the fundamental constants $\hbar$ and k_B in the last step to emphasize the universality of the result.

In the limit $T \ll \Delta$ we are now able to completely specify the form of the scaling function Φ_I in (10.39). The behavior of Φ_I is characterized by the concept of a *reduced scaling function*, which is determined entirely by classical physics; we have several occasions to use this concept later in this book. Note that the original function Φ_I had three arguments: the scales of space and time relative to T and the ratio Δ/T. For $T \ll \Delta$ the last argument disappears, and we find that the scales of space and time are determined by the large classical scales ξ_c and τ_φ, respectively. By an analysis of (10.69) we find that the correlations can be written in the following reduced classical scaling form:

$$C(x, t) = N_0^2 \Phi_R\left(\frac{x}{\xi_c}, \frac{t}{\tau_\varphi}\right), \tag{10.72}$$

where clearly $R(x,t)$ satisfies the scaling form

$$R(x,t) = \Phi_R\left(\frac{x}{\xi_c}, \frac{t}{\tau_\varphi}\right). \tag{10.73}$$

These scaling forms are valid only for $T \ll \Delta$, and they must be consistent with the fully quantum Φ_I in (10.39), which is valid for all Δ/T. This requirement implies that the scales ξ_c and τ_φ must be *universal* functions of Δ/T, as we have already seen in the expressions (10.63) and (10.71). Evaluating (10.69) we can obtain an explicit closed form expression for Φ_R:

$$\ln \Phi_R(\bar{x}, \bar{t}) = -\bar{x}\,\mathrm{erf}\left(\frac{\bar{x}}{\bar{t}\sqrt{\pi}}\right) - \bar{t}e^{-\bar{x}^2/(\pi\bar{t}^2)}. \tag{10.74}$$

Note that the characteristic time τ_φ and length ξ_c both diverge as $\sim e^{\Delta/T}$, and so we can define an effective classical dynamic exponent $z_c = 1$ (there is no fundamental reason why z_c and z should have the same value). We also note here that the classical dynamic scaling function obtained above is unrelated to the dynamic scaling functions associated with a popular classical statistical model for dynamics of Ising spins – the Glauber model [169]. The present underlying quantum dynamics leads to a rather different effective classical model in which energy and momentum conservation play a crucial role, the time evolution is deterministic, and the average is over the set of initial conditions.

All of the results above have been compared with exact numerical computations and the agreement is essentially perfect. We show a typical comparison in Fig. 10.4. This

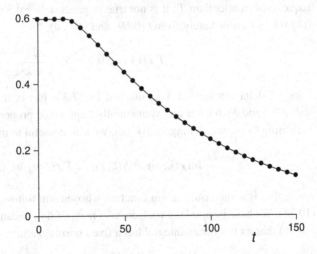

Fig. 10.4 Theoretical and numerical results from [442] for the real part of the correlator $\langle\hat{\sigma}^z(x,t)\hat{\sigma}^z(0,0)\rangle$ of H_I at $x = 20$ with $J = 1$, $g = 0.6$ (therefore $\Delta = 0.8$), $T = 0.3$; the system is thus in the low-T region on the magnetically ordered side of Fig. 10.2. The numerical data (shown in circles) were obtained for a lattice size $L = 512$ with free boundary conditions. This is compared with the theoretical prediction in (10.69). The imaginary part of the correlator was numerically found to be negligibly small, and the semiclassical theoretical prediction is that it vanishes.

agreement gives us confidence that the physical, "hand-waving," quasi-classical particle approach to dynamical properties outlined above is in fact exact.

The Fourier transform of (10.72) and (10.74) yields the portion of the dynamic structure factor, $S(k, \omega)$ (defined in (10.4)), describing the $T > 0$ broadening of the $T = 0$ delta function in (7.33). We expect this broadening to occur on a frequency scale of order $1/\tau_\varphi$, and so the predominant weight in $S(k, \omega)$ is at frequencies $\omega \ll T$. Under this condition, some simplifications occur in the relationships between the response functions introduced in the opening of this chapter. In particular, for $\omega \ll T$, the fluctuation–dissipation theorem (7.12) reduces to its simpler, "classical" form

$$S(k, \omega) = \frac{2T}{\omega} \text{Im} \chi(k, \omega). \tag{10.75}$$

As $\text{Im} \chi(k, \omega)$ is always an odd function of ω, in the limit that (10.75) is obeyed, $S(k, \omega)$ becomes an even function of ω (the reader should keep in mind that this last fact is not generally true). Applying (10.75) to (10.5), we see that the equal-time structure factor is simply T times the static susceptibility,

$$S(k) = \int \frac{d\omega}{\pi} \frac{\text{Im} \chi(k, \omega)}{\omega}$$
$$= T \chi(k), \tag{10.76}$$

where the second equation relies on the Kramers–Kronig transform in (7.20). So we see that the static, zero-frequency response to an external field contains information on the equal-time spin correlations; however, it must be remembered that this is only true for effectively classical systems in which the predominant weight in the spectral density is at frequencies smaller than T; it is not true in general. For the present situation, the value of $S(k)$ follows immediately from (10.69) and (10.70):

$$T \chi(k) = S(k) = N_0^2 \frac{2\xi_c}{1 + k^2 \xi_c^2}. \tag{10.77}$$

Thus the delta function in $S(k)$ implied by (7.33) has been broadened on a momentum scale ξ_c^{-1}, and $S(0)$ takes an exponentially large value proportional to $\xi_c \sim e^{\Delta/T}$.

Turning to the broadening in $S(k, \omega)$, we find it useful to introduce the parameterization

$$\frac{2T}{\omega} \text{Im} \chi(k, \omega) = S(k, \omega) = T \chi(k) \tau_\varphi \Phi_{\text{Sc}}(k\xi_c, \omega\tau_\varphi), \tag{10.78}$$

where Φ_{Sc} is a universal scaling function whose form follows from a Fourier transform of (10.74). We have inserted the prefactors in front of Φ_{Sc} because then it follows easily from (10.76) that its frequency integral has a fixed normalization

$$\int \frac{d\bar{\omega}}{2\pi} \Phi_{\text{Sc}}(\bar{k}, \bar{\omega}) = 1. \tag{10.79}$$

We use scaling forms such as (10.78) at several other occasions in this book. We perform a numerical Fourier transform of (10.74) and the result for Φ_{Sc} is shown in Fig. 10.5. We see that the dynamic structure factor has a large peak of order $N_0^2 \xi_c \tau_\varphi \sim e^{2\Delta/T}$ and decays

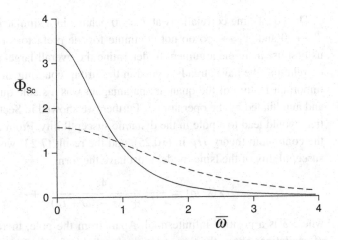

Fig. 10.5 The scaling function $\Phi_{Sc}(\bar{k}, \bar{\omega})$, appearing in (10.78), as a function of $\bar{\omega}$ at $\bar{k} = 0$ (full line) and $\bar{k} = 1.5$ (dashed line). This describes the broadening of the delta function in the dynamic structure factor in (7.33) at $0 < T \ll \Delta$.

monotonically to zero on a frequency scale $\sim \tau_\varphi^{-1}$ and on a momentum scale $\sim \xi_c^{-1}$. The frequency width of Φ_{Sc} broadens with increasing wavevector, but its maximum remains at $\bar{\omega} = 0$.

The existence here of a classical reduced scaling function describing relaxation of the order parameter reflects an important underlying physical property: the clear separation of scales at which quantum and thermal fluctuations are dominant. Quantum fluctuations are paramount at distance scales up to c/Δ and these cause a reduction in the ordered moment from unity to N_0. The influence of thermal fluctuations is not felt until the much larger scale ξ_c, where the excitations behave classically except during collisions.

10.4.2 Low T on the quantum paramagnetic side, $\Delta < 0$, $T \ll |\Delta|$

In a study of the model of Chapter 13, Chakravarty, Halperin, and Nelson [75] called the analogous regime "quantum disordered" [75]. However, as we have already noted and as we show below, this nomenclature does not capture the long-range time correlations associated with the exponentially large τ_φ in this regime.

We begin by describing the equal-time correlations. We need to take the $s \to -\infty$ limit of the functions $F_I(s)$, $G_I(s)$; from these limits we find

$$C(x, 0) = \frac{ZT}{|\Delta|^{3/4}} e^{-|x|/\xi}, \quad |x| \to \infty \text{ at fixed } 0 < T \ll |\Delta|, \tag{10.80}$$

with the correlation length ξ given by

$$\xi^{-1} = \frac{|\Delta|}{c} + \left(\frac{2|\Delta|T}{\pi c^2}\right)^{1/2} e^{-|\Delta|/T}. \tag{10.81}$$

Hence correlations decay exponentially on a scale $\sim c/|\Delta|$, and there is no long-range order.

The equal-time correlations at $T = 0$ behave in a similar manner, although the limits $T \to 0$ and $|x| \to \infty$ do not commute for the prefactor of the exponential decay. Let us first use a simple argument to determine the overall large-x dependence of the $T = 0$ correlation. We have already argued in the strong-coupling analysis of Section 5.2 that an important feature of the quantum paramagnet was a stable quasiparticle that was created and annihilated by the operator $\hat{\sigma}^z$. Further, we showed in Section 7.2.1 that this quasiparticle would lead to a pole in the dynamic susceptibility. From the relativistic invariance of the continuum theory H_F in (10.25), and the result (7.23) we conclude that the dynamic susceptibility of the Ising model must have the form

$$\chi(k, \omega) = \frac{\mathcal{A}}{c^2 k^2 + \Delta^2 - (\omega + i\delta)^2} + \cdots, \quad T = 0, \tag{10.82}$$

where δ is a positive infinitesimal. Apart from the pole, there will also be a continuum of excitations above the three-particle threshold of $\omega = 3\Delta$, as indicated by (7.27); these are represented by the ellipses in (10.82). The scale factor $\mathcal{A}$ is the quasiparticle residue, and we will obtain its value shortly. First, we use (10.82) to deduce the $T = 0$ equal-time correlations. This is most simply done by first analytically continuing (10.82) to imaginary frequencies ω_n, and then using the inverse of the definition (7.3). This gives us

$$C(x, 0) = \mathcal{A} \int \frac{d\omega}{2\pi} \int \frac{dk}{2\pi} \frac{e^{-ikx}}{\omega^2 + c^2 k^2 + \Delta^2}$$
$$= \frac{\mathcal{A}}{\sqrt{8\pi c |\Delta||x|}} e^{-\Delta|x|/c}, \quad |x| \to \infty \text{ at } T = 0. \tag{10.83}$$

Comparing this result with (10.80) and (10.81), we see that the two results differ in the power of $|x|$ that appears in the prefactor of the exponential. This is acceptable because the two cases involve different orders of limits of $T \to 0$ and $|x| \to \infty$, and there is no mathematical requirement that the orders of limit commute – in (10.83) we have sent $T \to 0$ first, whereas the limit $|x| \to \infty$ was taken first in (10.80).

To complete the description of the equal-time correlators we need to specify the value of $\mathcal{A}$. This requires a microscopic lattice calculation of the type considered in Section 10.3; an analysis of the large-n limit of T_n at $T = 0$ was carried out by McCoy [327] and Pfeuty [378], and comparing their results with (10.83) we can deduce that

$$\mathcal{A} = 2cZ|\Delta|^{1/4}, \tag{10.84}$$

where we recall that $Z = J^{-1/4}$ for the nearest-neighbor model H_I in (10.1). Hence the residue vanishes at the critical point $\Delta = 0$, where the quasiparticle picture breaks down, and we have a completely different structure of excitations. The relationship (10.84) also defines the value of Z on the quantum paramagnetic side in terms of the observables $\mathcal{A}$ and Δ; this complements the result (10.61), which defined Z on the magnetically ordered side.

The above is an essentially complete description of the correlations and excitations of the quantum paramagnetic ground state. We turn to the dynamic properties at $T > 0$. At nonzero T, there will be a small density of quasiparticle excitations that will behave classically for the same reasons as in Section 10.4.1: their mean spacing is much larger

than their de Broglie wavelength. The collisions of these thermally excited quasiparticles
lead to a broadening of the delta function pole in (10.82). The form of this broadening can
be computed exactly in the limit $T \ll |\Delta|$ using a semiclassical approach similar to that
employed for the ordered side [442]. The argument again employs a semiclassical path-
integral approach to evaluating the correlator in (10.2). The key observation is that we may
consider the operator $\hat{\sigma}^z$ to be given by

$$\hat{\sigma}^z(x, t) = (2cZ|\Delta|^{1/4})^{1/2}(\psi(x, t) + \psi^\dagger(x, t)) + \cdots, \tag{10.85}$$

where $\psi^\dagger$ is the operator that creates a single-particle excitation from the ground state, and
the ellipses represent multiparticle creation or annihilation terms, which are subdominant
in the long-time limit. This representation may also be understood from the $g \gg 1$ picture
discussed earlier, in which the single-particle excitations were $|-\rangle$ spins: the $\hat{\sigma}^z$ operator
flips spins between the $\pm x$ directions, and therefore creates and annihilates quasiparticles.

Because the computation of the nonzero T relaxation is best done in real space and time,
let us first write down the $T = 0$ correlations in this representation. We define $K(x, t)$ to
be the $T = 0$ correlator of the order parameter:

$$
\begin{aligned}
K(x, t) &\equiv \langle \hat{\sigma}^z(x, t)\hat{\sigma}^z(0, 0)\rangle_{T=0} \\
&= \int \frac{dk}{2\pi} \frac{cZ|\Delta|^{1/4}}{\varepsilon_k} e^{i(kx-\varepsilon_k t)} \\
&= \frac{Z|\Delta|^{1/4}}{\pi} K_0(|\Delta|(x^2 - c^2 t^2)^{1/2}/c),
\end{aligned} \tag{10.86}
$$

where K_0 is the modified Bessel function. This result is obtained by the Fourier transform
of (10.82) and (7.20). Note that for $t > |x|/c$, the Bessel function has imaginary argument
and is therefore complex and oscillatory. Indeed, (10.86) has a simple interpretation as the
spacetime Feynman propagator of a single relativistic particle in one dimension; this can
be made more evident by looking at the nonrelativistic limit of (10.86) well within the light
cone $x \ll ct$; in this case (10.86) reduces to

$$K(x, t) = Z|\Delta|^{1/4} e^{i\Delta t}\left(\frac{1}{2\pi i \Delta t}\right)^{1/2} \exp\left(i\frac{|\Delta|x^2}{2c^2 t}\right), \tag{10.87}$$

which is the familiar Feynman propagator of a nonrelativistic particle of mass $|\Delta|/c^2$; the
leading oscillatory term $\sim e^{i\Delta t}$ represents the common "rest mass" energy of all the particles. Well outside the light cone, $x \gg ct$, (10.86) reduces to the equal-time correlator
obtained earlier in (10.83); here the correlations become exponentially small. Our primary
interest is the $T > 0$ properties of the correlations within the light cone, where the correlations are large and oscillatory (corresponding to the propagation of real particles) and
display interesting semiclassical dynamics.

Now we consider the $T \neq 0$ evaluation of (10.2) in the same semiclassical path-integral
approach that was employed earlier in Section 10.4.1. Again we are dealing with semiclassical particles, although the physical interpretation of these particles is quite different: they are quasiparticle excitations above a quantum paramagnet, and not domain walls
between magnetically ordered regions. As in Section 10.4.1 the path-integral representation of (10.2) leads to *two* sets of paths: one forward and the other backwards in time.

However, there is a special trajectory that moves only forward in time: This is the trajectory representing the particle that is created by the first $\hat{\sigma}_0^z$ and annihilated at the second $\hat{\sigma}_i^z$. The inverse process in which the first $\hat{\sigma}_0^z$ annihilates a pre-existing thermally excited particle can be neglected because the probability of finding such a particle at a given location is exponentially small. Also, as in the semiclassical limit, the forward and backward trajectories of the thermally excited particles are expected to be the same; the particle on the trajectory created by the $\hat{\sigma}_0^z$ must be annihilated at the $\hat{\sigma}_i^z$, for otherwise the initial and final states in the trace in (10.2) will not be the same. This reasoning leads to a spacetime snapshot of the trajectories that is the same as in Fig. 10.3, but its physical interpretation is very different. The dashed line represents the trajectory of a particle created at $(0, 0)$ and annihilated at (x, t), and $\pm$ signs in the domains should be ignored. In the absence of any other particles this dashed line would contribute the $T = 0$ Feynman propagator above, $K(x, t)$, to $\langle \hat{\sigma}^z(x, t)\hat{\sigma}^z(0, 0)\rangle$. The scattering of the background thermally excited particles (represented by the full lines in Fig. 10.3 (which are not domain walls)) introduces factors of the S matrix element $S_{k_1 k_2}$ in (5.14) at each collision; as the dashed line only propagates forward in time, the S matrix elements for collisions between the dashed and full lines are not neutralized by a complex conjugate partner. All other collisions occur both forward and backward in time, and therefore they contribute $|S_{k_1 k_2}|^2 = 1$. Using the low momentum value $S_{k_1 k_2} = -1$, we see that the contribution to $\langle \hat{\sigma}^z(x, t)\hat{\sigma}^z(0, 0)\rangle$ from the set of trajectories in Fig.10.3 equals $(-1)^j K(x, t)$, where j is the number of full lines intersecting the dashed line. Remarkably, the $(-1)^j$ factor is precisely the term that appeared in the analysis at low T on the magnetically ordered side in Section 10.4.1, although for very different physical reasons. We can carry out the averaging over all trajectories as in the analysis leading to (10.69), thereby obtaining one of our main results for low-T dynamic correlations on the paramagnetic side [442]:

$$C(x, t) = K(x, t)R(x, t), \tag{10.88}$$

with $K(x, t)$ given by (10.86), and $R(x, t)$ again specified by the second result of (10.70). Now note that in going from the magnetically ordered to the quantum paramagnetic side the only change in parameters has been the change in sign of Δ. The dispersion spectrum ε_k is invariant under this change of sign, and so we can use precisely the same expressions for the relaxation function $R(x, t)$ as before: the result (10.70) still applies, and we can continue to use the expression (10.74) for the scaling function Φ_R. Furthermore, the characteristic space and time scales, ξ_c and τ_φ, on which R varies are still given by (10.63) and (10.71), respectively. (Note that we were careful to insert the absolute value $|\Delta|$ in these expressions, even though that was not needed for the magnetically ordered side.)

An interesting feature of the result (10.88) is that it clearly displays the separation in scales at which quantum and thermal effects act. Quantum fluctuations determine the oscillatory, complex function $K(x, t)$, which gives the $T = 0$ value of the correlator. Exponential relaxation of spin correlations occurs at longer scales $\sim \xi_c, \tau_\varphi$ and is controlled by the classical motion of particles and a purely real relaxation function. This relaxation is expected to lead to a broadening of the quasiparticle pole with widths of order $\xi_c^{-1}, \tau_\varphi^{-1}$ in momentum and energy space. We can consider the presence of a quasiparticle delta function in the spectral density of excitations above the ground state as a representation of the

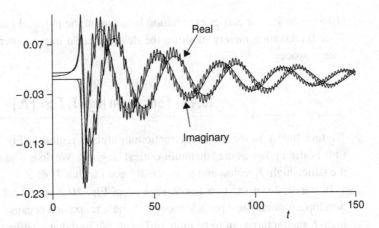

Theoretical and numerical results from [442] for the correlator $\langle \hat{\sigma}^z(x, t)\hat{\sigma}^z(0, 0)\rangle$ of H_I at $x = 30$ with $J = 1$, $g = 1.1$ (therefore $\Delta = -0.2$), $T = 0.1$; thus the system is in the low T region on the paramagnetic side of Fig. 10.2. The numerical data were obtained for a lattice size $L = 512$ with free boundary conditions; it has a "ringing" at high frequency owing to the lattice cutoff. The theoretical prediction is from the continuum theory prediction in (10.88) and is represented by the smoother curve. The envelope of the numerical curve fits the theoretical prediction well.

perfect quantum coherence in the ground state, and so for $T > 0$ its width in energy is a natural measure of the inverse phase relaxation time $1/\tau_\varphi$. In Fig. 10.6 we compare the predictions of (10.88) with numerical results on a lattice of size $L = 512$. As expected, there is a rapid oscillatory part representing the Feynman propagator of a single particle and an envelope that decays exponentially at a much slower rate. The theoretical curve was determined from the continuum expression for $K(x, t)$, but the full lattice form for ε_k was used. The theory agrees well with the numerics; some differences are visible for small x, outside the light cone, but this is outside the domain of validity of (10.88).

We can also compute the structure factor $S(k, \omega)$ from (10.88) by taking the Fourier transform as in (10.4). This will mainly have weight at positive frequencies $\omega \approx \varepsilon_k \approx |\Delta| + c^2 k^2/(2|\Delta|)$, corresponding to the creation of a quasiparticle by the external probe. It is not possible to analytically perform the Fourier transform in general, but the leading term in an asymptotic expansion in $T/|\Delta|$ can be obtained in closed form. For reasons discussed in [107], it turns out that because $\xi_c/c\tau_\varphi = (2T/\pi|\Delta|)^{1/2} \ll 1$, the slower relaxation in time dominates the Fourier transform, and we can simply evaluate the Fourier transform while ignoring the x dependence of R:

$$S(k, \omega) \approx \int dt \int dx\, K(x, t)R(0, t)e^{-i(kx - \omega t)}$$

$$= \left(\frac{2cZ|\Delta|^{1/4}}{\varepsilon_k}\right) \frac{1/\tau_\varphi}{(\omega - \varepsilon_k)^2 + (1/\tau_\varphi)^2}. \tag{10.89}$$

This result holds for k close enough to the band minimum, with $|k| \ll \sqrt{T\Delta}/c$; for larger k there is no alternative to complete numerical evaluation of the Fourier transform. The result

(10.89) verifies our earlier expectation based upon the physical interpretation of $1/\tau_\varphi$: the $T > 0$ relaxation merely modifies the delta function into a Lorentzian of width $1/\tau_\varphi$ in energy space.

10.4.3 Continuum high T, $T \gg |\Delta|$

We turn finally to the universal continuum high-T region of Fig. 10.2, where $T \gg |\Delta|$ (this is also known as the "quantum-critical" region). We do not have anything to say about the lattice high-T region and so implicitly assume that $T \ll J$.

In our study of the two low-T regions of Fig. 10.2 we found that it was possible to develop a semiclassical particle picture of phase relaxation because $1/\tau_\varphi \ll T$. The present high-T region turns out to be quite different. We find that no effective classical model can provide an adequate description of the dynamics because the phase relaxation time is quite short. In particular we find that $1/\tau_\varphi \sim T$, so that, as noted earlier, this regime is maximally incoherent. The de Broglie wavelength of the effective excitations is of the same order as their spacing; this holds whether we consider the excitations to be the domain walls of the magnetically ordered phase or the flipped spins of the quantum paramagnet. Consequently, it is difficult to disentangle quantum and thermal effects because they both play an equally important role. The large class of classical models discussed in the review of Hohenberg and Halperin [223] cannot, therefore, be applied in the present context. This novel regime of dynamics was first discussed in [440] and [86] in the context of the model of Chapter 13 and was dubbed *quantum relaxational*. We find it more convenient to introduce it in this book in the simpler context of the Ising chain.

As in the previous subsections, we begin by understanding the structure of the equal-time correlations. Right at the critical point, $\Delta = 0$, $g = g_c$, this high-T regime extends all the way down to $T = 0$. At the $T = 0$, $g = g_c$ quantum critical point, we can deduce the form of the correlator by a simple scaling analysis. As the ground state is scale invariant at this point, the only scale that can appear in the equal-time correlator is the spatial separation x; from the scaling dimension of $\hat{\sigma}^z$ in (10.38), we then know that the correlator must have the form

$$C(x, 0) \sim \frac{1}{(|x|/c)^{1/4}}, \qquad \text{at } T = 0, \Delta = 0. \tag{10.90}$$

Actually, we can also include time-dependent correlations at this level without much additional work. We know that the continuum theory (10.28) is Lorentz invariant, and so we can easily extend (10.90) to the imaginary-time result

$$C(x, \tau) \sim \frac{1}{(\tau^2 + x^2/c^2)^{1/8}}, \qquad \text{at } T = 0, \Delta = 0. \tag{10.91}$$

This result can also be understood by referring to the classical $D = 2$ Ising model in (3.1). In this context, (10.91) is simply the statement that correlations are isotropic with all D dimensions, and so the long-distance correlations depend only upon the Euclidean distance between two points.

We extend the result (10.91) to $T > 0$ by a trick that we quote without proof. (Later in Chapter 20 we note an explicit derivation using the bosonization method.) The basic point is that the $\Delta = 0$ continuum theory (10.28), in addition to being scale and Lorentz invariant, is also invariant under conformal transformations of spacetime [244]. Turning on a $T > 0$ is equivalent, in imaginary time, to placing the theory $\mathcal{L}_I$ on a spacetime manifold that is a cylinder of circumference $1/T$. However, it is known that one can conformally map the cylinder to the infinite plane, which allows one to obtain a remarkable and exact relationship between $T = 0$ and $T > 0$ correlators in imaginary time at the critical coupling $\Delta = 0$. This mapping was explicitly obtained in (10.47) where we simply noted it as an interesting property of a fermionic correlator we were able to obtain explicitly for $T > 0$. The implication of this discussion is that the same mapping can also be applied to (10.91), allowing us to obtain the correlators at $T > 0$:

$$C(x, \tau) \sim T^{1/4} \frac{1}{[\sin(\pi T(\tau - ix/c)) \sin(\pi T(\tau + ix/c))]^{1/8}} \qquad \Delta = 0. \qquad (10.92)$$

We can obtain an independent confirmation of this result by specializing to the equal-time case again and comparing to our earlier results in Section 10.3; we have from (10.92)

$$C(x, 0) \sim \frac{T^{1/4}}{[\sinh(\pi T|x|/c)]^{1/4}}$$

$$\sim T^{1/4} \exp\left(-\frac{\pi T|x|}{4c}\right), \qquad \text{as } |x| \to \infty. \qquad (10.93)$$

Compare this with the precise results for this regime quoted earlier in (10.48), where using the values $F_I(0) = \pi/4$ (from evaluation of (10.50)) and $G_I(0) = 0.858714569\ldots$ we have

$$\lim_{|x| \to \infty} C(x, 0) = Z T^{1/4} G_I(0) \exp\left(-\frac{\pi T|x|}{4c}\right) \quad \text{at } \Delta = 0. \qquad (10.94)$$

The two results, obtained by very different methods, are in perfect agreement. We can combine (10.94) with (10.92) to determine the prefactor in (10.92) and thus obtain our final closed-form result for the universal two-point correlator at $\Delta = 0$:

$$C(x, \tau) = Z T^{1/4} \frac{2^{-1/4} G_I(0)}{[\sin(\pi T(\tau - ix/c)) \sin(\pi T(\tau + ix/c))]^{1/8}}. \qquad (10.95)$$

As expected, this result is of the scaling form (10.39), and indeed it completely determines the function Φ_I for the case where its last argument is zero. It is the leading result everywhere in the continuum high-T region of Fig. 10.2. Note that this result has been obtained in imaginary time. Normally, as we have noted earlier, such results are not terribly useful in understanding the long real-time dynamics at $T > 0$ because the analytic continuation is ill-posed. However, in the present case, we have the *exact* expression, and so the analytic continuation is a useful tool.

Now let us turn to a physical interpretation of the main result (10.95). Consider first the case $T = 0$. By a Fourier transformation of (10.91), and using the normalization constant implied by (10.95), we obtain the dynamic susceptibility

$$\chi(k, \omega) = Z(4\pi)^{3/4} G_I(0) \frac{\Gamma(7/8)}{\Gamma(1/8)} \frac{c}{(c^2 k^2 - (\omega + i\delta)^2)^{7/8}}, \quad T = 0, \Delta = 0 \quad (10.96)$$

with δ a positive infinitesimal. Note that this function has a *branch cut* in the complex ω plane at $\omega = ck$; this is to be contrasted with the simple pole-like structure that appeared in the quantum paramagnet at $T = 0$ in (10.82). Instead the answer corresponds precisely to our expectations from Section 7.2.2, from where we see that the branch cut at the quantum critical point is a direct consequence of the anomalous dimension of $\hat{\sigma}^z$ in (10.38), which led to the noninteger powers in (10.91) and (10.96). We plot $\text{Im}\chi(k, \omega)$ in Fig. 10.7, which should be compared to Fig. 7.2. There are no delta functions in the spectral density, as there were in the quantum paramagnet in Section 7.2.1, indicating that the $\hat{\sigma}^z$ operator has negligible overlap with the single fermion quasiparticle state of Section 10.1. Instead, we have a critical continuum above a branch cut arising from a superposition of states with an arbitrary number of fermionic quasiparticles. However, the presence of sharp thresholds and singularities indicates that there is still perfect phase coherence, as there must be in the ground state. It is also interesting to think about how the $T = 0$ spectral density crosses over from the form of delta-function+multi-particle continua discussed in Section 7.2.1 as characteristic of the quantum paramagnet, to the critical continuum in Fig. 10.7. Consider, for instance, the case $k = 0$. In the quantum paramagnet, we have a quasiparticle delta function at $\omega = \Delta$, a continuum above the three-particle threshold at $\omega = 3\Delta$, another above the five-particle threshold at $\omega = 5\Delta$, and so on. As we approach the critical point with $\Delta \to 0$, all these continua come crashing down in energy and their limiting superposition leads to the critical form shown in Fig. 10.7.

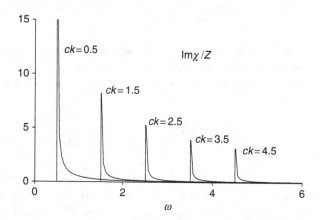

Fig. 10.7 Spectral density, $\text{Im}\chi(k, \omega)/Z$, of H_I at its critical point $g = 1$ ($\Delta = 0$) at $T = 0$, as a function of frequency ω, for a set of values of k.

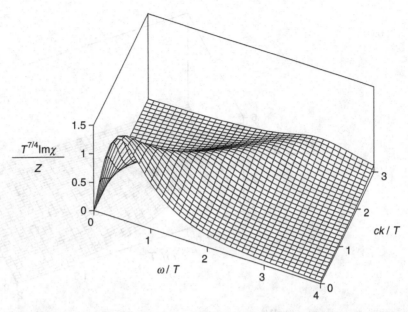

$\dfrac{T^{7/4}\mathrm{Im}\chi}{Z}$

ck/T

ω/T

Fig. 10.8   The same observable as in Fig. 10.7, $T^{7/4}\mathrm{Im}\chi(k,\omega)/Z$, but for $T \neq 0$. This is the leading result for $\mathrm{Im}\chi$ for $T \gg |\Delta|$ (i.e. in the high-T region of Fig. 10.2). All quantities are scaled appropriately with powers of T, and the absolute numerical values of both axes are meaningful.

Now let us turn to $T > 0$. We Fourier transform (10.95) to obtain $\chi(k,\omega_n)$ at the Matsubara frequencies ω_n and then analytically continue to real frequencies. This gives us the leading result for $\chi(k,\omega)$ in the high-T region:

$$\chi(k,\omega) = \frac{Zc}{T^{7/4}}\frac{G_I(0)}{4\pi}\frac{\Gamma(7/8)}{\Gamma(1/8)}\frac{\Gamma\left(\frac{1}{16}-i\frac{\omega+ck}{4\pi T}\right)\Gamma\left(\frac{1}{16}-i\frac{\omega-ck}{4\pi T}\right)}{\Gamma\left(\frac{15}{16}-i\frac{\omega+ck}{4\pi T}\right)\Gamma\left(\frac{15}{16}-i\frac{\omega-ck}{4\pi T}\right)}. \quad (10.97)$$

We show a plot of $\mathrm{Im}\chi$ in Fig. 10.8. This result is the finite-T version of Fig. 10.7. Note that the sharp features of Fig. 10.7 have been smoothed out on the scale T, and there is nonzero absorption at all frequencies. For $\omega, ck \gg T$ there is a well-defined "reactive" peak in $\mathrm{Im}\chi$ at $\omega \approx ck$ (Fig. 10.8) rather like the $T = 0$ critical behavior of Fig. 10.7. However, the low-frequency dynamics is quite different, and for $\omega, ck \ll T$ we cross over to the *quantum relaxational* regime [86]. This is made clear by an examination of the quantity $\mathrm{Im}\chi(k,\omega)/\omega$ as a function of ω/T and ck/T; note from (7.20) that for $\omega \ll T$ this quantity is proportional to the dynamic structure factor, $S(k,\omega)$ (defined in (10.4)), which is in turn proportional to the neutron scattering cross-section. (We prefer to work with $\mathrm{Im}\chi(k,\omega)/\omega$ rather than $S(k,\omega)$ because the former is an even function of ω, while the latter is not; in any case, the two are practically indistinguishable for low frequencies.) We show a plot of $\mathrm{Im}\chi(k,\omega)/\omega$ in Fig. 10.9 (notice that Fig. 10.9 is simply Fig. 10.8 divided by ω). Now the reactive peaks at $\omega \sim ck$ are just about invisible, and the spectral density is dominated by a large relaxational peak at zero frequency. We can understand the structure

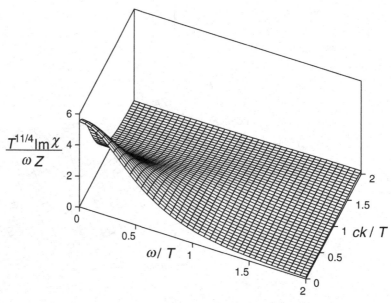

Fig. 10.9 The spectral density $T^{11/4} \mathrm{Im}\chi(k,\omega)/\omega Z$ as a function of ω/T and ck/T. Note that this is simply the quantity in Fig. 10.8 divided by ω. The reactive peaks at $\omega \approx ck$ in Fig. 10.8 are essentially invisible, and the plot is dominated by a large relaxational peak at zero wavevector and frequency.

of Fig. 10.9 by expanding the inverse of (10.97) in powers of k and ω; this expansion has the form

$$\chi(k,\omega) = \frac{\chi(0)}{1 - i(\omega/\omega_1) + k^2 \tilde{\xi}^2 - (\omega/\omega_2)^2}, \tag{10.98}$$

where $\omega_{1,2}$ and $\tilde{\xi}$ are parameters characterizing the expansion, and where we recall from (10.97) that $\chi(0) \sim T^{-7/4}$. For k not too large, the ω dependence in (10.98) is simply the response of a strongly damped harmonic oscillator: this is the reason we have identified the low-frequency dynamics as "relaxational." The function in (10.98) provides an excellent description of the spectral response in Fig. 10.9. We determine the best fit values of the parameters $\omega_{1,2}$ and $\tilde{\xi}$ by minimizing the mean square difference between the values of $\mathrm{Im}\chi(k,\omega)/\omega$ given by (10.98) and (10.97) over the range $0 < \omega < 2T$ and $0 < ck < 2T$ and obtain:

$$\omega_1 = 0.396\,T,$$
$$\omega_2 = 0.795\,T,$$
$$\tilde{\xi} = 1.280\,c/T. \tag{10.99}$$

The quality of the fit is shown in Figs. 10.10 and 10.11, where we compare the predictions of (10.97) and (10.98) for $\mathrm{Im}\chi(k,\omega)/\omega$ at $\omega = 0$ as a function of ck/T, and at $ck/T = 0, 1.5$ as a function of ω/T, respectively. For $k = 0$ ($\omega = 0$) there is a large overdamped

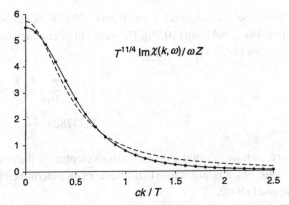

Fig. 10.10 Comparison of the predictions of (10.97) (dots) and (10.98) (solid line) for $\mathrm{Im}\chi(k,\omega)/\omega$ at $\omega = 0$ as a function of ck/T. The best-fit parameters in (10.99) were used. The function (10.98) yields the *square* of a Lorentzian as a function of k; a best fit by just a Lorentzian is also shown (dashed line) and is much poorer.

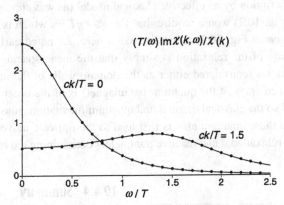

Fig. 10.11 Comparison of the predictions of (10.97) (dots) and (10.98) (line) for $(T/\omega)\mathrm{Im}\chi(k,\omega)/\chi(k)$ as a function of ω/T at $ck/T = 0$, 1.5. The dispersion relation (7.20) implies that the area under both curves for $-\infty < \omega < \infty$ is exactly π. Notice also the similarity of the quantity plotted to the scaling function considered in (10.78) and Fig. 10.5; however, in the present case $S(k) \neq T\chi(k)$ as the dynamics is not effectively classical – in particular $S(0) = 1.058T\chi(0)$. The overall magnitude of $\mathrm{Im}\chi$ at $ck/T = 1.5$ is smaller than this figure would suggest, as $\chi(k = 1.5)/\chi(0) = 0.216$.

peak at $\omega = 0$ ($k = 0$), but a weak reactive peak at $\omega \sim ck$ makes an appearance at larger wavevectors or frequencies.

For an alternative, and more precise, characterization of the relaxational dynamics we can introduce the relaxation rate Γ_R defined by

$$\Gamma_R^{-1} \equiv i\chi(0) \left.\frac{\partial \chi^{-1}(0,\omega)}{\partial \omega}\right|_{\omega=0} = \frac{S(0,0)}{2T\chi(0)}, \qquad (10.100)$$

where the second relation follows from (7.20). We have chosen this definition for the informative functional form (10.98), $\Gamma_R = \omega_1$, the frequency characterizing the damping. However, using (10.97) we determine:

$$\Gamma_R = \left(2 \tan \frac{\pi}{16}\right) \frac{k_B T}{\hbar}$$

$$\approx 0.397825 \frac{k_B T}{\hbar}, \tag{10.101}$$

where we have inserted physical units to emphasize the universality of the result. Note that the value of Γ_R is quite close to the value of ω_1 determined by the least square minimization discussed above.

The rate Γ_R is a satisfactory measure of how thermal effects have rounded out the sharp, $T = 0$ phase-coherent structure in the dynamic susceptibility in Fig. 10.7: we can therefore identify it with the phase coherence rate $1/\tau_\varphi$. At the scale of the characteristic rate Γ_R, the dynamics of the system involves intrinsic quantum effects that cannot be neglected. Description by an effective classical model (as was appropriate in both the low-T regions of Fig. 10.2) would require that $\Gamma_R \ll k_B T/\hbar$, which is thus not satisfied in the high-T region of Fig. 10.2 under discussion here. As noted earlier, the reason for the quantum nature of the relaxation is simply that the mean spacing between the thermally excited particles (considered either as the domain walls of the magnetically ordered state or the flipped spins of the quantum paramagnet) is of the order of their de Broglie wavelength, and so the classical thermal and quantum fluctuations must be treated on an equal footing. It is these quantum effects that lead to the intricate universal numerical relation between the relaxational and reactive parameters determining the response in (10.97) and (10.98).

10.4.4 Summary

Our detailed study of the $T > 0$ crossovers in the vicinity of the quantum critical point of the Ising chain has led to a rich variety of different physical regimes, and so it is useful to summarize their main properties. Such a summary is contained in our earlier Fig. 10.2 and in Fig. 10.12 and Table 10.1. At short enough times or distances in all three regions of Fig. 10.2, the systems display critical fluctuations characterized by the dynamic susceptibility (10.96). The regions are distinguished by their behaviors at the low frequencies and momenta. In both the low-T regimes of Fig. 10.2 (on the magnetically ordered and quantum paramagnetic sides), the long-time dynamics is relaxational and is described by effective models of quasi-classical particles; however, the physical interpretation of the particles is quite different between the two low-T regimes: they are domain walls on the magnetically ordered side but are flipped spins in the quantum paramagnet. The relaxation time, or equivalently, the phase-coherence time, is of order $(\hbar/k_B T)e^{\text{(energy gap)}/k_B T}$ and is therefore much longer than $\hbar/k_B T$; it is this condition which ensures that quantum thermal effects act at very different scales and allows for a semiclassical description of the low-frequency dynamics. In contrast, the dynamics in the high-T region is also relaxational, but it involves quantum effects in an essential way, as was described above. In this region,

Table 10.1 Summary of physical regimes.

Values of the correlation length, ξ (defined from the exponential decay of the equal-time correlations of the order parameter), and the phase coherence time, τ_φ (defined as discussed in the respective sections), in the regimes of Fig 10.2. The two low-T regimes have interpretations in terms of quasi-classical particles, but the physical interpretations of the two particles are very different, as indicated.

	Low T (magnetically ordered) quasi-classical particles –domain walls	T Continuum high (quantum critical)	Low T (quantum paramagnetic), quasi-classical particles –flipped spins
ξ	$\left(\dfrac{\pi c^2}{2\Delta T}\right)^{1/2} e^{\Delta/T}$	$\dfrac{4c}{\pi T}$	$\dfrac{c}{|\Delta|}$
τ_φ	$\dfrac{\pi}{2T} e^{\Delta/T}$	$\dfrac{\cot(\pi/16)}{2T}$	$\dfrac{\pi}{2T} e^{|\Delta|/T}$

Low T (magnetically ordered)

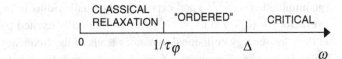

Continuum high T (quantum critical)

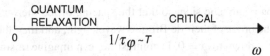

Low T (quantum paramagnetic)

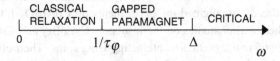

Fig. 10.12   Crossovers as a function of frequency for the Ising model in the different regimes of Fig. 10.2. The high frequency critical fluctuations are present in all regimes and are characterized by (10.96). The two classical relaxational regimes are described by multiple collisions of thermally excited quasi-classical particles; the physical correlations in these two regimes are quite different but are described by the same relaxation function R in (10.69). The quantum relaxation is described by (10.97) and the relaxation rate by (10.100). The "ordered" regime is in quotes, because there is no long-range order, and the system only appears ordered between spatial scales c/Δ and $c\tau_\varphi$. In the low-T regions $1/\tau_\varphi \sim Te^{-|\Delta|/T}$.

the spacing between the thermally excited particles is of the order of their de Broglie wavelength, and the phase relaxation time is of the order of $\hbar/k_B T$.

The ease with which our expressions for the phase coherence times τ_φ in (10.71) and (10.101) have been obtained belies their remarkable nature. Note that we have worked in a closed Hamiltonian system, evolving unitarily in time with the operator $e^{-iH_I t/\hbar}$, from an initial density matrix given by the Gibbs ensemble at a temperature T. Yet, we have obtained relaxational behavior at low frequencies and determined *exact* values for a dissipation constant. In contrast, in the theory of dynamics near classical critical points [223], a statistical relaxation dynamics is postulated in a rather ad hoc manner, and the relaxational constants are treated as phenomenological parameters to be determined by comparison with experiments. Our subsequent discussions of more complicated models in higher dimensions also only consider deterministic unitary evolution from an initial density matrix, but we are only able to obtain approximate values of dissipation constants.

It is also worth contrasting the small k, ω behavior of the dynamic structure factor, $S(k, \omega)$, in the three regimes of Fig. 10.2. At low T on the quantum paramagnetic side, there is a sharp quasiparticle peak at $\omega \sim |\Delta|$ whose frequency width is exponentially small ($\sim Te^{-|\Delta|/T}$); this peak arises from the creation of a "flipped spin" quasiparticle, and its width is a consequence of collisions with a dilute gas of pre-existing quasiparticles. It is strong for the case of energy absorption, $\omega > 0$, and has exponentially small weight on the energy emission side, $\omega < 0$. In the low-T regime on the magnetically ordered side, the peak in $S(k, \omega)$ is at $\omega, k = 0$, but it is now symmetric in ω and has an exponentially large amplitude ($\sim e^{2\Delta/T}$) and exponentially small widths in frequency ($\sim Te^{-\Delta/T}$) and wavevector ($\sim (c/\sqrt{T\Delta})e^{-\Delta/T}$). Now the thermally excited particles are domain walls, and their low-energy collisional dynamics leads to the frequency broadening of the peak. Finally, in the high-T regime, $S(k, \omega)$ is not a symmetric function of ω, but its dominant structure is near $\omega = 0$ and has a width of order T. There are two "dual" physical perspectives on the form of $S(k, \omega)$ at high T. First, we can imagine raising the T on the quantum paramagnetic side, so that the spacing of the thermally excited quasiparticles becomes of the order of their de Broglie wavelength, and then the quasiparticle peak at $\omega = \Delta$ broadens down to $\omega = 0$. In contrast, we can imagine raising T on the magnetically ordered side, so that the domain wall spacing is of the order of their de Broglie wavelength, in which case the quasiparticle motion has a relaxation rate of the order of T.

We conclude this chapter by mentioning some experimental applications. We have already highlighted the experiments on $CoNb_2O_6$ [90] in Section 1.3. We also mention here earlier studies on the insulators $CsCoBr_3$ and $CsCoCl_3$. The Co ions form chains of *antiferromagnetically* interacting Ising spins. Their effective Hamiltonian is not the Ising chain in a transverse field, but the dynamics and structure of the domain-wall excitations above the magnetically ordered ground state are essentially identical to those described in our discussion in Section 10.4.1. Neutron scattering studies [171, 355, 549] have focused on the temperature induced broadening of the $T = 0$ delta function in (7.33).

11

Quantum rotor models: large-N limit

This chapter turns to the $O(N)$ quantum rotor studied earlier in Chapter 6. We extend the earlier results to $T > 0$ aided by an exact solution obtained in the $N \to \infty$ limit.

The quantum Ising model studied in Chapter 10 had a discrete Z_2 symmetry. An important new ingredient in the rotor models is the presence of a continuous symmetry: the physics is invariant under a uniform, global $O(N)$ transformation on the orientation of the rotors, which is broken in the magnetically ordered state. Thus we have to use ideas on the *spin stiffness* which were introduced in Chapter 8. Apart from this, much of the technology and the physical ideas introduced earlier for the $d = 1$ Ising chain generalize straightforwardly, although we are no longer able to obtain exact results for crossover functions at finite N. The characterization of the physics in terms of three regions separated by smooth crossovers, the high-T and the two low-T regions on either side of the quantum critical point, continues to be extremely useful and is again the basis of our discussion. Because we consider models in spatial dimensions $d > 1$, it is possible to have a thermodynamic phase transition at a nonzero temperature, as in Fig. 1.2b. We are particularly interested in the interplay between the critical singularities of the finite-temperature transition and those of the quantum critical point.

The large-N expansion [60, 310, 311, 485] was developed earlier in the context of the classical model, and is extended here to the quantum rotor model at $T > 0$. This chapter largely confines itself to the results obtained at $N = \infty$. The results so obtained give an adequate description of gross features of the phase diagram and some static observables, but they are quite inadequate for dynamical properties at nonzero temperatures. The latter problems are addressed in subsequent chapters.

We examine here a slight extension of the quantum rotor model (6.1):

$$H_R = \frac{J\tilde{g}}{2} \sum_i \hat{\mathbf{L}}_i^2 - J \sum_{\langle ij \rangle} \hat{\mathbf{n}}_i \cdot \hat{\mathbf{n}}_j - \mathbf{H} \cdot \sum_i \hat{\mathbf{L}}_i. \tag{11.1}$$

Recall that the N-component vector operators $\hat{\mathbf{n}}_i$, with N geq 2, are of unit length, $\hat{\mathbf{n}}_i^2 = 1$, and represent the orientation of the rotors on the surface of a sphere in N-dimensional rotor space, while the operators $\hat{\mathbf{L}}_i$ are the $N(N-1)/2$ components of the angular momentum. We phrase our physical discussion using the physically important case $N = 3$, in which case these operators satisfy the commutation relations (1.21) on each site (the operators on different sites commute); the generalization to other values of N is immediate but will not be discussed explicitly for simplicity. The form (11.1) for H_R differs from that in (6.1) by a field $\mathbf{H}$, which couples to the total angular momentum; this field should not be confused with the field $\tilde{\mathbf{h}}$ in (6.46), which coupled to the rotor orientation $\mathbf{n}$. As we see later, the field **H** does not have a familiar analog upon inverting the mapping to (3.2). It is, however, an important perturbation of the quantum rotor model that arises in many experimental applications. The total angular momentum is conserved in zero field as it commutes with H_R at **H** = **0**, and we will see that this has important implications for its scaling properties.

The study of the quantum rotor model H_R in (11.1) occupies a substantial portion of Part III of this book. The motivation for this is primarily theoretical, but important experimental connections also exist. These were discussed in Section 1.4.2 and Chapters 9, and further connections are made in Chapter 19.

The large-N expansion is set up in Section 11.1; this is followed by descriptions of the $N = \infty$ solution for $T = 0$ and $T > 0$ in Sections 11.2 and 11.3, respectively.

Most of our results are expressed in terms of the dynamic susceptibility $\chi_{\alpha\beta}(\vec{k}, \omega)$ of the order parameter **n**. As in (7.1) and (7.3), this is defined most conveniently in imaginary time:

$$C_{\alpha\beta}(x, \tau) \equiv \langle n_\alpha(\vec{x}, \tau) n_\beta(0, 0) \rangle,$$

$$\chi_{\alpha\beta}(\vec{k}, \omega_n) \equiv \int_0^{1/T} \int d^d x \, C_{\alpha\beta}(x, \tau) e^{-i(\vec{k}\cdot\vec{x} - \omega_n \tau)}, \tag{11.2}$$

where $\mathbf{n}(x_i, i\tau)$ is the imaginary-time representation of the quantum operator $\hat{\mathbf{n}}_i$. The dynamic structure factor $S_{\alpha\beta}(\vec{k}, \omega)$ is then defined as in (10.4) and related to $\chi_{\alpha\beta}$ by a relationship analogous to (7.12). For the most part, we compute χ in zero field (**H** = **0**). Our analysis of the consequences of **H** is restricted here to determining its linear response susceptibility. For reasons that will become evident when we consider the relationship between quantum rotors and quantum antiferromagnets, we call this susceptibility the *uniform susceptibility*, χ_u. It is defined by the small-H expansion of the free energy density $\mathcal{F} = -T \ln \mathcal{Z}$:

$$\mathcal{F}(\mathbf{H}) = \mathcal{F}(\mathbf{H} = \mathbf{0}) - \frac{1}{2}\chi_{u\alpha\beta}H_\alpha H_\beta + \cdots \tag{11.3}$$

11.1 Continuum theory and large-N limit

Both the continuum analysis and the study of the large-N limit are most easily done in the imaginary time path integral. At **H** = **0**, the path integral can be derived by the inverse of the mapping discussed in Section 6.4 and indeed leads to the expression (2.12) already presented. The modification necessary for **H** $\neq$ **0** can be deduced by a simple trick that relies on the fact that **H** couples to the conserved total angular momentum. It is easy to see that the only effect of **H** is to cause a uniform Bloch precession of all the rotors and that this precession can be "removed" by transforming to a rotating reference frame. Because of a nonzero **H** each rotor acquires an additional precession $\delta\hat{n}_\alpha(t) = -\epsilon_{\alpha\beta\gamma}H_\beta\hat{n}_\gamma(t)\delta t$

in a small time δt. Including this extra precession in imaginary time in (2.12) we get the partition function

$$\mathcal{Z} = \text{Tr} \exp\left(-\frac{H_R}{T}\right) \approx \int \mathcal{D}\mathbf{n}(x,\tau)\delta(\mathbf{n}^2 - 1)\exp(-\mathcal{S}_n),$$

$$\mathcal{S}_n = \frac{N}{2cg}\int d^d x \int_0^{1/T} d\tau[(\partial_\tau \mathbf{n} - i\mathbf{H}\times\mathbf{n})^2 + c^2(\nabla_x \mathbf{n})^2]. \tag{11.4}$$

We have written the coupling to $\mathbf{H}$ in the form special for $N=3$, but it should be clear that for general N one writes a term that generates rotations of $O(N)$. Note also the i in the precession term, which therefore contributes a complex phase to the weights in the partition function. As a result the field $\mathbf{H}$ has no analog in classical statistical mechanics problems in D dimensions. We will be satisfied in this chapter by simply examining the linear response of the system to a small $\mathbf{H}$, as specified by the susceptibility χ_u in (11.3). Properties beyond linear response require examining the partition function (11.4) with a nonzero H, including the complex weights. This problem is of a class we examine only in Part IV, and we defer the analysis to Section 19.4. The coupling constant

$$g = N\sqrt{\tilde{g}}a^{d-1} \tag{11.5}$$

was defined in (6.51); it has the dimensions of $(\text{length})^{d-1}$ and is the primary coupling we change to vary the physical properties of the rotor model.

Because the above action is valid only at long distances and times, there is an implicit cutoff above momenta of order $\Lambda \sim 1/a$ and frequencies of order $c\Lambda$. Our main interest here is the universal physics at scales much smaller than Λ. The following large-N analysis makes it clear that such a universal regime does exist for $d < 3$ but that additional information on cutoff scale physics is necessary for $d \geq 3$. This identifies $d = 3$ as the so-called upper-critical dimension of the model. The large-N analysis is especially suited for describing the universal physics in $d < 3$, and we restrict our attention to these cases here. Properties in dimensions $d \geq 3$ are more easily analyzed by other methods and are discussed later.

The framework of the $N = \infty$ solution [58, 79, 85, 86, 100, 214, 385, 440, 475] is quite easy to set up, at least in the phase without long-range order in the order parameter $\mathbf{n}$; we consider the case with long-range order later in this chapter. We impose the $\mathbf{n}^2 = 1$ constraint by a Lagrange multiplier, λ. The action (11.4) then becomes at $\mathbf{H} = \mathbf{0}$ (which is assumed throughout the remainder unless explicitly stated otherwise)

$$\mathcal{Z} = \int \mathcal{D}\mathbf{n}(x,\tau)\mathcal{D}\lambda(x,\tau)\exp(-\mathcal{S}_{n1}),$$

$$\mathcal{S}_n = \frac{N}{2cg}\int d^d x \int_0^{1/T} d\tau[(\partial_\tau \mathbf{n})^2 + c^2(\nabla_x \mathbf{n})^2 + i\lambda(\mathbf{n}^2 - 1)]. \tag{11.6}$$

We rescale the $\mathbf{n}$ field to

$$\tilde{\mathbf{n}} = \sqrt{N}\mathbf{n}, \tag{11.7}$$

and, as (11.6) is quadratic in the $\tilde{\mathbf{n}}$ field, it can be integrated out to yield

$$\mathcal{Z} = \int \mathcal{D}\lambda(x, \tau) \exp\left[-\frac{N}{2}\left(\text{Tr}\ln\left(-c^2\nabla^2 - \partial_\tau^2 + i\lambda\right) - \frac{i}{cg}\int_0^{1/T} d\tau \int d^dx\,\lambda\right)\right].$$

(11.8)

(See [357] for further discussion on interpretation of the functional determinant above.) The action has a prefactor of N, and the $N = \infty$ limit of the functional integral is therefore given exactly by its saddle-point value. We assume that the saddle-point value of λ is space and time independent and given by $i\lambda = m^2$. The saddle-point equation determining the value of the parameter m^2 is

$$\int^\Lambda \frac{d^dk}{(2\pi)^d} T \sum_{\omega_n} \frac{1}{c^2k^2 + \omega_n^2 + m^2} = \frac{1}{cg},$$

(11.9)

where the sum over ω_n extends over the Matsubara frequencies $\omega_n = 2n\pi T$, n integer. It is also not difficult to evaluate the order parameter susceptibility at $N = \infty$ by inserting an appropriate source term in (11.6): as expected, the result is given simply by the propagator of the $\mathbf{n}$ field in (11.6) with λ replaced by its saddle-point value. The result obeys $\chi_{\alpha\beta} = \chi\delta_{\alpha\beta}$, where

$$\chi(k, \omega) = \frac{cg/N}{c^2k^2 - (\omega + i\delta)^2 + m^2}$$

(11.10)

is also the propagator of the $\mathbf{n}$ field. The large-N limit of the uniform susceptibility, χ_u, can also be evaluated by first expanding $\mathcal{F}$ in powers of H, and evaluating the resulting four- and two-point correlators of $\mathbf{n}$ at tree level using the propagator in (11.10). This gives

$$\chi_u = 2T \sum_{\omega_n} \int \frac{d^dk}{(2\pi)^d} \frac{c^2k^2 + m^2 - \omega_n^2}{\left(c^2k^2 + m^2 + \omega_n^2\right)^2}.$$

(11.11)

Equations (11.9)–(11.11) apply only when the system does not have long-range spatial order (at $T = 0$ or $T > 0$) and O(N) symmetry is preserved; they are the central results of the $N = \infty$ theory, and most of the remainder of this chapter is spent on analyzing their consequences. In spite of their extremely simple structure, these equations contain a great deal of information, and it takes a rather subtle and careful analysis to extract the universal information contained in them [85, 86, 440]. We begin by characterizing the $T = 0$ ground states and comparing the results to the strong- and weak-coupling analyses noted earlier. Then we turn to the finite-temperature crossovers.

11.2 Zero temperature

At $T = 0$, we can make use of the relativistic invariance of the action (11.4) to simplify our analysis. The summation over Matsubara frequencies in (11.9) turns into an integral, and after introducing spacetime momentum $p \equiv (k, \omega/c)$, the constraint equation (11.9) becomes

$$\int^{\Lambda} \frac{d^{d+1}p}{(2\pi)^{d+1}} \frac{1}{p^2 + (m/c)^2} = \frac{1}{g}.$$
(11.12)

The integral on the left-hand side increases monotonically with decreasing m; as $m \to 0$, it diverges as $\ln(1/m)$ in $d = 1$, and it has a maximum finite value at $m = 0$ in $d > 1$. It is then clear that it is always possible to find a solution for (11.12) in $d = 1$, and for $d > 1$ there is no solution to (11.12) for $g < g_c$ where

$$\int^{\Lambda} \frac{d^{d+1}p}{(2\pi)^{d+1}} \frac{1}{p^2} = \frac{1}{g_c}.$$
(11.13)

We have chosen the symbol g_c for the boundary point where the solution ceases to exist, following the discussion in Chapter 10. As we will see shortly, the regime where the solution exists describes a quantum paramagnetic ground state, and g_c is the quantum critical point for a transition to the $g < g_c$ magnetically ordered state. In $d = 1$ a solution exists for all g, and so the general d discussion for $g > g_c$ below can be applied to all g in $d = 1$. This indicates that the $d = 1$ ground state is always a quantum paramagnet. This is a large-N result and is manifestly incorrect for $N = 1$ as we saw in Chapter 10; it is also not true at $N = 2$, but we will see that the large-N theory leads to adequate results for all $N \geq 3$ in $d = 1$. For $g > g_c$ there is a unique solution of the saddle-point equation (11.12) describing a quantum paramagnetic ground state; we study its properties in the following subsection and find that they are quite similar to those of the quantum paramagnetic state of the Ising chain. The $d > 1$ critical point at $g = g_c$ is studied in the next subsection. Determination of the $d > 1$ ground state for $g \leq g_c$ requires a reanalysis of the derivation of the large-N saddle equation. This is done in Section 11.2.3, where we find a state with magnetic long-range order and spontaneous breakdown of the $O(N)$ symmetry.

11.2.1 Quantum paramagnet, $g > g_c$

For $d > 1$, subtract (11.12) from (11.13) and obtain

$$\frac{1}{g_c} - \frac{1}{g} = \int^{\Lambda} \frac{d^{d+1}p}{(2\pi)^{d+1}} \left(\frac{1}{p^2} - \frac{1}{p^2 + (m/c)^2} \right).$$
(11.14)

Now note that for $d < 3$ it is possible to send the upper cutoff Λ to infinity and still obtain a finite result. Thus, provided that we measure quantities in terms of deviations from their values at $g = g_c$, we see that observables are insensitive to the nature of the cutoff (i.e. they are universal). For $d \geq 3$ it is necessary to retain the upper cutoff, and observables have additional Λ dependence. (As briefly noted earlier, this identifies $d = 3$ as the upper-critical dimension.) The remaining analysis of this chapter is implicitly restricted to $d < 3$, and we examine $d \geq 3$ by other, more convenient, methods in subsequent chapters; in the language of the classical model (3.2), this restriction is equivalent to $D < 4$, where $D = 4$

is its upper-critical dimension. For $1 < d < 3$ we can evaluate the integral in (11.14) with an infinite cutoff and obtain

$$\frac{1}{g_c} - \frac{1}{g} = X_{d+1} (m/c)^{d-1}, \tag{11.15}$$

where the constant $X_d \equiv 2\Gamma((4-d)/2)(4\pi)^{-d/2}/(d-2)$. This equation can be easily solved to obtain the required value of m. In $d = 1$, we have $g_c = 0$, and evaluating (11.12) directly, we find for small m and small g

$$\frac{1}{2\pi} \ln \left(\frac{c\Lambda}{m} \right) = \frac{1}{g}, \tag{11.16}$$

which also has a simple solution for $m = c\Lambda e^{-2\pi/g}$. Apart from the difference in the expression in the value of m above, the remaining discussion in this subsection applies equally to $d = 1$ and $d > 1$.

A key step in the analysis of any ground state of a continuum theory is the determination of an energy scale that characterizes it. In this case, the quantum paramagnet has a gap, Δ_+, given by

$$\Delta_+ \equiv m(T = 0). \tag{11.17}$$

We emphasize that, by definition, the gap Δ_+ is a temperature-independent quantity; it equals the temperature-dependent value of m only at $T = 0$. The presence of a gap is apparent in the structure of the spectral density $\mathrm{Im}\chi(k, \omega)$, which from (11.10) is given by

$$\mathrm{Im}\chi(k, \omega) = \mathcal{A} \frac{\pi}{2\sqrt{c^2 k^2 + \Delta_+^2}} \left(\delta \left(\omega - \sqrt{c^2 k^2 + \Delta_+^2} \right) \right.$$
$$\left. - \delta \left(\omega + \sqrt{c^2 k^2 + \Delta_+^2} \right) \right), \tag{11.18}$$

which has weight only at frequencies greater than Δ_+. The spectral weight appears entirely in the form of delta functions, which indicate the presence of magnon quasiparticles; the quantity

$$\mathcal{A} = \frac{cg}{N} \tag{11.19}$$

is the quasiparticle residue. This magnon is obviously the same as the three-fold degenerate particle that appeared earlier in the strong-coupling analysis of the O(3) model in Section 6.1, and in Section 7.2.1. The spectral density (11.18) is also identical in form to the exact result for the quantum paramagnetic phase of the Ising chain obtained by taking the imaginary part of (10.82). The n-particle continua ($n \geq 3$, odd) are absent here in the $N = \infty$ theory, but they appear later when we study fluctuation corrections.

We can also evaluate the uniform susceptibility χ_u by converting the frequency summation in (11.11) to an integral and then evaluating the frequency integral. This gives the simple result

$$\chi_u = 0. \tag{11.20}$$

This result could have been anticipated. The ground state is a spin singlet; the lowest excited state is a triplet separated by a gap. In a small $\mathbf{H}$ field there is no change in the energy of the singlet, whereas one of the triplet states lowers its energy but remains above the singlet for $H < \Delta_+$. The ground state therefore remains unchanged and has vanishing uniform susceptibility.

Also justifying our identification of this phase as a quantum paramagnet is that equal-time $\mathbf{n}$ correlations decay exponentially in space:

$$\frac{1}{N} \langle \mathbf{n}(x,0) \cdot \mathbf{n}(0,0) \rangle = \frac{\mathcal{A}}{c} \int \frac{d^{d+1}p}{(2\pi)^{d+1}} \frac{e^{ip \cdot x}}{p^2 + (\Delta_+/c)^2}$$

$$= \frac{\mathcal{A}}{2c(2\pi)^{d/2}(\Delta_+/c)^{(2-d)/2}} \frac{e^{-x\Delta/c}}{x^{d/2}}, \qquad (11.21)$$

which identifies Δ/c as the inverse correlation length. Note again the precise agreement of this result with that for the quantum paramagnetic phase of the Ising chain in (10.83), where $2cZ\Delta^{1/4}$ played the role of the quasiparticle residue, $\mathcal{A}$.

11.2.2 Critical point, $g = g_c$

This subsection applies only for $1 < d < 3$. There is no critical point in $d = 1$, and there are violations of naive scaling hypotheses for $d \geq 3$.

As g approaches g_c from above, we see from (11.15) that the energy gap, Δ_+, vanishes as

$$\Delta_+ \sim (g - g_c)^{1/(d-1)}. \qquad (11.22)$$

The critical state at $g = g_c$ turns out to be scale invariant at scales much longer than Λ^{-1}, as expected by analogy with the Ising model. The coupling g is the parameter that tunes the system away from this scale-invariant point, and as Δ_+ is an energy (inverse time) scale, the definition (10.35), the definition of the exponent ν above it, and the result (11.22) identifies the exponent

$$z\nu = \frac{1}{d-1}. \qquad (11.23)$$

The equal-time correlations decay as

$$\langle \mathbf{n}(x,0) \cdot \mathbf{n}(0,0) \rangle \sim \int \frac{d^{d+1}p}{(2\pi)^{d+1}} \frac{e^{ikx}}{p^2}$$

$$\sim \frac{1}{x^{d-1}}, \qquad (11.24)$$

which is a power law, as expected for a scale-invariant theory; the decay as a function of time has the same exponent, and so

$$z = 1, \qquad (11.25)$$

as must be the case for a Lorentz-invariant theory. The application of the scaling transformation on (11.24) also tells us that as in (4.32)

$$\mathrm{dim}[\mathbf{n}] = \frac{d + z - 2 + \eta}{2}, \tag{11.26}$$

with η, the "anomalous dimension" of the field, vanishing in the $N = \infty$ theory. We see later in (13.51) that $1/N$ corrections induce a nonzero η, as in (4.37). The anomalous dimension also determines the scaling dimension of the quasiparticle residue $\mathcal{A}$: (11.26) implies that $\mathrm{dim}[\chi(k, \omega)] = -2 + \eta$, and demanding consistency of this with the expression (11.18), we conclude $\mathrm{dim}[\mathcal{A}] = \eta$. Therefore, as g approaches g_c from above,

$$\mathcal{A} \sim (g - g_c)^{\eta\nu}, \tag{11.27}$$

(i.e. in general, the quasiparticle residue vanishes as the system approaches the critical point). Again this scaling is consistent with the Ising model in which $\mathcal{A} = 2Z\Delta^{1/4} \sim (g - g_c)^{1/4}$ (see (10.84)). In the present $N = \infty$ theory, the quasiparticle residue $\mathcal{A}cg/N$ was nonzero all the way up to $g = g_c$. This is consistent with the $N = \infty$ result of $\eta = 0$: there is no dynamic scattering of the quasiparticle excitations at $N = \infty$ but such scattering appears upon including $1/N$ corrections, which also induce a nonzero η.

If there are no quasiparticles for $\eta \neq 0$, what do the excitations look like? As in the Ising chain, there is a critical continuum of excitations, whose spectral density is determined by η. Combining the Lorentz invariance of the theory with a simple analysis of scaling dimensions, we see that the dynamic susceptibility must have the form

$$\chi(k, \omega) \sim \frac{1}{(c^2 k^2 - \omega^2)^{1 - \eta/2}} \tag{11.28}$$

(compare (10.96)), and its imaginary part looks much like Fig. 10.7. The $\eta = 0$ case is of course special, in that the spectral density has a single delta function at $\omega = ck$, and the critical excitations have a particle-like nature. This is clearly an artifact of the $N = \infty$ theory and is one of its major failings.

We can also use simple scaling arguments to determine the exact scaling dimension of $\mathbf{H}$ and therefore from (11.3) that of χ_u. Note that in (11.4) $\mathbf{H}$ appears intimately coupled with a time derivative. As we discussed earlier, this is related to the fact that the only effect of $\mathbf{H}$ is to uniformly precess all the rotors, and this precession is not visible in a rotating reference frame. This is an exact property of theory, and therefore the precession angle must be invariant under scaling transformation. As a result the scaling dimension of $\mathbf{H}$ must be that of inverse time, which implies from (10.35) that

$$\mathrm{dim}[\mathbf{H}] = z, \tag{11.29}$$

and, using (10.37) and (11.3), that

$$\mathrm{dim}[\chi_u] = d - z. \tag{11.30}$$

11.2.3 Magnetically ordered ground state, $g < g_c$

This subsection necessarily applies only for $d > 1$, as there is no ordered state in $d = 1$.

Our analysis so far has shown no meaningful solution of the saddle-point equations in the large-N limit for $g < g_c$. The culprit for this shortcoming lies in the step before (11.8), where we indiscriminately integrated out all N components of the **n** field [61]. As we expect a magnetically ordered phase to appear for $g < g_c$, it seems sensible to allow for the possibility that fluctuations of **n** along the direction of the ordered ground state will be different from those orthogonal to it. So we write

$$\mathbf{n} = (\sqrt{N}r_0, \pi_1, \pi_2, \ldots, \pi_{N-1}), \tag{11.31}$$

where it is assumed that the order parameter is polarized along the 1 direction. Inserting this and (11.7) into (11.4), imposing the constraint with a Lagrange multiplier λ, and integrating out *only* the $\pi_{1,\ldots,N-1}$ fields, we find

$$\mathcal{Z} = \int \mathcal{D}\lambda \mathcal{D}r_0 \exp\left[-\frac{N-1}{2}\text{Tr}\ln\left(-c^2\partial_i^2 - \partial_\tau^2 + i\lambda\right) \right.$$
$$\left. + \frac{iN}{cg}\int_0^{1/T} d\tau \int d^dx\, \lambda \left(1 - r_0^2\right) \right]. \tag{11.32}$$

In the large-N limit, we can ignore the difference between $N-1$ and N and obtain the saddle-point equations with respect to variations in λ *and* r_0. As before, m^2 is taken to be the saddle-point value of $i\lambda$. The mean value of r_0 determines the spontaneous magnetization at $N = \infty$, which we denote by N_0; so

$$N_0 = \langle n_1 \rangle = \sqrt{N}r_0. \tag{11.33}$$

The saddle-point equations are

$$N_0^2 + g\int^\Lambda \frac{d^{d+1}p}{(2\pi)^{d+1}} \frac{1}{p^2 + (m/c)^2} = 1,$$
$$m^2 N_0 = 0, \tag{11.34}$$

where we have set $T = 0$. One solution of the second equation is $N_0 = 0$, but then the first equation for m becomes identical to the one considered earlier, which is known to fail for $g < g_c$. Therefore we choose the other solution, where

$$m = 0,$$

$$N_0^2 = 1 - g\int^\Lambda \frac{d^{d+1}p}{(2\pi)^{d+1}} \frac{1}{p^2}$$

$$= 1 - \frac{g}{g_c}. \tag{11.35}$$

It is satisfying to find that N_0 is nonzero precisely for $g < g_c$, reinforcing our belief in the correctness of our procedure in finding the saddle point. Note that N_0 vanishes as $(g_c - g)^{1/2}$ as g approaches g_c. It is conventional to define the critical exponent β by the dependence $N_0 \sim (g_c - g)^\beta$, and we therefore have $\beta = 1/2$ in the present $N = \infty$ theory.

More generally, the scaling dimension of N_0 must be the same as the scaling dimension of **n**, and we therefore have from (11.26) that (see also (4.47)

$$2\beta = (d + z - 2 + \eta)\nu,\tag{11.36}$$

an exponent relation that is satisfied by the $N = \infty$ theory.

The above approach also determines the two-point correlator of spin components orthogonal to the axis of the spontaneous magnetization. We denote the corresponding susceptibility by $\chi_\perp(k, \omega)$, and it is the Fourier transform of the n_2, n_2 correlator (say); we have at $N = \infty$

$$\chi_\perp(k, \omega) = \frac{cg/N}{c^2 k^2 - (\omega + i\delta)^2}.\tag{11.37}$$

Note that there is a quasiparticle pole at $\omega = ck$, and the energy of this excitation vanishes as $k \to 0$. These are the spin-wave excitations discussed earlier in the weak-coupling analysis. These spin waves survive fluctuation corrections as $k \to 0$, although the nature of the spectral density becomes different at larger k, as we discuss shortly.

We can now compute the spin-stiffness by combining the $N = \infty$ results ((11.35) and (11.37)) with (8.19):

$$\rho_s = cN \left(\frac{1}{g} - \frac{1}{g_c} \right).\tag{11.38}$$

In general, from (8.13), ρ_s is expected to vanish as $(g_c - g)^{(d-1)\nu}$, and the result (11.38) is consistent with the $N = \infty$ values of the exponents.

With the spin stiffness ρ_s in hand, we can now construct the energy scale, which we denote Δ_-, that characterizes the ground state for $g < g_c$. The requirement is that Δ_- should have scaling dimension z and physical units of $(\text{time})^{-1}$. Such an object has to be made out of powers of ρ_s, whose scaling dimension is in (8.13), and whose physical units are $(\text{length})^{2-d}(\text{time})^{-1}$, and the velocity c, whose scaling dimension is 0 and whose physical units are $(\text{length})(\text{time})^{-1}$; the unique combination is

$$\Delta_- \equiv (\rho_s/N)^{1/(d-1)} c^{(d-2)/(d-1)}.\tag{11.39}$$

The factor of N has been chosen for future convenience.

We conclude this subsection by remarking on two features of the response functions of the ordered ground state that depend upon having a nonzero η and are therefore absent in the $N = \infty$ theory. First, from (8.19), we deduce that the residue at the spin-wave pole (for $k \to 0$) is N_0^2/ρ_s; as g approaches g_c, this vanishes as $(g_c - g)^{\eta\nu}$, unlike the result (11.37) in which the spin-wave residue remains nonzero all the way up to g_c. Second, with energy scale Δ_- in hand, we can also define a corresponding length scale ξ_J:

$$\xi_J = \frac{c}{\Delta_-}.\tag{11.40}$$

This is known as the Josephson length. The forms (8.19) and (11.37), which are characteristic long-wavelength transverse responses of a phase with spontaneously broken continuous symmetry, remain valid at length scales larger than ξ_J and times longer than Δ_-^{-1}. At shorter scales, the responses cross over to the isotropic response of the critical points as in (11.28).

11.3 Nonzero temperatures

We have shown in the previous section that (for $d > 1$) there are two distinct ground states separated by a quantum critical point at $g = g_c$, and each ground state is characterized by a single energy scale Δ_+ or Δ_-, which vanishes as $|g - g_c|^{z\nu}$ near the critical point (in $d = 1$ we only have one phase characterized by Δ_+).

We can combine the insights gained from the solution of the Ising chain in Chapter 10 with some simple physical considerations, and also by partly anticipating some $N = \infty$ results to be discussed below, and sketch the $T > 0$ phase diagrams in Figs. 11.1–11.3. Compare these phase diagrams with those in Section 1.2.

First we show the phase diagram for $d = 1$ in Fig. 11.1 [250]. There is only one phase in $d = 1$: a quantum paramagnetic ground state with a gap Δ_+. The energy scale Δ_+ is the only one characterizing the universal physics, and therefore we expect a qualitative change in the nature of the physics at $T \sim \Delta_+ \sim \exp(-2\pi/g)$ (using (11.16)). We identify the region $T < \Delta_+$ as the low-temperature limit of the continuum theory, which will be similar to the low-T region on the quantum paramagnetic side of the Ising model in Fig. 10.2. The region $\Delta_+ < T < J$ is the high-temperature limit of the continuum theory; it differs from the high-T region of the Ising chain in Fig. 10.2, as we see in the next chapter, by the presence of logarithmic corrections that modify some key dynamic properties and their physical interpretation. Finally there is a lattice high-T region, $T > J$ (not shown in Fig. 11.1), where microscopic details matter. This region is not of interest to us here.

Turning next to $d = 2$, we show the anticipated large-N phase diagram in Fig. 11.2. The crossover phase boundaries and the physical interpretations of the regimes are essentially identical to those for the Ising chain in Fig. 10.2, both of which realize the phase diagrams in Fig. 1.2a and 1.3. There is an ordered magnetic state at $T = 0$, but the long-range order disappears at any nonzero T. This is similar to the Ising chain, but the physics behind the destruction of long-range order by thermal fluctuations is quite different and is discussed in more detail in subsequent chapters.

Finally, we also consider the large-N limit for $2 < d < 3$ in Fig. 11.3. Although these dimensions are unphysical, examining these cases is useful as we can deal with systems

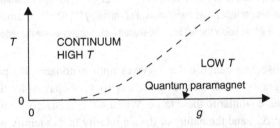

Fig. 11.1   Large-N phase diagram for the $O(N)$ rotor model in $d = 1$. This phase diagram applies for all $N \geq 3$. The dashed line is a crossover. Our interest is in the two universal regions, which are the low- and high-T limits of the continuum quantum field theory. The crossover boundary is at $T \sim \Delta_+ \sim \exp(-2\pi/g)$.

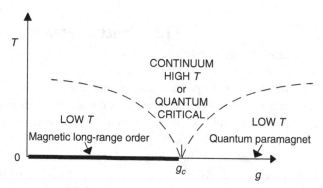

Fig. 11.2   Large-N phase diagram for the $O(N)$ rotor model for $d = 2$; compare with Figs. 1.2a and 1.3. As in Fig. 11.1, this is expected to apply for all $N \geq 3$, and the lower dashed lines are crossovers determined by the conditions $\Delta_{\pm} \sim T$. As g approaches the critical coupling g_c, $\Delta_+ \sim (g - g_c)^{z\nu}$ for $g > g_c$ and $\Delta_- \sim (g_c - g)^{z\nu}$ for $g < g_c$. The physical interpretation of the regimes is identical to those for the Ising chain in Fig. 10.2, and realizes the phase diagram in Fig. 1.2a. As in Fig. 10.2, there is also an additional, nonuniversal, lattice high-T region for $T > J$ (which is not shown here).

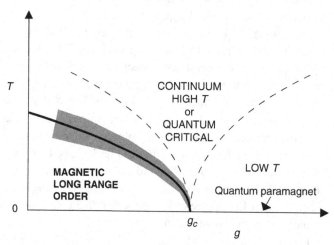

Fig. 11.3   Large-N phase diagram for the $O(N)$ rotor model with $2 < d < 3$, realizing the phase diagrams in Fig. 1.2b and 1.3. Qualitative features of the phase diagram apply for $N > 2$ and $2 < d < 3$, or $1 \leq N \leq 2$ and $2 \leq d < 3$. The dashed lines are crossovers determined by $\Delta_{\pm} \sim T (\Delta_{\pm} \sim |g - g_c|^{z\nu})$, while the full line is the locus of finite-temperature phase transitions with T_c given by (11.67). There is true magnetic long-range order at all temperatures below the full line. The shaded region shows where the reduced classical scaling functions apply.

whose long-range order survives until a nonzero temperature, as is required for phase diagrams like those in Fig. 1.2b. Also, the behavior for the physical cases $N = 1, 2, d = 2$ is quite similar to these large-N limits. The nonzero T phase transition is within the region $T < \Delta_-$, and the nature of the singularity in its vicinity is discussed below.

The crossovers in these phase diagrams can be described by scaling functions closely analogous to (10.39). It is more convenient to work in frequency and wavevector space, and we can obtain the scaling form by arguments similar to those used to obtain (10.39).

First, we can use the definition (11.2) and the scaling dimension (11.26) to conclude $\dim[\chi(k,\omega)] = 2\dim[\mathbf{n}] - d - z = -(2-\eta)$. Then recalling $\dim[T] = z$, we can obtain the scaling form

$$\chi(k,\omega) = \frac{Z}{T^{(2-\eta)/z}} \Phi_{\pm}\left(\frac{ck}{T^{1/z}}, \frac{\omega}{T}, \frac{\Delta_{\pm}}{T}\right), \tag{11.41}$$

where the upper (lower) sign applies for $g \geq g_c$ ($g \leq g_c$). Also, it should be clear that in $d = 1$ only the upper sign can apply. The functions $\Phi_{\pm}$ are completely universal and complex valued. They are chosen to have finite limits at all k and ω as $\Delta_{\pm} \to 0$ at fixed T (although there is an exception to this in $d = 1$, where, as we see in Chapter 12, the function Φ_{+} diverges logarithmically as $\Delta_{+}/T \to 0$; this logarithmic divergence is, however, absent in the present $N = \infty$ theory). There are strong restrictions that arise from the consistency of the two functions as they approach the common point $g = g_c$ from the two sides; not only their values must agree, but also the fact that $\chi(k,\omega)$ must be *analytic* as a function of g at $g = g_c$ for $T > 0$ places many additional restrictions (the reasons for this analyticity and its consequences are discussed in more detail in Section 14.2.1). For the Ising chain we were able to work with a single function by defining a $\Delta = \Delta_{+} > 0$ for $g \geq g_c$ and $\Delta = -\Delta_{-} < 0$ for $g \leq g_c$, but this is difficult to do in the present case as the definitions of $\Delta_{\pm}$ are quite different. Also, for the Ising chain, Δ was a simple, analytic linear function of g, and so the analyticity requirement was simply that Φ was analytic as a function of Δ at $\Delta = 0$.

The prefactor Z is a nonuniversal constant that is nonsingular at the $T = 0$ quantum critical point. It can be defined through (11.41) by relating it to some observable that depends upon the scale of the order parameter field. For $g > g_c$, by demanding that the form of $\chi(k,\omega)$ near the quasiparticle pole at $T = 0$ in (11.18) (which holds even beyond $N = \infty$, as we saw in the Ising chain) be consistent with the scaling form (11.41), we can specify

$$Z = (\text{constant}) \frac{\mathcal{A}}{\Delta_{+}^{\eta/z}}. \tag{11.42}$$

The constant can be chosen at our convenience and merely changes the definition of the $\Phi_{\pm}$. Alternatively, we could approach the critical point from $g < g_c$ and use (8.19) to define

$$Z = (\text{constant}) \frac{N_0^2 c^2}{\rho_s \Delta_{-}^{\eta/z}}. \tag{11.43}$$

A similar scaling form can be written for the uniform susceptibility from a knowledge of the scaling dimension in (11.30):

$$\chi_u = \frac{T^{d/z-1}}{c^d} \Phi_{u\pm}\left(\frac{\Delta_{\pm}}{T}\right). \tag{11.44}$$

Unlike (11.41), there is no nonuniversal prefactor such as Z in front. This is because the unknown field scale and the anomalous exponent η do not appear in the definition of χ_u; rather χ_u is related by (11.3) to the free energy density.

The remainder of this section presents explicit results for these scaling functions at $N = \infty$. In this limit, the expressions in (11.10) and (11.11) specify χ and χ_u, respectively.

These are consistent with the scaling forms (11.41) and (11.44) for $\eta = 0$ and $z = 1$, if the Lagrange multiplier m satisfies

$$m = T F_{\pm}\left(\frac{\Delta_{\pm}}{T}\right),\tag{11.45}$$

where $F_{\pm}$ are universal functions obtained from the solution of (11.9), as we show below. The resulting predictions for the physical properties at $T > 0$ are quite simple. By Fourier transforming (11.10), we see that m/c is the correlation length. The imaginary part of (11.10) also implies that there is a gap in the spectrum equal to m. This feature is an artifact of the $N = \infty$ limit: the response of any interacting system at $T > 0$ has a nonzero spectral density at all frequencies (in certain cases, the response could vanish above some large ultraviolet cutoff $\sim c\Lambda$), as there are essentially no restrictions on the set of frequencies at which all the possible thermally excited states can absorb energy. A particular objective of the remaining chapters in Part II is to describe a dynamical theory for the filling of this gap at finite temperatures.

The uniform susceptibility is obtained by evaluating the frequency summation in (11.11) by standard methods, which the reader can find in text books such as [136] and [314]; the result is

$$\chi_u = \frac{1}{2T}\int\frac{d^d k}{(2\pi)^d}\frac{1}{\sinh^2(\sqrt{c^2 k^2 + m^2}/2T)},\tag{11.46}$$

with m given by (11.45).

We now determine the universal functions $F_{\pm}$ and subsequently turn to a description of the physics in the various regions of Figs. 11.1–11.3. The method used here introduces a number of useful tricks for the extraction of universal, cutoff-independent crossover functions.

We present first the calculation on the disordered side where $g \geq g_c$. The first step is to subtract, from (11.9) the corresponding equation (11.12) at the same coupling constants at $T = 0$; this gives us

$$\int^{\Lambda}\frac{d^d k}{(2\pi)^d} T\sum_{\omega_n}\frac{1}{c^2 k^2 + \omega_n^2 + m^2} - \frac{1}{c}\int^{\Lambda}\frac{d^{d+1}p}{(2\pi)^{d+1}}\frac{1}{p^2 + (\Delta_+/c)^2} = 0,\tag{11.47}$$

where Δ_+ is the gap at the current value of g. A trick we often use is to subtract, from the summation over frequencies of any quantity, the integration over frequencies of precisely the same function; so we rewrite (11.47) as

$$\int^{\Lambda}\frac{d^d k}{(2\pi)^d}\left(T\sum_{\omega_n}\frac{1}{c^2 k^2 + \omega_n^2 + m^2} - \int\frac{d\omega}{2\pi}\frac{1}{c^2 k^2 + \omega^2 + m^2}\right)$$
$$+ \frac{1}{c}\int^{\Lambda}\frac{d^{d+1}p}{(2\pi)^{d+1}}\left(\frac{1}{p^2 + (m/c)^2} - \frac{1}{p^2 + (\Delta_+/c)^2}\right) = 0.\tag{11.48}$$

Now we use the general relation

$$T\sum_{\omega_n}\frac{1}{\omega_n^2 + a^2} - \int\frac{d\omega}{2\pi}\frac{1}{\omega^2 + a^2} = \frac{1}{a}\frac{1}{e^{a/T} - 1},\tag{11.49}$$

which is valid for any positive a (again this can be established by standard frequency summation methods [136, 314]). Note that the right-hand side falls off exponentially as a becomes large. This is a key property and was the reason for considering the combination in (11.49). Applying this identity to (11.48), we see that the first integration over k has an integrand that is exponentially small for large k, and hence it is quite insensitive to Λ, which can safely be sent to infinity. The integration over p in the second term is also ultraviolet convergent, again allowing Λ to be set to infinity. The resulting expression is then cutoff independent and hence universal; we obtain for $d > 1$

$$\int \frac{d^d k}{(2\pi)^d} \frac{1}{\sqrt{c^2 k^2 + m^2}} \frac{1}{e^{\sqrt{c^2 k^2 + m^2}/T} - 1} - \frac{X_{d+1}}{c^d}\left(m^{d-1} - \Delta_+^{d-1}\right) = 0, \qquad (11.50)$$

where the number X_d was defined below (11.15). When $d = 1$, this equation is modified to

$$\int \frac{dk}{(2\pi)} \frac{1}{\sqrt{c^2 k^2 + m^2}} \frac{1}{e^{\sqrt{c^2 k^2 + m^2}/T} - 1} - \frac{\ln(m/\Delta_+)}{2\pi c} = 0. \qquad (11.51)$$

The solution of these equations is clearly of the form (11.45); after rescaling momenta by c/T in (11.50), we find that the function $F_+(s)$ is determined implicitly by solution of the equation

$$\int \frac{d^d k}{(2\pi)^d} \frac{1}{\sqrt{k^2 + F_+^2}} \frac{1}{e^{\sqrt{k^2 + F_+^2}} - 1} - X_{d+1}\left(F_+^{d-1} - s^{d-1}\right) = 0, \qquad (11.52)$$

for $d > 1$, and similarly for $d = 1$. We discuss asymptotic features of the solution of these equations in the subsections below. We note here that precisely when $d = 2$, the equation (11.52) has a simple, explicit solution [86]:

$$F_+(s) = 2 \sinh^{-1}\left(\frac{e^{s/2}}{2}\right), \quad d = 2. \qquad (11.53)$$

Now we turn to the ordered side, $g \leq g_c$, which implicitly means that we have $d > 1$. We assume that T is large enough such that the magnetization is zero; the case of the magnetized state with $T \neq 0$ can be treated similarly and is referred to below. Subtract, from (11.9), the value of ρ_s/N in (11.38), and insert the value of $1/g_c$ in (11.13). Evaluating the frequency summation as above we find

$$\int \frac{d^d k}{(2\pi)^d} \frac{1}{\sqrt{c^2 k^2 + m^2}} \frac{1}{e^{\sqrt{c^2 k^2 + (m/c)^2}/T} - 1}$$

$$+ \frac{1}{c} \int \frac{d^{d+1} p}{(2\pi)^{d+1}} \left(\frac{1}{p^2 + m^2} - \frac{1}{p^2}\right) = \frac{\rho_s}{N c^2}. \qquad (11.54)$$

The solution of this is also in the form (11.45), and the function $F_-(s)$ is given by

$$\int \frac{d^d k}{(2\pi)^d} \frac{1}{\sqrt{k^2 + F_-^2}} \frac{1}{e^{\sqrt{k^2 + F_-^2}} - 1} - X_{d+1} F_-^{d-1} - s^{d-1} = 0. \qquad (11.55)$$

Again, there is a simple explicit solution when $d = 2$ [86]:

$$F_-(s) = 2\sinh^{-1}\left(\frac{e^{-2\pi s}}{2}\right), \quad d = 2. \tag{11.56}$$

With expressions for the crossover functions $F_\pm$ in hand, let us discuss the physical properties of the system in different regimes of the g–T plane for different values of d.

11.3.1 Low T on the quantum paramagnetic side, $g > g_c$, $T \ll \Delta_+$

The discussion here also applies when $d = 1$.

Properties of this phase are essentially identical to those of the low-T quantum paramagnetic region of the Ising model in Section 10.4.2. The ground state has a gap, and nonzero T induces an exponentially small density of thermally excited triplet magnons. For the parameter m we have

$$m = \Delta_+ + \mathcal{O}(e^{-\Delta_+/T}). \tag{11.57}$$

Hence there is a finite correlation length c/m that has exponentially small corrections from its $T = 0$ value c/Δ_+. The $N = \infty$ expression (11.10) has a quasiparticle peak that remains infinitely sharp at $T > 0$. This is clearly incorrect for finite N, as damping must be present, and is described in subsequent chapters. The uniform susceptibility can be computed from (11.46), and we find that it is exponentially small:

$$\chi_u = \mathcal{O}(e^{-\Delta_+/T}). \tag{11.58}$$

11.3.2 High T, $T \gg \Delta_+$, Δ_-

Again, properties are similar to those of the continuum high-T (or quantum critical) region of the Ising chain as discussed in Section 10.4.3. Now we have, for $d > 1$,

$$m = TF_+(0) = TF_-(0), \tag{11.59}$$

where $F_+(0)$ and $F_-(0)$ are pure numbers. This represents a correlation length $\sim c/T$. In $d = 1$, the correlation length has an additional logarithmic correction [250], as can be seen from the solution of (11.51),

$$m = \frac{\pi T}{\ln(\mathcal{C}T/\Delta_+)}, \tag{11.60}$$

where

$$\mathcal{C} = 4\pi e^{-\gamma} = 7.055507955\ldots \tag{11.61}$$

In a similar manner we find for the uniform susceptibility from (11.46) that when $d > 1$

$$\chi_u = \frac{T^{d-1}}{c^d}\Phi_{u+}(0) = \frac{T^{d-1}}{c^d}\Phi_{u-}(0), \tag{11.62}$$

where $\Phi_{u\pm}$ are universal pure numbers, which can be determined by solutions of (11.46) and (5.69); when $d = 2$ we have the simple result $\Phi_{u\pm}(0) = (\sqrt{5}/\pi) \ln((\sqrt{5} + 1)/2)$. Again, when $d = 1$ there are log corrections [250]

$$\chi_u = \frac{1}{\pi c} \ln(\mathcal{C}T/\Delta_+), \tag{11.63}$$

which will be better understood in the following chapter.

By analogy with the Ising chain, we expect that the dynamics is quantum relaxational with a phase coherence time $\sim 1/T$. However, damping and relaxation are completely absent at $N = \infty$ and are further discussed later.

11.3.3 Low T on the magnetically ordered side, $g < g_c$, $T \ll \Delta_-$

This section applies only for $d > 1$, as there is no such region for $d = 1$. The properties for $d = 1$ are analogous to the low-T ordered region of the Ising chain in Section 10.4.1, but there are important differences for $2 < d < 3$.

Let us assume first that T is large enough so that $\langle \mathbf{n} \rangle = 0$ and so (11.55) can be used to determine F_-. For $d = 2$, one finds that there is a solution of (11.55) for all T, and even as $T \to 0$ ($s = \Delta_-/T \to \infty$). We find that as $T \to 0$

$$m = T \exp(-2\pi \Delta_-/T) = T \exp(-2\pi \rho_s/NT). \tag{11.64}$$

Hence the correlation length $\sim c/m$ diverges as $T \to 0$ but remains finite for all nonzero T. This was exactly the situation as in the Ising chain, and the phase diagram for this model is therefore as shown in Fig. 11.2. We see in subsequent chapters that, as in the case of the Ising chain, because of the very large correlation length, it is possible to develop an effective classical dynamical model of the system and to express the result in terms of reduced scaling functions. Let us also note (from (11.46)) that the uniform susceptibility for $d = 2$ is given as $T \to 0$ by

$$\chi_u = \frac{2\Delta_-}{c^2} = \frac{2\rho_s}{Nc^2}. \tag{11.65}$$

This is actually an exact result even for finite N, as we see later.

Now let us consider the case $2 < d < 3$. Although there is no physical dimension in this region, the results obtained below will apply for $d = 3$ with cutoff-dependent logarithmic corrections we do not want to discuss here. Further, the physics of the quantum Ising model in $d = 2$ is expected to be similar to that of the large-N solution with $2 < d < 3$. The key observation in this case is that there is no solution of (11.55) for $F_-(s)$ above a critical value $s = s_c$, where $F_-(s_c) = 0$. The value of s_c is given by

$$s_c^{d-1} = \int \frac{d^d k}{(2\pi)^d} \frac{1}{k} \frac{1}{e^k - 1}$$

$$= \frac{2\Gamma(d-1)\zeta(d-1)}{\Gamma(d/2)(4\pi)^{d/2}}. \tag{11.66}$$

Just as was the case in the $T = 0$ analysis at the beginning of Section 11.2, the absence of a solution for the Lagrange multiplier m (related to $F_-(s)$ by (11.45)) implies that there must be magnetic order for $s > s_c$. This defines a critical temperature T_c given precisely by

$$T_c \equiv \Delta_- / s_c, \tag{11.67}$$

such that the system is in the paramagnetic phase only for $T > T_c$. The resulting phase diagram is shown in Fig. 11.3. There is a finite-temperature phase transition at $T = T_c$ and a magnetically ordered phase for $T < T_c$. As T approaches T_c, the conventional classical phase transition theory becomes applicable in the region $|T - T_c| \ll T_c$. The classical scaling functions of this transition emerge as reduced scaling functions of the quantum functions, in a manner very similar to the discussion on the quantum Ising chain in Section 10.4.1. One consequence of this behavior is that all the scale factors of the classical scaling functions, which are usually considered nonuniversal, are universally determined by the parameters Δ_-, c, and N_0 of the quantum crossover functions. We have already seen an example of this in (11.67), where T_c was universally determined by Δ_- [422].

Let us explicitly observe the collapse of the scaling function (11.45) in this classical region. Because the primary quantum crossover function has only one argument, the reduced function would have no arguments, that is, it is a pure power law. Indeed, solution of (11.55) for s close to but above s_c gives us

$$m = T_c \left[\left(\frac{T - T_c}{T_c} \right) \frac{(d-1)s_c^{d-1}}{X_d} \right]^{1/(d-2)}. \tag{11.68}$$

The correlation length c/m diverges with the classical exponent $v_c = 1/(d-2)$ with an amplitude that is universal.

The above is part of a very general lesson. Quantum critical scaling forms such as (11.41) hold everywhere in the vicinity of the quantum critical point, including at or close to any finite-temperature phase transition lines that may be approaching the quantum critical point. The classical critical singularities of these finite-temperature transitions appear as singularities of the quantum critical scaling function. Further, the amplitudes of the classical transitions, which are normally nonuniversal, become universal when expressed in terms of the arguments of the quantum critical scaling function.

11.4 Numerical studies

We close this chapter by briefly mentioning computer studies of quantum spin models which exhibit quantum phase transitions in the universality class of the O(3) quantum rotor model. As discussed in Section 1.4.3, a variety of dimerized (or "double layer") antiferromagnets have transitions from a Néel state to a quantum paramagnet which are described by the O(3) quantum rotor model. Quantum Monte Carlo studies of such antiferromagnets [129, 163, 323, 324, 443, 444, 509, 539], have yielded accurate estimates for critical exponent, with results in excellent agreement with theoretical expectations. Numerical results for the uniform susceptibility χ_u are in good agreement with (11.62) and its $1/N$ corrections [86].

Normand and Rice [365, 366] have proposed an interesting recent experimental realization of the quantum critical point of the $d = 3$ quantum rotor model in $LaCuO_{2.5}$. This is a spin-ladder compound in which the ladders are moderately coupled in three dimensions. By varying the ratio of the intraladder to interladder exchange it is possible to drive such an antiferromagnet across a $d = 3$ quantum critical point separating Néel ordered and quantum paramagnetic phases. The uniform susceptibility has a T^2 dependence at intermediate T, which is characteristic of the "high-T" dependence in (11.62) for $d = 3$. The entire T dependence of χ_u has been computed in Monte Carlo simulations of an $S = 1/2$ antiferromagnet on the $LaCuO_{2.5}$ lattice [511] and the results are in good agreement with quantum rotor model computations like those discussed here.

12 The $d = 1$, $\mathrm{O}(N \geq 3)$ rotor models

This and the following chapter are at a more advanced level, and some readers may wish to skip ahead to Chapter 14.

In Chapter 11 we studied the $\mathrm{O}(N)$ quantum rotor model in the large-N limit for a number of values of the spatial dimensionality, including $d = 1$. We noted that the results provided an adequate description of the static properties when $d = 1$ for $N \geq 3$. This is justified in the present chapter where we obtain a number of exact results for the same static observables. We also noted that the large-N limit did a very poor job of describing dynamical properties at nonzero temperatures. This is repaired in this chapter by simple physical arguments that lead to a fairly complete (and believed exact) description of the long-time behavior. Some of the discussion in this chapter is specialized to the $\mathrm{O}(N = 3)$ model, which is also the case of greatest physical importance; the properties of the $\mathrm{O}(N > 3)$ models are very similar, and many of our results are quoted for general N. Of the remaining cases, the $d = 1$, $N = 1$ model has already been considered in Chapter 10, and study of the $d = 1$, $N = 2$ model is postponed to Section 20.3.

The physical picture of the $T = 0$, $N = 3$ state that emerged in Chapter 11 was very simple. The ground state was a quantum paramagnet, which did not break any symmetries. There was an energy gap, Δ_+, above the ground state, and the low-lying excitations were a triplet of particles with dispersion $\varepsilon_k = \sqrt{c^2 k^2 + \Delta_+^2}$; this picture is verified here by a more complete renormalization group analysis in Section 12.1. These triplet particle excitations lead to a quasiparticle pole in the dynamic susceptibility $\chi(k, \omega)$, which has the form (11.18) near the pole. This form contains the quasiparticle residue, $\mathcal{A}$, which sets the overall scale of the order parameter field.

Turning next to nonzero temperatures, we obtained the crossover phase diagram shown in Fig. 11.1, a modified version of which is reproduced in Fig. 12.1. The primary purpose of this chapter is to give a fairly complete description of the dynamical properties in the two universal regions of Fig. 11.1 and Fig. 12.1: these are the low-T ($T \ll \Delta_+$) and high-T ($\Delta_+ \ll T \ll J$) regions of the continuum quantum field theory. As indicated in Fig. 12.1, the dynamics of the low-T region are described by an effective model of quasi-classical particles in Section 12.2, closely related to the particle model developed in Section 10.4.2 for the Ising chain. For the high-T region, we develop a new, "dual," description in a model of quasi-classical *waves*, which is introduced in Section 12.3. As indicated in Section 1.4.3, and discussed more extensively in Chapter 19, the $d = 1$, $\mathrm{O}(3)$ rotor model describes a large class of quantum spin chains. The low-T regime of Fig. 12.1 is applicable to all such spin chains, while the high-T, quasi-classical wave regime applies only if the continuum quantum field theory description for the lattice model holds at these elevated temperatures

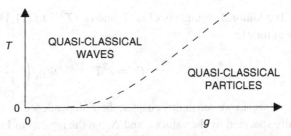

Fig. 12.1 Crossover phase diagram of the $d = 1$, $N \geq 3$ rotor model (11.1, 11.4) as a function of the temperature and the coupling g. The continuum theory description fails above some T, as in Fig. 10.2, but this has not been indicated. The quasi-classical particle model is developed in Section 12.2, while the quasi-classical wave model is discussed in Section 12.3.

(the precise restrictions this imposes are discussed in [65], but they will not be entered into here).

As we noted in Chapter 11, the dynamic susceptibility, $\chi(k, \omega)$, in the regions of Fig. 12.1 is completely determined by the parameters $\mathcal{A}$, c, and Δ_+ and obeys the scaling form (11.41) with $\eta = 0$, $z = 1$. The uniform susceptibility, χ_u, depends only on Δ_+ and c as shown in (11.44). We also examine here an important new observable that characterizes the transport of the conserved angular momentum of the rotor model in space: this is the *spin diffusion constant*, D_s. To compute this we need spacetime-dependent correlation functions of the angular momentum density $\mathbf{L}(x, t)$ (for a lattice model with spacing between sites, a, the continuum field $\mathbf{L}(x_i, t) = \hat{\mathbf{L}}_i(t)/a$); by analogy with (11.2) we define

$$C_{u,\alpha\beta}(x, \tau) \equiv \langle L_\alpha(x, \tau) L_\beta(0, 0) \rangle,$$

$$\chi_{u,\alpha\beta}(k, \omega_n) \equiv \int_0^{1/T} \int dx \, C_{u,\alpha\beta}(x, \tau) e^{-i(kx - \omega_n \tau)}. \tag{12.1}$$

Computations in this chapter show that $\chi_{u,\alpha\beta}$ has the following form at small k and ω:

$$\chi_{u,\alpha\beta}(k, \omega) = \delta_{\alpha\beta} \chi_u \frac{D_s k^2}{-i\omega + D_s k^2}. \tag{12.2}$$

For simplicity, we have set the external field $\mathbf{H} = 0$. This is done throughout this chapter, although it is not difficult to extend the results to a small $\mathbf{H} \neq 0$. The relationship (12.2) defines the value of the spin diffusion constant D_s. Actually the structure of (12.2) is a very general consequence of the conservation of $\mathbf{L}$, as we see in Chapter 15, and has been discussed in considerable detail in the book by Forster [150].

Note that the static uniform susceptibility is defined by

$$\chi_u \equiv \lim_{k \to 0} \lim_{\omega \to 0} \chi_u(k, \omega), \tag{12.3}$$

and the order of limits is important. It should also be clear that the full wavevector and frequency-dependent $\chi_u(k, \omega)$ obeys a scaling form quite analogous to (11.44); one simply adds additional arguments of ω/T and $ck/T^{1/z}$ to (11.44). This in turn implies a scaling form for D_s:

$$D_s = c^2 T^{-2/z+1} \Phi_{D_s} \left(\frac{\Delta_+}{T} \right); \tag{12.4}$$

therefore, as we see in this chapter, the T dependence of D_s is also completely and universally specified by the values c and Δ_+ in the regions of Fig. 11.1.

12.1 Scaling analysis at zero temperature

This section briefly reviews a well-known argument [61, 359, 384] that the large-N result for the $T = 0$ gap in (11.16), $\Delta_+ \sim c\Lambda \exp(-2\pi/g)$, is basically correct for all $N \geq 3$.

Our method is to examine the behavior of the coupling g under an RG transformation similar to that carried out in Chapter 4. However, rather than working with the field theory (2.11) with an unconstrained field ϕ_α, we work with the "nonlinear sigma model," in which the field $\mathbf{n}$ always obeys the fixed length constraint $\mathbf{n}^2 = 1$.

We work in a field theory with a momentum space cutoff Λ, and integrate out the degrees of freedom at momentum scales between $\Lambda e^{-\ell}$ and Λe^ℓ by the background field method of Polyakov [384, 385]. Let $\mathbf{n}_<(x, \tau)$ represent a "background" configuration of fields with wavevectors less than $\Lambda e^{-\ell}$. The fluctuations in the scales between $\Lambda e^{-\ell}$ and Λ must not violate the constraint $\mathbf{n}^2 = 1$, and they can therefore be parameterized by their $N - 1$ components along the directions orthogonal to $\mathbf{n}_<(x, \tau)$. Specifically, we write

$$\mathbf{n}(x, \tau) = \sqrt{1 - \pi_a \pi_a}\, \mathbf{n}_<(x, \tau) + \sum_{a=1}^{N-1} \pi_a \mathbf{e}_a(x, \tau), \tag{12.5}$$

where $\vec{\pi}$ is an $N - 1$ component field with wavevectors between $\Lambda e^{-\ell}$ and Λe^ℓ, and $\mathbf{e}_a(x, \tau)$, $\mathbf{n}_<(x, \tau)$ are N mutually orthogonal unit vectors in the N-dimensional rotor space. We insert (12.5) into (11.4) and expand the resulting action in powers of $\vec{\pi}$ at $\mathbf{H} = \mathbf{0}$. This gives the spatial gradient terms

$$\frac{cN}{2g} [(\nabla \mathbf{n}_<)^2 (1 - \pi_a \pi_a) + (\nabla \pi_a)^2 + \pi_a \pi_b \nabla \mathbf{e}_a \cdot \nabla \mathbf{e}_b + 2\pi_a \nabla \pi_b \mathbf{e}_b \cdot \nabla \mathbf{e}_a], \tag{12.6}$$

and also time derivative terms with an identical structure. Terms linear in π do not appear because they vanish upon spatial integration, as the momenta carried by the π_a are different from those of the background fields. Now the π_a fields are integrated out, and all terms containing up to two derivatives of the background fields are retained in the results. This results in an effective action for the fields $\mathbf{n}_<$ and $\mathbf{e}_a$; after using the orthonormality condition between these fields, all explicit dependence upon the $\mathbf{e}_a$ disappears, and the action for the $\mathbf{n}_<$ has precisely the form of (11.4) but with a modified coupling g'. Finally, we perform the rescaling (10.30) – this has no effect on the coupling g, which is dimensionless in

$d = 1$. We have now completed the required scaling transformation and it maps the original coupling g to a new coupling g' given by

$$\frac{1}{g'} = \frac{1}{g} - \frac{c(N-2)}{N} \int_{\Lambda e^{-\ell}}^{\Lambda} \frac{dk}{\pi} \int \frac{d\omega}{2\pi} \frac{1}{c^2 k^2 + \omega^2} + \mathcal{O}(g). \qquad (12.7)$$

The integrals in (12.7) can be easily carried out, and we can then represent the effects of successive application of this transformation (as in (4.22) and (4.24)) by the differential equation

$$\frac{dg}{d\ell} = \frac{N-2}{2\pi N} g^2 + \mathcal{O}(g^3). \qquad (12.8)$$

This is a key flow equation that helps us understand the properties of (11.4) at small g. By integrating (12.8) we can easily see that a system with a small initial value of g will flow into a system with a g of order unity at a scale

$$\ell = \frac{2\pi N}{(N-2)g} + \mathcal{O}(g^0), \qquad (12.9)$$

where the coefficient of the leading g^{-1} term does not depend upon the value of the order unity constant chosen, but that of the $\mathcal{O}(g)$ term does. We expect from the strong-coupling analysis of (6.1) that a system with a g of order unity will have a gap Δ_+ of the order of its cutoff $c\Lambda'$. Undoing the rescaling transformation (10.30), we know that the original cutoff Λ is related to the new cutoff by $\Lambda'/\Lambda = e^{-\ell} \sim \Delta/c\Lambda$, and therefore from (12.9)

$$\ln\left(\frac{c\Lambda}{\Delta_+}\right) = \frac{2\pi N}{(N-2)g} + \mathcal{O}(g^0), \qquad (12.10)$$

where again, the uncertainty in the precise value of Δ_+ relative to Λ' does not modify the leading g^{-1} term. This result has precisely the same form as the large-N result (11.16), establishing our earlier claims on the correctness of the large-N theory for static and equal-time properties – the only change in the present exact treatment has been the replacement of $\Delta_+ \sim c\Lambda \exp(-2\pi/g)$ by $\Delta_+ \sim c\Lambda \exp(-2\pi N/(N-2)g)$. This also shows that the large-N results break down badly at $N = 2$ but are quite reasonable for $N \geq 3$.

We have been rather sloppy in the above discussion about various constants of order unity. It is possible to be quite precise about these using a more sophisticated field-theoretic renormalization group analysis, which we discuss later in this chapter.

12.2 Low-temperature limit of the continuum theory, $T \ll \Delta_+$

This $T > 0$ region was shown in Figs. 11.1 and 12.1. All of the analysis of this section is specialized to $N = 3$, although the generalization to other $N \geq 3$ is straightforward.

The approach followed [107, 433] for $T \ll \Delta_+$ is very similar to that taken for the corresponding low-T region on the quantum paramagnetic side of the Ising chain in Section 10.4.2. The central difference here is that the quasiparticle excitations are triplets, and therefore they have an additional spin label, $m = -1, 0, 1$. This label is associated with the eigenvalues of the conserved total angular momentum and leads to important qualitative differences to be discussed below.

There are two key observations that allow our computation for $T \ll \Delta_+$. The first, as in the Ising chain, is that the density of thermally excited particles is so low that they can be treated, when well separated, as classical particles. In particular, as their density $\sim e^{-\Delta_+/T}$, their mean spacing $\sim e^{\Delta_+/T}$ is exponentially large at low T. Moreover, their thermal velocities are also small at low T, and so their typical wavelength becomes large; however, the divergence of the thermal de Broglie wavelength is only $\sim c/\sqrt{T\Delta_+}$ and is therefore much smaller than the particle spacing at low enough T. The density of particles with each spin m ($m = -1, 0, 1$ for $N = 3$), ρ_m, is given by expression (10.64), and the total density, ρ, therefore equals $\rho_1 + \rho_0 + \rho_{-1}$, which is

$$\rho = 3 \left(\frac{T\Delta_+}{2\pi c^2} \right)^{1/2} e^{-\Delta_+/T}. \tag{12.11}$$

This classical picture also allows us to simply obtain the value of the uniform susceptibility χ_u. In the presence of a field, the energy of a particle with spin component m simply acquires the Zeeman shift of $-mH$. This implies that in a field $\rho_m \to \rho_m e^{mH/T}$, expanding to linear order in the field we obtain [433, 510, 512]

$$\chi_u = \frac{2\rho}{3T} = \frac{1}{c} \left(\frac{2\Delta_+}{\pi T} \right)^{1/2} e^{-\Delta_+/T}. \tag{12.12}$$

Let us think about the dynamics of these classical particles. While well-separated particles behave classically, in one dimension these particles are forced to collide with their near neighbors and cannot avoid each other even in the extremely dilute limit. The collision must clearly be treated quantum mechanically, and we therefore need the two-particle S matrix. Because of the presence of the particle labels m, this S matrix can be a rather complicated object, and not simply a pure phase factor, as was the case in the Ising chain. Fortunately, we do not need the full S matrix, but only its value in the limit of vanishing momenta since the particles have thermal velocities that vanish, as noted above, in the low-T limit as $v_T = c(T/\Delta_+)^{1/2}$. Furthermore, this zero-momentum S matrix turns out to have a remarkably "super-universal" structure for $d = 1$. For the process shown in Fig. 12.2, the S matrix in the limit of vanishing momenta is

$$S_{m_1', m_2'}^{m_1 m_2} = (-1) \delta_{m_1 m_2'} \delta_{m_2 m_1'}. \tag{12.13}$$

In other words, the excitations behave like impenetrable particles that preserve their spin in a collision. As in the Ising chain, energy and momentum conservation in $d = 1$ require that

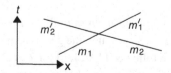

Fig. 12.2   Two-particle collision described by the S matrix (12.13). The momenta before and after the collision are the same, and so the figure also represents the spacetime trajectories of the particles.

these particles simply exchange momenta across a collision (Fig. 12.2). This result can be obtained in a variety of ways, which are explored in some detail in [107]. The simplest is to compute it in the strong-coupling expansion of Section 6.1: one solves the two-particle Schrödinger equation order-by-order in $1/g$ and finds that (12.13) holds at each order. Alternatively, one can take the low-momentum limit of the exact S matrix obtained by Zamolodchikov and Zamolodchikov [553] for the continuum theory (11.4) and find that (12.13) is valid. The first method shows that (12.13) holds even for lattice models and is thus not a special property of continuum relativistic theories. Indeed, (12.13) holds for practically every $d = 1$ model with a gap and excitations that have a quadratic dispersion at low momenta; exceptions arise only in specially fine-tuned cases when certain bound states happen to have exactly zero energy. The reasons for the "super-universality" are explored in more detail elsewhere [107], but the underlying physics can be seen to be a simple consequence of the arguments made below (5.15) in Section 5.2.2. We argued there that to a slowly moving particle, with a very long wavelength, any short-range repulsive potential can be approximated by an *impenetrable* delta function (i.e. a potential $u\delta(x)$ with $u \to \infty$). The wavefunctions of the two particles on either side of this potential therefore vanish as they approach $x = 0$. Exchange of spin requires actual overlap of the wavefunction, which we have shown becomes negligible in the low-momentum limit. Hence the spins of the two particles are preserved and we have the result (12.13).

We can now proceed to computation of correlation functions. As in Sections 10.4.1 and 10.4.2, we compute correlators as a "double time" path integral, and in the classical limit, stationary phase is achieved when the trajectories of the particles are time-reversed pairs of classical paths as shown in Fig. 12.3. Each trajectory has a spin label, m, which obeys (12.13) at each collision. The label, m, is assigned randomly at some initial time with equal probability but then evolves in time as discussed above (Fig. 12.3). We label the particles consecutively from left to right by an integer k; then their spins m_k are independent of t, and we denote their trajectories $x_k(t)$. The velocities of the particles are chosen independently at the initial time from the classical Boltzmann distribution $P(v)$:

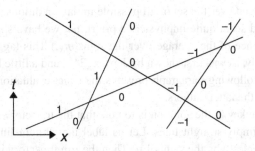

Fig. 12.3   A typical set of particle trajectories contributing to $C(x, t)$. Each trajectory represents paths moving both forward and backward in time, and the (-1) phase at each collision is neutralized by its time-reversed contribution. The particle coordinates are $x_k(t)$, with the labels k chosen so that $x_k(t) \leq x_l(t)$ for all t and $k < l$. Shown on the trajectories are the values of the particle spins m_k, which are independent of t in the low-T limit.

$$P(v) = \left(\frac{\Delta_+}{2\pi c^2 T}\right)^{1/2} \exp\left(-\frac{\Delta_+ v^2}{2c^2 T}\right). \tag{12.14}$$

We first discuss evaluation of the correlations of the conserved angular momentum density, $C_{u,\alpha\beta}$, defined in (12.1); this has no analog in the Ising case, as the latter model did not have a conserved charge associated with a continuous symmetry. In the absence of an external field $\mathbf{H}$, this correlator is rotationally invariant, and it is convenient to compute the correlator of the component of the angular momentum whose eigenstates we labeled in Fig. 12.3: we therefore compute $C_{u,33}$. The operator L_3 has a particularly simple effect on the particle trajectories in Fig. 12.3. It simply reports the azimuthal angular momentum of the particle it is operating on but does not create or annihilate any particles (this is evident from the strong-coupling expansion of Section 6.1). We therefore need only to sum over the trajectories shown in Fig. 12.3; for these every collision has a time-reversed pair, and therefore the -1s from the S matrix are completely neutralized. We are left then with a purely classical ensemble of point particles labeled with three "colors" (the azimuthal angular momentum). The observable $L_3(x, t)$ can be written in this ensemble as

$$L_3(x, t) = \sum_k m_k \delta(x - x_k(t)). \tag{12.15}$$

We have to determine its correlators under the average over a set of initial conditions of random, uncorrelated values of m_k and x_k, and over velocities given by the distribution (12.14). In particular we have

$$C_{u,33}(x - x', t - t') = \sum_{k,k'} \langle m_k m'_k \delta(x - x_k(t)) \delta(x' - x_{k'}(t')) \rangle$$

$$= \frac{2}{3} \sum_k \langle \delta(x - x_k(t)) \delta(x' - x_k(t')) \rangle. \tag{12.16}$$

In the second step (which is a crucial one), we have used the fact that the x'_k and m'_ks are uncorrelated and also that different m'_ks are mutually independent. We are now left with a well-defined problem in classical statistical mechanics. Place point particles independently and uniformly along an infinite line with a density ρ. Give each an initial velocity from the distribution (12.14). Tag a particle, k, and determine its position autocorrelation function, averaged over the set of all possible initial conditions. (Notice that such a particle tagging would seem quite unphysical a priori, but we have shown above how it is a natural consequence of the average over the spins m_k.) This tagged particle problem can be solved exactly, as was first shown by Jepsen [249] and a little later by Lebowitz and Percus [288]. The following paragraph presents the exact evaluation of (12.16) using a method drawn from the latter authors.

The key to the solution is to note that the trajectories in Fig. 12.3 are quite simple: they are simply straight lines. Let us label the straight line "trajectories" (as opposed to the "particles") by the symbol μ. Then the μth trajectory is simply

$$x_\mu(t) = x_\mu + v_\mu t, \tag{12.17}$$

where x_μ are the trajectory positions at $t = 0$, and v_μ are their velocities, and both of these have to be averaged over. Now, at a given time t, each trajectory μ will "belong" to a particle $k_\mu(t)$, where k_μ is a rather complicated integer-valued function of time. Its explicit expression is

$$k_\mu(t) = \sum_{\mu'=1}^{M} \theta(x_\mu(t) - x_{\mu'}(t)), \qquad (12.18)$$

where we have assumed there are a total of M trajectories (we will send $M \to \infty$ at a later stage), and $\theta(x)$ is the unit step function. It should be clear that (12.18) simply counts the trajectories to the left of a given trajectory at a time t, and this identifies the particle number. We can now rewrite (12.16) as a sum over trajectories, rather than particle number:

$$C_{u,33}(x,t) = \sum_{\mu,\mu'=1}^{M} \left\langle \delta(x - x_\mu(t))\delta(x_{\mu'})\delta_{k_\mu(t),k_{\mu'}(0)} \right\rangle$$

$$= \sum_{\mu,\mu'=1}^{M} \int_0^{2\pi} \frac{d\phi}{2\pi} \left\langle \delta(x - x_\mu(t))\delta(x_{\mu'}) \right.$$

$$\left. \times \exp\left[i \sum_{\mu''} (\theta(x_\mu(t) - x_{\mu''}(t)) - \theta(x_{\mu'} - x_{\mu''})) \right] \right\rangle, \qquad (12.19)$$

where in the second step we have introduced a Fourier representation of the Kronecker delta function. The average in (12.19) represents the multidimensional integral

$$\langle \cdot \rangle \equiv \prod_{\mu=1}^{M} \int_{-L/2}^{L/2} \frac{dx_\mu}{L} \int_{-\infty}^{\infty} dv_\mu P(v_\mu). \qquad (12.20)$$

We have assumed the particles are on a line of length L, and we are being quite sloppy about the boundary conditions. We ultimately want to take the limit $M \to \infty$ and $L \to \infty$ with the density $\rho = M/L$ fixed, and the result can be shown to be quite insensitive to the boundaries in this limit. Now the advantage of the Fourier representation in (12.19) should be quite evident: the $2M$-dimensional integral factorizes into products of M integrals. These integrals can be evaluated in closed form, and the subsequent limit $M \to \infty$, $L \to \infty$, $\rho = M/L$ fixed can be easily taken. We skip these intermediate steps and present the final results.

The final results satisfy the scaling forms discussed below (12.3), but they are, as expected, more usefully expressed in terms of reduced scaling forms that describe the semiclassical physics of the dilute gas of triplet magnons. The characteristic length and time scales of these reduced scaling functions are closely analogous to those found for the Ising chain in (10.63) and (10.71). In particular, we choose

$$\xi_c = \frac{1}{\rho} = \frac{1}{3}\left(\frac{2\pi c^2}{T\Delta_+} \right)^{1/2} e^{\Delta_+/T},$$

$$\tau_\varphi = \frac{\xi_c}{\sqrt{2}v_T} = \frac{\sqrt{\pi}}{3T}e^{\Delta_+/T}. \qquad (12.21)$$

Note that ξ_c is the mean spacing between the particles and τ_φ is a typical time between particle collisions, which is naturally identified also as phase coherence time. The final result for C_u is then

$$C_{u,\alpha\beta}(x, t) = \frac{2\rho^2}{3} F\left(\frac{|x|}{\xi_c}, \frac{|t|}{\tau_\varphi}\right) \delta_{\alpha\beta}, \qquad (12.22)$$

where F is a universal scaling function given by

$$F(\bar{x}, \bar{t}) = \left[\left(2G_1(u)G_1(-u) + e^{-u^2}/(\bar{t}\sqrt{\pi}) \right) I_0 \left(2\bar{t}\sqrt{G_2(u)G_2(-u)} \right) \right.$$

$$\left. + \frac{G_1^2(u)G_2(-u) + G_1^2(-u)G_2(u)}{\sqrt{G_2(u)G_2(-u)}} I_1 \left(2\bar{t}\sqrt{G_2(u)G_2(-u)} \right) \right]$$

$$\times \exp(-(G_2(u) + G_2(-u))\bar{t}), \qquad (12.23)$$

with $u \equiv \bar{x}/\bar{t}$, $G_1(u) = \mathrm{erfc}(u)/2$, and $G_2(u) = e^{-u^2}/(2\sqrt{\pi}) - uG_1(u)$. These expressions satisfy $\int_0^\infty d\bar{x} F(\bar{x}, \bar{t}) = 1/2$, which ensures the conservation of the total magnetization density with time and yields

$$\int dx\, C_{u,33}(x, t) = \frac{2\rho}{3} = T\chi_u, \qquad (12.24)$$

with the uniform susceptibility, χ_u, given by (12.12); this relationship between the spatial integral of C_u and χ_u follows from the conservation of total magnetization (which implies that the spatial integral of C_u is t independent), and the analog of the relation (10.76) (to be derived shortly) applied to correlators of the angular momentum density. For short times, F has the ballistic form

$$F(\bar{x}, \bar{t}) \approx e^{-\bar{x}^2/\bar{t}^2}/\bar{t}\sqrt{\pi}, \qquad (12.25)$$

which is the autocorrelator of a classical ideal gas for $d = 1$ and holds for $|\bar{t}| \ll |\bar{x}| \ll 1$. In contrast, for $|\bar{t}| \gg 1$, $|\bar{x}|$ it crosses over to the *diffusive* form

$$F(\bar{x}, \bar{t}) \approx \frac{e^{-\sqrt{\pi}\bar{x}^2/2\bar{t}}}{(4\pi \bar{t}^2)^{1/4}}. \qquad (12.26)$$

In the original dimensionful units, (12.21) and (12.26) imply a spin diffusion constant, D_s, given exactly by

$$D_s = \frac{c^2 e^{\Delta_+/T}}{3\Delta_+}. \qquad (12.27)$$

While this is an exact spin diffusion coefficient of the semiclassical model introduced above, it is not immediately clear that this result is also exact for the underlying quantum rotor model. There is a subtle question of orders of limits, which makes the above less than rigorous, and the reader is referred to [107] for further discussion. Also, let us note that the Fourier transform of (12.26) yields the diffusive form (12.2) with the susceptibility χ_u given by (12.12).

We turn to the correlations of the order parameter field $\mathbf{n}(x,t)$. These are very closely related to the computations of the $N=1$ case in Section 10.4.2. The basic observation is that, like $\hat{\sigma}^z$, the field $\mathbf{n}(x,t)$ is the creation and annihilation operator for magnon excitations above the ground state. In other words, a relationship analogous to (10.85) holds. This can be seen explicitly from the strong-coupling expansion in Section 6.1. Then, by arguments analogous to those in Section 10.4.2, we expect for the two-point correlator $C_{\alpha\beta} = C\delta_{\alpha\beta}$ in (11.2)

$$
\begin{aligned}
K(x,t) &\equiv C(x,t)|_{T=0} \\
&= \int \frac{dk}{2\pi} \frac{c\mathcal{A}}{2\varepsilon_k} e^{ikx - i\varepsilon_k t} \\
&= \frac{\mathcal{A}}{2\pi} K_0(\Delta_+ (x^2 - c^2 t^2)^{1/2}/c),
\end{aligned}
\tag{12.28}
$$

where $\mathcal{A}$ is the quasiparticle residue. The Bessel function is the Feynman propagator of a relativistic particle, and its properties were discussed below (10.86). The $T > 0$ computation proceeds as in Section 10.4.2. We have to augment the trajectories in Fig. 12.3 by an additional trajectory created and annihilated by the $\mathbf{n}$ fields. This is the only trajectory that moves only forward in time and hence picks up additional -1 signs at each of its collisions. The $T > 0$ modification is then a matter of averaging over these -1 signs. Unlike the Ising case, this cannot be done analytically, as the "colors" on the lines introduce additional complications. This problem and its numerical solution have been discussed elsewhere [107]; the answer has a structure closely analogous to that in Section 10.4.2. We find, as in (10.88), that

$$
C(x,t) = K(x,t) R(x,t),
\tag{12.29}
$$

where $R(x,t)$ is a relaxation function very similar, although not exactly equal, to that found in (10.88). It obeys a scaling form identical to (10.73), and so R decays exponentially on the spatial scale $\sim \xi_c$, and on the temporal scale $\sim \tau_\varphi$. As in Section (10.4.2), we can also Fourier transform (12.29) to obtain the structure factor $S(k,\omega)$. This has to be done numerically, and it is found that, for $|k| < \sqrt{\Delta T}/c$, the frequency dependence of the answer is reasonably well approximated by the following Lorentzian form:

$$
S(k,\omega) \approx \frac{\mathcal{A}}{\varepsilon_{||}} \frac{0.72/\tau_\varphi}{(\omega - \varepsilon_k)^2 + (0.72/\tau_\varphi)^2}.
\tag{12.30}
$$

This result is the analog of (10.89).

12.3 High-temperature limit of the continuum theory, $\Delta_+ \ll T \ll J$

If we continue to push the analogy with the Ising chain further, we would expect that the present region (Figs. 11.1 and 12.1) should be similar to the universal high-T region of the Ising chain discussed in Section 10.4.3. There, we found a novel regime of "quantum relaxational" dynamics for which no classical description was possible: the thermally excited particles had a spacing that was of the order of their de Broglie wavelength. The physics in the present region of the O(3) model is similar, but the presence of logarithms associated with the flow (12.8) does lead to a new twist. In particular, we find that logarithms of (T/Δ_+) make the classical thermal fluctuations marginally more important than the quantum fluctuations. If one is satisfied with results to leading logarithm accuracy (i.e. where one neglects all corrections of order $1/\ln(T/\Delta_+)$), then it is possible to develop an effective classical model of the dynamical properties. This classical model is quite different from that of the low-T region ($T \ll \Delta_+$), where we had a description in terms of classical particles. In contrast, the present description is in terms of *classical waves*. Our discussion here borrows heavily from the original analysis in [107].

There are a number of ways to make the basic argument. One is to note that the large-N result (11.60) predicts a correlation length for $\mathbf{n}$ correlations

$$\xi \sim (c/T) \ln(T/\Delta_+). \tag{12.31}$$

(We will shortly obtain the exact correlation length to leading logarithmic accuracy, and this has the same form as (12.31)). At distances of the order of, or shorter than, this correlation length we may crudely expect that the weak-coupling, spin-wave picture of Section 6.2 will hold, and the typical spin-wave excitations will have energy of order, or smaller than, $c\xi^{-1}$, which is logarithmically smaller than the thermal energy T; in other words

$$\frac{c\xi^{-1}}{T} \sim \frac{1}{\ln(T/\Delta_+)} < 1. \tag{12.32}$$

So the occupation number of these spin-wave modes will then be

$$\frac{1}{e^{c\xi^{-1}/T} - 1} \approx \frac{T}{c\xi^{-1}} > 1. \tag{12.33}$$

The last occupation number is precisely that appearing in a classical description of thermally excited spin waves, which is the approach we follow here.

Another way to state the dominance of classical effects is to run the flow equation (12.8) backwards: going to higher T means that we are exploring shorter scales and higher energies, at which (12.10) implies an effective coupling $g \sim 1/\ln(T/\Delta_+)$, which is small. The coupling g controls the strength of the quantum fluctuations, and these are therefore expected to be subdominant. This latter argument is made more precise in the following discussion.

We begin our analysis by first focusing on the static and thermodynamic correlations in this region. We shall use a method introduced by Luscher [307], and the same method is of considerable use to us in subsequent chapters. The main idea is to develop an effective action for only the *zero Matsubara frequency* ($\omega_n = 0$) components of $\mathbf{n}$ after integrating out all the $\omega_n \neq 0$ modes. We do this first for the correlation length in this and the following subsection. We turn to the thermodynamic uniform susceptibility in Section 12.3.2, and to the dynamical properties in Section 12.3.3.

The effective action for the zero-frequency modes can be obtained in the same background field method discussed in Section 12.1: we just identify the $\mathbf{n}_<$ modes with the zero-frequency components and the $\vec{\pi}$ fields with all finite-frequency components. Then it is easily seen that the effective action for $\mathbf{n}_<$ has precisely the same form as the $d = 1$ *classical ferromagnet* discussed in Section 6.4, with partition function (6.45) at $\tilde{\mathbf{h}} = 0$; for our purposes we write this as

$$\mathcal{Z} = \int \mathcal{D}\mathbf{n}(x)\delta(\mathbf{n}^2 - 1) \exp\left(-\frac{(N-1)\xi}{4}\int dx \left(\frac{d\mathbf{n}(x)}{dx}\right)^2\right), \quad (12.34)$$

where ξ is already known from Section 6.4 to be the spatial correlation length. (Actually we have usually reserved ξ to be the symbol for the equal-time correlations, whereas the present approach gives the correlation length for the zero-frequency correlations; as we will see in Section 12.3.3, these two lengths are asymptotically equal because of the dominance of classical thermal fluctuations.) Generalizing (12.7) to the present situation we have

$$\frac{(N-1)\xi T}{2} = \frac{cN}{g} - c^2(N-2)\int^\Lambda \frac{dk}{2\pi} T \sum_{\omega_n \neq 0} \frac{1}{c^2k^2 + \omega_n^2} + \cdots$$

$$\approx \frac{cN}{g} - \frac{c(N-2)}{2\pi}\ln\left(\frac{\Lambda}{T}\right), \quad (12.35)$$

where in the second equation we have ignored constants of order unity. Now we can use (12.10) to eliminate Λ, and we find

$$\xi = \frac{c(N-2)\ln(T/\Delta_+)}{T\pi(N-1)}, \quad (12.36)$$

in agreement with (11.60). Notably, dependence on g has also disappeared. This is not an accident – the renormalization group was designed to make this happen order-by-order in g, and all physical properties depend only upon the measurable ratio Δ_+/T.

   Actually, it is possible to be quite precise about the omitted constants of order unity in the argument of the logarithm in (12.36). To do this requires use of the field-theoretic renormalization group, and this is done in Section 12.3.1. The same method is applied to the uniform susceptibility, χ_u, in Section 12.3.2.

12.3.1 Field-theoretic renormalization group

We introduced the renormalization group in Chapter 4, and implemented it by the "momentum shell" method in Section 4.2. This method is adequate for most purposes. However, in a few cases, involving higher loop computations or the determination of universal amplitude ratios, a more formal field-theoretic approach is needed. While physically not as transparent, it does allow for efficient computations in which all nonuniversal features are automatically suppressed.

   A full description of this sophisticated approach is already available in a number of reviews in the literature [59, 244, 557] (we especially recommend the article by Brézin *et al.* [59] for a physical exposition), and the uninitiated reader is referred to these works for an in-depth treatment. Here we are satisfied by noting the essential points and quickly reviewing the computations necessary for our purposes.

To understand the low-energy and long-distance limit of the $d=1$ $O(N)$ rotor model, it is necessary to understand the behavior of the couplings under changes of the cutoff Λ. Computationally, it is advantageous to replace the cutoff by a new renormalization scale, μ, defined in the following manner. We define the coupling constants and the scale of the fields by relating them to the values of suitably chosen Green's functions (computed in the presence of a cutoff Λ) at external momenta proportional to μ. This statement is often shortened to "define the couplings at the scale μ." Now if we take an arbitrary observable, and re-express it in terms of couplings defined at the scale μ, we find that the resulting expressions are finite in the limit $\Lambda \to \infty$ (this is a consequence of the "renormalizability" of the field theory). So we take just this limit in all Green's function, and we are left with Λ-independent expressions in which we no longer have to deal with the (messy) details of the short-distance cutoff. As an added bonus, the independence of the underlying physics on the arbitrary scale μ also yields the required renormalization group equations. For the case of the $O(N)$ rotor model, only two redefinitions of coupling constants or field scales ("renormalizations") are necessary [61]: one renormalizing the coupling g to $g_R(\mu)$, and the other rescaling the overall field scale (related to the quasiparticle residue $\mathcal{A}$) by a factor $\bar{Z}$. Let us consider just the coupling constant renormalization for now. There is a multiplicative factor that relates g_R to the bare coupling constant, g, in the theory with a cutoff Λ; in an expansion in powers of g_R, this factor is a function of $\ln(\Lambda/\mu)$. However, it is advantageous to regulate the ultraviolet behavior by dimensional regularization (which means evaluating all momentum integrals in $d=1-\epsilon$ spatial dimensions), in which case the logarithms turn into poles in ϵ. The explicit relationship between the bare and renormalized coupling was shown by Brézin and Zinn-Justin [61] to be

$$g = g_R(\mu)\mu^{-\epsilon}\left(1 + \frac{N-2}{2\pi N}\frac{g_R(\mu)}{\epsilon} + \mathcal{O}(g_R^2)\right). \tag{12.37}$$

Similarly, the field rescaling factor is shown to be [61]

$$\bar{Z} = 1 - \frac{N-1}{2\pi N}g_R(\mu) + \mathcal{O}\left(g_R^2\right). \tag{12.38}$$

It is now possible to state the simple, field-theoretic recipe for computing correlators of (11.4) for $d=1$. First, obtain formal expressions for any rotationally invariant, physically observable correlator of the bare theory in an expansion in powers of g, and leave all the Feynman integrals as formal, unevaluated expressions. Next, perform the substitution (12.37) to replace g by g_R, and also multiply the correlator by a power of $\bar{Z}^{-1}$ for each power of the field $\mathbf{n}$ in the correlator. Now, evaluate all the integrals in $d=1-\epsilon$ dimensions, in powers of ϵ. The constants in (12.37) and (12.38) have been cleverly chosen so that all poles in ϵ cancel. The resulting expressions for the correlators of the theory are now expressed in terms of g_R and the momentum scale μ, with no explicit dependence on Λ.

It would seem that not much has been achieved with this rather sophisticated transformation. We began with a theory with a dimensionless coupling g and a cutoff Λ. This cutoff was rather hard to deal with in computing Feynman graphs, especially multiloop ones. We have ended up with a closely related theory with the same universal low-energy properties. This theory is expressed in terms of a dimensionless coupling g_R and a scale μ, which plays the physical role of an ultraviolet cutoff. The latter theory is much easier to compute with, and so it seems that all we have done is to devise a clever and convenient short-distance regularization that allows us to compute properties to a high order in g_R.

However, there is an additional advantage to the second approach: by using the independence of the original bare theory on μ, it is possible to easily derive an *exact* renormalization group equation for the flow of $g_R(\mu)$, and all observables, under rescalings of the "cutoff" $\mu \to \mu e^\ell$. Indeed, simply differentiating (12.37) with respect to μ at a fixed g gives us the flow equation

$$\frac{dg_R}{d\ell} = \frac{N-2}{2\pi N} g_R^2, \tag{12.39}$$

which is of course the same equation obtained earlier in (12.8). We are dealing with the coupling g_R rather than g, but this is physically innocuous as it is simply the consequence of trading the momentum cutoff Λ for a renormalization scale μ (which effectively plays the role of the cutoff). Similarly, the field-scale renormalization, $\bar{Z}$, also implies an exact statement on the behavior of correlation functions under changes of μ. (Again we are not terribly concerned with the physical consequences of changing the scale of correlators of $\mathbf{n}$ as we eventually set the overall amplitude of the structure factor using the physically measurable quasiparticle amplitude $\mathcal{A}$.) This is also discussed by Brézin and Zinn-Justin [61]; for the two-point correlator of $\mathbf{n}$ defined in (11.2) their result takes the form

$$\bar{Z}^{-1} C(x, t; g_R(\mu_1), \mu_1) = \left[\ln\left(\frac{\mu_1}{\mu_2}\right) \right]^{\frac{(N-1)}{(N-2)}} \bar{Z}^{-1} C(x, t; g_R(\mu_2), \mu_2). \tag{12.40}$$

We have several occasions to use this fundamental relation later.

Let us return to the physical problem of computing the correlation length using the present field-theoretic approach. The consequences of the above recipe are simple: we take the formal expression represented by the first equation in (12.35), perform the substitution in (12.37) to replace g by $g_R(\mu)$, and then evaluate the integrals in $d = 1 - \epsilon$ dimensions. Let us specify a few steps required in the latter evaluation:

$$T \sum_{\omega_n \neq 0} \int \frac{d^{1-\epsilon}k}{(2\pi)^{1-\epsilon}} \frac{1}{c^2 k^2 + \omega_n^2}$$

$$= \int \frac{d^{1-\epsilon}k}{(2\pi)^{1-\epsilon}} \left[T \sum_{\omega_n \neq 0} \frac{1}{c^2 k^2 + \omega_n^2} - \int \frac{d\omega}{2\pi} \frac{1}{c^2 k^2 + \omega^2 + T^2} \right]$$

$$+ c^{1-\epsilon} \int \frac{d^{2-\epsilon}p}{(2\pi)^{1-\epsilon}} \frac{1}{c^2 p^2 + T^2}$$

$$= \frac{1}{c} \left(\frac{T}{c}\right)^{-\epsilon} \left\{ \int \frac{d^{1-\epsilon}k}{(2\pi)^{1-\epsilon}} \left[\frac{1}{2k} \coth \frac{k}{2} - \frac{1}{k^2} - \frac{1}{2\sqrt{k^2+1}} \right] + \frac{\Gamma(\epsilon/2)}{(4\pi)^{1-\epsilon/2}} \right\}. \tag{12.41}$$

We are only interested in the poles in ϵ and the accompanying constants, and to this accuracy the first integral on the right-hand side can be evaluated directly at $\epsilon = 0$, while the

Γ function yields a $1/\epsilon$ term. Now inserting (12.41) and (12.37) into the first equation in (12.35), we find that all the poles in ϵ cancel in the resulting expression, and we get

$$\frac{(N-1)\xi T}{2c} = \frac{N}{g_R(\mu)} - \frac{(N-2)}{2\pi}\ln(\mu/T\sqrt{\mathcal{C}}), \tag{12.42}$$

where the constant $\mathcal{C}$ was defined in (11.61). Rather than leave this expression in terms of μ and $g_R(\mu)$, it is conventional to express the result in terms of the so-called renormalization group invariant $\Lambda_{\overline{MS}}$. This is a somewhat unfortunate conventional notation for this quantity, as it suggests that $\Lambda_{\overline{MS}}$ is some sort of cutoff. In fact it is not; it is really a quantity that is closely analogous to the momentum scale Δ_+/c, which is related to the energy gap, or the $T=0$ correlation length. In the language of Section 5.5.1, $\Lambda_{\overline{MS}}^{-1}$ is a "large" length scale, rather than a "short" scale. The basic idea behind the definition of $\Lambda_{\overline{MS}}$ is as follows. Choose any physically measurable length scale associated with a $d=1$ rotor model at $T=0$ you wish. By simple dimensional analysis, this scale must be of the form $(1/\mu) \times$ some function of $g_R(\mu)$. Because this scale is physically measurable, it must not depend upon the choice of μ, that is, the resulting combination should be invariant under the flow equation (12.37). This turns out to be a very strong restriction: up to an arbitrary overall numerical factor, it turns out there is only one such function. We choose this overall factor by convention and call the result $\Lambda_{\overline{MS}}$: by integration of the two-loop version of the flow equation, we define, following [307],

$$\Lambda_{\overline{MS}} = \mu\sqrt{\mathcal{C}}\left(\frac{(N-2)}{2\pi N}g_R\right)^{-1/(N-2)} \exp\left(-\frac{2\pi N}{g_R(N-2)}\right). \tag{12.43}$$

The constant $\mathcal{C}$ in the prefactor is purely for convenience and arbitrarily chosen.

Now the implication of the reasoning above is that *all* $T=0$ measurable length scales are universal numbers times $\Lambda_{\overline{MS}}^{-1}$, and they cannot depend separately upon μ and $g_R(\mu)$; similarly, all measurable length scales at $T>0$ are $\Lambda_{\overline{MS}}^{-1}$ times universal functions of the dimensionless ratio $c\Lambda_{\overline{MS}}/T$. It is easy to verify that this holds for our expression for the correlation length in (12.42). Solving (12.43) for $g_R(\mu)$ and substituting in (12.42) we find

$$\xi(T) = \frac{c(N-2)}{T\pi(N-1)}\left\{\ln\left[\frac{\mathcal{C}T}{c\Lambda_{\overline{MS}}}\right]\right.$$
$$\left. + \frac{1}{(N-2)}\ln\ln\frac{T}{c\Lambda_{\overline{MS}}} + \mathcal{O}\left(\frac{\ln\ln(T/c\Lambda_{\overline{MS}})}{\ln(T/c\Lambda_{\overline{MS}})}\right)\right\}. \tag{12.44}$$

As expected, the scale μ has completely dropped out.

However, the expression (12.44) is not very useful as it stands because it involves the scale $\Lambda_{\overline{MS}}$, which was defined by convention in the dimensional regularization scheme and is not a priori known for any physical system. To make it useful, we need to relate $\Lambda_{\overline{MS}}$ to some other physical observable. We have consistently been using the $T=0$ energy gap Δ_+ to characterize the ground state, and so it would be useful to know the universal dimensionless ratio $\Delta/c\Lambda_{\overline{MS}}$. This was computed recently by Hasenfratz and collaborators [205–207] using the Bethe ansatz solution of the σ-model; they obtained

$$\frac{\Delta_+}{c\Lambda_{\overline{MS}}} = \frac{(8/e)^{1/(N-2)}}{\Gamma(1 + 1/(N-2))}. \tag{12.45}$$

The results (12.44) and (12.45) constitute the more precise form of (12.36). Explicitly, for the case $N = 3$ we have the exact leading result for the correlation length

$$\xi(T) = \frac{c}{2\pi T} \ln\left(\frac{32\pi e^{-(1+\gamma)} T}{\Delta_+}\right), \tag{12.46}$$

where γ is Euler's constant.

12.3.2 Computation of χ_u

This section determines the uniform susceptibility, χ_u, by a strategy similar to that employed above in the computation of $\xi(T)$: place the system in an external magnetic field $\mathbf{H}$, integrate out the nonzero-frequency modes, and then perform the average over the zero-frequency fluctuations. We choose an $\mathbf{H}$ that rotates $\mathbf{n}$ in the 1–2 plane, and use (12.5) to integrate out the nonzero frequencies. Therefore the fields $\mathbf{n}_<$, $\mathbf{e}_a$ are independent of τ, while the π_a have no zero-frequency components. It is also clear that the fields $\mathbf{n}_<(x)$ are simply the $\mathbf{n}(x)$ fields appearing in (12.34). We expand the partition function to quadratic order in $\mathbf{H}$, drop all terms proportional to the spatial gradients of $\mathbf{n}(x)$ or $\mathbf{e}_a(x)$ (these can be shown to yield logarithmically subdominant contributions to χ_u), and find that the $\mathbf{H}$-dependent terms in the free energy density are

$$-\frac{N\mathbf{H}^2}{2cg}\left[\left(n_1^2 + n_2^2\right)\left(1 - \sum_a \langle\pi_a^2\rangle\right) + \sum_{ab}(e_{a1}e_{b1} + e_{a2}e_{b2})\langle\pi_a\pi_b\rangle\right.$$

$$-\frac{N}{cg}\sum_{abcd}(e_{a1}e_{b2} - e_{a2}e_{b1})(e_{c1}e_{d2} - e_{c2}e_{d1})$$

$$\left. \times \int dx d\tau \, \langle\pi_a\partial_\tau\pi_b(x,\tau); \pi_c\partial_\tau\pi_d(0,0)\rangle\right]. \tag{12.47}$$

Evaluating the expectation values of the π fields, and using orthonormality of the vectors $\mathbf{n}$, $\mathbf{e}_a$, we can simplify the expression (12.47) to

$$-\frac{N\mathbf{H}^2}{2cg}\left[\left(n_1^2 + n_2^2\right)\left(1 - \frac{c(N-2)g}{N}T\sum_{\omega_n \neq 0}\int\frac{dk}{2\pi}\frac{1}{c^2k^2 + \omega_n^2}\right)\right.$$

$$\left. + \frac{2cg}{N}\left(1 - n_1^2 - n_2^2\right)T\sum_{\omega_n \neq 0}\int\frac{dk}{2\pi}\frac{c^2k^2 - \omega_n^2}{(c^2k^2 + \omega_n^2)^2}\right]. \tag{12.48}$$

Finally, to obtain the susceptibility χ_u, we have to evaluate the expectation value of the zero-frequency field $\mathbf{n}$ under the partition function (12.34). This simply yields

$\langle n_1^2 \rangle = \langle n_2^2 \rangle = 1/N$. The first frequency summation is precisely the same as that evaluated earlier for ξ in the first equation in (12.35), while the second is explicitly finite in $d = 1$ and can be directly evaluated; in this manner we obtain our final result for χ_u:

$$\chi_u(T) = \frac{2}{N} \left[\frac{(N-1)T\xi}{2c^2} - \frac{(N-2)}{2\pi c} \right]$$

$$= \frac{(N-2)}{N\pi c} \ln \left(\frac{cT}{c\Lambda_{\overline{MS}}e} \right). \tag{12.49}$$

We have omitted the form of the subleading logarithms, which are the same as those in (12.44). Again, let us quote the explicit expression for χ_u for $N = 3$:

$$\chi_u(T) = \frac{1}{3\pi c} \ln \left(\frac{32\pi e^{-(2+\gamma)}T}{\Delta_+} \right). \tag{12.50}$$

It is useful to compare the $T \gg \Delta_+$ expression (12.50) for χ_u with the $T \ll \Delta_+$ result in (12.12): the two expressions are roughly equal for $T \approx \Delta$, suggesting that one or the other of the two asymptotic limits is always reasonable.

12.3.3 Dynamics

We have now assembled all the ingredients necessary for a complete description of the low-frequency dynamics. The key observation, made above (12.33), is that the energy, ω, of the characteristic excitation obeys $\omega \ll T$. We expect the spectral density, $\text{Im}\chi(k, \omega)$, to be dominated by weight at such frequencies, and the fluctuation–dissipation theorem (7.22) then takes its "classical" form in (10.75). We work here with an effective theory in which (10.75) is obeyed exactly, and so the equal-time structure factor, $S(k)$, is related to the static susceptibility, $\chi(k)$, by $S(k) = T\chi(k)$, as in (10.76). However, the static susceptibility is given by the two-point correlator of the $\omega_n = 0$ components of the $\mathbf{n}$ field, and these are determined by the effective action (12.34). Then we arrive at the important conclusion that (12.34) yields the *equal-time* correlators of $\mathbf{n}$ in the limit that the classical fluctuation–dissipation theorem in (10.75) is obeyed.

How do we extend (12.34) to unequal-time correlations? Recall that in classical statistical mechanics equal-time correlations are given by an integral over configuration space (as in $\int dq$), while an extension to dynamics requires an integral over phase space (as in $\int dp dq$). Furthermore, the integral over the conjugate momenta simply factorizes, and for equal-time correlations we can return to the configuration space formalism. So here, we need to extend (12.34) by finding the appropriate integral over conjugate momenta. The conjugate momentum of the rotor orientation $\mathbf{n}$ is clearly the rotor angular momenta $\mathbf{L}$. Therefore we treat $\mathbf{L}$ also as a classical variable and generalize (12.34) to a "$\int dq dp$"

integral of the form (we particularize the remainder of the discussion to the special case $N = 3$):

$$\mathcal{Z} = \int \mathcal{D}\mathbf{n}(x)\mathcal{D}\mathbf{L}(x)\delta(\mathbf{n}^2 - 1)\delta(\mathbf{L} \cdot \mathbf{n}) \exp\left(-\frac{\mathcal{H}_c}{T}\right),$$

$$\mathcal{H}_c = \frac{1}{2}\int dx\left[T\xi\left(\frac{d\mathbf{n}}{dx}\right)^2 + \frac{1}{\chi_{u\perp}}\mathbf{L}^2\right], \tag{12.51}$$

where $\mathbf{L}$ and $\mathbf{n}$ are classical commuting variables. The second term in $\mathcal{H}_c$ was absent in (12.34) and represents the kinetic energy of the classical rotors. Integrating out $\mathbf{L}$ we obtain (12.34) as we should. The value of the coupling $\chi_{u\perp}$ in the kinetic energy can be determined by a simple argument. It is clear that an external field $\mathbf{H}$ will couple to the total angular momentum and will therefore modify the classical Hamiltonian by

$$\mathcal{H}_c \to \mathcal{H}_c - \int dx\mathbf{H} \cdot \mathbf{L}. \tag{12.52}$$

Evaluating the linear response of (12.51) shows that

$$\chi_u = \frac{2}{N}\chi_{u\perp}, \tag{12.53}$$

with $N = 3$ (we have given, without proof, the expression for general N); the factor of $2/3$ comes from the constraint $\mathbf{L} \cdot \mathbf{n} = 0$. Using (12.49), we then have the value of $\chi_{u\perp}$. It should also be clear from this discussion that $\chi_{u\perp}$ has a simple physical interpretation: it is the susceptibility to a field oriented perpendicular to the local direction of the order parameter $\mathbf{n}$.

Finally, to proceed to unequal-time correlations, we need the equations of motion obeyed by the classical $\mathbf{n}$, $\mathbf{L}$ fields. A direct approach is to compute the quantum equations of motion and then to simply treat the quantum operators $\hat{\mathbf{n}}$ and $\hat{\mathbf{L}}$ as classical c-numbers. This is valid because the expectation value of any term will be dominated by large values as in (12.33), and any effects from noncommutativity will be suppressed. A quicker way to obtain the answer is to realize that the same result is obtained by replacing the quantum commutators by Poisson brackets and generating the Hamilton–Jacobi equations of the Hamiltonian $\mathcal{H}_c$. The required Poisson brackets here are the continuum classical limit of the commutation relations (1.21):

$$\{L_\alpha(x), L_\beta(x')\}_{PB} = \epsilon_{\alpha\beta\gamma}L_\gamma(x)\delta(x - x'),$$

$$\{L_\alpha(x), n_\beta(x')\}_{PB} = \epsilon_{\alpha\beta\gamma}n_\gamma(x)\delta(x - x'),$$

$$\{n_\alpha(x), n_\beta(x')\}_{PB} = 0. \tag{12.54}$$

From this, and (12.51), we obtain directly the equations of motion for the quasi-classical waves:

$$\frac{\partial \mathbf{n}}{\partial t} = \{\mathbf{n}, \mathcal{H}_c\}_{PB}$$

$$= \frac{1}{\chi_{u\perp}} \mathbf{L} \times \mathbf{n},$$

$$\frac{\partial \mathbf{L}}{\partial t} = \{\mathbf{L}, \mathcal{H}_c\}_{PB}$$

$$= (T\xi)\mathbf{n} \times \frac{\partial^2 \mathbf{n}}{\partial x^2}. \tag{12.55}$$

To compute the needed unequal-time correlation functions, pick a set of initial conditions for $\mathbf{n}(x)$, $\mathbf{L}(x)$ from the ensemble (12.51). Evolve these deterministically in time using the equations of motion (12.55). The value of the correlator is then the product of the appropriate time-dependent fields, averaged over the set of all initial conditions. We also note here that simple analysis of the differential equations (12.55) shows that small disturbances about a nearly ordered $\mathbf{n}$ configuration travel with a characteristic velocity $c(T)$ given by

$$c(T) = (T\xi(T)/\chi_{u\perp}(T))^{1/2}, \tag{12.56}$$

which is a basic relationship between thermodynamic quantities and the velocity $c(T)$. Note from (12.44) and (12.49) that to leading logarithms $c(T) \approx c$, but the second term in the first equation of (12.49) already shows that this result is not satisfied by the subleading terms.

Before relating the required correlators of the quantum model for $T \gg \Delta_+$ to the classical model defined above, we need to settle one final issue: that of the overall scale of the fields $\mathbf{n}$, $\mathbf{L}$. The scale of $\mathbf{L}$ is easy to set – it is specified completely by the coupling to the field $\mathbf{H}$ in (12.52), and by the first of the Poisson bracket relations in (12.54). These take the same values in the underlying quantum model and undergo no renormalization upon integrating out the finite-frequency degrees of freedom. We have therefore

$$C_{u,\alpha\beta}(x, t) = \langle L_\alpha(x, t) L_\beta(0, 0) \rangle_c, \tag{12.57}$$

where the subscript c represents the averaging procedure discussed below (12.55). The argument for the field scale of $\mathbf{n}$ is somewhat more subtle. So far the only parameter that has been sensitive to the scale of the order parameter has been the quasiparticle amplitude, $\mathcal{A}$, which was defined from the residue of the quasiparticle pole at $T = 0$. In contrast, we need the overall scale of $\mathbf{n}$ at a temperature $T \gg \Delta_+$. The matching between these two scales can, however, be performed with the aid of the renormalization group invariance equation (12.40) noted earlier. Now the quasiparticle amplitude $\mathcal{A}$ is naturally defined at a scale $\mu_1 \sim \Delta_+$, where the coupling g_R is of order unity. However, integrating out the finite-frequency modes and deriving the effective action for the zero-frequency modes is most easily done at $\mu_2 \sim T$, as the coupling $g_R \sim 1/(\ln(T/\Delta_+)$ and the perturbation theory will be free of large logarithms. The two scales can be related via (12.40), and in this way we obtain the required result

$$C_{\alpha\beta}(x, t) = \mathcal{A}\tilde{\mathcal{C}}\left[\ln\left(\frac{T}{\Delta_+}\right)\right]^{\frac{(N-1)}{(N-2)}} \langle n_\alpha(x, t)n_\beta(0, 0)\rangle_c. \tag{12.58}$$

The constant $\tilde{\mathcal{C}}$ is an unknown pure, universal number, which cannot be obtained by the present methods. It could, in principle, be obtained from the Bethe-ansatz solution.

Let us now examine the structure of the classical dynamics problem defined by (12.51) and (12.55). It obeys the crucial property of being free of all ultraviolet divergences. This is clear from the analysis of equal-time correlations in Section 6.4 and the unequal-time perturbation theory discussed in [65]. Consequently, we may determine its characteristic length and time scales by simple engineering dimensional analysis, as no short-distance cutoff scale is going to transform into an anomalous dimension. Indeed, a straightforward analysis shows that this classical problem is free of dimensionless parameters and is a *unique*, parameter-free theory. This is seen by defining

$$\bar{x} = \frac{x}{\xi},$$

$$\bar{t} = \frac{t}{\tau_\varphi},$$

$$\bar{\mathbf{L}} = \mathbf{L}\sqrt{\frac{\xi}{T\chi_{u\perp}}}, \tag{12.59}$$

where we have anticipated that the characteristic time, τ_φ, will be the phase coherence time, and it is given by

$$\tau_\varphi = \sqrt{\frac{\xi\chi_{u\perp}}{T}}; \tag{12.60}$$

then inserting these into (12.51) and (12.55), we find that all parameters disappear and the partition function and equations of motion acquire a unique, dimensionless form, given by setting $T = \xi = \chi_{u\perp} = 1$ in them.

The above transformations allow us to obtain scaling forms for the dynamic observables in terms of, as yet undetermined, universal functions.

First, consider the correlators of $\mathbf{n}$. The equal-time two-point correlations of (12.34) are known from Section 6.4 to decay simply as $e^{-|x|/\xi}/3$; from these and (12.58), we deduce that the equal-time structure factor $S(k)$ (defined in (10.5)) is given by

$$T\chi(k) = S(k) = \mathcal{A}\tilde{\mathcal{C}}\left[\ln\left(\frac{T}{\Delta_+}\right)\right]^2 \frac{2\xi/3}{(1 + k^2\xi^2)}. \tag{12.61}$$

For the dynamic structure factor, $S(k, \omega)$, (12.59) implies a scaling form similar to (10.78),

$$\frac{2T}{\omega}\mathrm{Im}\chi(k, \omega) = S(k, \omega) = S(k)\tau_\varphi\Phi_{\mathrm{Sc}}(k\xi, \omega\tau_\varphi), \tag{12.62}$$

where Φ_{Sc} is a universal scaling function, normalized as in (10.79). Also, because the equations of motion are classical, the relation (10.75) is obeyed exactly, and Φ_{Sc} is an even function of $\bar{\omega}$. For further information on the structure of Φ_{Sc} we refer to a recent paper [65], which used a combination of analytic and numerical methods. At sufficiently large $k\xi$, we expect a pair of broadened, reactive, "spin-wave" peaks at $\omega \approx c(T)k$ (with

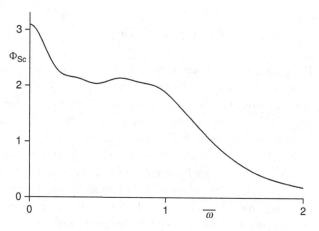

Numerical results of [65] for the scaling function $\Phi_{Sc}(0, \bar{\omega})$ appearing in (12.62).

$c(T)$ given in (12.56) and an exact expression for the linewidth given in [65]), which are similar to those found in the high-T limit of the quantum Ising chain in Fig. 10.11. For the opposite limit of small $k\xi$, we present the numerical results of [65] for $\Phi_{Sc}(0, \bar{\omega})$ in Fig. 12.4. There is a sharp relaxational peak at $\bar{\omega} = 0$, which is again similar to that found in the high-T limit of the quantum Ising chain in Fig. 10.11. However, there is now a well-defined shoulder at $\bar{\omega} \approx 0.7$, which was not found in the Ising case. This shoulder is a remnant of the large-N result (11.10), which predicts a delta function at $\omega = \pm m$, with m given by (11.60) in $d = 1$. So $N = 3$ is large enough for this finite-frequency oscillation to survive in the high-T limit.

There is an alternative, helpful way to view this oscillation frequency. The underlying degree of freedom in our dynamical field theory has a fixed amplitude, with $|\mathbf{n}| = 1$. However, correlations of $\mathbf{n}$ decay exponentially on a length scale ξ; consequently, if we imagine coarse-graining out to ξ, it is reasonable to expect significant *amplitude fluctuations* in the coarse-grained field. It is now useful to visualize an effective field ϕ_α with no length constraint, as we discussed in Section 2.1. On a length scale of order ξ, we expect the effective potential controlling fluctuations of ϕ_α to have a minimum at a nonzero value of $|\phi_\alpha|$ but to also allow fluctuations in $|\phi_\alpha|$ about this minimum. The finite frequency in Fig. 12.4 is due to the harmonic oscillations of ϕ_α about this potential minimum, while the dominant peak at $\omega = 0$ is due to angular fluctuations along the zero-energy contour in the effective potential. This interpretation is also consistent with the large-N limit, in which we freely integrate over all components of $\mathbf{n}$, and so angular and amplitude fluctuations are not distinguished. The above argument could also have been applied to the quantum Ising chain (in this case, angular fluctuations are replaced by low-energy domain wall motion, considered in Section 10.4.1), but the absence of such a reactive, finite-frequency peak at $k = 0$ in Fig. 10.11 indicates that $N = 1$ is too far from $N = \infty$ for any remnant of this large-N physics to survive. We meet related phenomena in our study of a quasi-classical wave model for the high-T limit for $d = 2$ in Section 14.3.

We turn next to the correlators of $\mathbf{L}$. The long-time behavior of these was examined numerically in [65], and it was found to be consistent with the diffusive form (12.2). We already know the value of the uniform susceptibility, χ_u. For the spin diffusion constant, D_s, we can deduce simply from the fact that it has dimension (length)^2/time, and from (12.59), that it must obey

$$D_s = \mathcal{B}\frac{T^{1/2}\xi^{3/2}}{\chi_{u\perp}^{1/2}}, \tag{12.63}$$

where $\mathcal{B}$ is a universal number. The numerical estimate [65] is $\mathcal{B} \approx 3.3$.

12.4 Summary

We summarize the basic properties of the two regimes in Figs. 11.1 and 12.1 and in Table 12.1. We also recall that in the low-T region, the dynamic structure factor, $S(k, \omega)$, has most of its weight in a frequency window about $\omega = \Delta_+$ of width $1/\tau_\varphi$. In the high-T region, $S(k, \omega)$ becomes an even function of ω and most of its weight is in a window of width $1/\tau_\varphi$ centered around $\omega = 0$.

Finally, we mention some application to experiments.

We have already seen in Section 1.4.3 that the $d = 1, O(3)$ quantum rotor model describes the so-called two-leg ladder antiferromagnets [33, 102]. There are materials, such as $SrCu_2O_3$ [33], that consist of two adjacent $S = 1/2$ spin chains, with neighboring spins

Table 12.1 Basic properties of low-T and high-T regimes.

Values of the correlation length, ξ (defined from the exponential decay of the equal-time correlations of **n**), the uniform spin susceptibility, χ_u, the phase coherence time, τ_φ, and the spin diffusion constant, D_s, for the two regimes in Figs. 11.1 and 12.1. Results are for $N = 3$, although many results for general $N \geq 3$ appear in the text. There is a large length scale, ξ_c, in the low-T region, which was given in (12.21) and does not appear below; this is the spacing between the thermally excited particles.

	Low-T, quasi-classical particles	High-T, quasi-classical waves
ξ	$\dfrac{c}{\Delta_+}$	$\dfrac{c}{2\pi T}\ln\left(\dfrac{32\pi e^{-(1+\gamma)}T}{\Delta_+}\right)$
χ_u	$\dfrac{1}{c}\left(\dfrac{2\Delta_+}{\pi T}\right)^{1/2}e^{-\Delta_+/T}$	$\dfrac{1}{3\pi c}\ln\left(\dfrac{32\pi e^{-(2+\gamma)}T}{\Delta_+}\right)$
τ_φ	$\dfrac{\sqrt{\pi}}{3T}e^{\Delta_+/T}$	$\left(\dfrac{3\xi\chi_u}{2T}\right)^{1/2}$
D_s	$\dfrac{c^2 e^{\Delta_+/T}}{3\Delta_+}$	$2.7\xi\left(\dfrac{T\xi}{\chi_u}\right)^{1/2}$

on the two chains coupled to each other like the rungs of a ladder; thus, they are modeled by (1.28) for the case where the sum over i, j extends a simple one-dimensional chain. Actually, as we see in Section 19.3.1, a much broader class of $d = 1$ antiferromagnets is described by the O(3) rotor model, including spin chains in which the individual spins have integer spin S. The mapping to the rotor model requires that all of these antiferromagnets have an energy gap above the ground state.

There have been a very large number of experimental studies of such one-dimensional antiferromagnets. For example, in a neutron scattering study of the $S = 1$ spin chain compound Y_2BaNiO_5, Xu *et al.* [544, 545] present clear evidence for a triplet particle in the low-T spectral density, along with the long phase coherence time associated with its presence [4, 544]. The most recent measurements [545] are in quantitative agreement with the results in Section 12.2. Thermodynamic and NMR measurements on $S = 1$ spin chains and spin ladders have been surveyed by Itoh and Yasuoka [239]: a striking feature of the data is that the energy gaps measured in activation plots of the NMR relaxation rate $1/T_1$ are about 1.5 times the measured gap in a thermodynamic measurement of the uniform susceptibility. It is argued in [107] that this feature can be quite generally explained by the picture of low-T spin diffusion developed in Section 12.2 and the value of the spin diffusivity in (12.27). Detailed comparisons [107] of the ballistic to diffusive crossover in (12.23) have been made against NMR experiments by Takigawa *et al.* [498] on the $S = 1$ spin chain compound $AgVP_2S_6$.

As we see in Chapter 19, the high-T analysis of Section 12.3 applies to spin chains with larger values of S, or to spin ladders with greater than two legs, at intermediate temperatures; the precise limits on experimental applicability are discussed in [65]. Explicit comparisons of the thermodynamic predictions in Section 12.3 have been made against Monte Carlo data for $S = 2$ chains by Kim *et al.* [269], with reasonable agreement. Experimental studies of $S = 2$ chains have also been undertaken [174], and there are interesting prospects for confrontation between theory and experiments on dynamical properties in future work. Dynamical measurements have been made on two-leg ladder compounds at higher temperatures [265] and the results have an interesting qualitative similarity to (12.63).

The large-N limit of quantum rotor models for $d = 2$ was examined in Chapter 11 and led to the phase diagram shown in Fig. 11.2. There we claimed that the large-N results provided a satisfactory description of the crossovers in the static and thermodynamic observables for $N \geq 3$. We establish this claim in this chapter and also treat the dynamic correlations of $\mathbf{n}$ at nonzero temperatures. The discussion of the dynamics takes place in a physical framework suggested by the modified version of Fig. 11.2 shown in Fig. 13.1. The low-T region on the quantum paramagnetic side can be described in an effective model of quasi-classical particles that is closely related to those developed in Sections 10.4.2 and 12.2. In the other low-T region on the magnetically ordered side, we obtain a "dual" model of quasi-classical waves, which is connected to that developed in Section 12.3. Finally, in the intermediate "quantum critical" or "continuum high-T" region, neither of these descriptions is adequate: quantum and thermal behavior, as well as particle- and wavelike behavior, all play important roles, and we use a melange of these concepts to obtain a complete picture in this and the following two chapters.

The results for the quasi-classical wave regime described in this chapter are obtained by a combination of analytical and numerical techniques, which become exact in the low-T limit. For the other two regions, we use the large-N expansion. This approximate approach is satisfactory for most purposes, but it fails in the very low-frequency regime, $\omega \ll T$. A proper description of the low-frequency dynamical correlators of $\mathbf{n}$ must await alternative techniques, which are developed in Chapter 14 and Section 14.3.

The cases $N = 1, 2$, $d = 2$ are special because they permit phase transitions at nonzero temperatures, and their crossover phase diagrams are of the form in Fig. 11.3. We do not treat the ordered phases or the vicinity of the nonzero-temperature transition in this chapter but defer their discussion to Chapter 14. In principle, the results obtained for the low-T region on the quantum paramagnetic side, and for the continuum high-T region (see Fig. 11.3), apply for all N, including $N = 1, 2$. However, the caveats mentioned in the previous paragraph on the failure of the large-N expansion at low frequencies apply even more strongly to $N = 1, 2$, and the dynamics for these cases is best understood using the methods of Chapters 14 and 15. Nevertheless, we quote our results in this chapter for these two regions for all values of N.

We do not consider time-dependent correlations of the angular momentum $\mathbf{L}$ in this chapter. The conservation of the total $\mathbf{L}$ implies that its low-frequency dynamics obeys the diffusive form (12.2). So the problem reduces to determination of a "transport coefficient" (the spin diffusion constant D_s), and we defer discussion of the transport problem to Chapter 15.

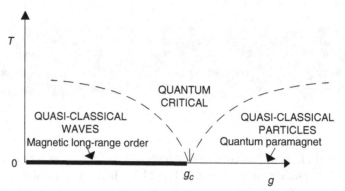

Fig. 13.1 Modified version of Fig. 11.2 for the crossovers for the rotor model (11.1, 11.4) for $d=2$, $N \geq 3$; see also Fig. 1.3. While quasi-classical descriptions of the dynamics and transport can be developed in the two low-T regions, the behavior in the "quantum critical" or "continuum high-T" region is more complex, with contributions from both thermal and quantum phenomena, and from both particle- and wave-like phenomena. We show in Section 14.3 that, to leading order in $\epsilon = 3 - d$, the low-frequency correlators of $\mathbf{n}$ in the quantum critical region are described by an effective quasi-classical wave model. In contrast, the transport of the conserved $\mathbf{L}$ in the quantum critical region is dominated by higher energy excitations and requires a particle-like description in a quantum Boltzmann equation, which is discussed in Chapter 15.

The main purpose of this chapter is a more complete description of the basic scaling forms for nonzero-temperature correlations of $\mathbf{n}$ introduced in Section 11.3. For $d=2$, on the magnetically ordered side ($g < g_c$), the scaling ansatz (11.41) is

$$\chi(k, \omega) = \frac{Z}{T^{(2-\eta)}} \Phi_- \left(\frac{ck}{T}, \frac{\omega}{T}, \frac{\rho_s}{NT} \right), \tag{13.1}$$

where we have set $z = 1$ and used the expression (11.39) for Δ_-, which for $d = 2$ is simply

$$\Delta_- = \rho_s/N; \tag{13.2}$$

that is, the $T = 0$ spin stiffness ρ_s is an energy that serves as a measure of the deviation of the magnetically ordered ground state from the quantum critical point; the factor of $1/N$ (13.2) is for future convenience, as ρ_s is naturally of order N in the large-N limit. Clearly ρ_s is defined only for the case of models with a continuous symmetry, and so (13.1) applies only for $N \geq 2$. For $g > g_c$ we have

$$\chi(k, \omega) = \frac{Z}{T^{(2-\eta)}} \Phi_+ \left(\frac{ck}{T}, \frac{\omega}{T}, \frac{\Delta_+}{T} \right), \tag{13.3}$$

characterizing the nonzero-temperature behavior on the quantum paramagnetic side for all N.

We begin in Section 13.1 by treating the low-T region on the magnetically ordered side of the $d=2$ phase diagram in Fig. 11.2; note that in this figure the magnetic long-range order disappears at any nonzero T. This is shown below to happen for all $N \geq 3$, and we only consider these cases. Section 13.2 then considers dynamical properties of the continuum high-T and quantum-paramagnetic low-T regions of Figs. 11.2 and 11.3 and describes the structure of the scaling function in (13.3); in principle, these results apply for all N.

13.1 Low T on the magnetically ordered side, $T \ll \rho_s$

As noted above, we only consider the case where magnetic order disappears at any nonzero T, and this happens (as shown below) for all $N \geq 3$. Recall that the quantum Ising chain, considered in Chapter 10, also had the feature of losing magnetic order at any nonzero T (compare the phase diagrams in Figs. 10.2 and 11.2). We find here that the static and dynamic properties of the $d = 2$ $N \geq 3$ rotor models in this low-T region are very similar to those discussed earlier for the corresponding region of the quantum Ising chain in Section 10.4.1. However, our analysis uses techniques that are very similar to those developed earlier in the "classical wave" description of the high-T region of the $d = 1$ rotor model in Section 12.3. The reader is urged to review these sections before proceeding.

The key property of this region is the very large value of the correlation length, obtained earlier in (11.64) in the large-N limit:

$$\xi_c \sim (c/T)\exp(2\pi\rho_s/NT). \tag{13.4}$$

We can use an argument similar to that following (12.31) for the $d = 1$ model, to establish the effective classical wave behavior of the system in this region; indeed the subscript c in (13.4) anticipates this. The typical wave excitations of the $\mathbf{n}$ field will have an energy $\sim c\xi_c^{-1}$ and hence a thermal occupation number

$$\frac{1}{e^{c\xi_c^{-1}/T} - 1} \approx \frac{T}{c\xi_c^{-1}} \approx \exp\left(\frac{2\pi\rho_s}{NT}\right) \gg 1. \tag{13.5}$$

Therefore, as in Section 12.3, we can treat these waves classically. Note that the classical thermal fluctuations are exponentially preferred, unlike the much weaker logarithmic preference in Section 12.3. The exponential preference is similar to that found for the quantum Ising chain in Section 10.4.1, although there the reason was the energy gap toward creation of domain walls.

This low-T region was studied in the influential paper of Chakravarty, Halperin, and Nelson [75], where they called it "renormalized classical," as seems natural from the reasoning above. We have not used this name here to prevent confusion with other types of effectively classical behavior that appear in different regions of the phase diagram.

As in the Ising case, we can expect that static and dynamic correlations obey a reduced scaling form of two arguments. The analog of the expression (10.72) turns out to be

$$C(x, t) = N_0^2 \left(\frac{T}{\rho_s}\right)^{\frac{(N-1)}{(N-2)}} \Phi_c\left(\frac{x}{\xi_c}, \frac{t}{\tau_\varphi}\right), \tag{13.6}$$

where N_0 is the ground state ordered moment, Φ_c is a completely universal function to be determined by some effective classical model, and, as before, τ_φ is a characteristic phase coherence time, which is determined below. Unlike the Ising case, it is not possible to determine Φ_c exactly, although most of its qualitative properties can be described. There is an additional prefactor of a power of T/ρ_s in (13.6) that is not present in (10.72); note that this is a rather weak prefactor on the scale of ξ_c as $\ln \xi_c \sim \rho_s/T$. As we discuss below, its origin is in the "wavefunction renormalization" of the $D = 2$ nonlinear σ-model, which also led to the logarithmic prefactor in (12.58) in the classical wave region of the $d = 1$ quantum rotor model. It is also easy to check from (11.43) that (13.6) is consistent with the global scaling form (13.1). Finally, note that (13.6) is consistent with the large-N result obtained from (11.10), (11.35), and (11.38). Also, by matching scaling forms at $N = \infty$ we obtain the value $\tau_\varphi = \xi_c/c$; however, we had no damping in the dynamic susceptibility (11.10) at $N = \infty$, and so τ_φ cannot even be interpreted as a phase coherence time. We shall find that the value of the phase coherence time at finite N is different: $\tau_\varphi \sim (\rho_s/T)^{1/2}\xi_c/c$; this result is actually quite similar to that for the quantum Ising chain, where, in (10.63) and (10.71), we obtained $\tau_\varphi \sim (\Delta/T)^{1/2}\xi_c/c$.

We describe the computation of the exact values of ξ_c and τ_φ in the following subsections. A description of the function Φ_c then follows.

13.1.1 Computation of ξ_c

The exact value of ξ_c, in the limit $T \ll \rho_s$, was obtained by Hasenfratz and Niedermayer [207] building upon foundations laid in [75]. Here we obtain the result by a different method, which has the advantage of connecting with results already obtained for the $d = 1$ case and also allowing for a streamlined discussion of dynamic properties in subsequent subsections.

We begin by precisely defining ξ_c: the definition (13.6) leaves it undefined up to an overall constant, which could be absorbed into a redefinition of Φ_c. It is then clear that demanding

$$\lim_{|\bar{x}| \to \infty} \Phi_c(\bar{x}, 0) \sim \frac{e^{-\bar{x}}}{\sqrt{\bar{x}}} \tag{13.7}$$

fixes ξ_c as the exponential decay rate of the long-distance equal-time correlations. The $\bar{x}$-dependent prefactor in (13.7) is the familiar "Ornstein–Zernicke" form expected in the long-distance decay of a classical, two-dimensional disordered system (the expression (11.21) is of this form in $D = d + 1$ dimensions). The missing coefficient in (13.7) is universal, but its value is not known exactly, although estimates have been made in the $1/N$ expansion [86] and in numerical simulations [75, 470].

As in Section 12.3, the inequality (13.5) suggests that we develop an effective action for the $\omega_n = 0$ component of $\mathbf{n}$ to describe its equal-time correlations. A simple argument then suggests the form of the effective action. Recall that we have true long-range order in $\mathbf{n}$ at $T = 0$, and we have denoted the exact spin stiffness of this ordered state as ρ_s (Section 11.2.3). The energy cost of any sufficiently slowly varying static deformation $\mathbf{n}(x)$

can be computed using this stiffness, and so we obtain the following partition function for the equal-time correlations:

$$\mathcal{Z} = \int \mathcal{D}\mathbf{n}(x)\delta(\mathbf{n}^2 - 1)\exp\left(-\frac{\rho_s}{2T}\int d^2x\,(\nabla\mathbf{n})^2\right). \tag{13.8}$$

This has the same form as the $d = 1$ classical wave model (12.34). The relationship of ρ_s to the couplings in the underlying quantum action (11.4) is not known exactly and in general is quite difficult to determine. At $N = \infty$ we obtained the relationship specified by (11.38) and (11.13). For our purposes here, it is useful to have an expression for ρ_s of the quantum model (11.4) in powers of g. Such an expansion can be obtained by a simple extension of the methods of Section 12.1. We take for $\mathbf{n}_<$ in (12.5) an externally imposed, long-wavelength, static deformation of $\mathbf{n}$ and then account for the quantum fluctuations by integrating out π fields at all wavevectors and frequencies at $T = 0$. The energy cost of such a deformation defines ρ_s, and this is obtained by generalization of (12.7):

$$\rho_s = \frac{cN}{g}\left[1 - \frac{(N-2)g}{N}\int^\Lambda \frac{d^3p}{(2\pi)^3}\frac{1}{p^2} + \mathcal{O}(g^2)\right], \tag{13.9}$$

where, as in Section 11.1, $p \equiv (\vec{k}, \omega/c)$, and the nature of the ultraviolet cutoff, Λ, was discussed below (11.4). We do not need to specify the precise form of this cutoff, for the scaling properties of the quantum critical point at $g = g_c$ imply that all observables become cutoff independent once expressed in terms of ρ_s and the ordered moment N_0, in place of the bare couplings in (11.4). Hence we shall really require the inverse of (13.9): a series for g in powers of $1/\rho_s$, which can, of course, be easily generated from (13.9). Note also that the large-N limit of (13.9) is consistent with (11.38) and (11.13).

Having determined ρ_s, let us return to the properties of the effective partition function (13.8) for the static fluctuations for $d = 2$. A little thought exposes a crucial difference from the corresponding model (12.34) for $d = 1$. In the latter case, the continuum theory (12.34) was ultraviolet finite and needed no short-distance regularization, and so the exact correlation length appeared as a coupling constant in (12.34) and completely specified the equal-time correlations of $\mathbf{n}$. In contrast, (13.8) is not well defined as it stands. Indeed, the action (13.8) has precisely the same form as the $d = 1$ *quantum* rotor model at $T = 0$ studied in Section 12.1, and it was shown there to require some short-distance regularization (see the expression (12.10) for the energy gap). In the present situation, we do not have the luxury of choosing the form of this regularization. The partition function (13.8) is only an effective classical theory and cannot be applied at distances so short that quantum effects become important. In particular, it cannot hold at wavevectors larger than where the energy of a spin wave $\sim ck$ becomes of order T. Thus quantum mechanics acts as an underlying high-momentum regularization of (13.8), at momenta of order $\Lambda_c \sim T/c$. We have added a subscript c to emphasize that this is a cutoff for the classical theory; Λ_c is completely unrelated to the cutoff of the quantum theory, Λ, noted in (13.9). The latter has a nonuniversal nature, while the cutoff at momenta of order Λ_c has a universal form, which is elucidated below.

An important property of the model (13.8) emerged in the renormalization group analysis of Chapter 12. We showed that its long-distance properties did not depend separately upon its coupling ρ_s/T and its cutoff $\sim \Lambda_c$, but only upon a single renormalization group invariant $\Lambda_{\overline{MS}}$. Therefore the central task facing us is determination of a precise expression for $\Lambda_{\overline{MS}}$ as a function of ρ_s/T and the momentum scale T/c. With this at hand, we obtain ξ_c by the analog of the Bethe-ansatz relation [205, 206] (12.45)

$$\xi_c^{-1} = \Lambda_{\overline{MS}} \frac{(8/e)^{1/(N-2)}}{\Gamma(1 + 1/(N-2))}, \qquad (13.10)$$

as the gap of the $d = 1$ quantum model at $T = 0$ becomes the exponential decay rate of correlations of the $d = 2$ classical model (13.8). We could also proceed, in principle, to use (13.8) to determine the entire function $\Phi_c(\bar{x}, 0)$.

One way to determine $\Lambda_{\overline{MS}}$ is to return to the underlying quantum model (11.4) and to directly compute the long-distance form of its equal-time correlators. This gives an expression for ξ_c in terms of c, g, and Λ; re-expressing g in terms of ρ_s using the inverse of (13.9), and matching against (13.10), we could then obtain the needed expression for $\Lambda_{\overline{MS}}$. This is clearly an intractable route, as it involves the physics of (13.8) in its strong-coupling regime. Instead, we use a simple trick that does the matching between the two theories in a weak-coupling regime.

Recall from our discussion in Chapter 12 that the theory (13.8) is strongly coupled at length scales longer than $\Lambda_{\overline{MS}}^{-1}$ and weakly coupled at shorter scales. Clearly, we should do the matching between (13.8) and (11.4) in the latter regime. To do this, imagine restricting the spatial coordinate, x, of both theories to an infinite cylinder of circumference L (the temporal direction of (11.4) remains unchanged). If we choose $L \ll \Lambda_{\overline{MS}}^{-1}$ then we are in the weak-coupling regime, and we can compute properties of both theories using perturbation theory. At the same time, we have to ensure that $L \gg c/T$ so that all length scales are longer than the inverse classical cutoff Λ_c^{-1}, allowing us to remain in the regime of effective classical behavior in model (11.4). Because $\Lambda_{\overline{MS}}^{-1} \sim \xi_c$ is exponentially large in $1/T$, these two conditions are easily compatible. Thus we have modified (13.8) to

$$\mathcal{Z} = \int \mathcal{D}\mathbf{n}(x)\delta(\mathbf{n}^2 - 1) \exp\left(-\frac{\rho_s}{2T} \int dx_1 \int_0^L dx_2 \, (\nabla \mathbf{n})^2\right), \qquad (13.11)$$

with periodic boundary conditions on $\mathbf{n}$ along the x_2 direction. However, precisely such a model, in the regime $L \ll \Lambda_{\overline{MS}}^{-1}$, was studied in Section 12.3: we simply have to identify x_2 with the imaginary-time direction τ of a fictitious $d = 1$ quantum model, and then L is just its inverse temperature. This model was analyzed by a further dimensional reduction in which we integrate out all modes of $\mathbf{n}$ that have a nonzero wavevector along the x_2 direction, obtaining an effective one-dimensional model:

$$\mathcal{Z} = \int \mathcal{D}\mathbf{n}(x_1)\delta(\mathbf{n}^2 - 1) \exp\left(-\frac{(N-1)\xi_c(L)}{4} \int dx_1 \left(\partial_{x_1}\mathbf{n}\right)^2\right). \qquad (13.12)$$

We have written the coefficient of the gradient coupling in a form such that $\xi_c(L)$ is precisely the correlation length of a two-point $\mathbf{n}$ correlator along the x_1 direction (this follows from (6.45). Indeed, we can read off the value of $\xi_c(L)$ as a function of $\Lambda_{\overline{MS}}$ and L from the result (12.44):

$$\xi_c(L) = \frac{L(N-2)}{\pi(N-1)} \ln\left(\frac{\mathcal{C}}{L\Lambda_{\overline{MS}}}\right) + \cdots, \tag{13.13}$$

where $\mathcal{C}$ is the constant defined in (11.61). We expect a universal scaling form $\xi_c(L) = LF(L\Lambda_{\overline{MS}})$ for general L, and (13.13) specifies the leading term in the small-u limit of $F(u)$. The $u \to \infty$ limit is the strong-coupling regime, with $\xi_c \equiv \xi_c(L \to \infty)$ given by the Bethe-ansatz result (13.10).

Let us compute the expression corresponding to (13.13) for the quantum model (11.4). We do this by performing the dimensional reduction to (13.12) in one step, in a perturbation theory in g (i.e. we will integrate out all modes with either a nonzero wavevector in the x_2 direction, or a nonzero frequency in the τ direction, but not both). By a simple generalization of (12.7) or (12.35) to this spacetime geometry, we get

$$\xi_c(L) = \frac{2}{(N-1)} \left[\frac{cNL}{gT} - \sum_{n,m}' \int \frac{dk}{2\pi} \frac{(N-2)}{k^2 + (2\pi m/L)^2 + (2\pi nT/c)^2} \right], \tag{13.14}$$

plus corrections of order g, where the prime indicates the sum is over all integers n, m excluding the single point $n = m = 0$. The integral and summation in (13.14) are badly divergent in the ultraviolet. However, expressing g in terms of ρ_s using (13.9) makes the resulting expression free of divergences, as we now show. The basic technical tool is to lift the denominators in the integrands of (13.9) and (13.14) up into exponentials using the simple identity

$$\frac{1}{a} = \int_0^\infty d\lambda\, e^{-\lambda a}. \tag{13.15}$$

Then the combination of (13.14) and (13.9) yields, after a suitable rescaling of λ,

$$\xi_c(L) = \frac{2L\rho_s}{(N-1)T} \left[1 - \frac{(N-2)T}{4\pi\rho_s} \int_0^\infty \frac{d\lambda}{\sqrt{\lambda}} \left(A(\lambda)A(\lambda v^2) - 1 - \frac{1}{\lambda v} \right) \right], \tag{13.16}$$

where $v = TL/c$ and the function $A(y)$ was defined in (6.30). By simple use of the identity (6.42) and (6.30) it is easy to show that the λ integral in (13.16) is convergent. As noted earlier, we are interested in the classical regime $L \gg c/T$ and therefore in the $v \to \infty$ limit of the integral in (13.16). It is not difficult to show that the integral $\sim \ln(v)$ in this limit; we determined the additive constant associated with this logarithm numerically and found

$$\xi_c(L) = \frac{2L\rho_s}{(N-1)T} \left[1 - \frac{(N-2)T}{2\pi\rho_s} \ln\left(\frac{LT}{\mathcal{C}c}\right) + \mathcal{O}\left(\frac{T}{\rho_s}\right)^2 \right], \tag{13.17}$$

where the constant $\mathcal{C}$ (given in (11.61), was again found to appear.

We are now prepared to perform the matching between the two approaches to computing $\xi_c(L)$. Comparing (13.13) and (13.17) we find that the L dependencies are consistent, as required, and that

$$\Lambda_{\overline{MS}} = \frac{T}{c} \exp\left(-\frac{2\pi\rho_s}{(N-2)T}\right). \tag{13.18}$$

This is the result to one-loop order. It is possible to improve this result to two-loop order by using the relationship (12.43) between $\Lambda_{\overline{MS}}$ and the coupling g_R at an arbitrary scale μ. Matching (13.18) with (12.43) by choosing $\mu = T/c$ ($\sim \Lambda_c$) we find that

$$\frac{1}{g_R(T/c)} = \frac{\rho_s}{NT} + \frac{(N-2)}{4\pi N}\ln\mathcal{C} + \mathcal{O}\left(\frac{T}{\rho_s}\right). \tag{13.19}$$

Inserting this back into (12.43), we obtain the final result [207]

$$\Lambda_{\overline{MS}} = \frac{T}{c}\left(\frac{2\pi\rho_s}{(N-2)T}\right)^{1/(N-2)}\exp\left(-\frac{2\pi\rho_s}{(N-2)T}\right)\left[1 + \mathcal{O}\left(\frac{T}{\rho_s}\right)\right]. \tag{13.20}$$

Combined with (13.10), we have the promised exact result for ξ_c.

13.1.2 Computation of τ_φ

We follow the same strategy employed in Section 13.1.1 for ξ_c. We extend to dynamical properties the static mapping of the model (11.4) on a cylinder of circumference L, $c/T \ll L \ll \Lambda_{\overline{MS}}^{-1}$, onto the effective one-dimensional classical rotor model. By exactly the same arguments as those leading to (12.51), we have to supplement the partition function (13.12) by an additional kinetic energy term for the classical rotors. We therefore consider

$$\mathcal{Z} = \int \mathcal{D}\mathbf{n}(x_1)\mathcal{D}\mathbf{L}(x_1)\delta(\mathbf{n}^2 - 1)\delta(\mathbf{L}\cdot\mathbf{n})\exp\left(-\frac{\mathcal{H}_c}{T}\right),$$

$$\mathcal{H}_c = \frac{1}{2T}\int dx_1 \left[\frac{(N-1)T\xi_c(L)}{2}\left(\frac{d\mathbf{n}}{dx_1}\right)^2 + \frac{1}{L\chi_{u\perp}(L)}\mathbf{L}^2\right], \tag{13.21}$$

where $L\chi_{u\perp}(L)$ is the uniform susceptibility *per unit length* of the model (11.4) on a cylinder of circumference L. The equations of motion of this one-dimensional classical rotor model follow from the Poisson brackets (12.54). The structure of these was analyzed in Section 12.3, and by (12.59) they imply a characteristic time

$$\tau_\varphi(L) \sim \left(\frac{\xi_c(L)L\chi_{u\perp}(L)}{T}\right)^{1/2}. \tag{13.22}$$

The value of $\tau_\varphi(L)$ is undetermined up to an overall constant, which we will choose later at our convenience.

It remains to compute $\chi_{u\perp}(L)$, and then to use scaling arguments to extrapolate perturbative results from the regime $L\Lambda_{\overline{MS}} \to 0$ to the required $L\Lambda_{\overline{MS}} \to \infty$.

The uniform susceptibility follows from a straightforward generalization of (12.48) to the present geometry. We obtain after a rotational average of the two terms in (12.48):

$$\chi_{u\perp}(L) = \frac{N}{cg}\left[1 - \frac{(N-2)g}{cN}\frac{T}{L}\sum_{\omega_n \neq 0}\sum_m \int \frac{dk}{2\pi}\frac{1}{k^2 + (2\pi m/L)^2 + (\omega_n/c)^2}\right.$$
$$\left. + \frac{(N-2)g}{cN}\frac{T}{L}\sum_{\omega_n \neq 0}\sum_m \int \frac{dk}{2\pi}\frac{k^2 + (2\pi m/L)^2 - (\omega_n/c)^2}{(k^2 + (2\pi m/L)^2 + (\omega_n/c)^2)^2}\right].$$

$$(13.23)$$

We can eliminate g in favor of ρ_s using (13.9) and obtain an expression for $\chi_{u\perp}$ in terms of ρ_s, c, T, and L. This expression can then be analyzed in a manner very similar to that used for (13.14) and (13.16). We will not describe the details of this but simply note that an important difference emerges from the structure of the earlier result (13.17): we find that there are no singular logarithmic terms in $\chi_{u\perp}(L)$ in the limit $TL/c \to \infty$. The dependence on TL/c is exponentially small in this limit, and we can therefore explicitly take the $L \to \infty$ limit already in the expression (13.23), by converting the summation over m into an integral. Taking this limit, and carrying out the summation over ω_n, we get

$$\chi_{u\perp}(L) = \frac{\rho_s}{c^2} - (N-2)\int \frac{d^2k}{(2\pi)^2}\left(\frac{1}{ck}\frac{1}{(e^{ck/T} - 1)} - \frac{T}{c^2 k^2}\right)$$
$$+ (N-2)\int \frac{d^2k}{(2\pi)^2}\left(\frac{1}{4T\sinh^2(ck/2T)} - \frac{T}{c^2 k^2}\right).$$

$$(13.24)$$

The two integrals in (13.24) are individually logarithmically divergent, but the combination is finite. This is a verification that the $L \to \infty$ limit was smooth, and that unlike (13.17), it was not necessary to keep L finite to obtain a finite answer. We can easily carry out the integral over the difference of the integrands in (13.24) and obtain a result $\chi_{u\perp}(L)$ that is independent of L to the accuracy we need:

$$\chi_{u\perp}(L) = \frac{\rho_s}{c^2}\left[1 + \frac{(N-2)T}{2\pi\rho_s} + \mathcal{O}\left(\frac{T}{\rho_s}\right)^2\right].$$

$$(13.25)$$

Note that, combined with $\chi_u = (2/N)\chi_{u\perp}$, this result agrees with our earlier large-N result (11.65).

We have assembled all the ingredients necessary to estimate τ_φ. Inserting the results (13.13) and (13.25) into (13.22) we get (ignoring numerical prefactors)

$$\tau_\varphi(L) \sim \left(\frac{\rho_s}{T}\right)^{1/2}\frac{L}{c}\left[\ln\left(\frac{c}{L\Lambda_{\overline{MS}}}\right)\right]^{1/2}$$

$$(13.26)$$

for small $L\Lambda_{\overline{MS}}$.

As the final step, we have to extrapolate the result (13.26) from $L\Lambda_{\overline{MS}} \to 0$ to $L\Lambda_{\overline{MS}} \to \infty$. This can be done by a relatively straightforward scaling argument. The phase relaxation time $\tau_\varphi(L)$ is expected to be given by a natural time scale times a dimensionless function of the ratio of the system width L to the only scale, $\Lambda_{\overline{MS}}^{-1}$, that characterized the two-dimensional nonlinear sigma model (13.11); in other words we expect

$$\tau_\varphi = \frac{AL}{c} G(L\Lambda_{\overline{MS}}), \tag{13.27}$$

where A is some prefactor and G is a universal scaling function. Clearly (13.26) is of this form, and the comparison allows us to fix the value of A. In the limit $L\Lambda_{\overline{MS}} \to \infty$ we expect $\tau_\varphi(L)$ to become independent of the system width L, and therefore we must have $G(u \to \infty) \sim 1/u$. Using this, we get our desired final result for $\tau_\varphi \equiv \tau_\varphi(L \to \infty)$ [75]:

$$\begin{aligned}\tau_\varphi &\sim \left(\frac{\rho_s}{T}\right)^{1/2} \frac{\Lambda_{\overline{MS}}^{-1}}{c} \\ &= \left(\frac{\rho_s}{T}\right)^{1/2} \frac{\xi_c}{c};\end{aligned} \tag{13.28}$$

in the last step we have arbitrarily chosen the prefactor and the relationship holds as an equality. This is the promised result for τ_φ. As we noted earlier, this result has an interesting similarity to that obtained in the corresponding low-T region of the quantum Ising chain in Section 10.4.1; there we found in (10.63) and (10.71) that $\tau_\varphi \sim (\Delta/T)^{1/2}\xi_c/c$.

13.1.3 Structure of correlations

We turn to a discussion of the structure of the reduced classical scaling function Φ_c in (13.6).

Equal-time correlations

For the equal-time case, $t=0$, it is possible to make exact analytic statements in certain asymptotic limits, which we now discuss (the full functional form of $\Phi_c(\bar{x}, 0)$ can be obtained in a $1/N$ expansion, as discussed in [86]). We have already noted the long-distance form in (13.7). We now discuss the behavior as $\bar{x} \to 0$. As we are restricting ourselves to the classical regime, we do not want to examine distances shorter than the thermal de Broglie wavelength of the spin waves – we are therefore examining the regime $c/T \ll x \ll \xi_c$. The overall dependence upon x in this regime follows immediately from the homogeneity relation (12.40); indeed by the precise analog of the argument used to obtain (12.58), but using distances rather than energies, we have

$$C(x, 0) \sim [\ln(\xi_c/x)]^{(N-1)/(N-2)}, \quad c/T \ll x \ll \xi_c. \tag{13.29}$$

We can also precisely fix the prefactor of the term in (13.29) by a simple argument. At the lower boundary, $x \sim c/T$ thermal fluctuations are no longer important, and the model crosses over into its quantum fluctuation-dominated ground state correlations. Because the ground state is ordered, the correlations are very simple: we must have $C(x, 0) = N_0^2/N$ for $x \sim c/T$ (the factor of N comes from the average over all orientations of the ground state magnetization). Demanding that (13.29) match smoothly with this criterion, using the value $\xi_c \sim \Lambda_{\overline{MS}}^{-1}$ in (13.20), and working to leading order in T/ρ_s, we find that the prefactor of the logarithm in (13.29) is uniquely determined. The resulting dependence of $C(x, 0)$ obeys the scaling form (13.6) (indeed, it requires the prefactor of $(T/\rho_s)^{(N-1)/(N-2)}$ in (13.6), and this is the reason for its presence) and gives us the small $\bar{x} = x/\xi_c$ limit of the scaling function:

$$\Phi_c(\bar{x} \ll 1, 0) = \frac{1}{N} \left[\frac{(N-2)}{2\pi} \ln(1/\bar{x}) \right]^{\frac{(N-1)}{(N-2)}}. \tag{13.30}$$

It is also useful to present these results in momentum space, in terms of the equal-time structure factor $S(k)$ defined in (10.5):

$$S(k) = \int d^2 x e^{-i\vec{k}\cdot\vec{x}} C(x, 0). \tag{13.31}$$

For small k, the scaling form (13.6) implies that

$$S(0) \sim N_0^2 \xi_c^2 \left(\frac{T}{\rho_s} \right)^{\frac{(N-1)}{(N-2)}}, \tag{13.32}$$

where the missing coefficient is a universal number given by the spatial integral of Φ_c at $t = 0$ (its numerical estimate [515] for $N = 3$ is ≈ 1.06). For larger k, we Fourier transform (13.30) and find that for $k\xi_c \gg 1$, but $ck \ll T$ [75, 86],

$$S(k) = \left(\frac{N-1}{N} \right) \frac{T N_0^2}{\rho_s k^2} \left[\frac{(N-2)T}{2\pi\rho_s} \ln(k\xi_c) \right]^{1/(N-2)}. \tag{13.33}$$

Note that for $k \sim c/T$, the term in the square brackets evaluates to $1 + \mathcal{O}(T/\rho_s)$, and so

$$S(k) = \left[\frac{N-1}{N} \right] \frac{T N_0^2}{\rho_s k^2}, \quad k \sim c/T. \tag{13.34}$$

This can be understood in terms of the response (8.19), with an additional factor of $(N-1)/N$ representing the fact that this response appears only in $N - 1$ directions transverse to the local ordered state.

It is instructive at this point to assemble all the known results for the equal-time correlator $C(x, 0)$ in the present low-T region, $T \ll \rho_s$. We have

$$C(x, 0) = \begin{cases} a_1 N_0^2 \left[\dfrac{c}{\rho_s x} \right]^{1+\eta}, & x \ll \dfrac{c}{\rho_s}, \\[3mm] \dfrac{N_0^2}{N}, & \dfrac{c}{\rho_s} \ll x \ll \dfrac{c}{T}, \\[3mm] \dfrac{N_0^2}{N} \left[\dfrac{(N-2)T}{2\pi\rho_s} \ln(\xi_c/x) \right]^{(N-1)/(N-2)}, & \dfrac{c}{T} \ll x \ll \xi_c, \\[3mm] a_2 N_0^2 \left(\dfrac{T}{\rho_s} \right)^{(N-1)/(N-2)} \dfrac{e^{-x/\xi_c}}{\sqrt{x/\xi_c}}, & \xi_c \ll x, \end{cases} \tag{13.35}$$

where a_1, a_2 are universal constants known only via $1/N$ expansion or numerical simulations. It is reassuring to note that all four asymptotic forms in (13.35) are perfectly compatible at the boundaries of their regions of applicability. The first result in (13.35) follows from a Fourier transform of (11.28), combined with prefactor constraints implied by (11.41), (11.43), and (11.39). In this region the correlations are those of the $T = 0$ quantum critical point at $g = g_c$, and the η is the anomalous dimension of the $(2 + 1)$-dimensional theory. The region $c/\rho_s \ll x \ll c/T$ is where the system appears to have the long-range order of the $g < g_c$ ground state (thermal fluctuations have not yet become apparent). A $T = 0$ quantum analysis of (11.4) is required to describe the crossover between these first two regimes. Finally, the last two regimes in (13.35) are those discussed in the present section and are contained in the reduced classical scaling function Φ_c.

Dynamic correlations

Let us turn to unequal-time correlations. A reasonable picture has been obtained through numerical simulations, combined with scaling arguments and matched to limiting weak-coupling regimes [515, 516]; these results are also supported by other analytic approaches [75, 86, 176]. By arguments similar to those in Section 12.3.3, the dynamics can be mapped onto the obvious two-dimensional generalization of the classical nonlinear wave problem defined by (12.51) and (12.55). This is a problem of classical rotors with orientation $\mathbf{n}(x, t)$ and angular momentum $\mathbf{L}(x, t)$. The equal-time correlations of $\mathbf{n}$, as already discussed, are given by the classical partition function (13.8). Those of $\mathbf{L}$ are defined, as in (12.51), by the kinetic energy term $\mathbf{L}^2/(2\chi_{u\perp})$ with $\chi_{u\perp}$ given by (13.25). An initial condition is chosen from this ensemble and then evolved deterministically under the equations of motion following from the Poisson brackets (12.54). This classical problem was numerically simulated by Tyc, Halperin, and Chakravarty [515] and we now describe their results.

It is convenient to express the results in terms of the dynamic structure factor $S(k, \omega)$ defined by (10.4). As in (10.78) and (12.62), we can incorporate the already determined information on the equal-time correlations and the scaling form (13.6) by writing

$$\frac{2T}{\omega}\mathrm{Im}\chi(k, \omega) = S(k, \omega) = S(k)\tau_\varphi \Phi_{\mathrm{Sc}}(k\xi_c, \omega\tau_\varphi), \tag{13.36}$$

where the first relation is the classical fluctuation–dissipation theorem (10.75), and the universal scaling function Φ_{Sc} is an even function of frequency and has a unit integral of frequency, as in (10.79). The function Φ_{Sc} was determined numerically by Tyc, Halperin, and Chakravarty [515]. They found that over a wide range of frequency and wavevectors, the frequency dependence of the results could be described by the simple functional form

$$\Phi_{\mathrm{Sc}}(\bar{k}, \bar{\omega}) = \frac{\gamma(\bar{k})}{(\bar{\omega} - \upsilon(\bar{k}))^2 + \gamma^2(\bar{k})} + \frac{\gamma(\bar{k})}{(\bar{\omega} + \upsilon(\bar{k}))^2 + \gamma^2(\bar{k})}, \tag{13.37}$$

where $\upsilon(\bar{k})$ and $\gamma(\bar{k})$ are functions of a wavevector that were determined numerically. This dynamic response consists of a peak at a spin-wave (rescaled) frequency $\upsilon(\bar{k})$ with a damping rate $\gamma(\bar{k})$.

For small $\bar{k}$, a best fit was obtained with a $\upsilon(\bar{k}) \to 0$ as $\bar{k} \to 0$, while $\gamma(\bar{k})$ approached a nonzero constant. Hence the spin waves are overdamped for $k\xi_c \ll 1$, and the dynamics is purely relaxational. There is no analog of the nonzero-frequency "shoulder" found in Fig. 12.4 for the classical wave dynamics for $d = 1$. Thus amplitude fluctuations are weaker for $d = 2$, and the relaxation is better considered as arising from angular fluctuations about an ordered state. This is physically sensible as it indicates that the **n** field is "more ordered" in the present $d = 2$, low-T region than it was in the $d = 1$ high-T region of Section 12.3.

For large $\bar{k}$ (more precisely, for $k\xi_c \gg 1$ and $ck/T \ll 1$) we expect that the system should cross over into the $T = 0$ spin-wave spectrum at $\omega = ck$. Using the values of ξ_c in (13.20), and that of τ_φ in (13.28), it is easy to see that this is consistent with the dimensionless frequency $\bar{\omega} = \nu(\bar{k})$ for $T \ll \rho_s$ only if

$$\nu(\bar{k} \to \infty) = \bar{k} \left[\frac{(N-2)}{2\pi} \ln \bar{k} \right]^{1/2}. \tag{13.38}$$

The large-$\bar{k}$ limit of the damping $\gamma(\bar{k})$ was examined in a self-consistent perturbation theory in [516] and it was found to be only logarithmically smaller than $\nu(\bar{k})$.

13.2 Dynamics of the quantum paramagnetic and high-*T* regions

We turn to the dynamical properties of the remaining two universal regions in Figs. 11.2 and 11.3. There is no signature of the ordered state in these regions at any length or time scale. Instead, the basic physics is of the critical ground state or of the quantum paramagnet eventually losing phase coherence at times longer than τ_φ, owing to the thermal effects. The qualitative nature of all the physics turns out not to be particularly sensitive to the precise value of N, and all of our results below apply to all N, including the cases $N = 1, 2$, which were excluded in the low-T discussion of Section 13.1. Indeed, the physical phenomena also turn out to be essentially identical to those in the corresponding regions of the $d = 1$, $N = 1$ quantum Ising chain, which were discussed in Sections 10.4.2 and 10.4.3. The dynamical properties of this latter model were summarized in Fig. 10.12, and the "high-T" and "low-T (quantum paramagnetic)" portions of this figure apply unchanged to the $d = 2$ models of interest here for all N. Exact dynamic response functions were obtained in Sections 10.4.2 and 10.4.3 for all the distinct dynamical regimes of the quantum Ising chain. The same response functions of the $d = 2$ models have a very similar form, but it is no longer possible to obtain exact results. In this section, we demonstrate how this structure emerges at first order in $1/N$. However, as we noted at the beginning of this chapter, the $1/N$ expansion breaks down at very low frequencies, and for this regime we provide an alternative approach in Section 14.3. In a sense, the purpose of this section is somewhat technical. The basic physical concepts are perhaps better appreciated in the simpler, and exact, discussion of Sections 5.2, 10.4.2, and 10.4.3, which the reader is urged to review.

We also note that the computations for $d = 1$ in Chapters 10 and 12, and for $d = 2$ here, treat interactions in opposing limits. In the $N = \infty$, $d = 2$ results of Chapter 11 we found a description in terms of N noninteracting massive particles with a self-consistently determined temperature-dependent energy gap; at first order in $1/N$ we find that these particles *weakly* scatter off each other with a T matrix that is of order $1/N$. In contrast, the collisions of the excitations for $d = 1$ are described by the S matrices (5.14), (12.13), describing full reflection of particles with a phase shift of π, and these are as far as one can get from the free-particle result $S = 1$, while being consistent with unitarity; the $d = 1$ case is therefore properly considered as a *strong* scattering limit. The qualitative similarity in $\chi(k, \omega)$ of the weak-coupling results below and the earlier strong-coupling results for $d = 1$ is reassuring and indicates that we have correctly understood the physics.

We begin by setting up the mechanics of the $1/N$ expansion for the dynamic susceptibility. The $N = \infty$ result was given in (11.10). At order $1/N$, it is necessary to include fluctuations in the λ field about the saddle point of (11.8), which is determined by the solution of (11.9). We insert a source term in the original action (11.6) for $\mathbf{n}$ and then expand the modified (11.8) up to cubic order in the deviation of λ about its saddle point (all higher order terms can be dropped at this order in $1/N$). The term purely quadratic in λ defines a propagator for the λ fluctuations, the structure of which is discussed in some detail below. We integrate out the λ fluctuations to order $1/N$, and this leads to the corrections to the $\mathbf{n}$ field correlator, $\chi(k, \omega)$, shown schematically in Fig. 13.2. This leads finally to the following expression for $\chi(k, \omega)$, which replaces (11.10) at order $1/N$ [86] (the reader can also consult [29, 504] for more explicit details on the mechanics of computing $1/N$ corrections for related models):

$$\chi(k, \omega) = \frac{cg/N}{c^2 k^2 - (\omega + i\epsilon)^2 + m^2 + \Sigma(k, \omega)}, \tag{13.39}$$

where the self-energy Σ is given by

$$\Sigma(k, \omega_n) = \tilde{\Sigma}(k, \omega_n) - \frac{1}{\Pi(0, 0)} T \sum_{\epsilon_n} \int \frac{d^2 q}{4\pi^2} G_0^2(q, \epsilon_n) \tilde{\Sigma}(q, \epsilon_n), \tag{13.40}$$

the two terms representing the contributions of the two graphs in Fig. 13.2; these graphs should be compared with the discussion in Section 7.2.1 and the perturbative expression

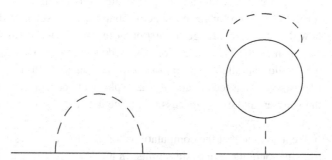

Fig. 13.2 Feynman diagrams which contribute to the self-energy of $\mathbf{n}$ at order $1/N$. The $\mathbf{n}$ propagator is a straight line, while the λ propagator, $1/\Pi$, is a dashed line.

(4.35). The frequency- and momentum-dependent contribution to the self-energy is $\tilde{\Sigma}$, which is given by

$$\tilde{\Sigma}(k, \omega_n) = \frac{2}{N} T \sum_{\epsilon_n} \int \frac{d^2q}{4\pi^2} \frac{G_0(\vec{k} + \vec{q}, \omega_n + \epsilon_n) - G_0(q, i\epsilon_n)}{\Pi(q, \epsilon)}, \tag{13.41}$$

with $1/\Pi$ the propagator of the λ field,

$$\Pi(q, \epsilon_n) = T \sum_{\Omega_n} \int \frac{d^2q_1}{4\pi^2} G_0(\vec{q} + \vec{q}_1, \epsilon_n + \Omega_n) G_0(q_1, \Omega_n), \tag{13.42}$$

and with G_0 proportional to the susceptibility of $\mathbf{n}$ at $N = \infty$,

$$G_0(k, \omega_n) \equiv \frac{1}{c^2k^2 + \omega_n^2 + m^2}. \tag{13.43}$$

The "mass" m in the propagators is the saddle-point value of the λ field, which was determined earlier in (11.53) to be

$$m = 2T \ln\left(\frac{e^{\Delta_+/2T} + (4 + e^{\Delta_+/T})^{1/2}}{2}\right) + \mathcal{O}(1/N), \tag{13.44}$$

where, as usual, Δ_+ represents the gap of the quantum-paramagnetic ground state. We also recall the important limiting forms, $m = \Delta_+$ for $T \ll \Delta_+$ (11.57) and $m = 2 \ln((\sqrt{5} + 1)/2)T$ for $T \gg \Delta_+$ (11.59). The value of m we are using in (13.44) is actually precisely the same as in the $N = \infty$ expression (11.53), when expressed in terms of the bare coupling constant g. The $1/N$ correction in (13.44) represents the change necessary because of the new value of the ground state energy gap Δ_+ at this order. At $N = \infty$, the gap Δ_+ was related to the bare coupling constant g in (11.15) and (11.17). The $1/N$ correction to the value of Δ_+ is obtained by solving the following equation for the location of the pole in the zero momentum $\mathbf{n}$ propagator in (13.39) at $T = 0$:

$$m^2 - \Delta_+^2 + \Sigma(0, \omega = \Delta_+). \tag{13.45}$$

The equation relating Δ_+ and the coupling g must then be inverted to express g in terms of Δ_+ and the result inserted into the expression for m. This leads to the corrections at order $1/N$ in the expression (13.44), and these are crucial in obtaining universal answers for the physical response function $\chi(k, \omega)$.

We study the properties of (13.39) at $T = 0$ in Section 13.2.1 and at nonzero temperatures in Section 13.2.2.

13.2.1 Zero temperature

The propagator of the λ field in (13.42) can be evaluated in closed form at $T = 0$. We find

$$\Pi(q, \omega) = \frac{1}{4\pi c^2 \sqrt{c^2q^2 - (\omega + i\epsilon)^2}} \tan^{-1}\left(\frac{\sqrt{c^2q^2 - (\omega + i\epsilon)^2}}{2\Delta_+}\right). \tag{13.46}$$

Note that Π is purely real for $|\omega| < (c^2 q^2 + 4\Delta_+^2)^{1/2}$ but acquires an imaginary part for larger $|\omega|$. The threshold corresponds to the minimum energy required to create two particles with total momentum q, in agreement with the expression (13.42) for Π as a two-particle propagator. We can insert (13.46) into (13.41) and determine the self-energy Σ. It is simpler to first consider only its imaginary part. This is obtained by using a spectral representation for $\mathrm{Im}(1/\Pi)$ and evaluating the summation over ϵ_n in the limit of zero T; taking the imaginary part of the result we obtain

$$\mathrm{Im}\,\Sigma(k, \omega) = \frac{1}{2\pi^2 N} \int \frac{d^2 q}{\varepsilon_{\vec{k}+\vec{q}}} \int_0^\infty d\Omega\, \mathrm{Im}\left(\frac{1}{\Pi(q, \Omega)}\right) \delta(\omega - \varepsilon_{\vec{k}+\vec{q}} - \Omega), \tag{13.47}$$

for $\omega > 0$ (for $\omega < 0$ we use the fact that $\mathrm{Im}\,\Sigma(k, \omega)$ is an odd function of ω), with

$$\varepsilon_{\vec{q}} \equiv \sqrt{c^2 q^2 + \Delta_+^2} \tag{13.48}$$

the energy spectrum of the quasiparticle. Actually, there is a little subtlety in obtaining (13.47) that we have glossed over: for large Ω, $\mathrm{Im}(1/\Pi(q, \Omega)) \sim \Omega$, and so its Kramers–Kronig transform is not well defined. This issue is discussed more carefully in [86], and it is shown there that for the imaginary part of Σ, the naive result obtained by simply ignoring this potential divergence is in fact correct. Now, the relativistic invariance of the $T = 0$ theory implies that (13.47) is a function only of $c^2 k^2 - \omega^2$, and so its general form can be deduced by evaluating it at $k = 0$. For this case, the q integral can be performed, and then changing variables from Ω to y with $y^2 = 2\omega\Omega - \omega^2 + \Delta_+^2$, we get our final expression for $\mathrm{Im}\,\Sigma$:

$$\mathrm{Im}\,\Sigma(k, \omega) = -\frac{4\pi}{N\sqrt{\omega^2 - c^2 k^2}} \int_{2\Delta_+}^{\sqrt{\omega^2 - c^2 k^2} - \Delta_+} dy\, y^2$$
$$\times [\pi^2 + \ln^2((y + 2\Delta_+)/(y - \Delta_+))]^{-1}, \tag{13.49}$$

for $\omega^2 > c^2 k^2 + 9\Delta_+^2$, and $\mathrm{Im}\,\Sigma$ is zero otherwise. Thus we have a threshold at the creation of *three* particles, above which $\mathrm{Im}\,\Sigma$ is nonzero: the $O(N)$ symmetry of the model only allows the N-fold degenerate particle with momentum k and energy ω to decay into a three-particle continuum if its energy is sufficiently large. We also note here the behavior of (13.49) for $\omega^2 - c^2 k^2 \gg \Delta_+^2$:

$$\mathrm{Im}\,\Sigma(k, \omega) = -\frac{\pi\eta}{2}(\omega^2 - c^2 k^2), \tag{13.50}$$

where η is

$$\eta = \frac{8}{3\pi^2 N}. \tag{13.51}$$

In fact, η is precisely the same critical exponent that appeared in (11.28), as becomes clear from the discussion below.

Inserting (13.49) into (13.39), we see that the resulting structure in $\mathrm{Im}\,\chi(k, \omega)$ is identical to that discussed in Section 7.2.1 and in Section 10.4.2 for the quantum Ising chain. Near the quasiparticle energy $\omega = \varepsilon_k$, Σ is purely real, there is no broadening of the quasiparticle spectral weight, and it remains a pure delta function. The real part of Σ does contribute a shift in the position of the pole, but this was already accounted for by our defining Δ_+ as the exact $T = 0$ energy gap in (13.45). The next nonzero spectral weight in χ arises at the three-particle threshold from the imaginary part of Σ discussed above, and also mentioned in (7.27). At next order in $1/N$ we will also find a threshold at $5\Delta_+$ and so on.

Let us return to the quasiparticle pole and consider the value of its residue at order $1/N$. For this we have to evaluate Σ at the pole position. This is most conveniently done by initially going to imaginary frequencies and explicitly using the relativistic invariance of the theory. In fact, if we define $K = (k^2 + \omega_n^2/c^2)^{1/2}$, the relativistic invariance implies that Σ is a function only of K. By an angular average of (13.41) in three-dimensional Euclidean spacetime, this function can be reduced to a one-dimensional integral

$$\Sigma(K) = \frac{1}{2\pi^2 cN} \int_0^\Lambda \frac{Q^2 dQ}{\Pi(Q)} \left[\frac{1}{2KQ} \ln\left(\frac{(K+Q)^2 + \Delta_+^2}{(P-Q)^2 + \Delta_+^2} \right) - \frac{2}{Q^2 + \Delta_+^2} \right], \quad (13.52)$$

where $\Pi(Q)$ is the relativistically invariant, imaginary frequency form of (13.46). A simple analysis shows that the integral is logarithmically divergent at large Q, and so we have introduced a relativistically invariant hard cutoff at momentum Λ; the same cutoff appears in other intermediate expressions below, but our final, universal, results are cutoff independent. Now, from (13.39), the quasiparticle residue $\mathcal{A}$ is given by

$$\mathcal{A} = \frac{cg}{N} \left(1 - \frac{d\Sigma(K^2 = -\Delta_+^2)}{dK^2} \right), \quad (13.53)$$

that is, we have to evaluate (13.52) and its derivatives at an imaginary $K = i\Delta_+$; this is quite easily done inside the integral in (13.52), and after a numerical evaluation of the resulting integrand we find

$$\mathcal{A} = \frac{cg}{N} \left(1 - \eta \ln\left(\frac{\Lambda}{\Delta_+} \right) + \frac{\mathcal{X}}{N} \right), \quad (13.54)$$

with the constant $\mathcal{X} = 0.481740823\ldots$, where the same η defined in (13.51) makes an appearance, and where we have omitted terms of order Δ_+/Λ, which can be neglected in the limit $\Lambda \to \infty$, which will eventually be taken. To order $1/N$, we can rewrite (13.54) as

$$\mathcal{A} = \frac{cg}{N} \left(\frac{\Delta_+}{\Lambda} \right)^\eta \left(1 + \frac{\mathcal{X}}{N} \right), \quad (13.55)$$

which indicates that the quasiparticle residue vanishes as Δ_+^η as the coupling g approaches g_c from above. That this is the correct form follows from the general scaling arguments made earlier in Section 11.2.2, which led to the result (11.27), and cannot be completely justified at any finite order in the $1/N$ expansion. The earlier arguments showed how such power laws appear as a general consequence of the vicinity of the system to a scale-invariant critical point. The exponents in the power laws can be expanded in powers of $1/N$, and so they are appearing here as logarithms in the computation of observables. We introduce the constant Z, which is precisely that appearing in the basic scaling form (13.3), by writing

$$A = Z\Delta_+^\eta \left(1 + \frac{\mathcal{X}}{N}\right), \tag{13.56}$$

so that by (13.55)

$$Z = \left(\frac{cg}{N}\right)\Lambda^{-\eta}, \tag{13.57}$$

and (13.56) corresponds to a particular choice of the numerical constant in (11.42). Note that Z is a nonuniversal constant, dependent upon the nature of the cutoff, and it is *nonsingular* as the coupling g goes through g_c. However, as neither g nor Λ is measurable, we should regard (13.56) as the definition of Z, where it is related to the $T = 0$ observables A and Δ_+. Indeed (13.56) is the analog of the relation (10.84) for the quantum Ising chain. In a similar manner, Z can also be related to observables of the ordered ground state (as in (10.61) for the Ising chain); we simply quote the result obtained in [86]:

$$\frac{N_0^2}{\rho_s} = Z\left(\frac{\rho_s}{N}\right)^\eta (1 + \eta \ln 16). \tag{13.58}$$

We reiterate that while the relations (13.56) and (13.58) relate Z to ground state observables that vanish in a singular manner at the critical point $g = g_c$, Z itself is nonsingular and finite.

One of the central implications of universal scaling forms like (13.3) is that when the overall scale of the susceptibility χ is expressed in terms of the quasiparticle residue A, or the closely related nonsingular constant Z, the remaining expression becomes universal. In particular, the cutoff dependence in the self-energy Σ in (13.52) must disappear. Using the value of Z above, we can rewrite (13.39) as

$$\chi(k, \omega) = ZT^\eta \left[c^2k^2 - \omega^2 + \lim_{\Lambda \to \infty}\left(m^2 + \Sigma(k, \omega) - \eta(c^2k^2 - \omega^2)\ln\left(\frac{\Lambda}{T}\right)\right)\right]^{-1}. \tag{13.59}$$

Provided the limit above exists, it is then evident that (13.59) is precisely of the scaling form (13.3) and defines the scaling function Φ_+. Conversely we can use the scaling arguments by which (13.3) was derived to argue that the limit must exist; indeed, it is not difficult to show explicitly that the limit exists at this order in $1/N$ at all T. Notice also that the subtraction in Σ affects only its real part, and this is why we saw no divergent terms in the computation earlier of its imaginary part. The constant m^2 has been included within the large-Λ limit because the $\mathcal{O}(1/N)$ corrections in (13.44) are Λ dependent, and these terms are required to obtain a finite limit.

A complete expression for χ at $T = 0$ is now available by combining (13.52) and (13.59). The integral over P cannot be simplified further, but explicit evaluation is however possible in the limit $\Delta_+ \to 0$, which we now consider. We can obtain this limit by approaching the critical point at fixed momenta and frequency, or by examining the large-energy regime $\omega^2 - c^2k^2 \gg \Delta_+$. From the former point of view, we have a picture of $1, 3, 5, \ldots$ particle continua in the spectral weight coming down in energy, and we can ask, what does their superposed spectral weight look like? From (13.52) and (13.46), we have in the limit $\Delta_+ \to 0$

$$
\lim_{\Lambda \to \infty} \left(\Sigma(K) - \eta \ln \left(\frac{\Lambda}{T} \right) \right)
$$

$$
= \frac{4}{\pi^2 N} \int_0^\infty dQ \left[\frac{Q^2}{K} \ln \left| \frac{K+Q}{K-Q} \right| - 2Q - \frac{2K^2 Q}{3(Q^2 + T^2)} \right]
$$

$$
= K^2 \left[\eta \ln \left(\frac{T}{K} \right) + \frac{8}{9\pi^2 N} \right]. \tag{13.60}
$$

Taking the imaginary part of (13.60) for real frequencies, we immediately get (13.50) for $\omega > ck$. This explains why we parameterized the spectral weight in terms of the exponent η. Inserting (13.60) into (13.59) we get

$$
\chi(k, \omega) = Z \left(1 - \frac{8}{9\pi^2 N} \right) \frac{1}{(c^2 k^2 - \omega^2)^{1 - \eta/2}}. \tag{13.61}
$$

Reassuringly T has dropped out. So as we move to the critical point at $T = 0$, the resultant of the superposition of the multiparticle continua is a single critical continuum characterized by the exponent η. This spectral weight has precisely the form sketched in Fig. 10.7 for the Ising chain (the latter model had $\eta = 1/4$). Indeed the entire structure of the $T = 0$ crossover from the quasiparticle pole and multiparticle continua to the critical continuum is essentially identical to that obtained earlier for the Ising model.

13.2.2　Nonzero temperatures

Turning on a nonzero temperature introduces additional thermal damping to the spectral functions computed above, and results in a finite phase coherence time, τ_φ. We find that the structure of these effects is again remarkably similar to those studied earlier for the quantum Ising chain in Sections 10.4.2 and 10.4.3.

First, we note some intermediate steps associated with the mechanics of the computation. We are particularly interested in imaginary parts of Green's functions. From (13.42) we get at $T > 0$

$$
\operatorname{Im}\left(\Pi(q, \omega) \right) = \int \frac{d^2 q_1}{16\pi \varepsilon_{\vec{q}_1 + \vec{q}} \varepsilon_{\vec{q}_1}} \left[\left| n(\varepsilon_{\vec{q}_1 + \vec{q}}) - n(\varepsilon_{\vec{q}_1}) \right| \delta\left(\omega - \left| \varepsilon_{\vec{q}_1 + \vec{q}} - \varepsilon_{\vec{q}_1} \right| \right) \right.
$$

$$
\left. + \left(1 + n(\varepsilon_{\vec{q}_1 + \vec{q}}) + n(\varepsilon_{\vec{q}_1}) \right) \delta\left(\omega - \varepsilon_{\vec{q}_1 + \vec{q}} - \varepsilon_{\vec{q}_1} \right) \right], \tag{13.62}
$$

where $n(\varepsilon)$ is the Bose function

$$
n(\varepsilon) = \frac{1}{e^{\varepsilon/T} - 1}, \tag{13.63}
$$

and the dispersion spectrum $\varepsilon_{\vec{q}}$ is given by

$$
\varepsilon_{\vec{q}}^2 = c^2 q^2 + m^2. \tag{13.64}
$$

Note that the T dependence of (13.62) arises from two sources: that contained in the Bose function (13.63), reflecting the T-dependent occupation of the modes, and that due to the

T dependence of the "mass" m in (13.44), which changes the quasiparticle dispersion. We also need the generalization of (13.47) to finite temperature, where it becomes

$$\text{Im}\Sigma(k, \omega) = \frac{1}{2\pi^2 N} \int \frac{d^2q}{\varepsilon_{\vec{k}+\vec{q}}} \int_0^\infty d\Omega \text{Im} \left(\frac{1}{\Pi(q, \Omega)} \right)$$
$$\times \left[|n(\varepsilon_{\vec{k}+\vec{q}}) - n(\Omega)| \delta(\omega - |\varepsilon_{\vec{k}+\vec{q}} - \Omega|) \right.$$
$$\left. + (1 + n(\varepsilon_{\vec{k}+\vec{q}}) + n(\Omega))\delta(\omega - \varepsilon_{\vec{k}+\vec{q}} - \Omega) \right]. \qquad (13.65)$$

We first discuss the physical properties of the above results in the limit $T \ll \Delta_+$ (i.e. in the low-T regions on the quantum paramagnetic side of Figs. 11.2 and 11.3). In this case, it is easy to see from (13.62) and (13.65) that all effects of temperature are exponentially suppressed (i.e. they are of order $e^{-\Delta_+/T}$ or smaller); also the "mass" $m \approx \Delta_+$ in this region. This is easy to understand since there is a gap Δ_+ to all excitations above the ground state, and all thermal effects are exponentially suppressed. One of the most important consequences of a nonzero T is broadening of the quasiparticle pole in $\chi(k, \omega)$. We explicitly describe the nature of this broadening at $k = 0$. The $T = 0$ pole is then at $\omega = \varepsilon_0 = \Delta_+$, and for $\omega \approx \Delta_+$ we can write χ as

$$\chi(0, \omega) = \frac{\mathcal{A}}{2\varepsilon_0} \frac{1}{(\varepsilon_0 - \omega - i/\tau_\varphi)}, \qquad (13.66)$$

where

$$\frac{1}{\tau_\varphi} = -\frac{1}{2\varepsilon_0} \text{Im}\Sigma(0, \varepsilon_0). \qquad (13.67)$$

Note the similarity of (13.66) to the Ising chain result (10.89) and the $d = 1$ rotor model result (12.30). As in the previous cases we have chosen to define the inverse phase coherence time, $1/\tau_\varphi$, as the width of the quasiparticle pole. We have included only the T-dependent corrections to $\text{Im}\Sigma$ and neglected those to $\text{Re}\Sigma$. This is because the former are much more important for broadening at $\omega \approx \Delta_+$, whereas the latter contribute only a negligible correction to the overall spectral weight of the quasiparticle feature. Evaluating $1/\tau_\varphi$ from (13.46), (13.62), and (13.65), we find for $T \ll \Delta_+$

$$\frac{1}{\tau_\varphi} = \frac{2\pi T e^{-\Delta_+/T}}{N} \left[1 + 2 \int_0^\infty dy \frac{e^{-y}}{\pi^2 + \ln^2(8\Delta_+/Ty)} \right]. \qquad (13.68)$$

Comparing this with the exact result (10.71) in the corresponding region of the quantum Ising chain (our definition of τ_φ there was slightly different) we see that the T dependence is essentially identical, and only the numerical prefactors differ. The latter need not agree, of course, as we are comparing models in different dimensions, and the prefactor in (13.68), unlike that in (10.71), is not exact and contains only the leading term in a $1/N$ expansion. There is also a subleading term with a $1/\ln^2(\Delta_+/T)$ dependence in (13.68). This logarithm is due to the T-matrix structure of a dilute Bose gas in two dimensions (which the thermally excited quasiparticles form), and its origin will be understood better in Chapters 15 and 16.

Finally, let us turn to the high-T region, $T \gg \Delta_+$. In this case T becomes the most important energy scale and controls the entire structure of the response functions. This is already apparent from the value of m in this limit: from (13.44) we have

$$m = \Theta T, \tag{13.69}$$

where $\Theta = 2\ln[(\sqrt{5} + 1)/2] \approx 0.962424\ldots$ Thus the two energy scales that determined spectral functions such as (13.62) and (13.65), the mass m and the temperature T in the Bose function, *both* become of order T. As a result, it is evident by a simple rescaling of the variables of integration in (13.62) that the propagator Π satisfies

$$\Pi(k, \omega) = \frac{1}{T} \Phi_\Pi \left(\frac{ck}{T}, \frac{\omega}{T} \right). \tag{13.70}$$

Determination of the scaling function Π requires complete evaluation of (13.62), and it is not possible to make any further simplifications. We therefore have to resort to numerical computation. In the limit $ck, \omega \gg T$, however, it is clear that Π reduces to the $\Delta_+ = 0$ limit of (13.46). Very similar considerations also apply to the expression for $\mathrm{Im}\Sigma$ in (13.65). The case of $\mathrm{Re}\Sigma$ is, however, somewhat more subtle. We already saw this in the computation at $T = 0$ where we encountered a logarithmic cutoff dependence. This was cured by expressing χ in terms of the quasiparticle residue $\mathcal{A}$, or the amplitude Z, which led to the result (13.59) with a subtraction that cancelled the cutoff dependence in $\mathrm{Re}\Sigma$. Indeed, we can use (13.59) also to evaluate χ for $T > 0$. Precisely the same subtraction is still adequate to cancel the cutoff dependence. The expression for (13.59) has to be evaluated numerically, and we do not present the details of this here; they may be found in [86]. The result satisfies the scaling form (13.3) and yields numerical values for the complex-valued scaling function Φ_+ at $\Delta_+/T = 0$.

We show the results of such a numerical evaluation in Fig. 13.3. Note the strong similarity to the corresponding result for the quantum Ising chain in Fig. 10.8, for which we had the exact expression (10.97). There are quasiparticle-like peaks with a width of order T. The typical excitation has an energy of order T and also a width of order T, and so the quasiparticles are, strictly speaking, not well defined. At very large $\omega, ck \gg T$, the spectrum crossovers to the $T = 0$ result in (13.61), whose form was sketched in Fig. 10.7.

It should also be clear from the above discussion that the phase coherence rate, $1/\tau_\varphi$, is of order T, as it is the only energy scale around. We want to choose a definition that yields $\tau_\varphi = \infty$ at $N = \infty$ as there is no damping in this limit. Indeed, as quasiparticles are well defined at large, but finite N, even in the high-T limit, we may continue to use (13.67) as our definition of τ_φ. Numerical evaluation yields

$$\frac{1}{\tau_\varphi} = 0.904 \frac{k_B T}{N\hbar}, \tag{13.71}$$

where we have inserted a factor of $k_B/\hbar$ to emphasize that this result depends only on fundamental constants.

Finally, we attempt to use the same expansion above to understand the low frequency behavior of the spectral density $\mathrm{Im}\chi(k, \omega \to 0)$, as was done in Fig. 10.9 for the quantum Ising chain. On general grounds, for an interacting system at nonzero temperatures that

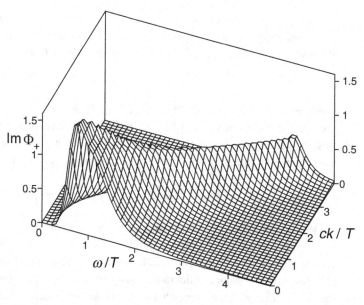

Fig. 13.3 Imaginary part of the scaling function Φ_+ in (13.3) as a function of ck/T and ω/T evaluated in the high-temperature limit $\Delta_+/T = 0$. The function was computed in the $1/N$ expansion and evaluated at $N = 3$. Compare with the exact answer for the $d = 1$ Ising model in Fig. 10.8.

has internal relaxational dynamics, we expect that $\chi(k, \omega)$ is analytic as a function of ω at $\omega = 0$; so the odd function $\mathrm{Im}\chi(k, \omega) \sim \omega$, and $\lim_{\omega \to 0} \mathrm{Im}\chi(k, \omega)/\omega$ is nonzero. This was found to be the case for the Ising chain in Fig. 10.9. However, the present large-N expansion does not obey this requirement; evaluation of (13.65) shows that a low-frequency spectral density comes only from collisions of particles with very high momenta, and their contributions are suppressed by exponentially small thermal factors. Specifically, we find [440] $\mathrm{Im}\chi(k, \omega \to 0) \sim \mathrm{sgn}(\omega) \exp(-c/|\omega|)$, for some constant c. This result is an artifact of the $1/N$ expansion, which places undamped intermediate states in the decay rate computation in (13.65). Alternatively stated, *even though the quasiparticles scatter weakly in the large-N limit, the low frequency relaxational dynamics of the order parameter* **n** *is strongly coupled.* This dynamics is discussed by alternative methods in Section 14.3.

13.3 Summary

As in previous chapters, we summarize the physical properties of the regions of Fig. 11.2 and Fig. 13.1 in Table 13.1. The evolution of the dynamic structure factor $S(k, \omega)$ between the three regimes is quite similar to that discussed for the $d = 1$ Ising model in Section 10.4.4. In the quasi-classical particle regime, we have a narrow peak of width $1/\tau_\varphi$ at a frequency $\omega \approx \Delta_+$. Conversely, in the quasi-classical wave regime, $S(k, \omega)$ becomes a symmetric function of ω and is sharply peaked near $k = 0$, $\omega = 0$ with an exponentially

Table 13.1 Physical propertiesis in three regimes.

Values of the correlation length, ξ (defined from the exponential decay of the equal-time correlations of **n**), the uniform spin susceptibility, χ_u, the phase coherence time, τ_φ, and the spin diffusion constant, D_s, for the two regimes in Figs. 11.2 and 13.1. The results in the quasi-classical wave regime are quoted only for $N=3$ and are asymptotically exact as $T/\rho_s \rightarrow 0$; other results are obtained in a $1/N$ expansion and are applicable in principle to all N. The $1/N$ corrections to the values for χ_u and ξ in the high-T region were not explicitly computed here and are taken from [86]. The values for D_s in the quantum critical and quasi-classical particle regimes anticipate results from Chapter 15, in particular from (15.11), (15.64), and (15.68). The order of magnitude of D_s in the quasi-classical wave regime follows from the general scaling arguments in Section 13.1. Finally, χ_u in the quasi-classical particle regime anticipates (15.14).

	Low T (magnetically ordered), quasi-classical waves	Continuum high T (quantum critical)	Low T (quantum paramagnetic), quasi-classical particles
ξ	$\dfrac{ec}{16\pi\rho_s}e^{2\pi\rho_s/T}$	$\left[2\ln\left(\dfrac{\sqrt{5}+1}{2}\right)\left(1+\dfrac{0.2373}{N}\right)\dfrac{c}{T}\right]^{-1}$	$\dfrac{c}{|\Delta|}$
χ_u	$\dfrac{2\rho_s}{3c^2}\left[1+\dfrac{T}{2\pi\rho_s}+\cdots\right]$	$\dfrac{\sqrt{5}}{\pi}\ln\left(\dfrac{\sqrt{5}+1}{2}\right)\left(1-\dfrac{0.6189}{N}\right)\dfrac{T}{c^2}$	$\dfrac{\Delta_+}{\pi c^2}e^{-\Delta_+/T}$
τ_φ	$\left(\dfrac{\rho_s}{T}\right)^{1/2}\dfrac{\xi}{c}$	$\dfrac{N}{0.904T}$	$\dfrac{N}{2\pi T}e^{\Delta_+/T}$
D_s	$\sim\dfrac{\xi^2}{\tau_\varphi}$	$\dfrac{0.1077N}{\chi_u}$	$\sim\dfrac{(\ln(\Delta_+/T))^2}{\chi_u}$

large height and an exponentially small width. In the high-T regime there is an interesting structure for ω, ck of order T, and this is discussed in Chapter 14.

We close with a discussion of experimental applications.

The primary application of the $d=2$ O(3) quantum rotor model has been as a continuum theory of the square lattice Heisenberg antiferromagnet. The connection between these models becomes clearer in Chapter 19, but the link between antiferromagnets and quantum rotors has already been motivated in Section 1.4.3.

In the low-T region, $T \ll \rho_s$, careful tests of the exact results (13.10) and (13.20) for the correlation length have been made. The agreement with neutron scattering measurements on the square lattice insulating antiferromagnets La$_2$CuO$_4$ [263] and Sr$_2$CuO$_2$Cl$_2$ is impressive. Much higher precision comparisons can be performed against state-of-the-art quantum Monte Carlo simulations and these have been discussed in [44] and [267]. The low-T dynamical properties discussed in Sections 13.1.2 and 13.1.3 were applied to NMR relaxation rates in [77] and compared against measurements in La$_2$CuO$_4$ in [234].

We turn next to the "high-T" region of the continuum quantum rotor model. Reference [85] argued for the existence of such a region in the intermediate temperature properties of the $S=1/2$ square lattice antiferromagnet. The high-T computations discussed in Section 13.2 were used to compute NMR relaxation rates [86, 88] and found to be in good agreement with measurements on La_2CuO_4 [234, 235]. It is not expected that the crossover from low T to high T would be visible in experimental measurements of the correlation length, as was pointed out early on in [85]; this has been discussed further in [179] and [267]. The issue of the "low-T" to "high-T" crossover in square lattice antiferromagnets was also examined in series expansion studies by Sokol, Glenister, and Singh [478] and Elstner *et al.* [128], and evidence was obtained for its existence in a number of static correlators for spin $S = 1/2$. Interestingly, no such evidence was found for the $S = 1$ case, which (as expected) is clearly too far from the quantum critical coupling.

The results of this chapter can also be applied, along with suitable reinterpretation, to the superconducting states of the doped cuprates. For a recent discussion of this, see [431]. As discussed in this chapter, the low-T quantum paramagnet has a sharp, N-fold degenerate, quasiparticle pole given by (13.66), with an exponentially small linewidth specified by (13.68). This is interpreted [431] as the "resonance peak" observed in neutron scattering experiments on high-temperature superconductors [53, 149, 345, 411]. Also, near the magnetic ordering transition, inelastic neutron scattering measurements [210, 263] have observed an ω/T scaling in their frequency-dependent susceptibility, which is consistent with the general scaling forms (13.1) and (13.3) for a vanishing value for their third arguments.

14 Physics close to and above the upper-critical dimension

We discussed the concept of the upper-critical dimension in Chapters 3 and 4 in the context of the classical models in D dimensions. There we showed that the critical physics for $D > 4$ was accessible in a perturbative analysis, while the $D < 4$ case required a renormalization group analysis. The latter case required an expansion in the parameter ϵ defined in (4.25). Here we discuss application of the ϵ expansion to physical issues associated with the quantum phase transition, where now

$$\epsilon = 3 - d. \tag{14.1}$$

The physics described by this perturbative method can, in most cases, also be elucidated by the large-N expansion we have developed in the previous chapters. However, there are a number of instances where the underlying principles are most transparently illustrated by studies close to and above $d = 3$. Our specific reasons for undertaking such an analysis are:

- As we have noted earlier, the quantum critical point at $T = 0$, $g = g_c$ extends out to a line of finite-temperature phase transitions for the cases $d = 2$, $N = 1, 2$. The ϵ expansion offers a controlled method for obtaining the structure of the crossovers in the vicinity of this line.
- We have not yet found a successful description of the low-frequency dynamics of the order parameter ($\mathbf{n}$ or σ^z) in the high-T regime for $d = 2$, although we did succeed for $d = 1$ in Chapters 10 and 12. We show in Section 14.3 that the ϵ expansion leads to an appealing quasi-classical wave description of this dynamics.
- For the quantum rotor models being studied here, the crossovers above the upper-critical dimension, with $d > 3$, are obviously in a physically inaccessible dimension. However, the basic structure that emerges is quite generic to quantum critical points above their upper-critical dimension. The results are therefore useful in Part IV, where we consider other models with a lower upper-critical dimension, so that dimensions above the upper-critical can be experimentally studied.
- The following Chapter 15 studies transport properties of the quantum rotor models in the high-T and quantum-paramagnetic low-T regions for $d = 2$. While it is possible to do this within the $1/N$ expansion, the computation is simplest, and most physically transparent, using the ϵ expansion we develop here.

The study in this chapter uses the "soft-spin" formulation of the continuum theory of the vicinity of the quantum critical point that was noted in Section 2.1, and also used in Section 6.6 and Chapter 7.

The theory is expressed in terms of an N-component field $\phi_\alpha(x, \tau)$ $(\alpha = 1 \ldots N)$, which is related to the lattice quantum rotor field $\mathbf{n}_i$ by the coarse-graining transformation (3.18); for $N = 1$, a similar relationship holds between the Ising spin $\hat{\sigma}_i^z$ and a one-component field ϕ. We study the quantum mechanics of the ϕ_α field as specified by the Hamiltonian in (6.52), or the imaginary-time path integral in (3.25), which is reproduced here for completeness:

$$\mathcal{Z} = \int \mathcal{D}\phi_\alpha(x, \tau) \exp(-\mathcal{S}_\phi),$$

$$\mathcal{S}_\phi = \int d^d x \int_0^{\hbar/k_B T} d\tau \left\{ \frac{1}{2} \left[(\partial_\tau \phi_\alpha)^2 + c^2 (\nabla_x \phi_\alpha)^2 + r\phi_\alpha^2(x, \tau) \right] + \frac{u}{4!} \left(\phi_\alpha^2(x, \tau) \right)^2 \right\}. \qquad (14.2)$$

The structure of this field theory is similar to that of the continuum theory (2.12) or (11.4) for the fixed-length $\mathbf{n}$ field, with the main difference being that the fixed-length constraint has been dropped and replaced instead by a quartic self-interaction u. The equivalence of the universal properties of these two formulations is a well-established principle in the theory of classical critical phenomena [59, 61]. This equivalence can be expected on general universality grounds, as the two models display a quantum critical point between a magnetically ordered and a quantum-paramagnetic phase with precisely the same symmetry structures and spectrum of low-lying excitations. We also explicitly see examples of the equivalence in our computations with (14.2) in this chapter. In practical terms, this equivalence means that the susceptibility $\chi(k, \omega)$, defined as the two-point correlator of the field ϕ_α, satisfies, for $d < 3$, the scaling forms (13.1) and (13.3), with precisely the same scaling function $\Phi_\pm$. We compute here some features of these scaling functions in an expansion in ϵ, whereas they were computed in a $1/N$ expansion in Chapter 13. The approaches have been compared in the overlapping region of validity where both ϵ and $1/N$ are small, and exact agreement is found (although this is not shown explicitly here).

We restrict ourselves in this chapter to results to the leading order in ϵ or u. The structure of the quantum/classical crossovers is quite complicated at higher orders, and the reader is referred to [422] for a discussion of this subtle issue; alternative approaches are also available [156, 287, 367]. Further, we limit our discussion to regions of the phase diagram where there is no spontaneous magnetization and complete $O(N)$ symmetry is preserved (the extension to ordered phases is straightforward). We are therefore approaching the finite-temperature phase boundaries from their high-temperature side.

We begin in Section 14.1 by a discussion of the $T = 0$ properties of (14.2); these are simply related to those obtained in Chapters 3 and 4 for the classical problem in $D = d + 1$ dimensions. Section 14.2 then provides a description of the ϵ expansion for the crossovers in the static properties of (14.2) at $T > 0$; although this expansion gives a useful qualitative picture, it is not particularly accurate for $d = 2$ and also fails for low-frequency dynamical properties. These deficiencies are repaired in Section 14.2, where we use the ϵ expansion to motivate an effective model for statics and dynamics that is solved exactly for $d = 2$.

We use units in which the velocity $c = 1$ throughout the remainder of this chapter only.

14.1 Zero temperature

We work in imaginary time throughout this section. We express the response functions in terms of a $D = (d + 1)$-dimensional wavevector $Q = (i\omega, \vec{q})$. At $T = 0$, all correlators of the action (14.2) are invariant under D-dimensional rotations in Euclidean space; they are therefore only functions of $Q^2 = q^2 + (i\omega)^2 = q^2 - \omega^2$. Dynamic quantum response functions are obtained by analytically continuing to negative Q^2. For positive Q^2 the responses are, of course, those associated with interpreting (14.2) as a classical statistical mechanics problem.

Recall our discussion in Section 3.3.2 of the computation of the two-point correlator of ϕ_α in ordinary perturbation theory in u, where we found that the results were adequate for $D > 4$ but suggested that higher orders had to be re-summed for $D < 4$. Below we need the explicit re-summation of this series for small u: this is done using the $1/N$ expansion in Subsection 14.1.1, where we also introduce the concept of the so-called tricritical crossover functions. Finally, in Subsection 14.1.2 we present a very concise review of the field-theoretic renormalization group approach to re-summing the perturbation theory in u.

14.1.1 Tricritical crossovers

Our main perturbative result for the susceptibility was presented in (3.57). Extended to nonzero wavevectors, and adapted to our present notation, this result is

$$\chi^{-1}(Q) = Q^2 + s\left[1 - \left(\frac{N+2}{6}\right)\frac{2\Gamma((4-D)/2)}{(D-2)(4\pi)^{D/2}}\frac{u}{s^{(4-D)/2}}\right]. \tag{14.3}$$

We are now interested in estimating the structure of the higher order corrections in u, especially for $D < 4$. For $D < 4$, the structure of (14.3) suggests that we can express the most important terms at higher order in u for the static susceptibility in the form

$$\chi^{-1}(Q) = s\Psi_D\left(\frac{Q}{s^{1/2}}, \frac{u}{s^{(4-D)/2}}\right), \tag{14.4}$$

where $\Psi_D(q, v)$ is a universal crossover function. This form is consistent with naive dimensional analysis and the expectation that it is permissible to send $\Lambda \to \infty$ in all the singular terms at higher orders. The result (14.3) gives us the small-v behavior of $\Psi_D(q, v)$:

$$\Psi_D(q, v) = q^2 + 1 - \left(\frac{N+2}{6}\right)\frac{2\Gamma((4-D)/2)}{(D-2)(4\pi)^{D/2}}v + \mathcal{O}(v^2). \tag{14.5}$$

To get the critical properties of the model for $D < 4$, however, we need its large-v behavior.

The function $\Psi_D(q, v)$ is the "tricritical crossover" function of [369] and [68]. (This terminology is motivated by considerations unrelated to those of interest here and will not be explained.) Computation of $\Psi_D(q, v)$ by various methods is described in the literature [64, 360]. We simply treat $\Psi_D(q, v)$ as a known function and find that some key properties of the $T > 0$ crossovers near the quantum critical point can be expressed in terms of it. For completeness, we note how $\Psi_D(q, v)$ may be computed in the large-N limit, with

vN fixed. The computation proceeds by a familiar approach: we decouple the quartic term in (14.2) by a Hubbard–Stratanovich field λ so that

$$\mathcal{Z} = \int \mathcal{D}\lambda \mathcal{D}\phi_\alpha(x, \tau) \exp(-\tilde{\mathcal{S}}_\phi),$$

$$\tilde{\mathcal{S}}_\phi = \int d^D x \left\{ \frac{1}{2} \left[(\partial_\tau \phi_\alpha)^2 + c^2(\nabla_x \phi_\alpha)^2 + (r + i\lambda)\phi_\alpha^2(x) \right] + \frac{3\lambda^2}{2u} \right\}. \tag{14.6}$$

Now, integrate out the ϕ_α fields, which appear only as quadratic terms in (14.6); evaluating the integral over λ as a saddle point, we obtain the required large-N limit. Expressing the result in terms of s using (3.54), we can easily show that $\Psi_D(q, v)$ is given by

$$\Psi_D(q, v) = q^2 + \Pi_D(v), \tag{14.7}$$

where the function $\Pi_D(v)$ is given by the solution of the following nonlinear equation:

$$\Pi_D(v) + Nv \frac{\Gamma((4 - D)/2)}{3(D - 2)(4\pi)^{D/2}} [\Pi_D(v)]^{(D-2)/2} = 1. \tag{14.8}$$

Note that as $v \to \infty$, $\Pi_D(v) \sim v^{-2/(D-2)}$, and therefore (14.7) and (14.4), imply that $\chi^{-1}(0) \sim s^{2/(D-2)}$ as $s \to 0$. This result agrees with our earlier large-N result in (11.22) and the $N = \infty$ relation $\chi^{-1}(0) \sim \Delta_+$.

14.1.2 Field-theoretic renormalization group

While most of our renormalization group computations can be performed using the "momentum shell" method introduced in Section 4.2, some results involving crossover functions are obtained by a field-theoretic approach. This is physically not as transparent, but is technically elegant and simpler to compute in. The basic ideas behind this approach have already been presented in Section 12.3.1 in the context of the $d = 1$ quantum rotor model, along with suggestions for further reading in the literature. Readers who skipped Chapter 12 should now read Section 12.3.1 up to (12.40).

As before, it is advantageous to replace the cutoff Λ by a renormalization scale, μ, at which various observable parameters are defined. At the scale μ we introduce renormalized couplings, which then replace the bare couplings in all expressions for observable quantities. Once this substitution has been performed, it is possible to send the cutoff $\Lambda \to 0$, order-by-order in an expansion in the nonlinearities. In practice, one never needs to introduce Λ at intermediate stages as all integrals are performed in dimensional regularization in $D = 4 - \epsilon$ dimensions. We work only to first order in ϵ here, in which case only two renormalized couplings are necessary: s_R, a renormalized measure of the deviation of the system from the quantum critical point, and u_R, a renormalized four-point interaction. The explicit relationships between the bare and renormalized couplings are [59]

$$u = u_R \frac{\mu^\epsilon}{S_D} \left(1 + \frac{N + 8}{6\epsilon} u_R \right),$$

$$s = s_R \left(1 + \frac{N + 2}{6\epsilon} \right). \tag{14.9}$$

A factor of μ^ϵ has been scaled out of u so that u_R is dimensionless, and $S_D = 2/[\Gamma(D/2)$ $(4\pi)^{D/2}]$ is a standard phase-space factor, introduced for notational convenience.

We can state the simple, field-theoretic recipe for computing correlators of (14.2). First, obtain formal expressions for the bare theory in terms of s and u, leaving integrals uneval-uated. Then, perform the substitution in (14.9) to expressions in terms of u_R and s_R. Now, evaluate all the integrals in $D = 4 - \epsilon$ dimensions, in powers of ϵ. The constants in (14.9) have been cleverly chosen so that all poles in ϵ cancel. The resulting expressions for the correlators of the theory are expressed in terms of s_R, u_R, and the momentum scale μ.

Exact renormalization group equations for all observables can be obtained by the fact that no physical quantity can depend upon the value of μ. By studying the behavior of the first equation in (14.9) under $\mu \rightarrow \mu e^\ell$ we obtain the flow equation

$$\frac{du_R}{d\ell} = \epsilon u_R - \left(\frac{N+8}{6}\right) u_R^2. \tag{14.10}$$

This equation is, of course, the present field-theoretic form of the flow equation obtained earlier in (4.24) by the simpler momentum-shell method. A simple analysis of this dif-ferential equation shows that, at long distances ($\ell \rightarrow \infty$), the coupling u_R flows to the attractive fixed point at

$$u_R^* = \frac{6\epsilon}{(N+8)}, \tag{14.11}$$

as also found in (4.26). This implies that a theory with $u_R = u_R^*$ and $s = 0$ does not flow under rescaling transformations and is therefore scale invariant. This specifies the univer-sal quantum critical point of theory. Turning on an $s > 0$ induces flow along the leading relevant direction and therefore determines the $T = 0$ energy gap. Deviations in u from u_R correspond to allowing corrections associated with the leading irrelevant operator; these can therefore be ignored in computations of the universal scaling functions.

Let us apply the above approach explicitly to the computation of $\chi(Q)$. We begin with the expression (3.55) and make the substitution in (14.9). Working to linear order in u_R, and evaluating the integrals in an expansion in $\epsilon = 4 - D$, we can write the result in the form

$$\chi(Q) = \frac{1}{Q^2 + \Delta_+^2}, \tag{14.12}$$

where Δ_+ is the $T = 0$ gap of the quantum paramagnetic phase with $s > 0$. The explicit expression for Δ_+ is

$$\Delta_+^2 = s_R \left[1 + u_R \frac{(N+2)}{12} \ln\left(\frac{s_R}{\mu^2}\right)\right], \tag{14.13}$$

where there is no additive term of order u_R associated with the logarithm. Precisely at $u_R = u_R^*$ the scale invariance of the theory implies that it is permissible to re-exponentiate the logarithm (as was done in the large-N expansion in (13.55)), in which case we can write

$$\Delta_+ = \mu \left(\frac{s_R}{\mu^2}\right)^\nu, \tag{14.14}$$

where ν is the usual correlation length exponent, defining how the gap vanishes at $s_R = 0$ (recall that this theory has $z = 1$); it is given to this order in ϵ by

$$\nu = \frac{1}{2} + \frac{(N+2)}{4(N+8)}\epsilon, \tag{14.15}$$

a result which agrees with (4.29) and (4.41). These results imply, from (14.4), that $\Psi_D(v \to \infty) \sim v^{-2(2\nu-1)/(4-D)} \sim v^{-(N+2)/(N+8)}$ to leading order for $4 - D$.

14.2 Statics at nonzero temperatures

This section describes the results of the ϵ expansion on the nonzero-temperature properties of (14.2). The results are helpful in exposing the general structure of the theory, but they are not expected to be very accurate for $d = 2$ ($\epsilon = 1$). An improved and quantitatively more accurate treatment appears in Section 14.2.1, which also considers dynamic properties.

The description of the $T > 0$ static correlators proceeds [422] by a method adapted from an approach developed by Luscher [307] (related methods have also been applied by others to study classical systems in finite geometries [62, 413]). Readers of Chapters 12 and 13 will recall that a similar method was used in Sections 12.3 and 13.1. The main idea is to integrate out the components of $\phi_\alpha(x, \tau)$ with a nonzero frequency along the imaginary-time direction by a straightforward ϵ expansion to the vicinity of the quantum critical point. This will result in an effective action for the zero-frequency component $\phi_\alpha(x)$ (which is independent of τ), which must subsequently be analyzed more carefully. The correlators of this zero-frequency effective action yield the static susceptibility, $\chi(k)$. It must be noted that, unlike the situation in Section 10.4.1, this static susceptibility does not yield the equal-time correlations, as the relationship (10.75) does not hold in general.

As we are only interested in the universal crossovers in the vicinity of the point $s = 0$, $T = 0$, for $D < 4$, we can set $u_R = u_R^*$ at the outset; further, as $u_R \sim \epsilon$, the derivation of the effective action for $\phi_\alpha(x)$ can be performed in an expansion in powers of the nonlinear coupling u_R. For $D > 4$ the mean-field behavior of the system at $T = 0$ suggests that an expansion in powers of u should be adequate for $T > 0$, and we indeed find that this is the case. A simple, one-loop, perturbative calculation then gives the following effective action for the static correlators:

$$\mathcal{Z} = \int \mathcal{D}\phi_\alpha(x)\exp(-\mathcal{S}_{\phi,\text{eff}}),$$

$$\mathcal{S}_{\phi,\text{eff}} = \frac{1}{T}\int d^d x \left\{ \frac{1}{2}\left[(\nabla_x \phi_\alpha)^2 + \tilde{R}\phi_\alpha^2(x)\right] + \frac{U}{4!}\left(\phi_\alpha^2(x)\right)^2 \right\}. \tag{14.16}$$

The couplings $\tilde{R}$ and U can be expressed in terms of the bare couplings in the quantum action (14.2):

$$\tilde{R} = r + u\left(\frac{N+2}{6}\right) T \sum_{\epsilon_n \neq 0} \int \frac{d^d k}{(2\pi)^d} \frac{1}{\epsilon_n^2 + k^2 + r},$$

$$U = u - u^2 \left(\frac{N+8}{6}\right) T \sum_{\epsilon_n \neq 0} \int \frac{d^d k}{(2\pi)^d} \frac{1}{(\epsilon_n^2 + k^2 + r)^2}. \tag{14.17}$$

The result for $\tilde{R}$ arises from the first diagram in Fig. 4.1, and that for U from the second diagram in Fig. 4.1, where the internal lines carry only nonzero Matsubara frequencies. We discuss the evaluation of these expressions shortly. For the moment, let us simply retain the formal expressions in (14.17) and proceed a bit further. Now note that the effective action (14.16) has precisely the same form as the original theory (14.2) at $T = 0$. The only, and crucial, difference is that the spacetime dimension D has been replaced by the spatial dimension d. Therefore, the theory (14.16) can be analyzed by the perturbative method of Section 3.3.2, or by the tricritical formulation of Section 14.1.1 simply by performing the replacement $D \rightarrow d$, and by a relabeling of the coupling constants. Using these methods, it is easy to get formal expressions for the equal-time correlators resulting from $\mathcal{S}_{\phi,\mathrm{eff}}$. We first define a shift in the value of the mass, as in (3.53) and (3.54):

$$R = \tilde{R} + U\left(\frac{N+2}{6}\right) \int \frac{d^d k}{(2\pi)^d} \frac{T}{k^2}. \tag{14.18}$$

Then the response function (14.4) of (14.2) tells us that the static susceptibility, defined in (10.7), is given by

$$\chi(k) = \frac{1}{R} \Psi_d^{-1}\left(\frac{k}{R^{1/2}}, \frac{TU}{R^{(4-d)/2}}\right). \tag{14.19}$$

As noted earlier, we regard Ψ_d as a known function, and so (14.19) construes the final solution of the crossovers of the static observables (14.2) at finite temperature in the region without long-range order. We emphasize again that Ψ_d has to be computed in the spatial dimension d, and not the spacetime dimensions $D = d + 1$ considered in Section 14.1. The large-N solution of $\Psi_d(q, v)$ was given in (14.7) and (14.8) and is valid for all values of v; however, we find below that the exact perturbative result (14.5) (valid for small v) is in fact sufficient over a substantial portion of the phase diagram.

The transition to the phase with long-range order is signaled by a divergence in $\chi(k=0)$. The general structure of (14.19) tells us that this will happen at a value $R = R_c$, with $R_c \sim (TU)^{2/(4-d)}$ (the missing coefficient is a universal number determined by the function Ψ_d). The $N = \infty$ result (14.8) has $R_c = 0$, and this is also found to leading order in the $4 - d$ expansion for tricritical crossovers. In our discussion in this section below, we assume $R_c = 0$ and that corrections due to a nonzero R_c are higher order in ϵ. So the result (14.19) is valid provided $R > 0$, and the condition $R = 0$ gives the boundary of the finite-temperature phase transition to the ordered phase.

It remains to compute the values of the couplings R, U to complete our description of static correlations and the associated phase diagram of (14.2) in the r–T plane. We consider the cases $d < 3$ and $d > 3$ separately, as the results are substantially different.

14.2.1 $d < 3$

We first determine the value of R for $d < 3$. The expression for R is given in (14.17) and (14.18), and to evaluate it in the scaling limit, we use precisely the same prescription discussed earlier in Section 14.1.2 for the $T = 0$ computation: the spatial integrals are evaluated in $d = 3 - \epsilon$ dimensions, the couplings are expressed in terms of the renormalized parameters as defined in (14.9), an expansion is made in powers of ϵ, and finally the resulting expression is evaluated at the fixed-point value (14.11). Just as at $T = 0$, the poles in ϵ cancel also at $T > 0$, and to first order in ϵ, the result is

$$R = s_R \left[1 + \epsilon \left(\frac{N+2}{N+8} \right) \ln \left(\frac{T}{\mu} \right) \right] + \epsilon T^2 \left(\frac{N+2}{N+8} \right) G \left(\frac{s_R}{T^2} \right), \qquad (14.20)$$

where the function $G(y)$ is given by

$$G(y) = \frac{y \ln y}{2} + 4 \int_0^\infty k^2 dk \left(\frac{1}{\sqrt{k^2 + y}} \frac{1}{e^{\sqrt{k^2+y}} - 1} - \frac{1}{k^2 + y} + \frac{1}{k^2} \right). \qquad (14.21)$$

We have obtained this expression assuming that $s_R > 0$ (and therefore $y = s_R/T^2 > 0$), and the result for $G(y)$ appears to have some singularity at $y = 0$. We shall shortly establish that this is not the case: a crucial property of the function $G(y)$ is that it is analytic at $y = 0$ and can therefore be analytically continued to $y < 0$. There is an important physical reason for this analyticity, and it is a key step in our analysis. Recall that, at $T = 0$, there was a quantum phase transition in (14.2) at $s_R = 0$ ($r = r_c$ from (3.54)), and so all response functions are certainly nonanalytic at $s_R = 0$. However, we are considering the case $T > 0$, and we expect that there is no thermodynamic singularity at $r = r_c$. The critical fluctuations surely get quenched at a nonzero T, and all observables should have a smooth, well-behaved dependence on r at $r = r_c$ for $T > 0$, as we saw in the case of the Ising chain in Chapter 10. There is eventually a nonzero T phase transition for some $s_R < 0$ ($r < r_c$) as in Fig. 11.3, and so there should be a thermodynamic singularity at this point. However, the latter singularity is a property of the scaling function Ψ_d in (14.19), and *not* a singularity in the value of the coupling R. Hence if our physical interpretation is correct, $G(y)$ should be analytic at $y = 0$, and it should be possible to analytically continue $G(y)$ to all $y < 0$ until the point when we hit the transition to the ordered phase where R, as defined in (14.20), first vanishes.

We explicitly demonstrate that the expectation above is indeed satisfied by (14.21) (indeed, our entire analysis of the crossover problem was carefully designed so that this would occur). After an integration by parts under the integral in (14.21), and some elementary manipulations, it can be shown that $G(y)$ can be transformed into the following:

$$G(y) = - \int_0^\infty dk \left[4 \ln \left(k \frac{\sinh(\sqrt{k^2 + y}/2)}{\sqrt{k^2 + y}/2} \right) - 2k - \frac{y}{\sqrt{k^2 + 1/e}} \right]. \qquad (14.22)$$

In this form, it is not difficult to see that $G(y)$ is analytic at $y = 0$; the function $\sinh(\sqrt{z})/\sqrt{z}$ is a smooth function of z near $z = 0$, and equals $\sin(\sqrt{|z|})/\sqrt{|z|}$ for $z < 0$, and so there is no singularity in the integrand as y goes through zero. Indeed $G(y)$ is smooth for all

$y > -2\pi$, with the singularity arising at -2π when the argument of the logarithm first turns negative. We find below that the transition to the ordered phase occurs for $y \sim -\epsilon$; thus the singularity at $y = -2\pi$ occurs well within the ordered phase, where the present results cannot be used, and is therefore of no physical consequence. We also note here some limits of (14.22) that will be useful later:

$$G(y) = \begin{cases} 2\pi^2/3 + 2.45381y, & |y| \ll 1 \\ (y/2)\ln y + 2\pi\sqrt{y} + y^{1/4}\sqrt{8\pi}e^{-\sqrt{y}}. & y \gg 1 \end{cases} \quad (14.23)$$

While we have a fairly complete picture of the function $G(y)$, the result (14.20) for R is still not ready to be used as it involves the unknown momentum scale μ. To remedy this, we recall a basic strategy used throughout this book: all correlators should be expressed in terms of observable parameters characterizing the $T = 0$ ground state. In the present situation we should clearly replace s_R as a measure of the deviation from the $T = 0$ quantum critical point at $s_R = 0$ by the energy scales $\Delta_\pm$ discussed in Chapter 11. We do this here only for $s_R > 0$ (the case $s_R < 0$ is discussed in [422]). The relationship between s_R and the energy gap of the quantum paramagnet, Δ_+, was obtained in (14.14). Substituting (14.14) into (14.20) we find

$$R = \Delta_+^2 \left[1 + \epsilon \left(\frac{N+2}{N+8} \right) \ln \left(\frac{T}{\Delta_+} \right) \right] + \epsilon T^2 \left(\frac{N+2}{N+8} \right) G \left(\frac{\Delta_+^2}{T^2} \right). \quad (14.24)$$

The dependence on the arbitrary scale μ has disappeared, and we have the required universal dependence of R on Δ_+ and T for $s_R > 0$ ($r > r_c$). A similar relationship exists between the scale Δ_- and R for $r < r_c$ [422].

A closely related computation can be performed for the quartic coupling U in (14.16) using the expression in (14.17). At the fixed point $u_R = u_R^*$ we again find that the μ dependence disappears:

$$U = \frac{6\epsilon T^\epsilon}{S_D(N+8)} \left[1 + \epsilon \frac{20 + 2N - N^2}{2(N+8)^2} + \epsilon G' \left(\frac{s_R}{T^2} \right) \right], \quad (14.25)$$

where $G'(y)$ is the derivative of $G(y)$, and we have actually used the expression for u_R^* to order ϵ^2 to obtain the complete result above. For $s_R > 0$ we can simply substitute $s_R = \Delta_+^2$ in the argument of G' to get the universal expression for U.

We have assembled all the ingredients to obtain the full crossover structure for the static susceptibility at $T > 0$. We use the expressions (14.20) or (14.24) for R and the expression (14.25) for U, and substitute them into (14.19), with results for the tricritical crossover function Ψ_d obtained in Section 14.1.1. A straightforward examination of the resulting expressions yields the phase diagram shown in Fig. 14.1, which is closely related to the phase diagram obtained earlier in the large-N limit in Fig. 11.3. The physical properties of the regimes have already been discussed in Section 11.3, and we note their properties for small ϵ in turn below.

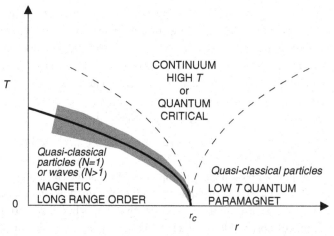

Fig. 14.1 Phase diagram of the theory (14.2) for $d < 3$ (compare with the large-N phase diagram in Fig. 11.3, and Figs. 1.2b and 1.3). The qualitative features are expected to apply to $d > 1$ for $N = 1$, $d \geq 2$ for $N = 2$, and $d > 2$ for $N \geq 3$. The quantum critical point is at $T = 0$ with coupling $r = r_c$ (this is also the coupling where $s = s_R = 0$). All properties are however analytic as a function of r at $r = r_c$ for $T > 0$. The dashed lines are crossovers at $T \sim |r - r_c|^{z\nu}$, as is the full line, which is the locus of finite-temperature phase transitions at $T_c(r)$. The shaded region is where the reduced classical scaling functions apply. The region $T > T_c(r)$, but $r < r_c$, is accessed in our calculation by analytic continuation from $r > r_c$, $T > 0$. The simple perturbative expression in (14.3) can be used in (14.19) for the static susceptibility everywhere in the paramagnetic region, except for the shaded portion. The low-T region for $r > r_c$ has a quasi-classical particle description as in Section 10.4.2 and is discussed in Chapter 15. In the magnetically ordered low-T region for $r < r_c$ and $N \geq 2$, the long-wavelength spin waves about the ordered state behave classically, while for $N = 1$, the amplitude oscillations in ϕ_α about its nonzero mean value lead to a quasi-classical particle. As we noted in Fig. 13.1, "the continuum high-T" or "quantum critical" region is more complex, with both thermal and quantum phenomena, and both particle- and wave-like phenomena, playing equal roles. In Section 14.3 we show that, to leading order in $\epsilon = 3 - d$, the low-frequency correlators of ϕ_α in this region are described by an effective quasi-classical *wave* model; however, the transport of the conserved angular momentum is dominated by higher energy excitations, and requires a particle-like description in a quantum Boltzmann equation which is discussed in Chapter 15.

The low-T regime on the quantum paramagnetic side was discussed in Section 11.3.1: it is present for $T \ll \Delta_+ \sim (r - r_c)^\nu$. Using (14.23)–(14.25), we have for this case

$$R \sim \Delta_+^2,$$

$$U \sim \epsilon \Delta_+^\epsilon,$$

$$TU/R^{(4-d)/2} \sim \epsilon T/\Delta_+ \ll 1. \tag{14.26}$$

The last quantity is that appearing in the argument of the tricritical scaling function, Ψ_d, in (14.19). As it is small, it is evident that a simply perturbative evaluation of Ψ_d in (14.5) is adequate for analyzing static properties in this regime. Using (14.5) and (14.23)–(14.25) in (14.19) we get

$$\chi^{-1}(k) = k^2 + \Delta_+^2 + \epsilon \left(\frac{N+2}{N+8} \right) T (8\pi T \Delta_+)^{1/2} e^{-\Delta_+/T}. \tag{14.27}$$

Thus there is only a correction of the order $e^{-\Delta_+/T}$ to the $T=0$ response; similar results were obtained in the large-N limit in Section 11.3.1. This exponentially small correction arises from the small density of pre-existing thermally excited particles. For the same reasons as those discussed in Section 10.4.2 (and also in Section 12.2), we expect that these particles form a Boltzmann gas, whose static and dynamic properties can be described by standard classical methods. We see this in our discussion of transport properties in Chapter 15.

Next we turn to the high-T regime of the continuum theory $T \gg |r - r_c|^\nu$. Now, the analogs of the estimates (14.26) are

$$R \sim \epsilon T^2,$$
$$U \sim \epsilon T^\epsilon,$$
$$TU/R^{(4-d)/2} \sim \sqrt{\epsilon} \ll 1. \tag{14.28}$$

So again, the second argument of Ψ_d is small, and a perturbative evaluation is permissible. Using (14.5) and (14.23)–(14.25) in (14.19) we get

$$\chi^{-1}(k) = k^2 + \epsilon \left(\frac{N+2}{N+8} \right) \frac{2\pi^2 T^2}{3} \tag{14.29}$$

to leading order in ϵ, which implies a correlation length $\xi \sim 1/\sqrt{\epsilon}T$. The almost free nature of this static result suggests that thermal fluctuations are noncritical and can be treated in an effectively Gaussian theory. However, when the present perturbative approach is extended to dynamical properties, one finds that it fails in the low-frequency limit [422] (just as we found for the $1/N$ expansion in Section 13.2.2). The strongly coupled dynamical problem is treated in Section 14.3, and associated transport properties are detailed in Chapter 15.

Finally, we turn to a novel part of the analysis using the ϵ expansion: the region of the phase without long-range order for $r < r_c$. Here $s_R < 0$, and it is possible for R to vanish. Using (14.20), we find that this happens at $s_R = s_{Rc}$ given by

$$s_{Rc} = -\epsilon \left(\frac{N+2}{N+8} \right) \frac{2\pi^2 T^2}{3} \tag{14.30}$$

to leading order in ϵ; this relationship can be translated into a universal proportionality between T_c and Δ_-, but we will not discuss that here. The value of s_{Rc} determines the phase transition line $T = T_c(r)$ shown in Fig. 14.1. The order of magnitude estimates of the couplings in (14.28) remain valid for $T > T_c(r)$ except that the omitted coefficient in the first expression of R vanishes as one approaches $T_c(r)$ from above. A simple estimate of the dimensionless coupling in the argument of Ψ_d then shows that the perturbative computation of Ψ_d fails when $T - T_c(r) \sim \epsilon T_c(r)$. This condition delineates the boundary of the shaded region shown in Fig. 14.1. Within this region the well-understood classical physics of a finite-temperature phase transition in d spatial dimensions applies. It is described by the appropriate classical singularity of Ψ_d discussed in Section 14.1.2. (We note again that these latter results have to be used in d rather than D dimensions; thus this emergence of classical statistical mechanics is completely unrelated to the $\mathcal{QC}$ mapping of Section 2.2, which mapped d-dimensional quantum mechanics to D-dimensional classical statistical mechanics.) From the perspective of the global quantum scaling functions such as (11.41), the shaded regime is where the reduced classical scaling functions apply.

Although (14.30) contains the leading prediction of the ϵ expansion for the value of the critical temperature $T_c(r)$, the result is not satisfactory in one important respect. Note that we find a $T_c > 0$ for *all* N. This is the correct result for $2 < d < 3$, but it is incorrect precisely when $d = 2$, the dimensionality of physical interest. For $d = 2$ we should find $T_c = 0$ for all $N \geq 3$, as we found in the large-N expansion in Chapters 11 and 13. This failure suggests that the estimate (14.30) for T_c is not very accurate for $d = 2$, $N = 1, 2$. We rectify this failure in Section 14.3, where we treat the effective theory (14.16) *directly in $d = 2$*. This can be done by a variety of analytical and numerical methods [426], which lead to quite accurate results for $d = 2$, $N = 1, 2$.

14.2.2 $d > 3$

This is obviously an unphysical regime, but we discuss it briefly to note the physics of models above their upper-critical dimension. We later meet models whose quantum critical points have a lower value for the upper-critical dimension, and their properties are quite similar to those found here. We assume here that $d < 4$, so that the classical finite-temperature transition remains below its upper-critical dimension. (There is little physical interest in discussing the case where both the quantum and classical transition are above their respective upper-critical dimensions.)

The basic results are already contained in the expression (3.53) for the position of the $T = 0$ critical point, the definition (3.54), and the values (14.17) and (14.18) for the effective coupling R. It is always sufficient to use just the first-order result $U = u$ for the nonlinear coupling. It is not necessary to renormalize the values of any coupling, and we can simply express the results in terms of bare parameters. The expressions also have a dependence upon the nonuniversal upper-cutoff Λ, and the main subtlety in the evaluation of the results is the separation of this nonuniversality from the T dependence, which we find is universal. Furthermore, this separation of Λ dependence must be done in a manner which maintains analyticity in s at $s = 0$ for $T > 0$.

The first step is evaluation of the frequency summation in the expressions noted above for R. This leads to a form for R closely related to expressions (14.20) and (14.21) for $d < 3$:

$$R = s + u \left(\frac{N+2}{6} \right) \left[\int \frac{d^d k}{(2\pi)^d} \left(\frac{1}{\sqrt{k^2 + s}} \frac{1}{e^{\sqrt{k^2 + s}/T} - 1} - \frac{T}{k^2 + s} + \frac{T}{k^2} \right) \right.$$

$$\left. + \int^{\Lambda} \frac{d^D K}{(2\pi)^D} \left(\frac{1}{K^2 + s} - \frac{1}{K^2} \right) \right]. \tag{14.31}$$

We observe that the cutoff dependence is isolated entirely in the second integrand, which is a property of the $T=0$ theory: this is why the T dependence, which depends only upon the low-energy excitations, is universal. We can remove this ultraviolet divergence by subtracting s/K^4 from the second integrand. (Note that this correction term is smooth in s so that the analyticity properties of the expression for R in terms of s will not be spoiled.) The correction term leads to a cutoff-dependent term that is also linear in s, and the remaining integral is convergent at high momenta. In this way we get our final result:

$$R = s\left(1 - c_1 u \Lambda^{d-3}\right) + u T^{d-1} \left(\frac{N+2}{6}\right) \tilde{G}_d \left(\frac{s}{T^2}\right), \tag{14.32}$$

where c_1 is the same nonuniversal constant that appeared in (3.56), and the universal function $\tilde{G}_d(y)$ is given by

$$\tilde{G}_d(y) = S_d \int_0^\infty k^{d-1} dk \left(\frac{1}{\sqrt{k^2+y}} \frac{1}{e^{\sqrt{k^2+y}} - 1} - \frac{1}{k^2+y} + \frac{1}{k^2}\right.$$

$$\left. + \frac{1}{2\sqrt{k^2+y}} - \frac{1}{2k} + \frac{y}{4k^3}\right). \tag{14.33}$$

Despite appearances, this function is analytic as a function of y at $y=0$. This can be established by studying the small-k behavior of the integrand and using the fact that the function $1/(e^x - 1) - 1/x + 1/2$ has an expansion about $x=0$ that involves only positive, odd powers of x. Consequently, (14.33) can also be analytically continued to $y < 0$, but we will not present the details of this.

With the result for R available in (14.32), and the value $U=u$, we obtain the static susceptibility by simply evaluating (14.19) and thence obtain the nonzero-T crossovers near the quantum critical point $T=0$, $s=0$. The structure of the results is very similar to those obtained in Section 14.2.1, and so we only state the main conclusions. Provided that there is no long-range order the static susceptibility takes the form

$$\chi^{-1}(k) = k^2 + \xi^{-2}, \tag{14.34}$$

where ξ is the correlation length. For $s > 0$ and $T \ll \sqrt{s}$ we have

$$\xi^{-2} = s\left(1 - c_1 u \Lambda^{d-3}\right) + u \left(\frac{N+2}{6}\right) \left(\frac{T}{2\pi}\right)^{d/2} s^{(d-2)/4} e^{-\sqrt{s}/T}; \tag{14.35}$$

thus the T-dependent correction to the correlation length is exponentially small, as expected for a system with an energy gap. At higher temperatures, $T \gg \sqrt{|s|}$, we have the limiting behavior

$$\xi^{-2} = s\left(1 - c_1 u \Lambda^{d-3}\right) + u T^{d-1} \left(\frac{N+2}{6}\right) S_d \Gamma(d-1) \zeta(d-1), \tag{14.36}$$

where $\zeta(x)$ is the Reimann zeta function; so for $s > 0$ and $\sqrt{s} \ll T \ll (s/u)^{1/(d-1)}$ the first $T=0$ term in (14.36) dominates, while for higher T the second T-dependent term takes over. For $s < 0$, setting $\xi^{-2} = 0$ gives us the condition for the transition to the ordered phase, $T_c \sim (|s|/u)^{1/(d-1)}$, which is analogous to the result (14.30) for $d < 3$.

We draw the reader's attention to an important property of the above results. Note that in the high-T limit, $T \gg \sqrt{|s|}$, the correlation length does *not* obey the relation $\xi \sim T^{-1/z}$ that might be expected from general scaling arguments; instead we have the result of (14.36) where $\xi \sim T^{-(d-1)/2}$, which agrees with the naive scaling estimate only in the upper-critical dimension $d = 3$. The violations of scaling are a consequence of the prefactor of the irrelevant coupling u in the T-dependent term in (14.36). In the strict scaling limit, we should set this irrelevant coupling to zero, but then we would have a T-independent correlation length. So, unlike the case for $d < 3$, irrelevant couplings have to be included to obtain the leading T dependence. Such couplings, which cannot be neglected even though they are formally irrelevant, are called *dangerously irrelevant*.

Apart from the above explicit computation, the power of temperature appearing in the correlation length in the quantum critical region can be easily deduced from a knowledge of the scaling dimension of the irrelevant coupling, u. We know from the structure of the perturbation theory that ξ^{-2} depends linearly on u. So let us assume

$$\xi^{-2} \sim u T^p. \tag{14.37}$$

Now using $\dim[\xi] = -1$, $\dim[u] = 3 - d$, and $\dim[T] = 1$, and matching scaling dimensions in (14.37), we conclude $p = d - 1$, which is consistent with (14.36).

It should also be evident (we briefly discuss this issue further in the following section) that such violations of scaling also appear in the characteristic time for dynamic fluctuations in the high-T regime. They are no longer simply universal numbers times $\hbar/k_B T$ but are proportional to higher powers of $1/T$ times a prefactor involving the nonuniversal bare value of the coupling u.

14.3 Order parameter dynamics in $d = 2$

We begin by formulating an effective theory for the low-energy, long-wavelength fluctuations of the order parameter ϕ_α. This model is then used to compute the behavior of $\mathrm{Im}\chi(k, \omega)$ at small k and small ω. An important limitation of the resulting model is that it cannot be used to compute universal transport properties (i.e. correlators of $\mathbf{L}$, the uniform susceptibility χ_u, and the spin diffusion constant D_s). These turn out to be dominated by larger k and ω, because although the small k and ω fluctuations of ϕ_α have large amplitudes, they carry very little angular momentum current. A separate model for transport properties is developed in Chapter 15.

We mostly limit our attention here to dynamics in the continuum high-T ("quantum critical") region of Fig. 14.1 (which applies to $N = 1, 2$ for $d = 2$) and Fig. 11.2 (which applies to $N \geq 3$ for $d = 2$; see also Fig. 13.1) but consider all values of N. Dynamical properties in this region were studied by the large-N expansion in Section 13.2 and led to Fig. 13.3 for $\mathrm{Im}\chi(k, \omega)$ (which is the analog of Fig. 10.8 for the Ising chain). The large-N expansion was found to be adequate near the position of the quasi-particle pole ($\omega \approx \varepsilon_{\vec{k}}$), but it failed badly for $\omega \ll T$, $ck \ll T$. It is this failure we can rectify here; our aim is to obtain the analog of the Ising chain results for $\mathrm{Im}\chi(k, \omega)/\omega$ in Figs. 10.9–10.11 for dimension $d = 2$.

As we shall see, there are some significant, qualitative physical differences between $d = 1$ and $d = 2$.

The basis of our approach depends upon the ϵ-expansion values of the coupling constants in the effective theory $\mathcal{S}_{\phi,\text{eff}}$, (14.16), for the static ϕ_α fluctuations. In particular, this theory is characterized by a "mass," R (defined in (14.18)), which in the high-T region is (from (14.20) and (14.29))

$$R = \epsilon \left(\frac{N+2}{N+8} \right) \frac{2\pi^2 T^2}{3}. \tag{14.38}$$

The characteristic wavevector and energy of the dominant ϕ_α fluctuations both equal $\sqrt{R}$ (in units with $c = 1$, which we are using in this chapter). Observe that from small ϵ, $\sqrt{R} \ll T$, and so the occupation number of ϕ_α modes with this energy is large:

$$\frac{1}{e^{\sqrt{R}/T} - 1} \approx \frac{T}{\sqrt{R}} \sim \frac{1}{\sqrt{\epsilon}} \gg 1. \tag{14.39}$$

The second term above is the classical equipartition value and suggests that predominant fluctuations are *classical waves* in the magnitude and orientation of ϕ_α.

How can we extend the model (14.16) to describe the dynamics of these classical waves? (The reasoning here is almost identical to that presented in Section 12.3.3 for the high-T regime of the $d = 1$ rotor model; readers who have skipped Chapter 12 may wish to read Section 12.3.3 until (12.55), but this is not essential.) The predominance of fluctuations with energy smaller than T implies that the classical fluctuation–dissipation theorem (10.75) applies, and (10.76) allows us to obtain the equal-time correlations of ϕ_α from the static susceptibility of (14.16). For unequal-time correlations, we need to account for the kinetic energy of the ϕ_α fluctuations. To this end, we introduce a canonically conjugate momentum, π_α, so that we have the following standard Poisson bracket relations between the ϕ_α, π_α:

$$\left\{ \phi_\alpha(x), \pi_\beta(x') \right\}_{PB} = \delta_{\alpha\beta} \delta(x - x'),$$

$$\left\{ \phi_\alpha(x), \phi_\beta(x') \right\}_{PB} = 0,$$

$$\left\{ \pi_\alpha(x), \pi_\beta(x') \right\}_{PB} = 0. \tag{14.40}$$

The Hamiltonian implied by (14.16) contains only "potential energy" terms, and it has to be extended to include the kinetic energy. At the low order in ϵ that we are working with here, there are no renormalizations of the gradient terms in (14.2), and so the kinetic energy is simply the standard $\int d^d x \pi_\alpha^2/2$ implied by the Hamiltonian form of the quantum Lagrangean in (14.2); in this respect, the present situation is simpler than that in Section 12.3.3, where a careful computation of the temperature dependence of the uniform susceptibility was necessary to obtain the proper kinetic energy. In this manner we are led to the following classical phase space integral (as in "$\int dq dp$"), which generalizes the configuration space integral in (14.16):

$$\mathcal{Z} = \int \mathcal{D}\phi_\alpha(x)\mathcal{D}\pi_\alpha(x) \exp\left(-\frac{\mathcal{H}_c}{T}\right),$$

$$\mathcal{H}_c = \int d^d x \left\{ \frac{1}{2}\left[\pi_\alpha^2 + (\nabla_x \phi_\alpha)^2 + \tilde{R}\phi_\alpha^2\right] + \frac{U}{4!}\left(\phi_\alpha^2(x)\right)^2 \right\}. \qquad (14.41)$$

Observe that we can freely integrate out the π_α in a Gaussian integral, and the functional integral over the ϕ_α and its equal-time correlations then reduce to those implied by (14.16), as they should. This argument shows that the couplings $\tilde{R}$ and U above are precisely those computed in Section 14.2 in the ϵ expansion.

For unequal-time correlations, we compute the classical Hamilton–Jacobi equations implied by (14.41) and the Poisson brackets (14.40):

$$\frac{\partial \phi_\alpha}{\partial t} = \{\phi_\alpha, \mathcal{H}_c\}_{\mathrm{PB}}$$

$$= \pi_\alpha,$$

$$\frac{\partial \pi_\alpha}{\partial t} = \{\pi_\alpha, \mathcal{H}_c\}_{\mathrm{PB}}$$

$$= \nabla_x^2 \phi_\alpha - \tilde{R}\Phi_\alpha - \frac{U}{6}(\phi_\beta^2)\phi_\alpha. \qquad (14.42)$$

Determination of the dynamic correlations now reduces to a problem of the form also discussed in Section 12.3.3. Pick a set of initial conditions for ϕ_α, π_α from the ensemble implied by (14.41). Then evolve forward in time, according to the deterministic equations (14.42). Finally, compute unequal-time correlations by averaging products of fields at different times over the set of initial conditions in (14.41).

(Readers familiar with the theory of classical critical dynamics [223] should be warned that the equations above look deceptively similar to those found in the classical model considered by Halperin, Hohenberg, and Ma [194, 195]. However, in the present situation, (14.42) is a set of deterministic Hamiltonian equations that preserve total energy (and other quantities). In contrast, the equations of [194] and [195] are *stochastic* equations with a damping coefficient inserted by hand.)

The scaling structure of the deterministic continuum dynamical problem defined above has been discussed carefully in [426], and the central result is simple and quite easy to understand. First, let us discuss the equal-time correlations computed in Section 14.2 in a slightly different language. The continuum statistical mechanics problem defined by the functional integral in (14.16) requires some consideration of the dependence of correlators on the short-distance cutoff, Λ^{-1}. For $d < 3$, the answer is very simple: introduce a new renormalized "mass" R as in (14.18), and then send the ultraviolet cutoff, Λ, to infinity; a finite, universal, continuum answer is obtained, which is, of course, that specified in the tricritical crossover function in (14.19). Notice that the integral in (14.18) is divergent in the ultraviolet for d close to 3. What we are saying is that this is the only short-distance singularity in the problem for $\epsilon = 3 - d > 0$ and small, and that this singularity can be removed by a simple, linear shift in the value of the mass R. After such a shift, the continuum limit is well defined, and we can then deduce the form of all correlators by a simple engineering analysis of dimensions. The claim of [426] (which we accept here without proof) is that this same shift is also sufficient for unequal-time correlations; readers who have read Chapter 12 will recall that a very similar claim was made in Section 12.3.3.

Let us, then, perform the straightforward engineering analysis of dimensions. We transform to dimensionless spatial, time, and field variables by the substitutions

$$\bar{x} = xR^{1/2},$$

$$\bar{t} = tR^{1/2},$$

$$\bar{\phi}_\alpha = R^{(2-d)/4}T^{-1/2}\phi_\alpha,$$

$$\bar{\pi}_\alpha = R^{-d/4}T^{-1/2}\pi_\alpha, \tag{14.43}$$

in (14.41) and (14.42). All dimensional couplings in (14.41) and (14.42) disappear, and they acquire a universal form dependent only on a single dimensionless parameter multiplying the quartic (cubic) term in (14.41) ((14.42)), which is

$$\mathcal{G} \equiv \frac{TU}{R^{(4-d)/2}}. \tag{14.44}$$

This is, of course, precisely the dimensionless parameter that appeared in crossover functions such as (14.19); we have now given it a separate symbol, which represents its role as a "Ginzburg parameter" [312], controlling the strength of thermal fluctuations. The scaling form (14.19) can now be immediately deduced from the mappings (14.43), but the present approach also allows us to write down the form of the dynamic structure factor in a manner similar to that used in (10.78), (12.62), and (13.36):

$$\frac{2T}{\omega}\operatorname{Im}\chi(k, \omega) = S(k, \omega) = \frac{T\chi(k)}{\sqrt{R}}\Phi_{\text{Sc}}\left(\frac{k}{R^{1/2}}, \frac{\omega}{R^{1/2}}, \mathcal{G}\right). \tag{14.45}$$

The first relation is the classical fluctuation–dissipation theorem (10.75), and Φ_{Sc} is a universal scaling function. The prefactor of the static susceptibility, $\chi(k)$, satisfies (14.19) and has already been computed in Section 14.2.1; it has been inserted so that the Kramers–Kronig relation (10.75) implies that Φ_{Sc} has a fixed integral over frequency,

$$\int \frac{d\bar{\omega}}{2\pi}\Phi_{\text{Sc}}(\bar{k}, \bar{\omega}, \mathcal{G}) = 1, \tag{14.46}$$

as in Fig. 10.11 and (10.79). The value of $\Phi_{\text{Sc}}(0, 0, \mathcal{G})$ is of particular importance because it determines the relaxational rate, Γ_R, of long-wavelength order parameter fluctuations, as in (10.100).

Our task is now clear. Solve the continuum equations of motion (14.42) for initial conditions specified by (14.41), and so determine the scaling function Φ_{Sc}. For the high-T region, this solution should be obtained at the value of $\mathcal{G}$ determined in Section 14.2.1, which is

$$\mathcal{G} = \frac{48\pi\sqrt{3}}{\sqrt{2(n+2)(n+8)}}\sqrt{\epsilon}, \tag{14.47}$$

and this is small for small ϵ. In general, as implied by the discussion in Section 14.2.1, $\mathcal{G}$ is a smooth, dimensionless function of $s_R/T^{1/z\nu}$ (and can be rewritten as a universal function of Δ_+/T on the quantum paramagnetic side, and similarly for the magnetically ordered side); it decreases (increases) from the high-T value in (14.47) as we decrease T toward the quantum paramagnetic (magnetically ordered) region. For $\mathcal{G} = 0$, the dynamic problem is one of linear waves and can be easily solved. For small $\mathcal{G}$, equal-time correlations can be obtained in perturbation theory, and this was discussed in Section 14.2.1. However, as we have already noted, perturbation theory fails for dynamic properties in the low-frequency limit [422] for any nonzero $\mathcal{G}$.

The only remaining possibility is to numerically solve the strong-coupling dynamical problem specified by (14.41) and (14.42). Formally, we are carrying out an ϵ expansion, and so the numerical solution should be obtained for d just below 3. However, it is naturally much simpler to simulate *directly in $d = 2$*, which is also the dimensionality of physical interest. Therefore, the approach to the solution of the dynamic problem in the quantum critical region breaks down into two systematic steps: (i) use the $\epsilon = 3 - d$ expansion to derive an effective classical nonlinear wave problem characterized by the couplings R and $\mathcal{G}$; (ii) obtain the exact numerical solution of the classical nonlinear wave problem at these values of R and $\mathcal{G}$ directly for $d = 2$. This division of the problem into two rather disjointed steps is also physically reasonable. The dimensionality $d = 2$ plays a special role, primarily for the classical thermal fluctuations, and the cases $N = 1, 2$ and $N \geq 3$ are strongly distinguished, and so it is important to treat these exactly; in contrast, the $\epsilon = 3 - d$ expansion provides a reasonable treatment of the quantum fluctuations down to $d = 2$ for all N.

Before embarking on a description of this numerical solution, let us make some peripheral remarks. First, the relationship (14.18) between $\tilde{R}$ and R cannot be used when $d = 2$ because there is now an *infrared* divergence in the momentum integral. To cure this, we replace (14.18) by

$$\tilde{R} = R - U \left(\frac{N+2}{6} \right) \int \frac{d^d k}{(2\pi)^d} \frac{T}{k^2 + R}, \qquad (14.48)$$

which is a nonlinear relationship between R and $\tilde{R}$. The change in the propagator makes no difference at large momenta, and so the cancellation of ultraviolet divergences goes through as before [303]. The value of R as computed in the ϵ expansion is now different, but the leading order results in (14.38) and (14.47) remain unchanged. The new relationship (14.48) does have some important consequences for the structure of the static properties at large $\mathcal{G}$, but we will not go into this here. In particular, the present approach can be used to reliably obtain physically important crossovers for $d = 2$ (such as those in the static susceptibility $\chi(k, \omega = 0)$ and in the superfluid density ρ_s for $N = 2$) between the high-T and low-T regions of Fig. 14.1; this is discussed elsewhere [426] in some detail.

Figures 14.2–14.4 contain the results of a recent numerical computation of the scaling functions in (14.45) at $k = 0$ (i.e. for $S(0, \omega)$). These results are the analog of Fig. 10.11 for the Ising chain (and Fig. 12.4 for the $d = 1$, $N = 3$ rotor model, for those who have read Chapter 12). Because (14.47) evaluates to a moderately large value of $\mathcal{G}$ for $d = 2$, the perturbation theory results of Section 14.2.1 for $\chi(k)$ are no longer accurate, and we also numerically computed the exact values of $\Psi_d(0, \mathcal{G})$ for $d = 2$ (see (14.19)), and these are reported in the captions to the figures. The results show a consistent trend from smaller values of $\mathcal{G}$ and larger values of N to larger values of $\mathcal{G}$ and smaller values of N, and we discuss the physical interpretation of the two limiting cases in turn.

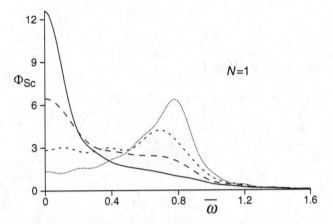

Fig. 14.2 Numerical results from [426] for the zero-momentum scaling function $\Phi_{Sc}(0, \bar{\omega}, \mathcal{G})$ appearing in (14.45) for $N = 1$. Results are shown for $\mathcal{G} = 25$ (dots), $\mathcal{G} = 30$ (short dashes), $\mathcal{G} = 35$ (long dashes), and $\mathcal{G} = 40$ (full line). The static susceptibility takes the values (see (14.19)) $R\chi(k) = \Psi_2^{-1}(0, \mathcal{G}) = 1.95, 2.45, 3.37$, and 4.78 at $\mathcal{G} = 25, 30, 35$, and 40, respectively. The high-T limit value of $\mathcal{G}$ in (14.47) evaluates to $\mathcal{G} = 35.5$ at $\epsilon = 1$ and $N = 1$.

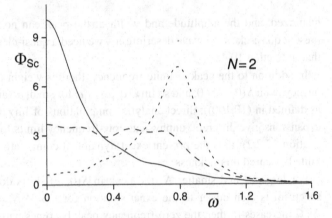

Fig. 14.3 Numerical results as in Fig. 14.2 but for $N = 2$. The values of $\mathcal{G}$ are now $\mathcal{G} = 20$ (short dashes), $\mathcal{G} = 30$ (long dashes), and $\mathcal{G} = 40$ (full line). The static susceptibility takes the values (see (14.19)) $R\chi(k) = \Psi_2^{-1}(0, \mathcal{G}) = 1.67, 2.65$, and 4.73 at $\mathcal{G} = 20, 30$, and 40, respectively. The high-T limit value of $\mathcal{G}$ in (14.47) evaluates to $\mathcal{G} = 29.2$ at $\epsilon = 1$ and $N = 2$.

For smaller $\mathcal{G}$ and larger N, we observe a peak in $S(0, \omega)$ at a nonzero frequency. This peak is the remnant of the delta function in the large-N result (11.10), where the value of m was given by (11.45), (11.53), and (11.56) (the same peak also appears in the $1/N$ computation in Fig. 13.3 in Chapter 13). In the present computation, it is clear that the peak is due to *amplitude fluctuations* as ϕ_α oscillates about the minimum in its effective potential at $\phi_\alpha = 0$. As $\mathcal{G}$ is reduced, we move out of the high-T region into the low-T region on the quantum paramagnetic side (see Fig. 14.1), and this finite-frequency, amplitude fluctuation peak connects smoothly with the quantum paramagnetic quasiparticle peak. Of course, once we are in the quantum paramagnetic region, the wave oscillations get

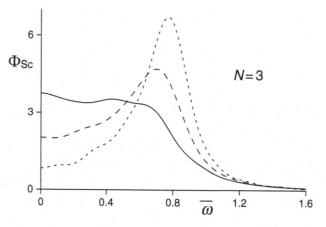

Fig. 14.4 Numerical results as in Fig. 14.2 but for $N = 3$. The values of $\mathcal{G}$ are now $\mathcal{G} = 20$ (short dashes), $\mathcal{G} = 25$ (long dashes), and $\mathcal{G} = 30$ (full line). The static susceptibility takes the values (see (14.19)) $R\chi(k) = \Psi_2^{-1}(0, \mathcal{G}) = 1.73, 2.17,$ and 2.75 at $\mathcal{G} = 20, 25,$ and 30, respectively. The high-T limit value of $\mathcal{G}$ in (14.47) evaluates to $\mathcal{G} = 24.9$ at $\epsilon = 1$ and $N = 3$.

quantized, and the amplitude and width of the peak can no longer be computed by the present quasi-classical wave description – we need a quasi-classical particle approach, like that in Section 10.4.2.

In addition to the peak at finite frequency, there is weight in $S(0, \omega)$ down to zero frequency, with $S(0, 0) > 0$, and so $\mathrm{Im}\chi(0, \omega) \sim \omega$ for small ω (and a finite relaxation rate Γ_R as defined in (10.100)); direct analytic computations of $\mathrm{Im}\chi(0, \omega)$ in either the ϵ or $1/N$ expansions give different, unphysical, low-frequency limits for $\mathrm{Im}\chi(0, \omega)$ (as was seen in Section 13.2.2); thus the present exact dynamical computation directly for $d = 2$ with N finite has cured this sickness.

For larger $\mathcal{G}$ and smaller N, the peak in $S(0, \omega)$ shifts down to $\omega = 0$. The resulting spectrum is then closer to the exact solution for $d = 1$, $N = 1$ presented in Fig. 10.11. As $\mathcal{G}$ increases further, the zero-frequency peak becomes narrower and taller. How do we understand the dominance of this low-frequency relaxation? For $N \geq 2$, there is a natural direction for low-energy motion of the order parameter: by angular or phase fluctuations of ϕ_α in a region where the value of $|\phi_\alpha|$ is nonzero. Of course, the *fully* renormalized effective potential controlling fluctuations of ϕ_α has a minimum only at $\phi_\alpha = 0$, as we are examining a region with no long-range order. However, for these values of $\mathcal{G}$, there is a significant intermediate length scale over which the local effective potential has a minimum at a $|\phi_\alpha| \neq 0$, and the predominant fluctuations of ϕ_α consist of a relaxational phase dynamics. This interpretation is also consistent with an analysis of the limit $\mathcal{G} \to \infty$ for $N = 3$ that was carried out in [426]. Then we expect to be well into the low-T region on the magnetically ordered side in Fig. 11.2, and so the analysis of Section 13.1.3 should apply. Indeed, it was shown that in the limit $\mathcal{G} \to \infty$, the three-argument scaling form (14.45) reduces to precisely the two-argument scaling form in (13.36). As readers of Chapter 13 will recall, this latter scaling form is described by a classical wave model in which $|\phi_\alpha|$ is fixed to be unity and its zero momentum dynamic structure factor has a large relaxational peak at zero frequency. The same "phase fluctuation"-dominated peak evidently survives at the moderate values of $\mathcal{G}$ we are considering here.

All of the above reasoning has been for cases with continuous symmetry, $N \geq 2$. However, closely related arguments can also be made for $N = 1$. In this case, in a region where $|\phi_\alpha|$ locally takes a nonzero value, there are low-energy modes corresponding to motions of *domain walls* between oppositely oriented magnetic phases. Indeed, precisely such a domain wall motion was considered for the $d = 1$, $N = 1$ case in Section 10.4.1, and it led to the zero-frequency relaxational peak in the structure factor in Fig. 10.5.

Even in a region dominated by angular (or domain wall) fluctuations about a locally nonzero value of $|\phi_\alpha|$, there could still be higher frequency amplitude fluctuations of $|\phi_\alpha|$ about its local potential minimum. This would be manifested by peaks in $S(0, \omega)$, both at $\omega = 0$ and at a nonzero frequency. A glance at Figs. 14.2–14.4 shows that this never happens in a well-defined manner. However, for $N = 1$, we do observe a nonzero frequency shoulder in $S(0, \omega)$ at $\mathcal{G} = 35$, along with a prominent peak at $\omega = 0$, indicating the simultaneous presence of domain wall relaxational dynamics and amplitude fluctuations in $|\phi_\alpha|$. Readers of Chapter 12 will also recognize the similarity of this with the shoulder in Fig. 12.4 describing the high-T limit of the $d = 1$, $N = 3$ case. For the other cases in $d = 2$, we do not see a clear signal of the concomitant amplitude and angular fluctuations. It seems, therefore, that once angular fluctuations appear with increasing $\mathcal{G}$, the nonlinear couplings between the modes reduce the spectral weight in the amplitude mode to a negligible amount.

It is interesting to examine the above results at the value of the high-T limit for $\mathcal{G}$ in (14.47) evaluated directly in $\epsilon = 1$. We find $\mathcal{G} = 35.5, 29.2, 24.9$ for $N = 1, 2, 3$, and these values are very close to the position where the crossover between the above behaviors occurs. The $N = 1$ case has a clear maximum in $S(0, \omega)$ at $\omega = 0$ (along with a finite frequency shoulder), while there is a more clearly defined finite frequency peak for $N = 3$.

We add a final parenthetic remark. Readers may recognize a passing resemblance between the above crossover in dynamical properties as a function of $\mathcal{G}$ and a well-studied phenomenon in dissipative quantum mechanics [294, 295, 537] – the crossover from "coherent oscillation" to "incoherent relaxation" in a two-level system coupled to a heat bath. However, here we do not rely on an arbitrary heat bath of linear oscillators, and the relaxational dynamics emerges on its own from the underlying Hamiltonian dynamics of an interacting many-body, quantum system. Our description of the crossover has been carried out in the context of a quasi-classical wave model here, but, as we noted earlier, the "coherent" peak connects smoothly to the quasiparticle peak in the low-T paramagnetic region; here the wave oscillations get quantized into discrete lumps, which must then be described by a "dual" quasi-classical particle picture.

14.4 Applications and extensions

Bitko, Rosenbaum, and Aeppli [49] have studied the vicinity of the quantum phase transition in the Ising spin system $LiHoF_4$ in the presence of a transverse magnetic field.

Thus Ising spins have long-range dipolar exchange, and this puts the quantum critical point above its upper-critical dimension because long-range forces tend to make the mean-field approximation better. The resulting exponents are therefore similar to those of the mean field, and physical properties can be computed in a manner very similar to Section 14.2.2.

We argued in Section 14.2.1 that for systems below the upper-critical dimension with a finite-temperature phase transition (the cases $N = 1, 2$; $d = 2$), the critical temperature of the transition is universally related to the ground state energy scale Δ_-. For the case $d = 2$, $N = 2$, we may choose $\Delta_- = \rho_s$, the ground state spin stiffness, and so T_c/ρ_s is a universal constant. Indeed the universality applies to the entire temperature dependence of ρ_s, and thus

$$\rho_s(T) = \rho_s \Phi_\rho \left(\frac{T}{\rho_s} \right), \tag{14.49}$$

where $\rho_s \equiv \rho_s(0)$, and Φ_ρ is a universal function, which can be computed by the methods of Section 14.2.1 in an ϵ expansion or by a direct numerical computation in $d = 2$ [426]. The value of T_c is determined by the argument at which Φ_ρ first vanishes. In $d = 2$, the function Φ_ρ will display a discontinuity at $T = T_c$ to allow for the Nelson–Kosterlitz jump [358] in the superfluid density. For experimental comparisons, it is easy to see that (14.49) implies the slightly weaker result

$$\frac{\rho_s(T)}{\rho_s(0)} = \tilde{\Phi}_\rho \left(\frac{T}{T_c} \right), \tag{14.50}$$

with the function $\tilde{\Phi}_\rho$ computable from Φ_ρ. The numerically exact computations for $d = 2$ discussed in Section 14.3 have been used to obtain explicit computations of the functions Φ_ρ, $\tilde{\Phi}_\rho$ for the model (14.2), and these results contain the jump in superfluid density. It is quite intriguing that the data [131, 202, 518] on high-temperature superconductors satisfy the relation (14.50), suggesting the proximity of a quantum critical point at which the superfluid density vanishes at $T = 0$.

We have not explicitly considered the case of the upper-critical dimension $d = 3$ in our discussion in this chapter. In this case there are logarithmic corrections, involving a nonuniversal upper cutoff scale in the argument of the logarithm, which can be computed using renormalization group arguments similar to those considered in Sections 12.3 and 13.1.1. We assume the system begins with a positive coupling u of order unity at a microscopic scale Λ. Then, as in Sections 12.3 and 13.1.1, it pays to use the renormalization group invariance to renormalize to a scale $\mu = T/c$; from the flow equation (14.10) we see that for $\epsilon = 0$ and $T \ll \Lambda$

$$u_R = \left(\frac{6}{N + 8} \right) \frac{1}{\ln(c\Lambda/T)}. \tag{14.51}$$

Hence the quartic coupling u is logarithmically small. The $\omega_n \neq 0$ modes can be integrated out by precisely the methods of Section 14.2 to derive an effective action for the $\omega_n = 0$ modes. This step should be carried out at the scale $\mu = T/c$, as this ensures that u_R is small and also that there are no large logarithms generated in the perturbation theory in u_R (the only scale running through the Feynman diagrams is T, and so dimensionally all logarithms must be order $\ln(c\mu/T)$, which is small). The subsequent analysis of the action for the $\omega_n = 0$ modes then proceeds as before. To lowest order, the physical results can be obtained from the $d < 3$ computation by the replacement $\epsilon \to 1/(\ln(c\Lambda/T))$.

Irkhin and Katanin [237] have applied the methods of this chapter to crossovers in anisotropic magnetic systems and have made comparisons to experimental and numerical data.

We have discussed the realization of the $d = 3$, $N = 3$ rotor quantum critical point in TlCuCl$_3$ by Ruegg et al. [414] in Sections 1.3 and 6.6. In particular, neutron scattering observations have reported a low-lying amplitude fluctuation mode (or the "Higgs particle") in the spectrum of the ordered phase, in addition to the usual spin-wave modes. The universal properties of such a mode can be obtained directly from the approach developed in this chapter; we can examine (14.2) for $r < r_c$ in an expansion in the nonlinear coupling u, which will be logarithmically small for $d = 3$ as in (14.51). The longitudinal fluctuations in $|\phi_\alpha|$ about a nonzero mean value lead to a mode whose energy vanishes as $|r - r_c| \to 0$. It should also be mentioned that such an amplitude fluctuation mode is not expected to be visible as $T \to 0$ for $r < r_c$ for $d = 2$, as is clear from the large-$\mathcal{G}$ computations in Section 14.3, and from the large-N expansion at $T = 0$ in [97] and [439] because the cross-section for the decay of the amplitude mode into multiple spin-wave excitations is too large in low dimensions.

This chapter turns to a systematic analysis of transport of conserved charges in the quantum rotor model. We introduced some general concepts in Section 8.3, and these are illustrated here by explicit computations at higher orders.

For $d = 1$, we considered time-dependent correlations of the conserved angular momentum, $\mathbf{L}(x, t)$, of the O(3) quantum rotor model in Chapter 12. We found, using effective semiclassical models, that the dynamic fluctuations of $\mathbf{L}(x, t)$ were characterized by a *diffusive* form (see (12.26)) at long times and distances, and we were able to obtain values for the spin diffusion constant D_s at low T and high T (see Table 12.1). The purpose of this chapter is to study the analogous correlations in $d = 2$ for $N \geq 2$; the case $N = 1$ has no conserved angular momentum, and so there is no possibility of diffusive spin correlations. Rather than thinking about fluctuations of the conserved angular momentum in equilibrium, we find it more convenient here to consider instead the response to an external space- and time-dependent "magnetic" field $\mathbf{H}(x, t)$ and to examine how the system transports the conserved angular momentum under its influence.

In principle, it is possible to address these issues in the high-T region using the nonlinear classical wave problem developed in Section 14.3 in the context of the $\epsilon = 3 - d$ expansion. However, an attempt to do this quickly shows that the correlators of $\mathbf{L}$ contain ultraviolet divergences when evaluated in the effective classical theory. Physically, this is a signal that, for small ϵ, transport properties are *not* dominated by excitations with energy $\ll T$ (while the order parameter fluctuations, considered in Section 14.3, were), and it is necessary to include fluctuations with higher energy, which must then be treated quantum mechanically. It is this quantum transport problem we address here. We show that it is necessary to solve a quantum transport equation for the quantized particle excitations to describe diffusion in the high-T and in the low-T quantum paramagnetic regions. We note, in passing, a proposal [439] applying classical wave models to transport directly for $d = 2$.

It is useful to begin by introducing some basic formalism. We mostly adopt the "soft-spin" approach to the quantum critical point discussed in Chapter 14, and so it is useful to set up the machinery of transport theory using its notation. For general N, the magnetic field $\mathbf{H}$ has $N(N - 1)/2$ components (as in (1.20)); this field generates rotations of the ϕ_α order parameter, and the number of components equals the number of ways of choosing independent planes of rotation in the N-dimensional order parameter space. Only for the case $N = 3$ considered earlier do the order parameter and magnetic field have the same number of components. We denote this generalized field by H^a, with $a = 1 \ldots N(N - 1)/2$. Note that for $N = 2$, a has only one allowed value and is therefore redundant. (We later apply the $N = 2$ model to the superfluid–insulator transition and see there that H^a represents the electrostatic potential.) In (11.4), we have already seen the form of the imaginary time effective action for the $N = 3$ rotor model in the presence of a nonzero H^a in the fixed-length $\mathbf{n}$ field formulation. By the precise analog of the arguments made in Section 11.1 in deriving (11.4), we may conclude that the generalization of the soft-spin theory (14.2) in the presence of an external field H^a is

$$\mathcal{Z} = \int \mathcal{D}\phi_\alpha(x, \tau) \exp(-\mathcal{S}_\phi),$$

$$\mathcal{S}_\phi = \int d^d x \int_0^{1/T} d\tau \left\{ \frac{1}{2} \left[\left(\partial_\tau \phi_\alpha - i H^a T_{\alpha\beta}^a \phi_\beta \right)^2 + c^2 (\nabla_x \phi_\alpha)^2 + r \phi_\alpha^2(x) \right] \right.$$
$$\left. + \frac{u}{4!} \left(\phi_\alpha^2(x) \right)^2 \right\}. \tag{15.1}$$

Note that H^a merely causes a precession of the ϕ_α field. The T^a are $N \times N$ real antisymmetric matrices that generate the O(N) rotation (they are i times the generators of the Lie algebra of O(N)). There are $N(N - 1)/2$ such linearly independent matrices. It is convenient to choose the basis in which all but two matrix elements of a given T^a are zero, with the nonvanishing elements equaling ± 1. In this functional language, the observable corresponding to the conserved angular momentum density is given by

$$L^a(x, t) = -\frac{\delta \mathcal{S}_\phi}{\delta H^a(x, t)}. \tag{15.2}$$

Note that there are $N(N - 1)/2$ components of this density, and their spatial integrals are all constants of the motion.

Many of the physical arguments are actually clearer in a Hamiltonian formalism. For the fixed-length model (11.4), we had the lattice Hamiltonian in (11.1). We can apply the inverse of the transformation used in going from (11.1) to (11.4) to obtain the Hamiltonian form of (15.1). If we interpret ϕ_a as the coordinate of a "particle" with unit mass moving in N dimensions (no longer constrained to move on a sphere), as discussed in Section 6.6, we obtain the following continuum Hamiltonian:

$$\mathcal{H} = \int d^d x \left\{ \frac{1}{2} \left[\pi_\alpha^2 + c^2 (\nabla_x \phi_\alpha)^2 + r \phi_\alpha^2(x) \right] \right.$$
$$\left. + \frac{u}{4!} \left(\phi_\alpha^2(x) \right)^2 - H^a(x, t) T_{\alpha\beta}^a \pi_\alpha \phi_\beta \right\}. \tag{15.3}$$

Here $\pi_\alpha(x, t)$ is the canonical momentum to the field ϕ_α, which therefore satisfy the equal-time commutation relations

$$[\phi_\alpha(x, t), \pi_\beta(x', t)] = i \delta_{\alpha\beta} \delta(x - x'). \tag{15.4}$$

Equations (15.3) and (15.4) are, of course, the quantized Hamiltonian versions of the classical Hamiltonian problem defined by (14.40) and (14.41). The operator representation of the angular momentum density is obtained by the analog of (15.2), and therefore

$$L^a(x, t) = T_{\alpha\beta}^a \pi_\alpha(x, t) \phi_\beta(x, t). \tag{15.5}$$

It can be verified that the fields L^a, ϕ_α satisfy commutation relations that are the continuum limit of the relations (1.21) for the fixed-length rotor model.

The basic purpose of this chapter is an analysis of the time evolution of the expectation value of $L^a(x, t)$ in situations close to thermal equilibrium. Let us first examine the exact Heisenberg equation of motion of the operator in (15.5) under the Hamiltonian in (15.3); an elementary computation using the commutation relation (15.4) gives a result that can be written in the form

$$\frac{\partial L^a}{\partial t} = -\vec{\nabla} \cdot \vec{J}^a + f_{abc} H^b L^c, \tag{15.6}$$

where f_{abc} are the structure constants of the Lie algebra of O(N) defined by the commutation relations

$$[T^a, T^b] = f_{abc} T^c. \tag{15.7}$$

These structure constants are totally antisymmetric in a, b, c; for $N = 3$, $f_{abc} = \epsilon_{abc}$, while for $N = 2$, $f_{abc} = 0$. The term proportional to f_{abc} in (15.6) represents the Bloch precession of the angular momentum about the external field. The quantity $\vec{J}^a$ in (15.6) is the angular momentum current. An expression for $\vec{J}^a$ can easily be obtained by generating the equation of motion as noted above; however, as the equations of motion involve only the divergence of $\vec{J}^a$, this expression will be uncertain up to the curl of an arbitrary vector. It is customary to choose this arbitrary vector to obtain the *transport* current $\vec{J}^a_{\text{tr}}$ so that the expectation value $\langle \vec{J}^a_{\text{tr}} \rangle$ vanishes in thermal equilibrium. In this case a nonzero $\langle \vec{J}^a_{\text{tr}} \rangle$ will describe bulk transport of the angular momentum density $\langle L^a \rangle$ across macroscopic distances when the system is driven out of equilibrium by an external perturbation.

Let us introduce some phenomenological considerations for a system close to thermal equilibrium. We imagine that weak perturbations from unspecified sources deform a thermal equilibrium state into one characterized by a nonzero, space-dependent angular momentum density $\langle L^a \rangle$. In addition a slowly varying "magnetic" field $H^a(x, t)$ is also present. Both these perturbations tend to induce a nonzero transport current $\langle \vec{J}^a_{\text{tr}} \rangle$, and provided that the perturbations are weak and very slowly varying, we can write down the following phenomenological expression for the current:

$$\langle \vec{J}^a_{\text{tr}}(x, t) \rangle = \sigma \vec{\nabla} H^a(x, t) - D_s \vec{\nabla} \langle L^a(x, t) \rangle. \tag{15.8}$$

We have introduced two transport coefficients in the above equation. The first, σ, is the conductivity: a uniform H^a is expected to induce only a nonzero magnetization density, and so any induced current can only be due to gradients of H^a. The second, D_s, is the spin diffusion constant we met in the discussion of the fluctuations of $L^a(x, t)$ for $d = 1$ in Chapter 12: the combination of (15.6) and (15.8) shows that for the external field $H^a = 0$, $\langle L^a(x, t) \rangle$ satisfies the diffusion equation

$$\frac{\partial \langle L^a(x, t) \rangle}{\partial t} = D_s \nabla^2 \langle L^a(x, t) \rangle, \tag{15.9}$$

and identifies D_s as the diffusion constant.

Continuing our phenomenological analysis, we discuss the important Einstein relation between σ and D_s. Imagine we are considering a closed system in which H^a is time independent but a slow function of x. Eventually the system reaches thermal equilibrium in which the local angular momentum density is simply given by the equilibrium response to a uniform field,

$$\langle L^a \rangle = \chi_u H^a(x), \tag{15.10}$$

where χ_u was defined in (11.3) (we have arranged the initial conditions so that this result is compatible with conservation of the total angular momentum). Under this condition of equilibrium the transport current should also vanish. This is compatible with the defining transport relation (15.8) only if

$$\sigma = \chi_u D_s. \tag{15.11}$$

This is the basic Einstein relation between the diffusion coefficient characterizing fluctuations and the conductivity representing the response of the system to an external field.

Using (15.11), we can obtain the basic scaling properties of the conductivity σ. Recall from (11.30) that the scaling dimension of χ_u is $d - z$ (we henceforth use the value $z = 1$ for the rotor models in the remainder of this chapter), and χ_u satisfies the scaling forms (11.44):

$$\chi_u = \frac{T^{d-1}}{c^d} \Phi_{u\pm} \left(\frac{\Delta_\pm}{T} \right) \tag{15.12}$$

on the two sides of the quantum critical point (recall that $\Delta_\pm$ are energy scales measuring the deviation of the ground state from the $T = 0$ quantum critical point). For $d = 2$, the $N = \infty$ result for the scaling functions $\Phi_{u\pm}$ can be obtained by inserting (11.45), (11.53), and (11.56) into (11.46). We note here some asymptotic limits, also mentioned in Table 13.1: on the ordered side, where $\Delta_- = \rho_s$, the ground state spin stiffness, we have the exact result for $T \ll \rho_s$ obtained in (13.25), which shows that χ_u reaches a nonzero value as $T \to 0$:

$$\chi_u = \frac{2\rho_s}{Nc^2} \left[1 + \frac{(N-2)T}{2\pi\rho_s} + \mathcal{O}\left(\frac{T}{\rho_s} \right)^2 \right]; \tag{15.13}$$

on the quantum paramagnetic side, we expect an exponentially small uniform susceptibility for $T \ll \Delta_+$, and we can again obtain an exact result for $d = 2$ by the same dilute gas of quasiparticles argument that led to (12.12):

$$\chi_u = \frac{\Delta_+}{\pi c^2} e^{-\Delta_+/T}; \tag{15.14}$$

finally, in the high-T limit, $T \gg \Delta_\pm$, we must rely on the large-N expansion to obtain the value of the constant $\Phi_{u\pm}(0)$, and the $N = \infty$ result is

$$\chi_u = \frac{T}{c^2} \frac{\sqrt{5}}{\pi} \ln \left(\frac{\sqrt{5}+1}{2} \right). \tag{15.15}$$

Turning to the diffusion constant, no explicit scaling results have yet been obtained (indeed, that is the primary purpose of this chapter), but we can deduce its scaling dimension by a simple argument. A glance at (15.9) shows that D_s has the dimensions of (length)^2/time.

There is no field scale that appears in the definition of D_s, and as the scaling dimension has to respect the conservation law for L^a, we have the scaling dimensions

$$\dim[D_s] = z - 2,$$

$$\dim[\sigma] = d - 2, \tag{15.16}$$

where the second relation follows from (11.30). Using $\dim[T] = z$, and matching engineering dimensions with those of c and T, we then see that D_s must equal c^2/T times a universal function of $\Delta_\pm/T$. Combining this with (15.11) and (15.12) we have the main scaling form for the conductivity [70, 106, 145, 147, 417, 538]:

$$\sigma(\omega) = \frac{Q^2}{\hbar} \left(\frac{k_B T}{\hbar c} \right)^{d-2} \Phi_{\sigma\pm} \left(\frac{\hbar\omega}{k_B T}, \frac{\Delta_\pm}{k_B T} \right), \tag{15.17}$$

where $\Phi_{\sigma\pm}$ are completely universal scaling functions. We have momentarily returned to physical units by reinserting factors of $\hbar$, k_B, and the charge of the carriers, Q (this had been absorbed into our definition of H^a; $Q = 2e$ for the superfluid–insulator transition); we do this occasionally below when quoting results for σ. For future use, we have generalized the conductivity to a dynamical frequency-dependent conductivity $\sigma(\omega)$ representing the response in the current at frequency ω to an external field at the same frequency ($\sigma \equiv \sigma(0)$); the scaling dependence on ω/T then follows from now familiar arguments. We focus in this chapter mainly on the high-T regime, $T \gg \Delta_\pm$, and therefore on the value of $\Phi_{\sigma\pm}(\omega/T, 0)$. Also note an important and remarkable property of (15.17): in spatial dimension $d = 2$, the prefactor of the power T disappears, and the conductivity is entirely given by the scaling function $\Phi_{\sigma\pm}$ times the fundamental constants $Q^2/\hbar$. In the high-T limit, we are then left with the dimensionless scaling function $\Phi_{\sigma\pm}(\omega/T, 0)$, which depends on no system parameters at all.

We work, throughout this chapter, in the high-T and quantum paramagnetic low-T regions of Fig. 14.1. We do not study the crossover in the shaded classical region of Fig. 14.1 near the finite-temperature transition for $N = 2$. Although transport properties in this region are of considerable practical interest, the methods developed here are not adequate to describe them. One consequence of restricting ourselves out of the shaded region is that we are always in a regime where the perturbative expansion for the tricritical crossover function in (14.5) is adequate.

15.1 Perturbation theory

We begin our computation of σ by a simple perturbative evaluation of the leading order term in both the $\epsilon = 3 - d$ and $1/N$ expansions [106].

First, let us specify more carefully the configuration of the system. We begin, at some time in the remote past, with an infinite d-dimensional quantum rotor system in thermal equilibrium at a temperature T. A small "magnetic field" with a uniform spatial gradient, and oscillating with a frequency ω, is turned on, also in the remote past. We are interested in the eventual steady state in which there is a spatially uniform angular momentum current present, also oscillating with the frequency ω. The proportionality constant between the current and the field gradient defines the conductivity $\sigma(\omega)$. As the current is spatially uniform, the magnetization density is zero at all times (so the term proportional to D_s in (15.8) is not present).

It should be evident that this physical situation has a translational symmetry. However, in considering the response to the field, we need a uniform gradient, and therefore it appears necessary to consider a response at a nonzero wavevector. Because this is slightly inconvenient, we use an alternative method that should be familiar to most readers in the context of discussion of the Kubo formula in many-body systems [136]. The basic point is to note that H^a appears in (15.1) in the same form as the time component of an $O(N)$ non-Abelian gauge field. It is then useful to generalize (15.1) to also introduce a fictitious spatial component of this gauge field, denoted $\vec{A}^a$, by changing only the gradient terms in (15.1) to

$$\frac{1}{2}\left[\left(\partial_\tau\phi_\alpha + iH^a T^a_{\alpha\beta}\phi_\beta\right)^2 + c^2\left(\vec{\nabla}\phi_\alpha - \vec{A}^a T^a_{\alpha\beta}\phi_\beta\right)^2\right]. \tag{15.18}$$

One advantage of introducing $\vec{A}^a$ is that we can check that the current $\vec{J}^a$ appearing in (15.6) is given by the simple expression

$$\vec{J}^a = -\frac{\delta \mathcal{S}_\phi}{\delta \vec{A}^a}; \tag{15.19}$$

this result is the analog of (15.2). The action $\mathcal{S}_\phi$ is seen to be invariant under the non-Abelian gauge transformation

$$\phi_\alpha \rightarrow \phi_\alpha + \Omega^a T^a_{\alpha\beta}\phi_\beta,$$

$$\vec{A}^a \rightarrow \vec{A}^a + \vec{\nabla}\Omega^a,$$

$$H^a \rightarrow H^a + i\partial_\tau\Omega^a, \tag{15.20}$$

where Ω^a is an arbitrary infinitesimal function of space and time. We can use this gauge invariance to transform away the field H^a appearing with the time component, leaving us with only a nonzero $\vec{A}^a$. From (15.20) we see that a system with a nonzero H^a and $\vec{A}^a = 0$ is equivalent to a system with $H^a = 0$ and $\vec{A}^a = \vec{\nabla}\Omega^a$, where $\partial_\tau\Omega^a = iH^a$. So if we have a uniform, time-dependent, spatial gradient in H^a, we can define the space-independent $\vec{E}^a = \vec{\nabla}H^a$ and we see, in imaginary frequencies, that $\vec{A}^a(\omega_n) = \vec{E}^a(\omega_n)/\omega_n$ in the gauge-transformed system.

The above mapping allows us to present a simple prescription to compute $\sigma(\omega_n)$. Work with an $\mathcal{S}_\phi$ with $H^a = 0$ and a nonzero space-independent $\vec{A}^a(\omega_n)$. (Note that the external source $\vec{A}^a$ is explicitly at zero momentum.) Compute the expectation value of (15.19) under this $\mathcal{S}_\phi$. Then the conductivity is given by the linear response to a nonzero $\vec{A}^a$ by

$$\sigma(\omega_n) = -\frac{1}{\omega_n}\frac{\delta}{\delta\vec{A}^a}\left\langle\frac{\delta\mathcal{S}_\phi}{\delta\vec{A}^a}\right\rangle\bigg|_{\vec{A}^a=0}. \tag{15.21}$$

It is a simple matter to use (15.21) to compute $\sigma(i\omega_n)$, either to first order in u or in the $N = \infty$ limit (in this case one simply uses the self-consistent large-N propagator in

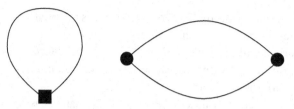

Fig. 15.1 Feynman diagrams leading to the two terms in (15.22). The dark circle represents the term linear in $\vec{A}^a$ in (15.18), while the dark square is the term quadratic in $\vec{A}^a$.

(11.10), and otherwise ignores interactions). In both cases the answer can be written in the following form:

$$
\sigma(\omega_n) = -\frac{2c^2 T}{\omega_n} \sum_{\epsilon_n} \int \frac{d^d k}{(2\pi)^d} \left[\frac{2c^2 k_x^2}{(\epsilon_n^2 + c^2 k^2 + m^2)((\epsilon_n + \omega_n)^2 + c^2 k^2 + m^2)} - \frac{1}{\epsilon_n^2 + c^2 k^2 + m^2} \right].
$$
(15.22)

The first term is the "paramagnetic" contribution, while the second is the "diamagnetic" term, and these arise from the diagrams shown in Fig. 15.1. Here we have taken the gradient of H^a along the x direction and k_x is the x component of the d-dimensional momentum **k**. The "mass" m has been computed in earlier chapters: the ϵ and large-N results differ only in their T-dependent values for m. At $N = \infty$ we have the result in (13.44). The ϵ expansion was considered in Chapter 14, and for the high-T and quantum paramagnetic low-T regions of interest here, we have from (14.19), (14.5), and (14.25) that

$$
m^2 = R - \epsilon \left(\frac{N+2}{N+8} \right) 2\pi T \sqrt{R},
$$
(15.23)

where R is given in (14.24); in the high-T limit we have $m^2 = \epsilon((N+2)/(N+8))2\pi^2 T^2/3$ to leading order in the ϵ expansion.

Now insert $1 = \partial k_x / \partial k_x$ in front of the diamagnetic term in (15.22) and integrate by parts. The surface terms vanish in dimensional or lattice regularization, and the expression for the conductivity becomes

$$
\sigma(\omega_n) = -\frac{2c^2}{\omega_n} T \sum_{\epsilon_n} \int \frac{d^d k}{(2\pi)^d} \frac{2c^2 k_x^2}{\epsilon_n^2 + c^2 k^2 + m^2}
$$
$$
\times \left[\frac{1}{(\epsilon_n + \omega_n)^2 + c^2 k^2 + m^2} - \frac{1}{\epsilon_n^2 + c^2 k^2 + m^2} \right].
$$
(15.24)

We evaluate the summation over Matsubara frequencies and analytically continue to real frequencies. The resulting $\sigma(\omega)$ is complex, and we decompose it into its real and imaginary parts, $\sigma(\omega) = \sigma'(\omega) + i\sigma''(\omega)$. We only present results for the real part $\sigma'(\omega)$; the imaginary part $\sigma''(\omega)$ can be obtained via the standard dispersion relation.

We find that the result for $\sigma'(\omega)$ has two distinct contributions [106, 522] of very different physical origin. We separate these by writing

$$\sigma'(\omega) = \sigma'_I(\omega) + \sigma'_{II}(\omega). \tag{15.25}$$

The first part, $\sigma'_I(\omega)$, is a delta function at zero frequency:

$$\sigma'_I(\omega) = 2\pi c^4 \delta(\omega) \int \frac{d^d k}{(2\pi)^d} \frac{k_x^2}{\varepsilon_k^2} \left(-\frac{\partial n(\varepsilon_k)}{\partial \varepsilon_k} \right), \tag{15.26}$$

where $n(\varepsilon)$ is the Bose function in (13.63) and the excitations have the energy momentum relation ε_k given in (13.64). We discuss the physical meaning of the delta function in (15.26) in Section 9.1.1, and obtain a separate and more physical derivation of its weight in Section 15.2. The second part, $\sigma'_{II}(\omega)$, is a continuum above a threshold frequency of 2σ:

$$\sigma'_{II}(\omega) = \pi c^4 \int \frac{d^d k}{(2\pi)^d} \frac{k_x^2}{2\varepsilon_k^3} (1 + 2n(\varepsilon_k)) \delta(|\omega| - 2\varepsilon_k)$$

$$= \frac{\pi S_d}{d} \theta(|\omega| - 2m) \left(\frac{\omega^2 - 4m^2}{4\omega^2} \right)^{d/2} [1 + 2n(\omega/2)] \left| \frac{\omega}{c} \right|^{d-2}, \tag{15.27}$$

where S_d was defined below (14.9). It can be verified that the above results for $\sigma(\omega)$ obey the scaling form (15.17).

We now discuss the physical and scaling properties of the two components of the conductivity in turn; the results are also sketched in Fig. 15.2.

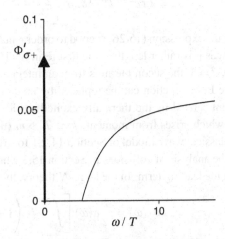

The real part, $\Phi'_{\sigma+}$, of the universal scaling function $\Phi_{\sigma+}$ in the high-T limit ($T \gg \Delta_\pm$) (see (15.17)) at the one-loop level. The numerical values are obtained from (15.26) and (15.27) with $d = 2$ ($\epsilon = 1$). There is a delta function precisely at $\omega/T = 0$ represented by the heavy arrow. The weight of this delta function is given in (15.28) and (15.29). The delta function contributes to σ_I, and the higher frequency continuum to σ_{II}.

15.1.1 σ_I

This is a zero-frequency delta function and is present only for $T > 0$. It is interpreted as the contribution of thermally excited particles that propagate ballistically without any collisions with other particles. This becomes evident when we re-derive this delta function contribution later in Section 15.2 using a transport equation formalism. Indeed, to first order in ϵ (Chapter 14), or at $N = \infty$ (Section 13.2), the excitations are simply undamped particles with an infinite lifetime and energy momentum relation ε_k. As we saw in Section 13.2.2, it is necessary to go to first order in $1/N$, to include collisions that give the quasiparticles a finite lifetime and lead to a finite phase coherence time τ_φ; a similar analysis in the ϵ expansion shows that such effects appear at order ϵ^2. We show in Section 15.3 that these collisions also broaden the delta function in σ_I. The magnitude of the broadening is expected to be determined by the inverse lifetime of the quasiparticles; in the high-temperature limit, this inverse lifetime is of order $\epsilon^2 T$ [422] in the ϵ expansion, or of order T/N in the large-N theory (see (13.68)). The typical energy of a quasiparticle at the critical point is of order T, and so the quasiparticles are well defined, at least within the ϵ or $1/N$ expansion. Note, however, that the quasiparticle interpretation breaks down at the physically important values of $\epsilon = 1$, $N = 2, 3$.

Let us evaluate the expression (15.26) for σ_I in its limiting regimes.

First, consider the high-T region, looking first at the ϵ expansion. The coefficient of the delta function is a function of the ratio m/T, but note from (15.23) and below that $m \ll T$ for small ϵ. Evaluating (15.26) in this limit we find for ϵ small

$$\sigma_I'(\omega) = 2\pi c^{2-d} T^{d-1} \delta(\omega) \left[\frac{1}{18} - \frac{m}{8\pi T} + \cdots \right]$$
$$= 2\pi c^{2-d} T^{d-1} \delta(\omega) \left[\frac{1}{18} - \frac{\sqrt{\epsilon}}{8} \left(\frac{2(N+2)}{3(N+8)} \right)^{1/2} + \cdots \right]. \quad (15.28)$$

Actually the expression (15.26) is good to order ϵ but we have refrained from displaying the next term as it is rather lengthy. The first term in (15.28) is obtained by evaluating (15.26) at $m = 0$, $d = 3$; the second term is from an integral dominated by small $ck \sim m \ll T$ and hence the Bose function can be replaced by its classical limit. It is important to note that the current carried by the thermally excited carriers is dominated by the leading term of (15.28), which arises from momenta $k \sim T \gg m$ (this is the reason we are not allowed to use the classical wave model of Section 14.2.1 for transport properties). This will be useful to us in the analysis of collisions in Section 15.3 where we will simply be able to set $m = 0$ to obtain the leading term. In the large-N theory, the corresponding expression for $d = 2$ is

$$\sigma_I'(\omega) = \frac{T}{2} \delta(\omega) \left[\int_\Theta^\infty d\varepsilon \left(1 + \frac{\Theta^2}{\varepsilon^2} \right) \frac{1}{e^\varepsilon - 1} \right]$$
$$= \frac{T}{2} \delta(\omega) \times 0.68940\ldots, \quad (15.29)$$

where $\Theta = 2\ln((\sqrt{5} + 1)/2)$. Notice that, as $m \sim T$, we have now been unable to approximate $\varepsilon_k \approx k$ to get the leading result, as was done in the ϵ expansion. The spectral weight of the delta function to leading order in the ϵ expansion is, from (15.28), $\pi T/9 = 0.3491 \ldots T$ whereas the $N = \infty$, $d = 2$ result is $0.3447 \ldots T$, which is remarkably close.

Next consider the low-T regime on the quantum paramagnetic side of the transition, $T \ll \Delta_+$. Here both the ϵ and large-N expansions give the following result for $d = 2$:

$$\sigma'_I(\omega) = T \left(\frac{\Delta_+ T}{2\pi c^2} \right)^{(d-2)/2} e^{-\Delta_+/T} \delta(\omega). \tag{15.30}$$

Hence the spectral weight of the delta function is exponentially small, since free quasi-particles are thermally activated.

15.1.2 σ_{II}

This is the continuum contribution to σ that vanishes for $\omega < 2m$. At this order in ϵ (or $1/N$) there is a sharp threshold at $\omega = 2m$, but we expect that this singularity will be rounded out when collisions are included at order ϵ^2 ($1/N$) (but we will not describe this rounding out here). Although collisions have a strong effect at the threshold, they are not expected to significantly modify the form of $\sigma'_{II}(\omega)$ at higher frequencies where the transport is predominantly collisionless. In particular, the $\omega \to \infty$ limit is precisely the $T = 0$ result [70]

$$\sigma'_{II}(\omega \to \infty) = \frac{\pi S_d}{2^d d} \left| \frac{\omega}{c} \right|^{d-2}. \tag{15.31}$$

15.2 Collisionless transport equations

The low-order perturbative result for $\sigma(\omega)$ in Section 15.1 is clearly not physically satisfactory. Owing to the absence of any collisions between the thermally excited particles, we found a singular delta function at $\omega = 0$ and a sharp threshold at $\omega = 2m$. Before we can repair these singularities, we present an alternative derivation of the delta function contribution at $\omega = 0$. This is carried out using an equation of motion analysis that clearly exposes the role of collisionless transport of thermally excited particles [106]. The advantage of this new approach is that subsequently we can readily include the effects of collisions.

We saw in the previous section that, at the one-loop level, the only effect of the $(\phi_\alpha^2)^2$ interaction was in inducing the T-dependent mass m in the propagators. This suggests that we perform our equation of motion analysis with the following simplification of the Hamiltonian in (15.3):

$$\mathcal{H}' = \mathcal{H}_0 + \mathcal{H}_{\text{ext}}. \tag{15.32}$$

The first term, $\mathcal{H}_0$, is the free particle part in (6.52), but with a renormalized mass m:

$$\mathcal{H}_0 = \frac{1}{2} \int d^d x \left[\pi_\alpha^2 + c^2 (\nabla_x \phi_\alpha)^2 + m^2 \phi_\alpha^2 \right], \tag{15.33}$$

and $\mathcal{H}_{\text{ext}}$ contains the coupling to the external magnetic field H^a:

$$\mathcal{H}_{\text{ext}} = -\int d^d x \, H^a(\mathbf{x}, t) T^a_{\alpha\beta} \pi_\alpha(x, t) \phi_\beta(x, t). \tag{15.34}$$

As noted earlier, we are interested only in the linear response of the current to the gradient, $\vec{E}^a = -\vec{\nabla}_x H^a(x, t)$, and it is assumed below that $\vec{E}^a$ is independent of x. Note that, unlike in Section 15.1, we are making the gauge choice of coupling to H^a rather than the vector potential $\vec{A}^a$; this is for convenience and should not change the final gauge-invariant results. Strictly speaking, the renormalized mass m that appears in $\mathcal{H}_0$ also depends upon $\vec{E}^a$; however, to linear order in $\vec{E}^a$, and for the case of a momentum-independent interaction, this "vertex correction" can be neglected, and we do so here without proof. The explicit form of the angular momentum current $\vec{J}^a$ can be obtained by computing the equation of motion for the angular momentum density and putting it in the form (15.6). In the present situation of an x-independent $\vec{E}^a$, the choice

$$\vec{J}^a = c^2 T^a_{\alpha\beta} \phi_\alpha \vec{\nabla}_x \phi_\beta \tag{15.35}$$

ensures that $\langle \vec{J}^a \rangle$ vanishes when $\vec{E}^a = 0$. Moreover, $\langle \vec{J}^a \rangle$ is independent of x for $\vec{E}^a$ nonzero. For completeness, let us also note here the expression for the total momentum density, $\vec{P}$, of the quantum field theory $\mathcal{H}$; this can be derived by studying the response of the action to translations, as is discussed in standard graduate texts [245]:

$$\vec{P} = \pi_\alpha \vec{\nabla}_x \phi_\alpha. \tag{15.36}$$

Notice that it is quite distinct from $\vec{J}^a$. In particular, in the absence of an external potential, $\vec{P}$ is conserved (i.e. it obeys an equation of the form $\partial_t \vec{P} + \vec{\nabla} \cdot \overleftrightarrow{Q} = 0$ for some local second rank tensor field $\overleftrightarrow{Q}$), whereas $\vec{J}^a$ is not.

The subsequent analysis is simplest in terms of the normal modes that diagonalize $\mathcal{H}_0$. Using the standard approach of diagonalizing harmonic oscillator Hamiltonians we make the mode expansions which generalize (6.54) to arbitrary time dependence:

$$\phi_\alpha(x, t) = \int \frac{d^d k}{(2\pi)^d} \frac{1}{\sqrt{2\varepsilon_k}} \left(a_\alpha(k, t) e^{i\vec{k}\cdot\vec{x}} + a^\dagger_\alpha(\vec{k}, t) e^{-i\vec{k}\cdot\vec{x}} \right),$$

$$\pi_\alpha(x, t) = -i \int \frac{d^d k}{(2\pi)^d} \sqrt{\frac{\varepsilon_k}{2}} \left(a_\alpha(\vec{k}, t) e^{i\vec{k}\cdot\vec{x}} - a^\dagger_\alpha(\vec{k}, t) e^{-i\vec{k}\cdot\vec{x}} \right), \tag{15.37}$$

where the $a(\vec{k}, t)$ operators satisfy the equal-time commutation relations in (6.55). It can be verified that (15.4) is satisfied, and $\mathcal{H}_0$ is given by the analog of (6.10)

$$\mathcal{H}_0 = \int \frac{d^d k}{(2\pi)^d} \varepsilon_k \left[a^\dagger_\alpha(\vec{k}, t) a_\alpha(\vec{k}, t) + 1/2 \right]. \tag{15.38}$$

We also need the expression for the current $\vec{J}^a$ in terms of the a and $a^\dagger$. We are only interested in the case where the system carries a position-independent current. For this case, inserting (15.37) into (15.35), we find

$$\vec{J}^a(t) = \vec{J}_I^a(t) + \vec{J}_{II}^a(t),$$

$$\vec{J}_I^a(t) = ic^2 L_{\alpha\beta}^a \int \frac{d^d k}{(2\pi)^d} \frac{\vec{k}}{\varepsilon_k} \langle a_\alpha^\dagger(\vec{k}, t) a_\beta(\vec{k}, t) \rangle,$$

$$\vec{J}_{II}^a(t) = -ic^2 L_{\alpha\beta}^a \int \frac{d^d k}{(2\pi)^d} \frac{\vec{k}}{2\varepsilon_k} \langle a_\alpha^\dagger(-\vec{k}, t) a_\beta^\dagger(\vec{k}, t) \rangle + \text{H.c.} \tag{15.39}$$

It should be evident that processes contributing to $\vec{J}_{II}^a$ require a minimum frequency of $2m$, and so $\vec{J}_{II}^a$ contributes only to $\sigma_{II}(\omega)$. We therefore drop the $\vec{J}_{II}^a$ contribution below and approximate $\vec{J}^a \approx \vec{J}_I^a$. The ease with which the high-frequency components of $\sigma(\omega)$ can be separated out is an important advantage of the present formulation of the quantum transport equations. Of course, at this simple free-field level it is not difficult to include $\vec{J}_{II}^a$ also and rederive the complete results for σ_I and σ_{II} obtained in Section 15.1. We do not do so in the interest of simplicity, but urge the reader to carry out this instructive computation.

The central object in our presentation of transport theory is the mean, time-dependent occupation number of the normal modes:

$$f_{\alpha\beta}(\vec{k}, t) = \langle a_\alpha^\dagger(\vec{k}, t) a_\beta(\vec{k}, t) \rangle, \tag{15.40}$$

in terms of which the expectation value of current is

$$\langle \vec{J}^a(t) \rangle = ic^2 T_{\alpha\beta}^a \int \frac{d^d k}{(2\pi)^d} \frac{\vec{k}}{\varepsilon_k} f_{\alpha\beta}(\vec{k}, t). \tag{15.41}$$

The corresponding expression for the momentum density is

$$\langle \vec{P}(t) \rangle = \int \frac{d^d k}{(2\pi)^d} \vec{k} f_{\alpha\alpha}(\vec{k}, t). \tag{15.42}$$

Note the difference in the structure of the $O(N)$ indices between (15.41) and (15.42).

For the subsequent analysis it is convenient to choose a definite orientation for the field H^a in the $O(N)$ space. As in Section 12.3, and the discussion above (12.47), we choose the field that generates rotations in the 1–2 plane, that is, $H^a = 0$ except for the component a, which couples to the generator of $O(N)$ with $T_{1,2}^a = -T_{2,1}^a = 1$. We henceforth denote this nonzero component simply by H (and $\vec{E} = \vec{\nabla} H$) and the index a is dropped. Similarly, the current $\vec{J}^a$ is nonzero only for this component and denoted $\vec{J}$. It is also not difficult to see that to linear order in $\vec{E}$, the distribution functions in (15.40) do not get modified for all components $\alpha > 2$. This is because any change in these components must be even in $\vec{E}$. This conclusion is also true to all orders in the interaction u in $\mathcal{H}$. We have therefore

$$f_{\alpha\beta}(\vec{k}, t) = \delta_{\alpha\beta} n(\varepsilon_k), \qquad \alpha > 2 \text{ or } \beta > 2, \tag{15.43}$$

where $n(\varepsilon)$ is the Bose function in (13.63).

The interesting transport phenomena all occur within the 1, 2 components of $f_{\alpha\beta}(\vec{k}, t)$. Within this subspace, it is helpful to transform to a basis where the external field is diagonal. We therefore define

$$a_\pm(\vec{k}, t) \equiv \frac{a_1(\vec{k}, t) \pm i a_2(\vec{k}, t)}{\sqrt{2}}, \tag{15.44}$$

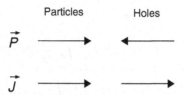

The contribution of the particle-like and hole-like excitations to the total momentum $\vec{P}$ and the angular momentum current $\vec{J}$. The particles are moving to the right, while the holes are moving to the left. Their contributions to $\vec{P}$ cancel out, while their contributions to $\vec{J}$ add.

and we occasionally refer to a_+ (a_-) as the annihilation operators for the particles (holes). The Hamiltonian $\mathcal{H}_0$ can also be expressed in terms of the $a_\pm$, and it remains diagonal, with the same form as in (15.33). The current becomes

$$\langle \vec{J} \rangle = \int \frac{d^d k}{(2\pi)^d} \sum_\lambda \lambda \frac{c^2 \vec{k}}{\varepsilon_k} \langle a_\lambda^\dagger(\vec{k}, t) a_\lambda(\vec{k}, t) \rangle$$

$$= \int \frac{d^d k}{(2\pi)^d} \sum_\lambda \lambda \frac{c^2 \vec{k}}{\varepsilon_k} f_\lambda(\vec{k}, t), \tag{15.45}$$

where the index λ is assumed here and below to extend over the values ± 1, and $f_\lambda \equiv f_{\lambda\lambda}$ are the particle distribution functions (the components of f that are off-diagonal in this λ space can easily be shown to vanish). Let us also note the expression for the momentum density

$$\vec{P} = \int \frac{d^d k}{(2\pi)^d} \sum_\lambda \vec{k} f_\lambda(\vec{k}, t). \tag{15.46}$$

An important difference between (15.45) and (15.46) is the λ inside the summation in (15.45), which is absent from (15.46). Thus the angular momentum current is proportional to the difference of the particle and hole number currents, while the momentum density is proportional to their sum, see Fig. 15.3.

We have introduced all the basic formalism necessary to formulate the transport equations, which are the equations of motion of the distribution functions $f_\lambda(\vec{k}, t)$. These are obtained by computing the Heisenberg equations of motion of $a_\pm$ under the Hamiltonian $\mathcal{H}'$ in (15.32). In deriving these equations we make approximations similar to those made for $\vec{J}$: we drop all terms involving the product of two as or $a^\dagger$s as these contribute only to the high frequency σ_{II} (the mixing of these modes with the f_λ can also be neglected to linear order in $\vec{E}$). A straightforward computation then gives the central result of this section:

$$\left(\frac{\partial}{\partial t} + \lambda \vec{E}(t) \cdot \frac{\partial}{\partial \vec{k}} \right) f_\lambda(\vec{k}, t) = 0. \tag{15.47}$$

Let us solve (15.45) and (15.47) in linear response. In the absence of $\vec{E}$, the distribution function has the equilibrium value given by the Bose function $f_\lambda(\vec{k}, t) = n(\varepsilon_k)$. We Fourier transform from time, t, to frequency, ω, and parameterize to linear order in $\vec{E}$:

$$f_\lambda(\vec{k}, \omega) = 2\pi\delta(\omega)n(\varepsilon_k) + \lambda\vec{k} \cdot \vec{E}(\omega)\psi(k, \omega), \qquad (15.48)$$

where we have used the fact that only $\vec{E}$ breaks spatial rotation invariance and $O(N)$ symmetry to conclude that ψ is independent of $\vec{k}/k$ and λ. Now inserting (15.48) in (15.47), and using $\partial\varepsilon_k/\partial\vec{k} = \vec{k}/\varepsilon_k$ it is simple to solve for ψ to leading order in $\vec{E}$:

$$\psi(k, \omega) = \frac{c^2}{-i\omega}\frac{1}{\varepsilon_k}\left(-\frac{\partial n(\varepsilon_k)}{\partial\varepsilon_k}\right). \qquad (15.49)$$

Finally, we insert this result in (15.45) and deduce the conductivity

$$\sigma(\omega) = \frac{2c^4}{-i\omega}\int\frac{d^dk}{(2\pi)^d}\frac{k_x^2}{\varepsilon_k^2}\left(-\frac{\partial n(\varepsilon_k)}{\partial\varepsilon_k}\right). \qquad (15.50)$$

The real part of this agrees with (15.26). Note that the leading factor of 2 comes from the sum over λ. The current is therefore carried equally by the thermally excited particles and holes; they move in opposite directions to create a state with vanishing momentum but nonzero charge current. We see in the next section that this charge current can be relaxed by collisions among the particles and holes.

15.3 Collision-dominated transport

We proceed to improve (15.47) by including collisions among the excitations. These collisions were previously considered in Section 13.2.2, where they led to a finite lifetime for the excitations. Here we study how the same collisions degrade the transport of angular momentum current.

A full analysis and derivation [104] of the collision contributions to the transport equation is quite lengthy and involved, and is beyond the scope of our discussion here. However, the physical interpretation of the final result is quite straightforward, and with the benefit of hindsight, it is possible to guess the collision terms by a simple application of Fermi's Golden Rule. We follow this latter route here and omit presentation of a complete, formal derivation. We begin, in Section 15.3.1, by using the ϵ expansion on the Hamiltonian $\mathcal{H}$ in (15.3). The large-N approach is considered later in Section 15.3.2.

15.3.1 ϵ expansion

The basic idea is to treat (15.47) as a rate equation for the occupation probability of particle states with momentum $\vec{k}$ and polarization λ. The terms present in (15.47) then represent the flow of particles with their momenta obeying "Newton's Law" $d\vec{k}/dt = \lambda\vec{E}$. Collisions can therefore be accounted for by including terms that represent the rate at which particles in state $\vec{k}$ collide with other particles (the "out" terms) and also the rate at which particles in other states scatter into the state $\vec{k}$ (the "in" terms). Thus, if there is a matrix element $\mathcal{M}$ for scattering of two particles with momenta and polarizations $\vec{k}$, λ and $\vec{k}_1$, λ_1 into states

$\vec{k}_2$, λ_2 and $\vec{k}_3$, λ_3, then Fermi's Golden Rule implies that the right-hand side of the transport equation will acquire the term

$$- |\mathcal{M}|^2 (2\pi) \delta(\varepsilon_k + \varepsilon_{k_1} - \varepsilon_{k_2} - \varepsilon_{k_3})$$

$$\times \left\{ f_\lambda(\vec{k}, t) f_{\lambda_1}(\vec{k}_1, t) [1 + f_{\lambda_2}(\vec{k}_2, t)] [1 + f_{\lambda_3}(\vec{k}_3, t)] \right.$$

$$\left. - f_{\lambda_2}(\vec{k}_2, t) f_{\lambda_3}(\vec{k}_3, t) [1 + f_\lambda(\vec{k}, t)] [1 + f_{\lambda_1}(\vec{k}_1, t)] \right\} \tag{15.51}$$

summed over momenta $\vec{k}_{1,2,3}$ and polarizations $\lambda_{1,2,3}$. The expression outside the curly brackets is clearly the collision rate as specified by Fermi's Golden Rule. Inside the curly brackets we have the factors associated with the out and in processes, respectively: particles entering into a collision are being annihilated and have associated with them the average Bose matrix element $\langle |\langle n_k - 1 | a_k | n_k \rangle|^2 \rangle = f(k)$ (where n_k is the occupation of state k in one realization of the thermal ensemble), while those emerging from a collision have the Bose factor $\langle |\langle n_k + 1 | a_k^\dagger | n_k \rangle|^2 \rangle = 1 + f(k)$.

Applying the rules discussed above, we pursue a lengthy, but straightforward computation to give us the following rather formidable transport equation [106]:

$$\left(\frac{\partial}{\partial t} + \lambda \vec{E} \cdot \frac{\partial}{\partial \vec{k}} \right) f_\lambda(\vec{k}, t)$$

$$= -\frac{u^2}{9} \int \frac{d^d k_1}{(2\pi)^d} \frac{d^d k_2}{(2\pi)^d} \frac{d^d k_3}{(2\pi)^d} \frac{1}{16 \varepsilon_k \varepsilon_{k_1} \varepsilon_{k_2} \varepsilon_{k_3}}$$

$$\times (2\pi)^d \delta(\vec{k} + \vec{k}_1 - \vec{k}_2 - \vec{k}_3) 2\pi \delta(\varepsilon_k + \varepsilon_{k_1} - \varepsilon_{k_2} - \varepsilon_{k_3})$$

$$\times \left\{ 4 \{ f_\lambda(\vec{k}, t) f_{-\lambda}(\vec{k}_1, t) [1 + f_\lambda(\vec{k}_2, t)][1 + f_{-\lambda}(\vec{k}_3, t)] \right.$$

$$- [1 + f_\lambda(\vec{k}, t)][1 + f_{-\lambda}(\vec{k}_1, t)] f_\lambda(\vec{k}_2, t) f_{-\lambda}(\vec{k}_3, t) \}$$

$$+ 2 \{ f_\lambda(\vec{k}, t) f_\lambda(\vec{k}_1, t) [1 + f_\lambda(\vec{k}_2, t)][1 + f_\lambda(\vec{k}_3, t)]$$

$$- [1 + f_\lambda(\vec{k}, t)][1 + f_\lambda(\vec{k}_1, t)] f_\lambda(\vec{k}_2, t) f_\lambda(\vec{k}_3, t) \}$$

$$+ (N - 2) \{ f_\lambda(\vec{k}, t) n(\varepsilon_{k_1}) [1 + f_\lambda(\vec{k}_2, t)] [1 + n(\varepsilon_{k_3})]$$

$$- [1 + f_\lambda(\vec{k}, t)] [1 + n(\varepsilon_{k_1})] f_\lambda(\vec{k}_2, t) n(\varepsilon_{k_3}) \}$$

$$+ \frac{(N - 2)}{2} \{ f_\lambda(\vec{k}, t) f_{-\lambda}(\vec{k}_1, t) [1 + n(\varepsilon_{k_2})][1 + n(\varepsilon_{k_3})]$$

$$\left. - [1 + f_\lambda(\vec{k}, t)][1 + f_{-\lambda}(\vec{k}_1, t)] n(\varepsilon_{k_2}) n(\varepsilon_{k_3}) \} \right\}, \tag{15.52}$$

with $\lambda = \pm 1$. Fortunately, interpreting the individual terms is quite simple. The first two pairs of collision terms represent processes within those with polarizations in the 1–2 plane, while the last two pairs (proportional to $(N - 2)$) represent collisions with particles with polarizations with $\alpha > 2$; in linear response, the latter have their distribution function given simply by the Bose function, as was noted earlier in (15.43).

In writing down (15.52), we have omitted terms associated with collisions that involve creation or annihilation of particle–hole pairs, as they have a negligible contribution in both the high- and low-T limits in the ϵ expansion (such processes are included in our later discussion of the $1/N$ expansion). Thus a collision in which, for example, a positively charged particle of momentum $\vec{k}$ turns into two positively charged particles and a negatively charged hole with momenta $\vec{k}_1$, $\vec{k}_2$, and $\vec{k}_3$, respectively, is permitted by the symmetries of the problem. However, it remains to evaluate the phase space over which such collisions conserve total energy and momentum. In the low-T quantum paramagnetic region, we need a particle with energy of at least $3\Delta_+$ to have sufficient energy to emit a particle–hole pair, and such particles are exponentially rare. In the opposite high-T region, note that the "mass" m of the particles/holes is of order $\sqrt{\epsilon}T$ (below (15.23)), whereas their momentum is of order T. Consequently, to leading order in ϵ we may just replace the energy momentum relation (13.64) by $\varepsilon_k = ck$ (see also the discussion below (15.28)). The particle–hole pair-creation collision requires that $\vec{k} = \vec{k}_1 + \vec{k}_2 + \vec{k}_3$ and $k = k_1 + k_2 + k_3$. This is only possible if all three momenta are collinear, and this process therefore has vanishing phase space in the high-T limit. More generally, for a nonzero m, the phase space vanishes as $\epsilon \to 0$.

We analyze the solutions of (15.52) separately in the high-T and low-T paramagnetic regions of Figs. 11.2, 11.3, and 14.1.

High T, $T \gg \Delta_+$

To obtain the $T \gg \Delta_+$ limit of the scaling results for conductivity as encapsulated in (15.17), it is sufficient to replace the interaction strength u on the right-hand side of (15.52) by the fixed point value discussed in Chapter 14 and in (14.11). For the result to leading order in ϵ, we can set

$$u = \frac{48\pi^2 c^3}{(N+8)}\epsilon\mu^\epsilon. \tag{15.53}$$

The prefactor of c^3 has been deduced by dimensional analysis; it did not appear in Chapter 14 because we used units with $c = 1$ there. We expect that in the high-T limit, $\mu \sim T/c$, as that is the only natural scale in the problem; in any case, to leading order in ϵ, the precise value of μ is not needed.

The next step is to linearize the transport equation (15.52) by using the ansatz (15.48) and to examine the structure of its solution in the limit of small ϵ. First, we find that the λ dependencies in (15.48) and (15.52) are completely compatible, in that the linearized equation for the unknown function $\psi(k, \omega)$ is independent of λ. Then we perform a simple dimensional analysis of the linear integral equation satisfied by ψ. The dependencies on ϵ (for small ϵ), T, and c, can all be scaled out, and it is not difficult to show that the solution of the linear integral equation can be written in the form

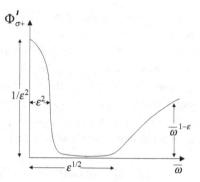

Fig. 15.4 Structure of the real part, $\Phi'_{\sigma+}(\bar{\omega}, 0) = \Phi_{\sigma-}(\bar{\omega}, 0)$, of the universal scaling functions $\Phi_{\sigma\pm}$ in (15.17) in the high-T region, $T \gg \Delta_{\pm}$, as a function of $\bar{\omega} = \omega/T$ in the limit of small ϵ. The peak at small $\bar{\omega}$ has a width of order ϵ^2 and a height of order $1/\epsilon^2$; this feature of the conductivity is denoted by σ_I. The collisionless contribution (denoted σ_{II}) begins at $\bar{\omega}$ of order $\epsilon^{1/2}$; as $\bar{\omega} \to \infty$, this contribution is a number of order unity times $\bar{\omega}^{1-\epsilon}$.

$$\psi(k, \omega) = \frac{c^2}{\epsilon^2 T^3} \Psi\left(\frac{\omega}{\epsilon^2 T}, \frac{ck}{T}\right), \tag{15.54}$$

where the dimensionless complex function Ψ satisfies a parameter-free and universal linear integral equation. This equation has to be solved numerically [106], and we will not discuss the details of the numerical analysis here. Finally, computing the current by using (15.45) and (15.48) we see that the conductivity σ_I can be written in the form

$$\sigma_I(\omega) = \frac{(T/c)^{d-2}}{\epsilon^2} \Phi_{\sigma I}\left(\frac{\omega}{\epsilon^2 T}\right), \tag{15.55}$$

where the scaling function $\Phi_{\sigma I}$ is simply related to Ψ. This result is clearly compatible with the scaling form (15.17) for the total conductivity. Note that the natural frequency scale in (15.54) and (15.55) is of order $\epsilon^2 T$: this is the scale over which the delta function in σ'_I was expected to be broadened. Furthermore, the peak value of the zero frequency conductivity, which diverged at the one-loop level, is seen to be of order $1/\epsilon^2$. (These features are sketched in the schematic of the frequency-dependent conductivity in the high-T limit in Fig. 15.4.)

The function $\Phi_{\sigma I}$ therefore defines the smoothing of the delta function in (15.26) and has the same total spectral weight. From (15.28) we see that it satisfies

$$\int_0^\infty d\tilde{\omega} \text{Re}\Phi_{\sigma I}(\tilde{\omega}) = \frac{\pi}{18} \tag{15.56}$$

in the high-T limit. It should be noted that this sum rule is special to the leading order in ϵ being considered here. For ϵ of order unity, there is no sharp distinction between σ_I and σ_{II} and there is no sum rule. Indeed the integral in (15.56) when carried out over the total σ will be divergent. For any realistic lattice model there is a large microscopic energy scale ($\sim J$) beyond which the universal scaling results do not apply, and the entire spectral weight (including frequencies beyond J) is not divergent; this latter spectral weight satisfies a sum rule related to nonuniversal microscopic quantities and is unrelated to the universal result (15.56).

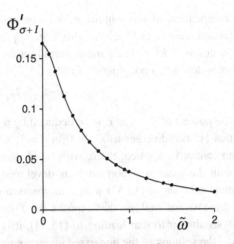

Fig. 15.5 The real part of the universal function $\Phi_{\sigma I}$ as a function of $\tilde{\omega} = \omega/\epsilon^2 T$, defined in (15.55) in the high-T limit $\Delta_+/T = 0$. This function describes the inelastic collision-induced broadening of the $\omega = 0$ delta function in Fig. 15.2 at a frequency scale of order $\epsilon^2 T$. The conductivity has an additional continuum contribution ($\sigma_{//}(\omega)$) at frequencies larger than $\omega \sim \epsilon^{1/2} T$, which is not shown above (see Fig. 15.4).

A complete numerical solution for the function $\Phi'_{\sigma I}$ has been carried out in the high-T limit in [106] for the case $N = 2$, and the solution is sketched in Fig. 15.5. Most important is the value of $\Phi'_{\sigma I}(0)$, which gives the value of the d.c. conductivity [106]

$$\sigma(0) = \frac{Q^2}{\hbar}\left(\frac{T}{c}\right)^{d-2}\frac{0.1650}{\epsilon^2}, \qquad N = 2, T \gg \Delta_{\pm}. \tag{15.57}$$

As noted earlier, the result is a pure number times $Q^2/\hbar$ for $d = 2$ (recall that Q is the charge of the carriers, which we usually absorb into the definition of H). These are among the main results for low-frequency transport in this chapter.

Low T, $T \ll \Delta_+$

First, we have to determine the value of u that must be used in the transport equation (15.52). Now Δ_+ is the largest scale in the problem, and so the generalization of the result (15.53) suggests that

$$u \sim \epsilon c^3 (\Delta/c)^\epsilon. \tag{15.58}$$

However, this result is not adequate, as we now argue that the limits $T \to 0$ and $\epsilon \to 0$ do not commute. For $T \ll \Delta_+$, as in Sections 10.4.2 and 12.2, all the thermally excited particles are at energies just above the gap, and so we can approximate their dispersion by

$$\varepsilon_k = \Delta_+ + \frac{c^2 k^2}{2\Delta_+}. \tag{15.59}$$

The typical value of the particle momentum is $k \sim \sqrt{\Delta_+ T}/c$. The coupling (15.58) would imply that these quadratically dispersing, slowly moving particles scatter with a T matrix that is independent of momentum at low momentum. However, we know from elementary quantum mechanics [502] that this Born approximation result is incorrect; the full T matrix scales as $\sim k^{d-2}$ as the momentum transfer $k \to 0$, and so we should really use a momentum-dependent coupling u of order

$$u \sim k^{d-2} \Delta_+^{5-2d} c^{2d-2}, \tag{15.60}$$

where the powers of Δ_+ and c were deduced by a dimensional comparison with (15.58). (Note that (15.60) diverges as $k \to 0$ in $d = 1$, where the present perturbative transport equation cannot be applied; there, we should instead use the exact S matrix in (12.13), along with the exact transport analysis developed in Section 12.2.) We do not attempt a complete solution of (15.52) with a momentum-dependent u here but will be satisfied with a dimensional analysis that exposes the T dependence of physical observables. By an analysis similar to that leading to (15.54), it is not difficult to show that in the limit $T \ll \Delta_+$, the solution of the linearized integral equation satisfied by ψ takes the form

$$\psi(k, \omega) = \frac{c^2 \tau_\varphi e^{-\Delta_+/T}}{\Delta_+ T} \Psi\left(\omega \tau_\varphi, \frac{ck}{\sqrt{\Delta_+ T}}\right), \tag{15.61}$$

where the particle scattering time, τ_φ, is deduced by a dimensional analysis of the collision term in (15.52) with the momentum-dependent coupling u in (15.60):

$$\frac{1}{\tau_\varphi} \sim T\left(\frac{T}{\Delta_+}\right)^{2(d-2)} e^{-\Delta_+/T}. \tag{15.62}$$

Note that this result for τ_φ is consistent with the $d = 2$ result in (13.68). Now we can compute the current by inserting (15.61) into (15.45) and (15.48), and the result for σ_I takes the form

$$\sigma_I(\omega) = T\tau_\varphi \left(\frac{\Delta_+ T}{2\pi c^2}\right)^{(d-2)/2} e^{-\Delta_+/T} \Phi_{\sigma+I}(\omega \tau_\varphi). \tag{15.63}$$

This is consistent with (15.30) in the collisionless limit $\tau_\varphi \to \infty$. The scaling function $\Phi_{\sigma+I}$ is expected to be a constant when its argument vanishes, and so the d.c. conductivity can be obtained from (15.62) and (15.63):

$$\sigma_I(0) \sim \left(\frac{T}{\Delta_+}\right)^{-3(d-2)/2} \left(\frac{\Delta_+}{c}\right)^{(d-2)}. \tag{15.64}$$

This result is valid for $d > 2$, where we see that the d.c. conductivity actually diverges as $T \to 0$. The total spectral weight in the "Drude" peak of the d.c. conductivity, σ_I, is exponentially small, $\sim e^{-\Delta_+/T}$, but the weak inelastic scattering between the thermally excited particles is also exponentially rare; the two exponential factors cancel each other out, and we get a power-law divergent conductivity. For $d = 2$, because of the logarithmic factors obtained in (13.68), we expect $\sigma_I(0)$ to diverge as $(\ln(\Delta_+/T))^2$; recall that this logarithmic divergence was absent in the high-T limit ($T \gg \Delta_+$), where the d.c. conductivity was a completely universal constant in $d = 2$. Finally, as we have already noted, these methods do not apply for $d = 1$, but it is interesting to note that the Einstein relation (15.11), when combined with our earlier results (12.12) and (12.27), gives us a d.c. conductivity, $\sigma \sim T^{-1/2}$, which also diverges as $T \to 0$.

15.3.2 Large-N limit

A closely related analysis of collisions can also be carried out in the large-N limit. It has the advantage of working directly for $d = 2$ at all stages, and so we briefly discuss its formulation here [425].

The central simplification of the large-N limit is apparent by a glance at the right-hand side of (15.52): the $\vec{E}$ field changes the distribution of particles only with polarization $\alpha = 1, 2$, but their scattering is dominated completely by collisions with particles with polarization $\alpha > 2$ (note the prefactor of $(N - 2)$ in some of the collision terms). The collisions with these particles actually appear in the form of interactions with the fluctuations of the λ field, which were considered in Section 13.2. The propagator, Π, of this λ field was given in (13.42), and the upshot of the result (15.43) is that this propagator remains unchanged in the presence of the $\vec{E}$ field. To leading order in $1/N$ we can then simply consider the Gaussian fluctuations of the λ field as an infinite set of harmonic oscillators with density of modes given by the imaginary part of $1/\Pi$. These harmonic oscillators are coupled to the normal modes of the order parameter $\mathbf{n}$ by the $\lambda \mathbf{n}^2$ vertex in (11.6). The collision terms arise entirely from this vertex, and their form can be deduced from Fermi's Golden Rule as discussed earlier. The resulting generalization of (15.47) is then

$$
\left(\frac{\partial}{\partial t} + \lambda \vec{E} \cdot \frac{\partial}{\partial \vec{k}} \right) f_\lambda(\vec{k}, t)
$$

$$
= -\frac{2}{N} \int_0^\infty \frac{d\Omega}{\pi} \int \frac{d^2 q}{(2\pi)^2} \operatorname{Im} \left(\frac{1}{\Pi(\vec{q}, \Omega)} \right)
$$

$$
\times \left\{ \frac{(2\pi)\delta(\varepsilon_k - \varepsilon_{|\vec{k}+\vec{q}|} - \Omega)}{4\varepsilon_k \varepsilon_{|\vec{k}+\vec{q}|}} \left[f_\lambda(\vec{k}, t)(1 + f_\lambda(\vec{k} + \vec{q}, t))(1 + n(\Omega)) \right. \right.
$$

$$
\left. - f_\lambda(\vec{k} + \vec{q}, t)(1 + f_\lambda(\vec{k}, t))n(\Omega) \right]
$$

$$
+ \frac{(2\pi)\delta(\varepsilon_k - \varepsilon_{|\vec{k}+\vec{q}|} + \Omega)}{4\varepsilon_k \varepsilon_{|\vec{k}+\vec{q}|}} \left[f_\lambda(\vec{k}, t)(1 + f_\lambda(\vec{k} + \vec{q}, t))n(\Omega) \right.
$$

$$
\left. - f_\lambda(\vec{k} + \vec{q}, t)(1 + f_\lambda(\vec{k}, t))(1 + n(\Omega)) \right]
$$

$$
+ \frac{(2\pi)\delta(\varepsilon_k + \varepsilon_{|-\vec{k}+\vec{q}|} - \Omega)}{4\varepsilon_k \varepsilon_{|-\vec{k}+\vec{q}|}} \left[f_\lambda(\vec{k}, t) f_{-\lambda}(-\vec{k} + \vec{q}, t)(1 + n(\Omega)) \right.
$$

$$
\left. \left. - (1 + f_{-\lambda}(-\vec{k} + \vec{q}, t))(1 + f_\lambda(\vec{k}, t))n(\Omega) \right] \right\}, \tag{15.65}
$$

where the function Π is defined by analytic continuation from (13.42). Note that this equation is formulated directly in $d = 2$ and is entirely free of parameters, other than the energy scales T and Δ_+ (through the value of ε_k in (13.64)). Hence it is already in the scaling limit, and its solution leads to a σ consistent with the scaling form (15.17).

Equation (15.65) can of course also be formulated for arbitrary d, and it is reassuring to verify that in their overlapping regions of validity (N large and ϵ small) the results (15.52) and (15.65) are in precise agreement with each other. However, there are important differences between the $d = 2$, large-N analysis of (15.65) and the small-ϵ analysis of (15.52) discussed earlier. Now we have to use the full dispersion $\varepsilon_k = (c^2 k^2 + m^2)^{1/2}$ in (13.64), and in no regime is it possible to approximate it by $\varepsilon_k = ck$. Also, unlike (15.52), (15.65) does contain terms corresponding to collisions that cause production of new particle–hole pairs.

As in the case of the ϵ expansion, it is useful to scale out the small parameter $1/N$ from the transport equation (15.65). Using the ansatz (15.48), and obtaining the linear integral equation for ψ, it can be shown that its solution can be written in the form

$$\psi(k, \omega) = \frac{Nc^2}{T^3} \Psi\left(\frac{N\omega}{T}, \frac{ck}{T}, \frac{\Delta_+}{T}\right), \tag{15.66}$$

where again the dimensionless complex function Ψ satisfies a parameter-free and universal linear integral equation. If we compute the current by using (15.45) and (15.48), it follows that the analog of (15.55) and (15.63) is in $d = 2$

$$\sigma_I(\omega) = N\Phi_{\sigma+I}\left(\frac{N\omega}{T}, \frac{\Delta_+}{T}\right). \tag{15.67}$$

The natural frequency scale in (15.66) and (15.67) is of order T/N – this is the scale, from (13.71), over which the delta function in σ_I' was expected to be broadened. A schematic of the large-N frequency-dependent conductivity in the high-T limit is shown in Fig. 15.6.

The sum rule on $\Phi'_{\sigma+I}$ corresponding to (15.56) is specified by (15.29).

A complete numerical solution for the function $\Phi'_{\sigma+I}$ has been carried out in the high-T limit in $d = 2$ and the solution is shown in Fig. 15.7.

The large-N value of $\Phi'_{\sigma+I}(0, 0)$, which gives the value of the d.c. conductivity, was obtained as

$$\sigma(0) = \frac{Q^2}{\hbar} 0.1077N, \qquad d = 2, T \gg \Delta_\pm. \tag{15.68}$$

These results complement similar results discussed earlier in the ϵ expansion.

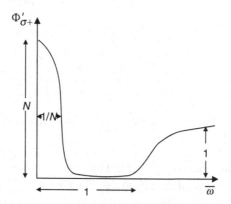

Fig. 15.6   The analog of Fig. 15.4 for the large-N limit in $d = 2$.

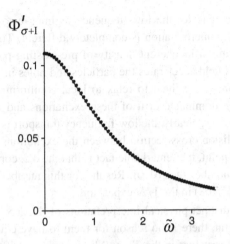

Fig. 15.7 The real part of the large-N universal function $\Phi_{\sigma+I}$ as a function of $\tilde{\omega} = N\omega/T$, defined in (15.67) in the high-T limit $\Delta_+/T = 0$. This function describes the inelastic collision-induced broadening of the $\omega = 0$ delta function in Fig. 15.2 at a frequency scale of order T/N. The conductivity has an additional continuum contribution ($\sigma_{//}(\omega)$) at frequencies larger than $\omega \sim T$, which is not shown above (see Fig. 15.6).

15.4 Physical interpretation

This is a convenient point to emphasize some interesting physical features of the above computations of the universal behavior of the conductivity near a quantum critical point in $d = 2$. The central property is the basic scaling form (15.17) and the above computations have all been aimed at describing the structure in the scaling function $\Phi_{\sigma+}$. A remarkable property of the result emerges in the high-T region of Fig. 14.1. Here the dynamical conductivity in $d = 2$ depends upon no material parameters at all and is given by the pure universal function $\Phi_{\sigma+}(0, \omega/T)$.

We focus on two limiting regions of this result in the high-T region.

First, consider the high-frequency regime, $\omega \gg T$. Here we found that the perturbative analysis considered in Section 15.1 gave an adequate description of the physics. The main result is contained in (15.31) and has a simple physical interpretation. The system is in its ground state, and the oscillating external field creates a particle–hole pair. The conductivity is then determined by the subsequent motion of this particle–hole pair. As we are effectively at the critical coupling, there is a gapless spectrum, and this particle–hole pair will also create a cascade of lower energy particle–hole pairs. Such processes lead to corrections to (15.31) that are higher order in ϵ (computed in [134]) or $1/N$ (computed in [70]). It is clear, however, that all these processes are essentially coherent. The system was originally in its phase-coherent ground state, and the particle–hole pairs created move coherently in response to the external field. This coherent transport is characterized by the universal number $\Phi_{\sigma+}(\infty, 0)$.

Now, consider the low-frequency regime, $\omega \ll T$, which also includes the d.c. case. Here, the interpretation is completely different. The system is initially at finite temperature, with an incoherent density of pre-existing particle–hole pairs already present. The external field accelerates the particles and holes in opposing directions, but their repeated collisions cause them to relax to local equilibrium. The transport is therefore due to a collision-dominated drift of these excitations and is controlled entirely by inelastic processes. Now, clearly, the low-frequency transport is entirely incoherent. However, because the collision cross-section between the excitations has a universal form near a quantum critical point, the remarkable fact is that the d.c. conductivity remains universal: it is given by the number $\Phi_{\sigma+}(0, 0)$. Results for this number appear in (15.57) in the ϵ expansion, and in (15.68) in the $1/N$ expansion.

The distinct physical interpretations of $\Phi_{\sigma+}(\infty, 0)$ and $\Phi_{\sigma+}(0, 0)$ make it clear that, in general, there is no reason for them to have equal values. This difference leads to an unusual structure in the $T \to 0$ limit of the conductivity for $d = 2$. In Fig. 15.8 we show the universal value of $\sigma(\omega, T \to 0)$. For all $\omega > 0$ we have a frequency-independent conductivity given by the number $\Phi_{\sigma+}(\infty, 0)$ describing coherent transport; however, only the single point $\omega = 0$ is given by the value $\Phi_{\sigma+}(0, 0)$, which characterizes incoherent transport. For laboratory measurements, we note that a degree Kelvin in temperature converts approximately to 20 GHz frequency by the factor k_B/h; so even a radio frequency measurement is usually comfortably in the regime $\hbar\omega \ll k_B T$, and will therefore measure $\Phi_{\sigma+}(0, 0)$, given by the isolated $\omega = 0$ point in Fig. 15.8.

In the first edition of this book it was noted that we could not rule out the existence of exotic models or symmetries that may cause the collision-dominated and collisionless limits (specified by $\Phi_{\sigma+}(0, 0)$ and $\Phi_{\sigma+}(\infty, 0)$, respectively) to be equal. Remarkably, since then, precisely such an "exotic" model has been found [219]: it is the $N \to \infty$ limit of a SU(N) gauge theory with maximal possible supersymmetry in 2+1 dimensions (this is one of the phases of M theory). The transport properties of this model are obtained via the AdS/CFT correspondence. We briefly review these recent developments in Section 15.5 below.

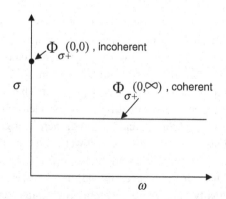

Fig. 15.8   The value of $\sigma(\omega, T \to 0)$ in $d = 2$ at the quantum critical coupling $s = 0$ (in the notation of Chapter 14) or $g = g_c$ (in the notation of Chapter 11). The value $\Phi_{\sigma+}(0, 0)$ characterizes the single point $\omega = 0$.

In physical applications of the $N = 2$ transport analysis of this chapter (discussed a bit more explicitly in Section 15.6), we interpret σ as the electrical conductivity of carriers of charge Q ($Q = 2e$ for the superconductor–insulator transition); then, in laboratory units, the conductivity is the quantum unit of conductance, $Q^2/\hbar$, times the scaling functions computed here. So, when $d = 2$, the high-T region has a d.c. conductivity that is $Q^2/\hbar$ times a universal number. The reader may be familiar with other physical situations in which universal conductances of order e^2/h have been discussed previously. These include Landauer transport in one-dimensional microstructures [19], universal conductance fluctuations [18, 291], or critical points in noninteracting electron models of transitions between quantum Hall plateaus [78, 227, 229]. However, in all these cases, the transport is *phase coherent* and the phase-breaking length is assumed to be larger than the sample size. In contrast, the conductivity studied here near an interacting quantum critical point is dominated entirely by *inelastic* processes; it is therefore quite remarkable that the d.c. conductivity is universal despite being entirely incoherent.

15.5 The AdS/CFT correspondence

All our explicit computations so far of quantum critical transport have begun from a quasiparticle picture. Then we have included interactions to destroy the quasiparticles, and moved towards a generic quantum critical description. One consequence of this weak-coupling approach was the parametrically strong frequency dependence of the conductivity, as illustrated in Fig. 15.6.

It would clearly be valuable to have a complementary description which avoids any reference to quasiparticle excitations: of the quantum-critical region of a strongly-coupled theory, these are not well-defined excitations at any scale. Remarkably, just such a description has emerged [219] from what is known as the AdS/CFT correspondence in string theory.

Before describing these results, we rephrase the general discussion of transport presented at the beginning of this chapter. First, let us generalize the uniform susceptibility χ_u appearing in (15.10) to $\chi_u(k, \omega)$, describing the response to a space- and time-dependent magnetic field. Using the Kubo formula for the dynamic conductivity in (8.32) or (15.21), and relating correlations of the current to those of the conserved density via the equation of motion (15.6), we conclude that $\sigma(\omega)$, is given by

$$\sigma(\omega) = \lim_{k \to 0} \frac{-i\omega}{k^2} \chi_u(k, \omega). \tag{15.69}$$

The correlation function $\chi_u(k, \omega)$ allows us to concisely describe the essence of the crossover from "phase-coherent" or "collisionless" dynamics at high frequencies $\omega \gg T$, to "phase-incoherent" or "collision-dominated" dynamics at low frequencies, $\omega \ll T$. The ω and k dependence takes distinct forms in the two limits, as we now describe.

First, consider the collisionless regime, $\omega \gg T$. Here, we may as well set $T = 0$, and the form of $\chi_u(k, \omega)$ is then strongly restricted by scale and relativistic invariance. Here it pays to use a fully relativistic notation, and use spacetime indices (μ, $v = t, x, y$), and consider the general form of $\chi_{u,\mu v}(p_\mu)$, where $p_\mu = (\omega, ck)$ is a spacetime momentum; note that $\chi_u \equiv \chi_{u,tt}$. The two essential constraints on χ_u are (i) its scaling dimension $\dim[\chi_u] = d-z$ which follows from (15.16) and (15.11), and (ii) the conservation equation (15.6), which implies $p^\mu \chi_{u,\mu v} = 0$. There is a unique relativistically covariant function which satisfies these constraints (we restrict our attention to $d = 2$):

$$\chi_{u,\mu v} = \frac{Q^2}{\hbar c^2} K \sqrt{p^2} \left(\eta_{\mu v} - \frac{p_\mu p_v}{p^2} \right), \tag{15.70}$$

where $p^2 = \eta^{\mu v} p_\mu p_v$ with $\eta_{\mu v} = \mathrm{diag}(-1, 1, 1)$, and only the universal dimensionless constant K is undetermined. For $\chi_u \equiv \chi_{u,tt}$, we can write this as

$$\chi_u(k, \omega) = \frac{Q^2}{\hbar} K \frac{k^2}{\sqrt{c^2 k^2 - (\omega + i\eta)^2}}. \tag{15.71}$$

Thus collisionless transport at relativistic quantum critical points in $d = 2$ is described completely by the single number K. Combining (15.69) with (15.71), we obtain the conductivity

$$\sigma(\omega) = \frac{Q^2}{\hbar} K \; ; \quad \hbar\omega \gg k_B T, \tag{15.72}$$

and so the number K is that determined in (15.31) to leading order in the ϵ expansion [70].

Now let us turn to the collision-dominated regime, $\omega \ll T$. Here we may no longer use relativistic invariance, because thermal fluctuations have strongly broken the relativistic invariance of the ground state. Instead, the form of χ_u is now dictated by the constraints of "hydrodynamics": this is the idea that perturbations relax back to thermal equilibrium under the transport equation (15.8). We may now use (15.6) and (15.8) to directly compute the linear response in the density L^a to an applied field H^a: we assume that the field is oriented along a single direction in spin space, and so drop the index a and the structure constants f_{abc}. In this manner, we obtain the response function

$$\chi_u(k, \omega) = \chi_u \frac{D_s k^2}{D_s k^2 - i\omega}, \tag{15.73}$$

where χ_u is the static susceptibility, and D_s is the diffusion constant in (15.9). Note the strong contrast in the functional forms in (15.71) and (15.73) between the collisionless and collision-dominated behaviors. Applying (15.69) to (15.73), we can verify the Einstein relation in (15.11) for the conductivity at zero frequency:

$$\sigma(\omega) = \chi_u D_s, \quad \hbar\omega \ll k_B T. \tag{15.74}$$

We have already presented the scaling form for χ_u in (15.12); that for D_s follows similarly

$$D_s = \frac{c^2}{T} \Phi_{s\pm} \left(\frac{\Delta_\pm}{T} \right), \tag{15.75}$$

and then (15.74) implies (15.17). Note that the resulting value of $\sigma(0)$ has nothing to do with the constant K in the collisionless dynamics in (15.71).

15.5.1 Exact results for quantum critical transport

Now we are ready to describe the solvable model obtained from the AdS/CFT correspondence, which fully confirms the structure of $\chi_u(k, \omega)$ obtained in (15.71) and (15.73).

The solvable model may be viewed as a generalization of the gauge theory we consider in (19.80) when we discuss the criticality of antiferromagnets on the square lattice in Section 19.3.5. We take the same basic structure of critical matter fields coupled to a gauge field, and generalize it to a relativistically invariant model with a non-Abelian SU(N) gauge group and the maximal possible supersymmetry. The resulting supersymmetric Yang-Mills (SYM) theory has only one independent coupling constant g, which is the analog of the couplings g_μ in (19.80). The matter content is naturally more complicated than the complex scalar w_a in (19.80), and also involves relativistic Dirac fermions as in Chapter 17. However, all the terms in the action for the matter fields are also uniquely related by supersymmetry to the single coupling constant g. Under the renormalization group, it is believed that g flows to an attractive fixed point at a nonzero coupling $g = g^*$; the fixed point then defines a supersymmetric conformal field theory in 2+1 dimensions (a SCFT3). We are interested here in computing the transport properties of the SCFT, as a paradigm of quantum critical transport at a strongly interacting quantum critical point.

A remarkable recent advance has been the exact solution of this SCFT3 in the $N \to \infty$ limit using the AdS/CFT correspondence [243]. The solution proceeds by a dual formulation as a four-dimensional supergravity theory on a spacetime with uniform negative curvature: anti-de Sitter space, or AdS$_4$. The solution is also easily extended to nonzero temperatures, and allows direct computation of the correlators of conserved charges in real time. At $T > 0$ a black hole appears in the gravity theory, resulting in an AdS-Schwarzschild spacetime, and T is the Hawking temperature of the black hole; the real-time solutions also extend to $T > 0$.

A description of the derivation of the results is beyond the scope of the present treatment, and the reader is referred to the original paper [219] and reviews cited therein. The results of a full computation of the density correlation function, $\chi_u(k, \omega)$ are shown in Fig. 15.9 and 15.10. The most important feature of these results is that the expected limiting forms in the collisionless (15.71) and collision-dominated (15.73) are obeyed. Thus the results do display the collisionless to collision-dominated crossover at a frequency of order $k_B T/\hbar$, as we expected from the physical discussion in Section 15.4.

At this point, we describe some technical aspects of the results which turn out to have important physical implications. For this, let us generalize the arguments leading to $\chi_{u,\mu\nu}$ in (15.70) to $T > 0$. At $T > 0$, we do not expect $\chi_{\mu\nu}$ to be relativistically covariant, and so can only constrain it by spatial isotropy and density conservation. Setting the velocity $c = 1$ for the remainder of this subsection, these two constraints lead to the most general form

$$\chi_{u,\mu\nu}(\vec{k}, \omega) = \frac{Q^2}{\hbar}\sqrt{p^2}\Big(P_{\mu\nu}^T K^T(k, \omega) + P_{\mu\nu}^L K^L(k, \omega)\Big), \tag{15.76}$$

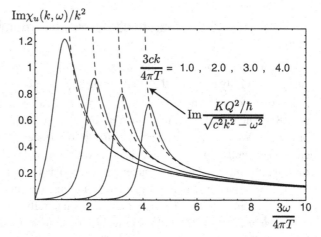

Fig. 15.9 Spectral weight of the density correlation function of the SCFT3 with $\mathcal{N} = 8$ supersymmetry in the collisionless regime.

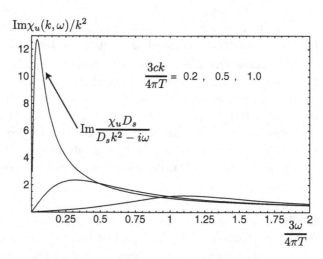

Fig. 15.10   As in Fig. 15.9, but for the collision-dominated regime.

where $P_{\mu\nu}^T$ and $P_{\mu\nu}^L$ are orthogonal projectors defined by

$$P_{00}^T = P_{0i}^T = P_{i0}^T = 0, \quad P_{ij}^T = \delta_{ij} - \frac{k_i k_j}{k^2}, \quad P_{\mu\nu}^L = \left(\eta_{\mu\nu} - \frac{p_\mu p_\nu}{p^2}\right) - P_{\mu\nu}^T, \quad (15.77)$$

with the indices i, j running over the two spatial components. Thus, in the general case at $T > 0$, the full density and current response functions as a function of k and ω are described in terms of two functions $K^{L,T}(k, \omega)$, representing current fluctuations longitudinal and transverse to the momentum. The results in (15.73) and (15.71) are obtained by taking suitable limits of these functions. These two functions are not entirely independent. At $T > 0$, we expect all correlations to be smooth functions at $k = 0$: this is because all correlations are expected to decay exponentially to zero as a function of spatial separation. However, this is only possible from (15.76) if we have the additional relation

$$K^T(0, \omega) = K^L(0, \omega). \tag{15.78}$$

The relations of the previous paragraph are completely general and apply to any theory. Specializing to the AdS-Schwarzschild solution of SYM3, the results were found to obey a simple and remarkable identity [219]:

$$K^L(k, \omega) K^T(k, \omega) = \mathcal{K}^2, \tag{15.79}$$

where $\mathcal{K}$ is a known pure number, independent of ω and k. Let us give some more details on the origin of (15.79). In the AdS/CFT correspondence, every globally conserved quantity in the CFT gets mapped onto a gauge field in AdS. Moreover, in the leading classical gravity theory on AdS, different global charges commute with each other, and so can be considered separately. In the end, we have a U(1) gauge field on AdS for every global conservation law of the CFT. The low-energy effective field theory on AdS$_4$ has the standard Einstein–Maxwell action for gravity+electromagnetism. The Maxwell action is in 3+1 dimensions, and this is well-known to have a self-dual structure corresponding to the exchange of electric and magnetic fields. Thus we have the important and key result that every global charge in a CFT3 maps onto a self-dual theory in the leading gravity approximation on AdS$_4$. The identity in (15.79) is a consequence of this emergent self-duality of CFT3s.

The combination of (15.79) and (15.78) now fully determines the response functions at zero momenta: $K^L(0, \omega) = K^T(0, \omega) = \mathcal{K}$. Then, from (15.69) we have for the conductivity

$$\sigma(\omega) = \frac{Q^2}{\hbar} \mathcal{K}. \tag{15.80}$$

This result is an important surprise: the conductivity of the classical gravity theory on AdS$_4$ is frequency-independent. Furthermore, its value is fixed by self-duality to be the constant $\mathcal{K}$ appearing in the self-duality relation (15.79). All these remarkable results are a direct consequence of the self-duality of the U(1) Maxwell theory on AdS$_4$.

It is also possible to go beyond the simplest Einstein–Maxwell theory on AdS4, which corresponds to the $N \to \infty$ limit of SYM3. In [354], Myers *et al.* argued that the leading corrections away from $N \to \infty$ could be incorporated into corrections to the Einstein–Maxwell action for the U(1) gauge field. They showed that such corrections did induce a universal frequency dependence in $\sigma(\omega)$, so that for a wide parameter regime the results acquired features bearing some similarity to Fig. 15.4.

It is also natural to ask whether anything like the self-duality relation (15.79) applies to the $O(N)$ rotor model considered in the rest of this chapter. For $N = 2$, there have been extensive discussions in the literature of a "particle–vortex" duality [109]. Under this duality, the theory of particles given by (2.11) for the $N = 2$ rotor model maps onto a relativistic theory of vortices. However this is not a *self*-duality: the theory of vortices has an emergent U(1) gauge field, and is described by the theory of quantum electrodynamics in the presence of a complex scalar field representing the vortex excitations. Reference [219] considered the action of this particle–vortex duality on the correlation functions in (15.76), and found the following interesting relations:

$$K^L(k, \omega)\widetilde{K}^T(k, \omega) = (2\pi)^2, \quad K^T(k, \omega)\widetilde{K}^L(k, \omega) = (2\pi)^2, \quad\quad (15.81)$$

where $\widetilde{K}^{L,T}$ determine the correlators of the vortex current as in (15.76), as opposed to the particle current determined by $K^{L,T}$. Because of the lack of self-duality the vortex current correlators do not equal the particle current correlators. Unlike (15.79), (15.81) does *not* fully determine the correlation functions at $\mathbf{k}=0$: it only serves to reduce the four unknown functions $K^{L,T}$, $\widetilde{K}^{L,T}$ to two unknown functions. Thus, in general, there is no exact self-duality, and we expect $\sigma(\omega)$ to be frequency dependent. We believe that the self-duality arises *only* in the classical gravity approximation of the dual theory on AdS$_4$.

15.5.2 Implications

Let us summarize the lessons we have learnt from the AdS theory of quantum critical transport in strongly interacting systems in 2+1 dimensions. This theory should be viewed as complementary to the quasiparticle-based theory, whose implications were discussed in Section 15.4.

The lessons are:

- There is a large class of strongly interacting 2+1 dimensional "nearly perfect" fluids which are able to relax back to thermal equilibrium in a time of order $\hbar/(k_B T)$, as we indicated in (1.5). They are "nearly perfect" because this relaxation time is the shortest possible, as noted in (2.13).
- The quasiparticle transport theory of Sections 15.2 and 15.3 starts from the free theory with an infinite thermal equilibration time, and includes the effect of weak interactions using the Boltzmann equation. Complementary to this is the quantum-critical transport theory applicable for the shortest possible equilibration time of order $\hbar/(k_B T)$, which is the classical Einstein-Maxwell theory on AdS$_4$.
- The Einstein–Maxwell theory exhibits collisionless dynamics for $\omega \gg T$, and collision-dominated dynamics for $\omega \ll T$, as shown in Figs. 15.9 and 15.10.
- All continuous global symmetries are represented by a self-dual Einstein–Maxwell theory.
- This emergent self-duality implies that $\sigma(\omega)$ is frequency-independent in the Einstein–Maxwell theory and equal to the self-dual value.
- A frequency-dependent conductivity is obtained [354] upon considering corrections to the effective Einstein–Maxwell theory.
- Such "nearly perfect" fluids also have universal momentum transport. By extending the scaling arguments to momentum transport we conclude that the ratio of the shear viscosity to the entropy density η/s should equal a universal number characterizing the collision-dominated regime. This number was computed in the Einstein–Maxwell theory by Kovtun, Son, and Starinets [281] and found to equal $\hbar/(4\pi k_B)$.

It is a remarkable fact that many experiments on the superfluid–insulator transition [490] do exhibit a critical conductivity very close to the self-dual value obtained by setting

$K^{L,T} = \widetilde{K}^{L,T}$ in (15.81): i.e. $\sigma(\omega) = Q^2/h$. We suggest that this indicates a reasonable description of the low-energy physics by the Einstein–Maxwell theory.

15.6 Applications and extensions

An important application of the transport results of this chapter is for the $N = 2$ case, which describes a superfluid to insulator transition in lattice models of bosons. This connection is clearer from the discussion in Chapter 9. An intuitive understanding can be gained by returning to the lattice Hamiltonian representation in (11.1) and interpreting it as an effective Hamiltonian for a regular two-dimensional array of mesoscopic superconducting quantum dots [20, 76, 83, 121, 125, 252, 450]. For $N = 2$, there is only one component of the angular momentum operator, $\hat{L}_i$ (see (6.33)). We interpret $\hat{L}_i$ as the number operator for bosonic Cooper pairs on a superconducting quantum dot at site i, minus a fixed integer that equals the number of Cooper pairs on an isolated island (the role of this integer should be clear from Chapter 9). The term proportional to gJ in (11.1) is a caricature of the additional Coulomb energy required for deviation in the number of Cooper pairs on a dot from its optimum value. The angle, θ, defining the orientation of $\hat{n}$ (as in (6.15)), is taken as the phase of the superconducting order parameter. Then the term proportional to J represents Josephson tunneling of Cooper pairs between neighboring dots. The phase of the rotor model with long-range order in $\mathbf{n}$ represents the superfluid, while the quantum paramagnet is the Mott insulator of Cooper pairs.

A large number of experiments have measured d.c. transport on granular films, Josephson junction arrays, and homogeneously disordered superconductors undergoing a zero-temperature transition from a superconductor to an insulator; one of the earliest such experiments was carried out by Strongin *et al.* [495] and there are reviews of more recent works in [211] and [301]. However, these experiments cannot be quantitatively modeled by the simple models we have considered here because all experimental systems have an appreciable amount of randomness, and this is surely a relevant perturbation on the simple, clean, quantum critical point we have studied here. (A $d = 1 + \epsilon$ expansion for finite-temperature transport in a disordinal system has been described recently by Herbut [215, 216].) Further, we have entirely ignored fermionic excitations [447, 508], and these could be important near the critical point, although there are indications in recent simulations [166] that neglect of fermionic excitations is justified. In some interesting experiments [405], a disordered superconducting film was coupled to a tunable, dissipative metallic bath, and some initial theoretical attempts to explain the result have also appeared [450, 534]. (We briefly consider the general consequences of fermionic excitations in Chapter 18.) Finally, the long-range part of the Coulomb interaction is probably also relevant at the superfluid–insulator transition. This issue has been addressed in [146], [547], and [546]. Nevertheless, the scaling forms for the conductivity, and our general discussion on the crossover between coherent and incoherent transport at a frequency scale of order $k_B T/\hbar$, is expected to apply to these more complex systems as well.

Dynamical measurements of the conductivity at frequencies of order $k_B T / \hbar$ in systems near a superfluid–insulator transition are not yet available. However, such measurements have been made for a system near a metal–insulator transition by Lee *et al.* [290], and they nicely exhibit scaling as a function of ω / T. Related measurements [132] have also been made near quantum Hall transitions in $d = 2$, and these are again consistent with scaling as a function of ω / T. As we discussed at the conclusion of Section 14.2.2, universal time scales of order $\hbar / k_B T$ require that the quantum critical theory have nonvanishing interactions between its thermal excitations, for otherwise the interactions are "dangerously irrelevant" and the characteristic times are higher powers of $1/T$. These quantum Hall measurements therefore indicate that the noninteracting electron models for these transitions [227, 481] have to be extended to include interactions.

Recently, exciting prospects have emerged for measuring the dynamic conductivity near the superfluid–insulator transition in ultracold atom systems [164, 175].

For these ultracold atom systems, and more generally, it is essential for the theory to move beyond the strict relativistic theories considered in the present chapter. We need to allow for generic chemical potentials, where, as we showed in Chapter 9, the theory acquires deviations from the $z = 1$ critical theory. It would also be useful to allow for weak applied magnetic fields, as well as scattering off impurities. A general theory for such perturbations in the quantum critical region was presented by Hartnoll *et al.* [203]: a number of universal results were obtained in the collision-dominated regime, and verified by exact solutions using the AdS/CFT correspondence.

PART IV

OTHER MODELS

16 Dilute Fermi and Bose gases

We consider a number of different models in this chapter, but they share some important unifying characteristics. They all have a global U(1) symmetry. We are particularly interested in the behavior of the conserved density, generically denoted as Q, associated with this symmetry. All the models exhibit a quantum phase transition between two phases with a specific $T = 0$ behavior in the expectation value of Q. In one of the phases, $\langle Q \rangle$ is pinned precisely at a quantized value and does not vary as microscopic parameters are varied. This quantization ends at the quantum critical point with a discontinuity in the derivative of $\langle Q \rangle$ with respect to the tuning parameter, and $\langle Q \rangle$ varies smoothly in the other phase; there is no discontinuity in the value of $\langle Q \rangle$, however.

We have already met a transition of the above type in Chapter 9: the Mott insulator to superfluid transition at points excluding the tips of the lobes in Fig. 9.1, where the coupling K_1 in (9.34) did not vanish. In this case Q was just the boson density $\hat{b}_i^\dagger \hat{b}_i / a^d$. We will find it convenient to shift the definition of Q by a constant so that the quantized value is zero: in this case, Q equals $(\hat{b}_i^\dagger \hat{b}_i - n_0(\mu/U))/a^d$. In this chapter, we study the universal properties of the continuum theory of this transition, which, following (9.34), we write in the following form:

$$
\mathcal{Z}_B = \int \mathcal{D}\Psi_B(x, \tau) \exp\left(-\int_0^{1/T} d\tau \int d^d x \mathcal{L}_B\right),
$$

$$
\mathcal{L}_B = \Psi_B^* \frac{\partial \Psi_B}{\partial \tau} + \frac{1}{2m}|\nabla \Psi_B|^2 - \mu |\Psi_B|^2 + \frac{u_0}{2}|\Psi_B|^4. \tag{16.1}
$$

We have dropped the second-order time derivative (proportional to K_2) from (9.34) and have not included any nonlinearity beyond the quartic, as these will all be shown to be irrelevant near the transition. We have rescaled Ψ_B by a factor of the square root of K_1 so that the first-order time derivative has coefficient unity. This sets the normalization of the continuum field Ψ_B, which is always consistently maintained in this chapter. This time-derivative term is the same as that arising in the coherent state path integral for canonical bosons, where it is the "Berry phase" associated with the adiabatic evolution of the coherent states, which were described in Section 9.2.1. Physically, the Berry phase term here accounts for the Josephson precession in the phase of a condensate of the bosons in the presence of an external chemical potential. So the normalization of Ψ_B is determined by its Berry phase, a feature we see in other models. With the above rescaling of Ψ_B it is easy to see from (9.37) and (9.34) that, close to the quantum critical point, the coefficient of $|\Psi_B|^2$ is the negative of the chemical potential, μ, up to an additive constant. We absorb this unimportant additive constant into a redefinition of μ, and this leads to the $|\Psi_B|^2$ term shown in (16.1). We can also identify the charge Q with $\Psi_B^* \Psi_B$ as

$$\langle Q \rangle = -\frac{\partial \mathcal{F}_B}{\partial \mu} = \langle |\Psi_B|^2 \rangle, \tag{16.2}$$

with $\mathcal{F}_B = -(T/V) \ln \mathcal{Z}_B$. With the form of the quadratic term in (16.1), we also see from the mean-field results in Chapter 9 that the quantum critical point is precisely at $\mu = 0$ and $T = 0$. We see in this chapter that there are *no* fluctuation corrections to this location from the terms in $\mathcal{L}_B$ (the K_2 term in (9.34) does lead to shifts in the position of the quantum critical point, but we have already set it to zero here as it is not important for the critical theory). So at $T = 0$, $\langle Q \rangle$ takes the quantized value $\langle Q \rangle = 0$ for $\mu < 0$, and $\langle Q \rangle > 0$ for $\mu > 0$; we are particularly interested in the nature of the onset at $\mu = 0$ and finite-T crossovers in its vicinity. We have also assumed here that $K_1 > 0$, and so, from (9.37) and (9.40), $\langle Q \rangle$ increases from its quantized value away from the quantum critical point. The opposite case of decreasing Q can be treated after a particle–hole transformation and has essentially identical properties.

After our study of $\mathcal{Z}_B$, we find it useful to introduce a closely related model that also displays a quantum phase transition with the same behavior in a conserved U(1) density $\langle Q \rangle$ and has many similarities in its physical properties. The model is exactly solvable and is expressed in terms of a continuum canonical spinless fermion field Ψ_F; its partition function is

$$\mathcal{Z}_F = \int \mathcal{D}\Psi_F(x, \tau) \exp\left(-\int_0^{1/T} d\tau \int d^d x \mathcal{L}_F\right),$$

$$\mathcal{L}_F = \Psi_F^* \frac{\partial \Psi_F}{\partial \tau} + \frac{1}{2m} |\nabla \Psi_F|^2 - \mu |\Psi_F|^2. \tag{16.3}$$

The functional integral is over fluctuations of an anticommuting Grassman field $\Psi_F(x, \tau)$ (see the discussion in [357] for an introduction to Grassman numbers and their functional integrals). Note that the terms in $\mathcal{L}_F$ are in one-to-one correspondence with those in $\mathcal{L}_B$ in (16.1), except that there is no quartic $|\Psi_F|^4$ term. Such a term vanishes because the square of a Grassman number is zero, which is just a mathematical representation of the Pauli exclusion principle. As a result, $\mathcal{L}_F$ is just a free-field theory. Like $\mathcal{Z}_B$, $\mathcal{Z}_F$ has a quantum critical point at $\mu = 0$, $T = 0$ and we discuss its properties in this chapter; in particular, we show that all possible fermionic nonlinearities are irrelevant near it. The reader should not be misled by the apparently trivial nature of the model in (16.3); using the theory of quantum phase transitions to understand free fermions might seem like technological overkill. We will see that $\mathcal{Z}_F$ exhibits crossovers that are quite similar to those near far more complicated quantum critical points, and observing them in this simple context leads to considerable insight.

In a general spatial dimension, d, the continuum theories $\mathcal{Z}_B$ and $\mathcal{Z}_F$ have different, though closely related, universal properties. However, we argue here that the quantum critical points of these theories are *exactly* equivalent in $d = 1$. This is one of the important results of this chapter. We will see that the bosonic theory $\mathcal{Z}_B$ is very strongly coupled in $d = 1$, and we present compelling evidence that the solvable fermionic theory $\mathcal{Z}_F$ is its exactly universal solution in the vicinity of the $\mu = 0$, $T = 0$ quantum critical point. We are also able to make a correspondence between the operators of the two theories, and this allows us to obtain certain exact results for experimentally measurable bosonic correlation functions of $\mathcal{Z}_B$, including some for the nonzero-temperature dynamical properties that are an important focus of this book. Of course, all fermionic correlators of $\mathcal{Z}_F$ are exactly known in arbitrary d, but these are not of significant practical interest.

The last major topic of this section is a discussion of the dilute spinful Fermi gas in Section 16.4. This generalizes $\mathcal{Z}_F$ to a spin $S = 1/2$ fermion $\Psi_{F\sigma}$, with $\sigma = \uparrow, \downarrow$. Now Fermi statistics do allow a contact quartic interaction, and so we have

$$\mathcal{Z}_{Fs} = \int \mathcal{D}\Psi_{F\uparrow}(x, \tau)\mathcal{D}\Psi_{F\downarrow}(x, \tau) \exp\left(-\int_0^{1/T} d\tau \int d^d x\, \mathcal{L}_{Fs}\right),$$

$$\mathcal{L}_{Fs} = \Psi_{F\sigma}^* \frac{\partial \Psi_{F\sigma}}{\partial \tau} + \frac{1}{2m}|\nabla \Psi_{F\sigma}|^2 - \mu|\Psi_{F\sigma}|^2 + u_0 \Psi_{F\uparrow}^* \Psi_{F\downarrow}^* \Psi_{F\downarrow} \Psi_{F\uparrow}. \qquad (16.4)$$

This theory conserves fermion number, and has a phase transition as a function of increasing μ from a state with fermion number 0 to a state with nonzero fermion density. However, unlike the above two cases of $\mathcal{Z}_B$ and $\mathcal{Z}_F$, the transition is not always at $\mu = 0$. The problem defined in (16.4) has recently found remarkable experimental applications in the study of ultracold gases of fermionic atoms. These experiments are also able to tune the value of the interaction u_0 over a wide range of values, extended from repulsive to attractive. For the attractive case, the two-particle scattering amplitude has a *Feshbach resonance* where the scattering length diverges, and we obtain the unitarity limit. This Feshbach resonance plays a crucial role in the phase transition obtained by changing μ, and leads to a rich phase diagram of the so-called "unitary Fermi gas."

Our treatment of $\mathcal{Z}_{Fs}$ in the important experimental case of $d = 3$ shows that it defines a strongly coupled field theory in the vicinity of the Feshbach resonance for attractive interactions. It therefore pays to find alternative formulations of this regime of the unitary Fermi gas. One powerful approach is to promote the two-fermion bound state to a separate canonical Bose field. This yields a model, $\mathcal{Z}_{FB}$, with both elementary fermions and bosons; i.e. it is a combination of $\mathcal{Z}_B$ and $\mathcal{Z}_{Fs}$ with interactions between the fermions and bosons. We define $\mathcal{Z}_{FB}$ in Section 16.4, and use it to obtain a number of experimentally relevant results for the unitary Fermi gas.

We begin in Section 16.1 by discussing a simple solvable model in $d = 1$: the spin-1/2 quantum XX chain. This allows us to motivate the physical origin of the fermionic theory $\mathcal{Z}_F$ and indicate the relationship between $\mathcal{Z}_B$ and $\mathcal{Z}_F$ in the context of a lattice model. Then Section 16.2 presents a thorough discussion of the universal properties of $\mathcal{Z}_F$. This is followed by an analysis of $\mathcal{Z}_B$ in Section 16.3, where we use renormalization group methods to obtain perturbative predictions for universal properties. The perturbation theory for $\mathcal{Z}_B$ becomes strongly coupled in $d = 1$, but we are able to obtain exact results for this case by the $d = 1$ mapping between $\mathcal{Z}_B$ and $\mathcal{Z}_F$. This is discussed in Section 16.3.3. This section also contains further discussion of the properties of the XX chain of Section 16.1. The spinful Fermi gas is discussed in Section 16.4.

16.1 The quantum *XX* model

This model is obtained by taking the $U \to \infty$ limit of the boson Hubbard model H_B in (9.4). This is then a model of "hard-core" bosons with an infinite on-site repulsion energy. The only states with a finite energy are those with $\hat{n}_{bi} = 0$ or 1 on every site of the lattice. The Mott insulating states in Fig. 9.1 with $n_0 > 1$ have therefore been expelled, and only the two Mott insulators with $n_0 = 0$ or $n_0 = 1$ are permitted. Precisely at $w = 0$, we have the $n_0 = 1$ Mott insulator for $\mu > 0$, while for $\mu < 0$ we have the $n_0 = 0$ Mott insulator, which is a fanciful term for a bare vacuum with no particles.

This model of hard-core bosons can also be written as a magnet of $S = 1/2$ spins with nearest-neighbor exchange interactions. The idea is to associate the two states on each site with the up and down states of an $S = 1/2$ spin degree of freedom. In operator language, we can identify

$$\hat{\sigma}_j^x = \hat{b}_j + \hat{b}_j^\dagger,$$

$$\hat{\sigma}_j^y = -i\left(\hat{b}_j - \hat{b}_j^\dagger\right),$$

$$\hat{\sigma}_j^z = 1 - 2\hat{b}_j^\dagger \hat{b}_j. \tag{16.5}$$

Then the boson commutation relations (9.1) and the hard-core restriction imply that the $\hat{\sigma}_j^{x,y,z}$ obey the commutation relations of the Pauli matrices and satisfy $\hat{\sigma}_j^{\alpha 2} = 1$ (no sum over α). We may therefore consider them to be the Pauli matrices. With this mapping, the fully polarized state with all spins up is the $n_0 = 0$ Mott insulator, while that with all spins down is the $n_0 = 1$ Mott insulator. Inverting (16.5), we see that the Hamiltonian H_B in (9.4) becomes (up to an uninteresting additive constant)

$$H_{XX} = -\frac{w}{2} \sum_{\langle ij \rangle} \left(\hat{\sigma}_i^x \hat{\sigma}_j^x + \hat{\sigma}_i^y \hat{\sigma}_j^y\right) + \frac{\mu}{2} \sum_i \hat{\sigma}_i^z. \tag{16.6}$$

This is the so-called XX model; it describes spin-1/2 degrees of freedom on the lattice sites with a nearest-neighbor ferromagnetic exchange $w/2 > 0$ confined to the x–y plane in spin space, and in a magnetic field $\mu/2$ in the $-z$ direction in spin space. We argue later that additional exchange in the z direction in spin space (this corresponds to nearest-neighbor interactions in the boson Hubbard model) does not modify the universal properties of the Mott insulator to superfluid transitions. Note also that both the simple fully polarized $n_0 = 0$ and $n_0 = 1$ Mott insulators are exact eigenstates of H_{XX} for arbitrary w. For the $n_0 = 1$ state this is a consequence of having sent $U \to \infty$, which eliminates virtual particle–hole pair fluctuations.

Now we specialize to the one-dimensional case, $d = 1$. In this case exact expressions for the thermodynamic properties of H_{XX} can be easily obtained. The basic tool is the Jordan–Wigner transformation introduced in Section 10.1 for the solution of the Ising chain in

$d = 1$. This transforms the spin-1/2 model into a model of spinless fermions. Inserting (10.8) and (10.9) into (16.6), we get

$$H_{XX} = -\sum_i \left(w(c_{i+1}^\dagger c_i + c_i^\dagger c_{i+1}) + \mu c_i^\dagger c_i \right). \tag{16.7}$$

Note that H_{XX} is simply a free spinless fermion Hamiltonian and its spectrum can therefore be easily determined. Adding non-on-site interactions to the original H_B would lead to fermion interactions in H_{XX}, which are shown below to be irrelevant. Fourier transforming as in (10.15) we get the simple diagonal form

$$H_{XX} = \sum_k \varepsilon_k c_k^\dagger c_k, \tag{16.8}$$

with the free fermion dispersion $\varepsilon_k = -2w\cos(ka) - \mu$. So for $\mu < -2w$, the energy of all the fermions is positive and the ground state has no fermions present: this is clearly the Mott insulator with $n_0 = 0$. For $\mu > 2w$ all the fermions have negative energy and every fermion state is occupied, leading to the Mott insulator with $n_0 = 1$. At intermediate values of μ there is partial occupation, which can be easily computed at $T = 0$:

$$\langle \hat{b}_i^\dagger \hat{b}_i \rangle = \frac{1}{2}\left(1 - \langle \hat{\sigma}_i^z \rangle\right) = \langle c_i^\dagger c_i \rangle = \begin{cases} 0, & \mu \leq -2w, \\ 1 - (1/\pi)\cos^{-1}(\mu/2w), & |\mu| \leq 2w, \\ 1, & \mu \geq 2w. \end{cases} \tag{16.9}$$

We show a plot of the boson number as a function of μ in Fig. 16.1. The state with intermediate occupation number has a nonzero superfluid stiffness but only "quasi-long-range order" in the superfluid order parameter in $d = 1$, as discussed in Section 16.3.3. We continue to refer to it as a superfluid, however. Hence the result (16.9) displays two superfluid–Mott insulator transitions: one at $\mu = -2w$ and the other at $\mu = 2w$. We focus

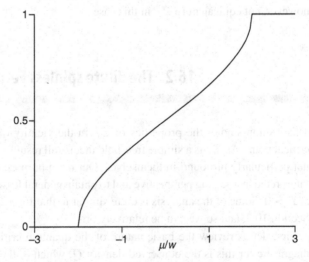

Fig. 16.1 The boson number per site as a function of the chemical potential μ for the $U \to \infty$ limit of the boson Hubbard model H_B in (9.4) in dimension $d = 1$. There are Mott insulator to superfluid transitions at $\mu = \pm 2w$.

on the one at $\mu = -2w$, where the transition is from a simple vacuum state with no particles (a Mott insulator with $n_0 = 0$) to a low-density superfluid.

For μ close to $2w$, the density of bosons is seen to vanish from (16.9) as $\sim (1+\mu/2w)^{1/2}$. We may identify the power of $1/2$ as an exact critical exponent of the quantum critical point at $\mu = -2w$. Compare this with the mean-field result (9.40), which has the value 1 for this critical exponent. We see that the mean-field result applies for $d > 2$.

We can derive a continuum theory for the quantum critical point at $\mu = -2w$, $T = 0$ using an analysis very similar to that in Section 10.2. The low-energy fermionic states that are occupied across the transition are near $k = 0$. Therefore we may take the continuum limit simply by taking spatial gradients of the fields. We define the continuum field Ψ_F as in (10.23), and we expand H_{XX} in spatial gradients. This leads to the Hamiltonian

$$H_F = \int dx \left(-\frac{1}{2m} \Psi_F^\dagger(x) \nabla^2 \Psi_F(x) - \mu \Psi_F^\dagger(x) \Psi_F(x) \right), \tag{16.10}$$

where the fermion mass $m = 1/(2wa^2)$. The coherent state path integral of H_F is, of course, the fermionic theory $\mathcal{Z}_F$ (16.3).

We have thus presented evidence that the critical theory of the transition in the XX model in $d = 1$ is given by $\mathcal{Z}_F$. A complete demonstration requires that there are no further relevant perturbations that can appear in H_F, and this is taken up in the following section. Recall also that H_{XX} was derived from the boson Hubbard model (9.4), which was shown to be related to $\mathcal{Z}_B$ in (16.1) in Chapter 9. These mappings therefore equate the universal critical properties of $\mathcal{Z}_B$, H_{XX}, and $\mathcal{Z}_F$ in $d = 1$. These universal correlators are described explicitly in the subsequent sections.

For dimensions $d > 1$ the analysis of this section and the arguments of Chapter 9 have established that H_{XX} has Mott insulator–superfluid transitions (or transitions between fully polarized and partially polarized spin states) that are described by $\mathcal{Z}_B$. These models are, however, not equivalent to $\mathcal{Z}_F$ in this case.

16.2 The dilute spinless Fermi gas

This section studies the properties of $\mathcal{Z}_F$ in the vicinity of its $\mu = 0$, $T = 0$ quantum critical point. As $\mathcal{Z}_F$ is a simple free-field theory, all results can be obtained exactly and are not particularly profound in themselves. Our main purpose is to show how the results are interpreted in a scaling perspective and to obtain general lessons on the nature of crossovers at $T > 0$. Some of the analysis is quite similar to that for a different free fermion theory in Section 10.2, and so we can be relatively brief.

First, let us review the basic nature of the quantum critical point at $T = 0$. A useful diagnostic for this is the conserved density Q, which in the present model we identify as $\Psi_F^\dagger \Psi_F$. As a function of the tuning parameter μ, this quantity has a critical singularity at $\mu = 0$:

$$\langle \Psi_F^\dagger \Psi_F \rangle = \begin{cases} (S_d/d)(2m\mu)^{d/2}, & \mu > 0, \\ 0, & \mu < 0, \end{cases} \tag{16.11}$$

where the phase space factor S_d was defined below (14.9). When $d = 1$, this result is clearly the universal continuum limit of (16.9).

We now proceed to a scaling analysis. Note that at the quantum critical point $\mu = 0$, $T = 0$, the theory $\mathcal{L}_F$ is invariant under the scaling transformations closely related to those in (10.30):

$$x' = xe^{-\ell},$$
$$\tau' = \tau e^{-z\ell},$$
$$\Psi_F' = \Psi_F e^{d\ell/2}, \tag{16.12}$$

provided that we make the choice of the dynamic exponent

$$z = 2. \tag{16.13}$$

The parameter m is assumed to remain invariant under the rescaling, and its role is simply to ensure that the relative physical dimensions of space and time are compatible (a role rather analogous to that of the velocity c in Section 10.2). The transformation (16.12) also identifies the scaling dimension

$$\dim[\Psi_F] = d/2. \tag{16.14}$$

Now turning on a nonzero μ, it is easy to see that μ is a relevant perturbation with

$$\dim[\mu] = 2. \tag{16.15}$$

There are no other relevant perturbations at this quantum critical point, and so by the definition of ν above (10.33) we have

$$\nu = 1/2. \tag{16.16}$$

We can now examine the consequences of adding interactions to $\mathcal{L}_F$. A contact interaction such as $\int dx (\Psi_F^\dagger(x)\Psi_F(x))^2$ vanishes because of the fermion anticommutation relation. (A contact interaction is however permitted for a spin-1/2 Fermi gas and is discussed in Section 16.4.) The simplest allowed term for the spinless Fermi gas is

$$\mathcal{L}_1 = \lambda \big(\Psi_F^\dagger(x, \tau)\nabla\Psi_F^\dagger(x, \tau)\Psi_F(x, \tau)\nabla\Psi_F(x, \tau) \big), \tag{16.17}$$

where λ is a coupling constant measuring the strength of the interaction. However, a simple analysis shows that

$$\dim[\lambda] = -d. \tag{16.18}$$

This is negative and so λ is irrelevant and can be neglected in the computation of universal crossovers near the point $\mu = T = 0$. In particular, it modifies the result (16.11) only by contributions that are higher order in μ. The arguments show the sense in which the fermionic theory $\mathcal{L}_F$ is the universal critical theory describing the phase transition in H_{XX} in $d = 1$. Additional exchange couplings in the z direction, or further neighbor interactions, can only lead to terms like that in (16.17), and all of these are irrelevant.

Turning to nonzero temperatures, we can write down scaling forms by the same arguments that led to (10.39). Let us define the fermion Green's function

$$G_F(x, t) = \langle \Psi_F(x, t) \Psi_F^\dagger(0, 0) \rangle; \tag{16.19}$$

then the scaling dimensions above imply that it satisfies

$$G_F(x, t) = (2mT)^{d/2} \Phi_{G_F}\left((2mT)^{1/2}x, Tt, \frac{\mu}{T}\right), \tag{16.20}$$

where Φ_{G_F} is a fully universal scaling function. For this particularly simple theory $\mathcal{L}_F$ we can of course obtain the result for G_F in closed form:

$$G_F(x, t) = \int \frac{d^d k}{(2\pi)^d} \frac{e^{ikx - i(k^2/(2m) - \mu)t}}{1 + e^{-(k^2/(2m) - \mu)/T}}, \tag{16.21}$$

and it is easy to verify that this obeys the scaling form (16.20). Similarly the free energy $\mathcal{F}_F$ obeys the scaling dimension (10.37), and we have

$$\mathcal{F}_F = T^{d/2+1} \Phi_{\mathcal{F}_F}\left(\frac{\mu}{T}\right), \tag{16.22}$$

with $\Phi_{\mathcal{F}_F}$ a universal scaling function; the explicit result is, of course,

$$\mathcal{F}_F = -T \int \frac{d^d k}{(2\pi)^d} \ln\left(1 + e^{(\mu - k^2/(2m))/T}\right), \tag{16.23}$$

which clearly obeys (16.22). The crossover behavior of the fermion density

$$\langle Q \rangle = \langle \Psi_F^\dagger \Psi_F \rangle = -\frac{\partial \mathcal{F}_F}{\partial \mu} \tag{16.24}$$

follows by taking the appropriate derivative of the free energy. Examination of these results leads to the now familiar crossover phase diagram of Fig. 16.2. We examine each of the regions of the phase diagram in turn, beginning with the two low-temperature regions.

16.2.1 Dilute classical gas, $k_B T \ll |\mu|$, $\mu < 0$

The ground state for $\mu < 0$ is the vacuum with no particles. Turning on a nonzero temperature produces particles with a small nonzero density $\sim e^{-|\mu|/T}$. The de Broglie wavelength of the particles is of order $T^{-1/2}$, which is significantly smaller than the mean spacing between the particles, which diverges as $e^{|\mu|/dT}$ as $T \to 0$. This implies that the particles behave semiclassically. These properties are quite similar to those of the low-T region on the quantum paramagnetic side of the Ising chain in Section 10.4.2. To leading order from (16.21), the fermion Green's function is simply the Feynman propagator of a single particle

$$G_F(x, t) = \left(\frac{m}{2\pi i t}\right)^{d/2} \exp\left(-\frac{imx^2}{2t}\right), \tag{16.25}$$

and the exclusion of states from the other particles has only an exponentially small effect. Note that G_F is independent of μ and T and (16.25) is the exact result for $\mu = T = 0$. The free energy, from (16.22) and (16.23), is that of a classical Boltzmann gas

$$\mathcal{F}_F = -T \left(\frac{mT}{2\pi}\right)^{d/2} e^{-|\mu|/T}. \tag{16.26}$$

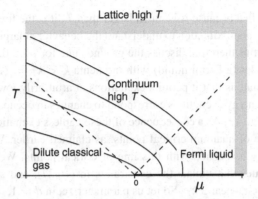

Fig. 16.2   Phase diagram of the dilute Fermi gas $\mathcal{Z}_F$, (16.3), as a function of the chemical potential μ and the temperature T. The regions are separated by crossovers denoted by dashed lines, and their physical properties are discussed in the text. The full lines are contours of equal density, with higher densities above lower densities; the zero density line is $\mu < 0, T = 0$. The line $\mu > 0, T = 0$ is a line of $z = 1$ critical points that controls the longest scale properties of the low-T Fermi liquid region. The critical end point $\mu = 0, T = 0$ has $z = 2$ and controls the global structure of the phase diagram. In $d = 1$, the Fermi liquid is more appropriately labeled a Tomonaga–Luttinger liquid. The hatched region marks the boundary of applicability of the continuum theory and occurs at $\mu, T \sim w$.

16.2.2 Fermi liquid, $k_B T \ll \mu, \mu > 0$

The behavior in this regime is quite complex and rich. As we see, and as noted in Fig. 16.2, the line $\mu > 0$, $T = 0$ is itself a line of quantum critical points. The interplay between these critical points and those of the $\mu = 0$, $T = 0$ critical end point is displayed quite instructively in the exact results for G_F and is worth examining in detail. It must be noted that the scaling dimensions and critical exponents of these two sets of critical points need not be (and indeed are not) the same. The concept of a *reduced scaling function*, used earlier (e.g. in Section 10.4.1 for the quantum Ising chain) to describe the emergence of effective classical models now comes in useful to obtain the critical behavior of the $\mu > 0$, $T = 0$ critical line out of the global scaling functions of the $\mu = 0$, $T = 0$ critical end point. Precisely the same structure is also present in the physically measurable bosonic correlators of $\mathcal{Z}_B$ in $d = 1$ (discussed in Section 16.3.3) but there the results are far more complicated and only available in restricted regimes. In this case the closed form results (16.21) and (16.23) contain all the structure, and so these are worth examining explicitly.

First it can be argued, for example, by studying asymptotics of the integral in (16.21), that for very short times or distances, the correlators do not notice the consequences of other particles present because of a nonzero T or μ and are therefore given by the single-particle propagator, which is the $T = \mu = 0$ result in (16.25). More precisely we have

$$G(x, t) \text{ is given by (16.25) for } |x| \ll (2m\mu)^{-1/2}, \quad |t| \ll \frac{1}{\mu}. \tag{16.27}$$

With increasing x or t, the restrictions in (16.27) are eventually violated and the consequences of the presence of other particles, resulting from a nonzero μ, become apparent.

Note that because μ is much larger than T, it is the first energy scale to be noticed, and as a first approximation to understand the behavior at larger x we may ignore the effects of T.

Let us therefore discuss the ground state for $\mu > 0$. It consists of a filled Fermi sea of particles (a Fermi liquid) with momenta $k < k_F = (2m\mu)^{1/2}$. An important property of this state is that it permits excitations at arbitrarily low energies (i.e. it is *gapless*). These low-energy excitations correspond to changes in occupation number of fermions arbitrarily close to k_F. As a consequence of these gapless excitations, the points $\mu > 0$ ($T = 0$) form a line of quantum critical points, as claimed earlier. We now derive the continuum field theory associated with this line of critical points. We are interested here only in x and t values that violate the constraints in (16.27), and so in the occupation of states with momenta near $\pm k_F$. So let us parameterize, in $d = 1$,

$$\Psi(x, \tau) = e^{ik_F x}\Psi_R(x, \tau) + e^{-ik_F x}\Psi_L(x, \tau), \tag{16.28}$$

where $\Psi_{R,L}$ describe right- and left-moving fermions and are fields that vary slowly on spatial scales $\sim 1/k_F = (1/2m\mu)^{1/2}$ and temporal scales $\sim 1/\mu$. A similar parameterization can be used for $d > 1$, and we discuss it in Section 18.1; most of the results discussed below hold, with small modifications, in all d. Inserting the above parameterization in $\mathcal{L}_F$, and keeping only terms of lowest order in spatial gradients, we obtain the "effective" Lagrangean for the Fermi liquid region, $\mathcal{L}_{FL}$ in $d = 1$:

$$\mathcal{L}_{FL} = \Psi_R^{\dagger}\left(\frac{\partial}{\partial\tau} - iv_F\frac{\partial}{\partial x}\right)\Psi_R + \Psi_L^{\dagger}\left(\frac{\partial}{\partial\tau} + iv_F\frac{\partial}{\partial x}\right)\Psi_L, \tag{16.29}$$

where $v_F = k_F/m = (2\mu/m)^{1/2}$ is the Fermi velocity. The Lagrangean $\mathcal{L}_{FL}$ also describes a massless Dirac field in one spatial dimension and (like (10.28) for $\Delta = 0$) is invariant under relativistic and conformal transformations of spacetime. (These facts are of some use to us later.) Now note that $\mathcal{L}_{FL}$ is invariant under a scaling transformation, which is rather different from (16.12) for the $\mu = 0$, $T = 0$ quantum critical point:

$$x' = xe^{-\ell},$$

$$\tau' = \tau e^{-\ell},$$

$$\Psi'_{R,L}(x', \tau') = \Psi_{R,L}(x, \tau)e^{\ell/2},$$

$$v'_F = v_F. \tag{16.30}$$

The above results imply

$$z = 1, \tag{16.31}$$

unlike $z = 2$ (16.13) at the $\mu = 0$ critical point, and

$$\dim[\Psi_{R,L}] = 1/2, \tag{16.32}$$

which actually holds for all d and therefore differs from (16.14). Further note that v_F, and therefore μ, are *invariant* under rescaling, unlike (16.15) at the $\mu = 0$ critical point. Thus v_F plays a role rather analogous to that of m at the $\mu = 0$ critical point: it is simply the physical units of spatial and length scales. The transformations (16.30) show that $\mathcal{L}_{LF}$ is scale invariant for each value of μ, and we therefore have a line of quantum critical points as claimed earlier. It should also be emphasized that the scaling dimension of interactions such as λ will also change; in particular not all interactions are irrelevant about the $\mu \neq 0$ critical points. These new interactions are, however, small in magnitude provided μ is small (i.e. provided we are within the domain of validity of the global scaling forms (16.20) and (16.22)), and so we neglect them here. Their main consequence is to change the scaling dimension of certain operators, but they preserve the relativistic and conformal invariance of $\mathcal{L}_{FL}$. This more general theory of $d = 1$ fermions is known as a Tomonaga–Luttinger liquid, and we discuss it in Chapter 20.

The action (16.29) and the scaling transformations (16.30) can be considered as defining scaling forms on their own, independent of any derivation from the original $\mathcal{L}_F$. By complete analogy with the arguments presented earlier, we may deduce that

$$G_{R,L}(x,t) = \left\langle \Psi_{R,L}(x,t)\Psi_{R,L}^{\dagger}(0,0)\right\rangle$$

$$= \left(\frac{T}{v_F}\right)\phi_{R,L}\left(\frac{Tx}{v_F}, Tt\right), \qquad (16.33)$$

where the powers of T follow from the scaling dimensions of G, x, and t; the factors of v_F merely keep track of physical units; and $\phi_{R,L}$ are universal scaling functions. The result is a reduced scaling form of (16.20) in the sense of the discussion in Section 10.4.1; the former has three arguments, and in the limit $\mu/T \to \infty$ it collapses into (16.33), which is itself described by the quantum critical theory (16.29).

Explicit expressions for $G_{R,L}$ can of course easily be obtained from the definition (16.33) and the theory $\mathcal{L}_{FL}$ in (16.29); however, let us proceed from an instructive derivation from the globally valid expression (16.21). For $|x| \gg (1/2m\mu)^{1/2}$, $|t| \gg 1/\mu$, and $T \ll \mu$, the integral in (16.21) is dominated by contributions near the Fermi points $k = \pm k_F$. So, near k_F let us parameterize $k = k_F + p$, expand terms in the integrand to linear order in p, and to leading order let the integral extend over all real p; a similar procedure can be carried out near $-k_F$. In this manner the expression (16.21) for G_F reduces to

$$G(x,\tau) = e^{ik_F x}\int_{-\infty}^{\infty}\frac{dp}{2\pi}\frac{e^{p(ix-v_F\tau)}}{1+e^{-v_Fp/T}} + e^{-ik_Fx}\int_{-\infty}^{\infty}\frac{dp}{2\pi}\frac{e^{p(-ix-v_F\tau)}}{1+e^{-v_Fp/T}}. \qquad (16.34)$$

The integrals over p can be evaluated exactly, and we obtain

$$G_F(x,t) = e^{ik_Fx}G_R(x,t) + e^{-ik_Fx}G_L(x,t), \qquad (16.35)$$

with

$$G_{R,L}(x,\tau) = \left(\frac{T}{v_F}\right)\frac{1}{2\sin(\pi T(\tau \mp ix/v_F))}. \qquad (16.36)$$

This result is clearly consistent with the scaling form (16.33). For $T > 0$, the equal-time $G_{R,L}$ decay exponentially with a correlation length $\xi = v_F/\pi T$, and the power of T is consistent with the $z = 1$ dynamic exponent of $\mathcal{L}_{FL}$. At $T = 0$, these fermionic

Green's functions take the scale-invariant power-law decay characteristic of the $\mu > 0$ critical ground state,

$$G_{R,L}(x, \tau) = \frac{1}{2\pi v_F(\tau \mp ix/v_F)}; \qquad (16.37)$$

note that this is consistent with the scaling transformations (16.30). Note also that the $T > 0$ result (16.36) and the $T = 0$ result (16.37) of $\mathcal{L}_{FL}$ are related by the mapping (10.47) (with the replacement $c \to v_F$) asserted to be a general property of conformally invariant theories with $z = 1$ in $d = 1$.

16.2.3 High-T limit, $k_B T \gg |\mu|$

This is the last, and in many ways the most interesting, region of Fig. 16.2. Now T is the most important energy scale controlling the deviation from the $\mu = 0$, $T = 0$ quantum critical point, and the properties will therefore have some similarities to the continuum high-T regions discussed in Part II. As always, it should be emphasized that while the value of T is significantly larger than $|\mu|$, it cannot be so large that it exceeds the limits of applicability for the continuum action $\mathcal{L}_F$: this implies that $T \ll w$.

We discuss first the behavior of the fermion density. In the high-T limit of the continuum theory $\mathcal{L}_F$, $|\mu| \ll T \ll w$, we have from (16.23) and (16.24) the universal result

$$\langle \Psi_F^\dagger \Psi_F \rangle = (2mT)^{d/2} \int \frac{d^d y}{(2\pi)^d} \frac{1}{e^{y^2} + 1}$$

$$= (2mT)^{d/2} \zeta(d/2) \frac{(1 - 2^{d/2})}{(4\pi)^{d/2}}. \qquad (16.38)$$

This density implies an interparticle spacing that is of order the de Broglie wavelength $= (1/2mT)^{1/2}$. Hence, thermal and quantum effects are equally important, and neither dominates, as we found in corresponding regions in Chapters 10, 11, and 13.

For completeness, let us also consider the fermion density for $T \gg w$ (the region above the hatched marks in Fig. 16.2), to illustrate the limitations on the continuum description discussed above. Now the result depends upon the details of the nonuniversal fermion dispersion: on a hypercubic lattice with dispersion $\epsilon_k - \mu$, we obtain

$$\langle \Psi_F^\dagger \Psi_F \rangle = \int_{-\pi/a}^{\pi/a} \frac{d^d k}{(2\pi)^d} \frac{1}{e^{(\varepsilon_k - \mu)/T} + 1}$$

$$= \frac{1}{2a^d} - \frac{1}{4T} \int_{-\pi/a}^{\pi/a} \frac{d^d k}{(2\pi)^d} (\varepsilon_k - \mu) + \mathcal{O}(1/T^2). \qquad (16.39)$$

The limits on the integration, which extends from $-\pi/a$ to π/a for each momentum component, had previously been sent to infinity in the continuum limit $a \to 0$. In the presence of lattice cutoff, we are able to make a naive expansion of the integrand in powers of $1/T$, and the result therefore only contains negative integer powers of T. Contrast this with the universal continuum result (16.38), where we had noninteger powers of T dependent upon the scaling dimension of Ψ.

We return to the universal high-T region, $|\mu| \ll T \ll w$, and describe the behavior of the fermionic Green's function G_F, given in (16.21). At the shortest scales we again have the free quantum particle behavior of the $\mu = 0$, $T = 0$ critical point:

$$G_F(x, t) \text{ is given by (16.25) for } |x| \ll (2mT)^{-1/2}, \quad |t| \ll \frac{1}{T}. \tag{16.40}$$

Note that the limits on x and t in (16.40) are different from those in (16.27), in that they are determined by T and not μ. At larger $|x|$ or t the presence of the other thermally excited particles becomes apparent, and G_F crosses over to a novel behavior characteristic of the high-T region. We illustrate this by looking at the large-x asymptotics of the equal-time G in $d = 1$ (other d are quite similar):

$$G_F(x, 0) = \int \frac{dk}{2\pi} \frac{e^{ikx}}{1 + e^{-k^2/2mT}}. \tag{16.41}$$

For large x this can be evaluated by a contour integration, which picks up contributions from the poles at which the denominator vanishes in the complex k plane. The dominant contributions come from the poles closest to the real axis, and give the leading result

$$G_F(|x| \to \infty, 0) = -\left(\frac{\pi^2}{2mT}\right)^{1/2} \exp\left(-(1 - i)(m\pi T)^{1/2} x\right). \tag{16.42}$$

Thermal effects therefore lead to an exponential decay of equal-time correlations, with a correlation length $\xi = (m\pi T)^{-1/2}$. Note that the T dependence is precisely that expected from the exponent $z = 2$ associated with the $\mu = 0$ quantum critical point and the general scaling relation $\xi \sim T^{-1/z}$. The additional oscillatory term in (16.42) is a reminder that quantum effects are still present at the scale ξ, which is clearly of order the de Broglie wavelength of the particles.

16.3 The dilute Bose gas

This section studies the universal properties of the quantum phase transition of the dilute Bose gas model $\mathcal{Z}_B$ in (16.1) in general dimensions. We begin with a simple scaling analysis that shows that $d = 2$ is the upper-critical dimension. The first subsection analyzes the case $d < 2$ in some more detail, while the next subsection considers the somewhat different properties in $d = 3$. Some of the results of this section were also obtained by Kolomeisky and Straley [273, 274].

We begin with the analog of the simple scaling considerations presented at the beginning of Section 16.2. At the coupling $u = 0$, the $\mu = 0$ quantum critical point of $\mathcal{L}_B$ is invariant under the transformations (16.12), after the replacement $\Psi_F \to \Psi_B$, and we have as before $z = 2$ and

$$\dim[\Psi_B] = d/2, \qquad \dim[\mu] = 2; \tag{16.43}$$

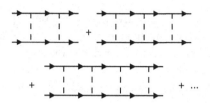

Fig. 16.3   The ladder series of diagrams that contribute to the renormalization of the coupling u in $\mathcal{Z}_B$ for $d < 2$.

these results are shortly seen to be exact in all d. We can easily determine the scaling dimension of the quartic coupling u at the $u = 0$, $\mu = 0$ fixed point under the bosonic analog of the transformations (16.12); we find

$$\dim[u_0] = 2 - d. \tag{16.44}$$

Thus the free-field fixed point is stable for $d > 2$, in which case it is suspected that a simple perturbative analysis of the consequences of u will be adequate. However, for $d < 2$, a more careful renormalization group-based resummation of the consequences of u is required, for reasons similar to those presented in Section 3.3.2 for the case of the quantum Ising/rotor models. This identifies $d = 2$ as the upper-critical dimension of the present quantum critical point.

Our analysis of the case $d < 2$ for the dilute Bose gas quantum critical point is very similar to that in Section 14.1.2. However, we find, somewhat surprisingly, that all the renormalizations, and the associated flow equations, can be determined exactly in closed form. We begin by considering the one-loop renormalization of the quartic coupling u_0 at the $\mu = 0$, $T = 0$ quantum critical point. It turns out that only the ladder series of Feynman diagrams shown in Fig. 16.3 need be considered (the T matrix). Evaluating the first term of the series in Fig. 16.3 for the case of zero external frequency and momenta, we obtain the contribution

$$-u_0^2 \int \frac{d\omega}{2\pi} \int \frac{d^d k}{(2\pi)^d} \frac{1}{(-i\omega + k^2/(2m))} \frac{1}{(i\omega + k^2/(2m))} = -u_0^2 \int \frac{d^d k}{(2\pi)^d} \frac{m}{k^2}, \tag{16.45}$$

(the remaining ladder diagrams are powers of (16.45) and form a simple geometric series). Note the infrared singularity for $d < 2$, which is cured, as in Section 3.3.2, by moving away from the quantum critical point, or by external momenta.

We can proceed further by a simple application of the momentum shell RG of Chapter 4, but extended here to a quantum problem. Note that we apply cutoff Λ only in momentum space. The RG then proceeds by integrating *all* frequencies, and momentum modes in the shell between $\Lambda e^{-\ell}$ and Λ. The renormalization of the coupling u_0 is then given by the first diagram in Fig. 16.3 (see also the second diagram in Fig. 4.1), and leads here to the analog of the flow equation in (4.24). First, it is useful to absorb some phase space factors by a redefinition of interaction coupling:

$$u_0 = \frac{\Lambda^{2-d}}{2m S_d} u, \tag{16.46}$$

and then we obtain [139, 148]

$$\frac{du}{d\ell} = \epsilon u - \frac{u^2}{2}.$$ (16.47)

Here $S_d = 2/(\Gamma(d/2)(4\pi)^{d/2})$ is the usual phase space factor, and

$$\epsilon = 2 - d.$$ (16.48)

Note that for $\epsilon > 0$, there is a stable fixed point at

$$u^* = 2\epsilon,$$ (16.49)

which controls all the universal properties of $\mathcal{Z}_B$.

We now state a very important and surprising feature of the above results, which is not shared by the corresponding calculations in Chapter 14. The flow equation (16.47), and the fixed-point value (16.49) are *exact* to all orders in u or ϵ, and it is not necessary to consider u-dependent renormalizations to the field scale of Ψ_B or any of the other couplings in $\mathcal{Z}_B$. This result is ultimately a consequence of a very simple fact: the ground state of $\mathcal{Z}_B$ at the quantum critical point $\mu = 0$ is simply the empty vacuum with no particles. So any interactions that appear are entirely due to particles that have been created by the external fields. In particular, if we introduce the bosonic Green's function (the analog of (16.21))

$$G_B(x, t) = \langle \Psi_B(x, t)\Psi_B^\dagger(0, 0)\rangle,$$ (16.50)

then for $\mu \leq 0$ and $T = 0$, its Fourier transform $G(k, \omega)$ is given exactly by the free-field expression

$$G_B(k, \omega) = \frac{1}{-\omega + k^2/(2m) - \mu}.$$ (16.51)

The field $\Psi_B^\dagger$ creates a particle that travels freely until its annihilation at (x, t) by the field Ψ_B; there are no other particles present at $T = 0$, $\mu \leq 0$, and so the propagator is just the free-field one. The simple result (16.51) implies that the scaling dimensions in (16.43) are exact. Turning to the renormalization of u, it is clear from the diagram in Fig. 16.3 that we are considering the interactions of just two particles. For these, the only nonzero diagrams are the ones shown in Fig. 16.3, which involve repeated scattering of just these particles. Formally, it is possible to write down many other diagrams that could contribute to the renormalization of u; however, all of these vanish upon performing the integral over internal frequencies for there is always one integral that can be closed in one half of the frequency plane where the integrand has no poles. This absence of poles is of course just a more mathematical way of stating that there are no other particles around.

We consider application of these renormalization group results separately for the cases below and above the upper-critical dimension of $d = 2$.

16.3.1 $d < 2$

The approach and analysis here are very similar to those carried out in Chapter 14 below the upper-critical dimension ($d < 3$) for the quantum rotor/Ising models.

First, let us note some important general implications of the theory controlled by the fixed-point interaction (16.49). As we have already noted, the scaling dimensions of Ψ_B and μ are given precisely by their free-field values in (16.43), and the dynamic exponent z also retains the tree-level value $z = 2$. All these scaling dimensions are identical to those obtained for the case of the spinless Fermi gas in Section 16.2. Further, the presence of a nonzero and universal interaction strength u^* in (16.49) implies that the bosonic system is stable for the case $\mu > 0$ because the repulsive interactions prevent condensation of an infinite density of bosons (no such interaction was necessary for the fermion case, as the Pauli exclusion was already sufficient to stabilize the system). These two facts imply that the formal scaling structure of the bosonic fixed point being considered here is identical to that of the fermionic one considered in Section 16.2, and that the scaling forms of the two theories are *identical*. In particular, G_B obeys a scaling form identical to that for G_F in (16.20) (with a corresponding scaling function Φ_{G_B}), while the free energy, and associated derivatives, obey (16.22) (with a scaling function $\Phi_{\mathcal{F}_B}$). The universal functions Φ_{G_B} and $\Phi_{\mathcal{F}_B}$ can be determined order by order in the present $\epsilon = 2 - d$ expansion, and this will be illustrated shortly.

Although the fermionic and bosonic fixed points share the same scaling dimensions, they are distinct fixed points for general $d < 2$. However, the arguments already presented in Section 16.1 suggest that these two fixed points are identical precisely in $d = 1$ [438]. Further evidence for this identity was presented in [105], where the anomalous dimension of the composite operator Ψ_B^2 was computed exactly in the ϵ expansion and was found to be identical to that of the corresponding fermionic operator. Assuming the identity of the fixed points, we can then make a stronger statement about the universal scaling function: those for the free energy (and all its derivatives) are identical, $\Phi_{\mathcal{F}_B} = \Phi_{\mathcal{F}_F}$ in $d = 1$. In particular, from (16.23) and (16.24) we conclude that the boson density is given by

$$\langle Q \rangle = \langle \Psi_B^\dagger \Psi_B \rangle = \int \frac{dk}{2\pi} \frac{1}{e^{(k^2/(2m)-\mu)/T} + 1}, \tag{16.52}$$

for $d = 1$ only. The operators Ψ_B and Ψ_F are still distinct and so there is no reason for the scaling functions of their correlators to be the same. (We compute numerous exact properties of the scaling function Φ_{G_B} for G_B in the following Section 16.3.3.) The crossover diagram of Fig. 16.2 also applies to $\mathcal{Z}_B$ in $d = 1$. The critical Fermi liquid state for $\mu > 0$, $T = 0$ is expected to become a critical superfluid state. As we show in Section 16.3.3, the bosonic correlation functions decay with a power law in space implying "quasi-long-range" superfluid order at $T = 0$. However, correlations decay exponentially at any nonzero T, implying the absence of any finite-T phase transition. This is again consistent with the $T > 0$ behavior of Fig. 16.2.

As not all observables can be computed exactly in $d = 1$ by the mapping to the free fermions, we now consider the $\epsilon = 2 - d$ expansion. We present a simple ϵ expansion calculation [437] for illustrative purposes. We focus on the density of bosons at $T = 0$. Knowing that the free energy obeys the analog of (16.22), we can conclude that a relationship like (16.11) holds:

$$\langle \Psi_B^\dagger \Psi_B \rangle = \begin{cases} \mathcal{C}_d (2m\mu)^{d/2}, & \mu > 0, \\ 0, & \mu < 0, \end{cases} \tag{16.53}$$

at $T = 0$, with $\mathcal{C}_d$ a universal number. The identity of the bosonic and fermionic theories in $d = 1$ implies from (16.11) or from (16.52) that $\mathcal{C}_1 = S_1/1 = 1/\pi$. We will show how to compute $\mathcal{C}_d$ in the ϵ expansion; similar techniques can be used for almost any observable.

Even though the position of the fixed point is known exactly in (16.49), not all observables can be computed exactly because they have contributions to arbitrary order in u. However, universal results can be obtained order-by-order in u, which then become a power series in $\epsilon = 2 - d$. As an example, let us examine the low-order contributions to the boson density. To compute the boson density for $\mu > 0$, we anticipate that there is a condensate of the boson field Ψ_B, and so we write

$$\Psi_B(x, \tau) = \Psi_0 + \Psi_1(x, t), \tag{16.54}$$

where Ψ_1 has no zero wavevector and frequency component. Inserting this into $\mathcal{L}_B$ in (16.1), and expanding to second order in Ψ_1, we get

$$\mathcal{L}_1 = -\mu |\Psi_0|^2 + \frac{u_0}{2} |\Psi_0|^4 - \Psi_1^* \frac{\partial \Psi_1}{\partial \tau} + \frac{1}{2m} |\nabla \Psi_1|^2$$
$$- \mu |\Psi_1|^2 + \frac{u_0}{2} \left(4 |\Psi_0|^2 |\Psi_1|^2 + \Psi_0^2 \Psi_1^{*2} + \Psi_0^{*2} \Psi_1^2 \right). \tag{16.55}$$

This is a simple quadratic theory in the canonical Bose field Ψ_1, and its spectrum and ground state energy can be determined by the familiar Bogoliubov transformation. Carrying out this step, we obtain the following formal expression for the free-energy density $\mathcal{F}$ as a function of the condensate Ψ_0 at $T = 0$:

$$\mathcal{F}(\Psi_0) = -\mu |\Psi_0|^2 + \frac{u_0}{2} |\Psi_0|^4$$
$$+ \frac{1}{2} \int \frac{d^d k}{(2\pi)^d} \left[\left\{ \left(\frac{k^2}{2m} - \mu + 2u_0 |\Psi_0|^2 \right)^2 - u_0^2 |\Psi_0|^4 \right\}^{1/2} \right.$$
$$\left. - \left(\frac{k^2}{2m} - \mu + 2u_0 |\Psi_0|^2 \right) \right]. \tag{16.56}$$

To obtain the physical free-energy density, we have to minimize $\mathcal{F}$ with respect to variations in Ψ_0 and to substitute the result back into (16.56). Finally, we can take the derivative of the resulting expression with respect to μ and obtain the required expression for the boson density, correct to the first two orders in u_0:

$$\langle \Psi_B^\dagger \Psi_B \rangle = \frac{\mu}{u_0} + \frac{1}{2} \int \frac{d^d k}{(2\pi)^d} \left[1 - \frac{k^2}{\sqrt{k^2(k^2 + 4m\mu)}} \right]. \tag{16.57}$$

To convert (16.57) into a universal result, we need to evaluate it at the coupling appropriate to the fixed point (16.49). This is most easily done by the field-theoretic RG employed in Chapter 14. So let us translate the RG equation (16.47) into this language. As in (14.9), we introduce a momentum scale $\tilde{\mu}$ (the tilde is to prevent confusion with the chemical potential) and express u_0 in terms of a dimensionless coupling u_R by

$$u_0 = u_R \frac{(2m)\tilde{\mu}^\epsilon}{S_d} \left(1 + \frac{u_R}{2\epsilon}\right). \tag{16.58}$$

The motivation behind the choice of renormalization factor in (16.58) is the same as that behind (14.9): the renormalized four-point coupling, when expressed in terms of u_R, and evaluated in $d = 2-\epsilon$, is free of poles in ϵ as can easily be explicitly checked using (16.45) and the associated geometric series. Then, we evaluate (16.57) at the fixed-point value of u_R, following the basic recipe as in Section 14.1.2: compute any physical observable as a formal diagrammatic expansion in u_0, substitute u_0 in favor of u_R using (16.58), and expand the resulting expression in powers of ϵ. All poles in ϵ should cancel, but the resulting expression depends upon the arbitrary momentum scale $\tilde{\mu}$. At the fixed-point value u_R^*, dependence upon $\tilde{\mu}$ then disappears and a universal answer remains. In this manner we obtain from (16.57) a universal expression in the form (16.53) with

$$\mathcal{C}_d = S_d \left[\frac{1}{2\epsilon} + \frac{\ln 2 - 1}{4} + \mathcal{O}(\epsilon)\right]. \tag{16.59}$$

16.3.2 $d = 3$

Although we only discuss the case $d = 3$ here, precisely the same manipulations and results hold for all $2 < d < 4$. The same methods can also be used to compute the logarithmic corrections in $d = 2$, along the lines of the discussion in Section 14.4; this was carried out elegantly by Prokof'ev' Ruebenacker, and Svistunov [388]. Related results, obtained through somewhat different methods, are available in the literature [139, 386, 387, 438].

The quantum critical point at $\mu = 0$, $T = 0$ is above its upper-critical dimension, and we expect mean-field theory to apply. The analog of the mean-field result in the present context is the $T = 0$ relation for the density

$$\langle \Psi_B^\dagger \Psi_B \rangle = \begin{cases} \mu/u_0 + \cdots, & \mu > 0, \\ 0, & \mu < 0, \end{cases} \tag{16.60}$$

where the ellipsis represents terms that vanish faster as $\mu \to 0$. Note that this expression for the density is not universally dependent upon μ; rather it depends upon the strength of the two-body interaction u_0 (more precisely, it can be related to the s-wave scattering length a by $u_0 = 4\pi a/m$).

We turn to the crossovers and phase transitions at $T > 0$. These are sketched in Fig. 16.4. These crossovers were computed by Rasolt *et al.* [394], Weichman *et al.* [536] and also addressed in earlier work [96, 473, 474]; we do not follow their approach here, however. Instead, we show that the results can be obtained by a direct application of the method used in Section 14.2.2 to study the quantum rotor/Ising models above their upper-critical dimension.

The basic approach is one used several times in this book: integrate out the modes with a nonzero Matsubara frequency to obtain an effective action for the static, time-independent

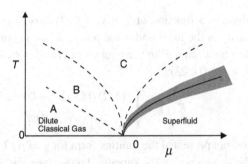

Fig. 16.4 Crossovers of the dilute Bose gas in $d = 3$ as a function of the chemical potential μ and the temperature T. The regimes labeled A, B, C are described in the text. The solid line is the finite-temperature phase transition where the superfluid order disappears; the shaded region is where the classical $D = 3, N = 2$ theory describes thermal fluctuations. The contours of constant density are similar to those in Fig. 16.2 and are not displayed.

modes. In the present situation it is clear that in the nonsuperfluid phase, the effective action again has the form of $\mathcal{S}_{\phi,\text{eff}}$ in (14.16) for the case $N = 2$ after the identification

$$\Psi(x) = \sqrt{m}(\phi_1(x) + i\phi_2(x))$$ (16.61)

of the static modes. The values of the couplings, R and U, can be obtained by a simple perturbation theory in u. From expressions analogous to (14.17), (14.18), and (14.31) we obtain

$$R = -2m\mu + 4mu_0 \int \frac{d^3k}{(2\pi)^3} \left[\frac{1}{e^{(k^2/2m-\mu)/T} - 1} - \frac{2mT}{k^2 - 2m\mu} + \frac{2mT}{k^2} \right],$$ (16.62)

and

$$U = 12m^2 u_0.$$ (16.63)

Armed with a knowledge of the values of R and U we can then proceed precisely as in Section 14.2.2: we simply insert these values into the form (14.19) involving the tricritical crossover function (the susceptibility χ in the present case is the boson Green's function G_B defined in (16.50)), and we use the results for the tricritical crossovers in Section 14.1.1 for the case $N = 2$. Hence a clear understanding of the functional form of R is useful, and we now discuss this.

Let us first rewrite R in the form analogous to (14.32):

$$R = -2m\mu + 4mu_0(2mT)^{3/2} K\left(\frac{\mu}{T}\right),$$ (16.64)

where the universal function $K(y)$ is given by (compare (14.21) and (14.33))

$$K(y) = \frac{1}{2\pi^2} \int_0^\infty k^2 dk \left(\frac{1}{e^{k^2 - y} - 1} - \frac{1}{k^2 - y} + \frac{1}{k^2} \right).$$ (16.65)

Note that the result for R depends explicitly on the bare value of the coupling u, as is expected for a system above its upper-critical dimension, and as we also found in Section 14.2.2. A crucial property of $K(y)$ (as was the case for (14.21) and (14.33)) is that it is analytic as a function of y at $y = 0$. This is clear from the fact that the only possible singularity of the integrand is the pole at $k^2 = y$, but its residue vanishes because of cancellation between the first two terms in (16.65). We quote some limiting forms for $K(y)$ analogous to (14.23):

$$K(y) = \begin{cases} \zeta(3/2)/(4\pi)^{3/2} - 0.0327826 y, & |y| \ll 1, \\ \sqrt{|y|}/(4\pi) + e^{-|y|}/(4\pi)^{3/2}, & y \ll -1. \end{cases} \tag{16.66}$$

We have not presented the limiting form for $y = \mu/T \gg 1$ as that is not needed. This limit puts the system within the superfluid phase (see Fig. 16.4) and the present results are valid only for the normal phase.

Inserting the above results into (14.19) and performing some straightforward analysis allows one to construct the phase diagram in Fig. 16.4. We can characterize the nonsuperfluid regions of Fig. 16.4 by the behavior of the zero-frequency limit of the boson Green's function G_B; following (14.34) we parameterize this as

$$G_B(k, i\omega_n = 0) = \frac{2m}{k^2 + \xi^{-2}}, \tag{16.67}$$

where ξ can be identified as the correlation length of the superfluid order parameter. An expression for ξ follows from (14.19) and (14.5) at $N = 2$:

$$\xi^{-2} = R - \frac{TU\sqrt{R}}{6\pi}. \tag{16.68}$$

As in Section 14.2 the condition for the boundary to the ordered superfluid phase is simply $R = 0$. Using (16.64)–(16.66) we therefore obtain, to leading order in u, the critical temperature

$$T_c = \frac{2\pi}{m} \left(\frac{\mu}{2u_0\zeta(3/2)} \right)^{2/3}, \tag{16.69}$$

which describes the phase boundary shown in Fig. 16.4; note that $\mu/T_c \sim \mu^{1/3}u_0^{2/3}m \ll 1$, and so the $\mu/T_c \gg 1$ case of (16.66) was not necessary. Before discussing the various normal state regimes in Fig. 16.4, however, we also obtain an expression for the free-energy density, $\mathcal{F}$; the boson density then follows immediately from the identity $\langle \Psi_B^\dagger \Psi_B \rangle = -\partial \mathcal{F}/\partial\mu$. The free energy is computed by adding the contribution of the $\omega_n \neq 0$ modes to that of the $\omega_n = 0$ modes as described by $\mathcal{S}_{\phi,\text{eff}}$ in (14.16). In this manner we obtain

$$\mathcal{F} = T \int \frac{d^3k}{(2\pi)^3} \ln\left(1 - e^{-(k^2/(2m)-\mu)/T}\right) + T \int_0^\Lambda \frac{d^3k}{(2\pi)^3} \ln\left(\frac{k^2 + \xi^{-2}}{k^2 - 2m\mu}\right). \tag{16.70}$$

The integral over the $\omega_n \neq 0$ terms yields the first logarithm and the denominator in the argument of the second logarithm. Note that this combination is well defined even for $\mu > 0$, and the singularity at $k^2 = 2m\mu$ is illusory; the expression (16.70) is analytic at $\mu = 0$ and can be straightforwardly numerically evaluated in the present form for all real values of μ. The integral over the ϕ_α modes in $\mathcal{S}_{\phi,\text{eff}}$ gives the numerator of the second logarithm. Note also that the second integral requires a large momentum cutoff Λ. It is natural to take this cutoff of order the de-Broglie wavelength $\sqrt{2mT}$, because the zero frequency theory (14.16) is valid only at such scales. However, more precisely, there can be no cutoff dependence because the problem of the dilute Bose gas is ultraviolet finite after the two-body interaction has been expressed in terms of the two-body scattering length, a (this connection is given later in (16.93)). Seeing this cutoff-independence requires a systematic expansion in powers of u_0 or in $1/N$ (for an N-component Bose gas), which we have not carried out here. From a knowledge of $\mathcal{F}$, and therefore of the boson density $\langle \Psi_B^\dagger \Psi_B \rangle$, we can, in principle, convert the μ–T phase diagram in Fig. 16.4 into a density–T phase diagram. Such a procedure also yields an expression for T_c in terms of the density: this was studied in [41, 42, 224], which also give cutoff-independent expressions using the $1/N$ expansion. The constant density contours in Fig. 16.4 have a shape quite similar to those in Fig. 16.2. However, the theoretical analysis and the manner in which the present problem fits into the general theory of crossovers near quantum phase transitions are much more transparent in the μ–T plane, and this representation continues to be the basis of our remaining discussion. We turn to a separate description of the normal state regions in turn (the discussion parallels that below (14.34)).

(A) $\mu < 0$, $T \ll |\mu|$, *Dilute classical gas*: We use the $y \ll -1$ limit of (16.66) in (16.64) and (16.68) to obtain

$$\xi^{-2} = 2m|\mu| + \frac{mu_0}{2} \left(\frac{2mT}{\pi} \right)^{3/2} e^{-|\mu|/T}. \tag{16.71}$$

So the correlation length is given by its $T = 0$ value and all T-dependent corrections are exponentially small. The density of bosons follows from the μ derivative of (16.70) and we obtain

$$\langle \Psi_B^\dagger \Psi_B \rangle = \left(\frac{mT}{2\pi} \right)^{3/2} e^{-|\mu|/T} + \cdots \tag{16.72}$$

The ellipsis represents small corrections that depend upon the strength of the weak interaction u_0, and we invite the reader to work them out from (16.70). This density is very small and, as in Section 16.2.1, the spacing between the particles is much larger than their thermal de Broglie wavelength. We therefore expect an effective classical Boltzmann gas description to apply. Although (16.67) and (16.71) give an adequate description of the static correlations, dynamic properties require further analysis following that presented in Chapters 13 and 15 for the quantum rotor models.

(B) $\mu < 0$, $|\mu| \ll T \ll (|\mu|/u_0)^{2/3}/m$: As in (A), the correlation length is dominated by its $T = 0$ value of $(2m|\mu|)^{-1/2}$ but the form of the T-dependent corrections differs from the exponentially small corrections in (A); we have instead, power-law corrections that follow from the $|y| \ll 1$ limit of (16.66) inserted in (16.64) and (16.68):

$$\xi^{-2} = 2m|\mu| + \frac{mu_0}{2} \left(\frac{2mT}{\pi} \right)^{3/2} \zeta(3/2). \tag{16.73}$$

The density is no longer exponentially small, and (16.70) gives

$$\langle \Psi_B^\dagger \Psi_B \rangle = \left(\frac{mT}{2\pi} \right)^{3/2} \zeta(3/2) + \cdots , \tag{16.74}$$

where again the ellipsis represents u_0-dependent corrections, which are somewhat messy but easy to compute from the expressions provided above. For this density the spacing between the particles is of order their thermal de Broglie wavelength, and in this respect this regime is similar to the high-T limit of the Fermi gas discussed in Section 16.2.3. Of course, there are nonuniversal u-dependent corrections here, which were absent for the Bose gas for $d < 2$ and for the spinless Fermi gas in all d. Again, a description of dynamics in this region (B) requires extension of the computations of Chapters 13 and 15.

(C) $T \gg (|\mu|/u_0)^{2/3}/m$, *High T*: This is of course the true high-T limit of the continuum theory $\mathcal{Z}_B$. Its physical properties are similar to those of (B) but with some significant differences. The expression (16.73) for the correlation length still applies, but it is clear that the second T-dependent term is the larger one. Thus the correlation length $\xi \sim T^{-3/4}$, which does not agree with the naive scaling estimate $\xi \sim T^{-1/z}$; as we discussed in Section 14.2.2, this is because the interaction u is dangerously irrelevant, and its bare value appears in the high-T limit of (16.73). The leading term in the density is also as in (16.74), but the omitted u-dependent corrections have a rather different structure.

16.3.3 Correlators of $\mathcal{Z}_B$ in $d = 1$

Readers not interested in the details of correlations in $d = 1$, may skip ahead to the next section.

The study of the bosonic correlators of $\mathcal{Z}_B$ is of some interest because they can be measured directly in neutron scattering or NMR experiments on spin systems that undergo a quantum phase transition of the type studied here. Explicit realizations include the XX chain of Section 16.1 or gapped antiferromagnets in a strong field (discussed in Chapter 19). In all of these cases, the bosonic field Ψ_B has a simple, local relationship to the spin operators (as in (16.5)), allowing its correlators to be simply related to measurable quantities. In contrast, the fermionic correlators of Ψ_F (discussed in Section 16.2) have no physical interpretation in such applications.

We have argued in Sections 16.1, 16.2, and 16.3.1 that the theories $\mathcal{Z}_B$ and $\mathcal{Z}_F$ are equivalent for small μ. The universal expression for the boson density was given in (16.52). Here we discuss how to map the two theories at the operator level. For the case of the transition from a Mott insulator with $n_0 = 0$, there are no background particles to account for, and we can derive the theory $\mathcal{Z}_B$ simply by the naive continuum limit of the lattice boson coherent state path integral (9.31). Such a procedure leads to the exact operator correspondence $\Psi_B = \hat{b}_i/a$. We know from Section 16.1 that $\hat{b}_i = (\hat{\sigma}_i^x + i\hat{\sigma}_i^y)/2$ and, further, that the Pauli matrices are related to the lattice fermion field by (10.9) and thence to the continuum Fermi field Ψ_F via (10.23). Combining these transformations, and taking the naive continuum limit, we obtain the formal operator correspondence

$$\Psi_B(x, t) = \exp\left(i\pi \int_{-\infty}^{x} dy \Psi_F^\dagger(y, t)\Psi_F(y, t)\right)\Psi_F(x, t). \qquad (16.75)$$

So our task is, in principle, well defined: all correlators of Ψ_F under $\mathcal{L}_F$ are known – use these to compute those of Ψ_B using the mapping (16.75). In practice, this evaluation cannot be carried out in the continuum as severe short-distance divergences appear. We have to return to the underlying lattice degrees of freedom, evaluate the expectation values under the lattice Hamiltonian, and then return to the continuum limit. A calculation such as this was discussed in Section 10.3 for equal-time correlators of the quantum Ising model. A very similar analysis can also be performed for the present XX model. We refer the reader to the literature for details [420] and present the main results.

We are interested here in the two-point bosonic correlation function G_B in (16.50). As discussed in Section 16.3.1, we know that this satisfies a scaling form identical to (16.20), but the bosonic scaling function Φ_{G_B} will be quite different from Φ_{G_F}. The large-distance limit of the equal-time case can be obtained by the methods of Section 10.3. We use the mapping

$$G_B(x,0) = \frac{1}{2a}\langle \hat{\sigma}_i^x \hat{\sigma}_0^x \rangle, \tag{16.76}$$

where $x = ia$ and the latter expectation value is evaluated under H_{XX} at a temperature T. This can be performed using essentially the same analysis as in Section 10.3, and we obtain for $T > 0$ that [420]

$$\lim_{|x| \to \infty} G_B(x,0) = \left(\frac{mT}{2}\right)^{1/2} G_X(\mu/T) \exp\left(-F_X(\mu/T)(2mT)^{1/2}|x|\right), \tag{16.77}$$

where the universal crossover functions $F_X(y)$ and $G_X(y)$ are given by

$$F_X(y) = \int_0^\infty \frac{ds}{\pi} \ln\coth\frac{|s^2 - y|}{2} + \theta(-y)\sqrt{-y}, \tag{16.78}$$

$$\ln G_X(y) = 2\int_{-\infty}^{-1} ds \left[\left(\frac{dF_X(s)}{ds}\right)^2 + \frac{1}{4s}\right] + 2\int_{-1}^y ds \left(\frac{dF_X(s)}{ds}\right)^2. \tag{16.79}$$

Note the similarity of these results to (10.50) and (10.51) for the Ising chain. As in the Ising case, both functions F_X and G_X are analytic (despite appearances) for all real values of y, as must be the case owing to the absence of thermodynamic singularities at nonzero T. We show a plot of these functions in Fig. 16.5. These results for F_X and G_X have also been obtained in [294], [279], [240], and [278] by the rather different, and far more sophisticated, quantum inverse scattering method.

Let us look at the physical implications of the above results for G_B in the different regimes of Fig. 16.2.

Dilute classical gas, $k_B T \ll |\mu|$, $\mu < 0$

We need the $y \to -\infty$ limits of the F_X and G_X scaling functions. From (16.78) and (16.79) we get $F_X(y \to -\infty) = \sqrt{-y}$ and $G_X(y \to -\infty) = 1/\sqrt{-y}$, and so we have for the equal-time correlator

$$G_B(x,0) = \frac{T}{2}\left(\frac{2m}{|\mu|}\right)^{1/2} \exp\left(-(2m|\mu|)^{1/2}|x|\right) \text{ as } |x| \to \infty. \tag{16.80}$$

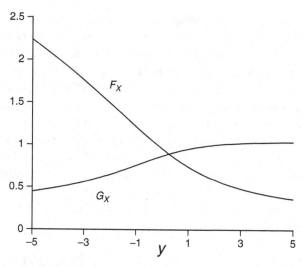

Fig. 16.5 The universal scaling functions $F_X(y)$ for the inverse correlation length and the amplitude $G_X(y)$ (defined in (16.77)), as a function of $y = \mu/T$.

This equal-time result has a very simple interpretation. It is precisely the Fourier transform of $TG_B(k, \omega_n = 0)$ with the G_B given in (16.67), the prefactor of T coming from the classical limit of the fluctuation–dissipation theorem as in (10.76), and with the leading low-temperature value for ξ from (16.71), $\xi^{-2} = 2m|\mu|$. Classical behavior is of course expected, because, as in Section 16.2.1, the spacing between the particles is much larger than their thermal de Broglie wavelength.

Long-time correlators can be obtained by a simple physical argument that relies on the similarity of this regime to the low-T regime on the paramagnetic side of the quantum Ising chain, discussed in Section 10.4.2. In that case, and here, we have an exponentially dilute concentration of particles and are interested in the single-particle boson Green's function. Semiclassical arguments to compute these were advanced in Section 10.4.2, and those led to the main result in (10.88). Its analog in the present case is

$$G_B(x, t) = G_F(x, t)R(x, t). \tag{16.81}$$

Here G_F is the result given in (16.25) with $d = 1$ (this result is also the Fourier transform of (16.51)) and is the Feynman propagator for a single particle moving quantum mechanically from $(0, 0)$ to (x, t). The factor R represents the consequence of collisions with the exponentially dilute background of thermally excited particles. As argued in Section 10.4.2, G_B picks up a (-1) from the S matrix of each collision, and the result of averaging over such collisions leads to $R(x, t)$ given in (10.69); the only change here is in the dispersion spectrum of the particles, $\varepsilon_k = k^2/(2m) + |\mu|$:

$$R(x, t) = \exp\left(-\int \frac{dk}{\pi} e^{-\varepsilon_k/T} \left| x - \frac{d\varepsilon_k}{dk}t \right| \right). \tag{16.82}$$

The explicit structure of the function R was described in Section 10.4.1: equal-time correlations decay exponentially in space with the length ξ_c, whereas equal-space correlations decay exponentially in time with the time τ_φ (see (10.70)), and the general function obeys the scaling from (10.73) with the scaling function given in (10.74). The only change is in the specific values of the characteristic scales ξ_c and τ_φ, which are given by

$$\xi_c = \left(\frac{\pi}{2mT}\right)^{1/2} e^{|\mu|/T},$$
$$\tau_\varphi = \frac{\pi}{2T} e^{|\mu|/T}.$$

Note that both scales are exponentially large at low T.

The dynamic structure factor can be obtained by a Fourier transform of (16.81). Its physical properties are very similar to those in Section 10.4.2, with a well-defined quasiparticle pole at $\omega = \varepsilon_k$, which is broadened by collisions with other particles on the spatial and temporal scales given in (16.83).

The results (16.81) and (16.82) have also been obtained in [242] and [278] and by a more rigorous and much lengthier method. The precise agreement gives us confidence that the simple semiclassical arguments used above are essentially exact.

Tomonaga–Luttinger liquid, $k_B T \ll \mu, \mu > 0$

This is the region labeled a Fermi liquid in Fig. 16.2. In $d = 1$ the generic state with interaction among the fermions away from the critical point is a Tomonaga–Luttinger liquid (as we discuss in Chapter 20), and we use this more general and standard terminology.

In our discussion of the correlators of Ψ_F in this region (Section 16.2.2) we showed that the long-distance properties were described by a line of $z = 1$ critical points at $\mu > 0$, $T = 0$ and that this manifested itself in a collapse of the fermion scaling functions into a reduced scaling form. A similar collapse must also occur for the G_B correlator, and indeed for all other observables. To describe this, we need the scaling dimension of Ψ_B under the continuum critical theory of this line of critical points, which was $\mathcal{L}_{FL}$ in (16.29).

This dimension can be easily obtained from the equal-time results above. Using, from (16.78) and (16.79), $F_X(y \to \infty) = \pi/4\sqrt{y}$ and $G_X(y \to \infty) = 1.042828\ldots$ we get for $T \ll \mu$

$$G_B(x, 0) = G_X(\infty) \left(\frac{mT}{2}\right)^{1/2} \exp\left(-\frac{\pi}{2}\frac{T}{v_F}|x|\right) \text{ as } |x| \to \infty. \tag{16.83}$$

The prefactor $\sim T^{1/2}$, along with quantities invariant under the scaling transformation (16.30) and the exponent $z = 1$, fixes

$$\dim[\Psi_B] = 1/4 \tag{16.84}$$

along the $\mu > 0$ critical line; recall that $\dim[\Psi_B] = 1/2$ at the $\mu = 0$ critical end point.

The results (16.83) and (16.84) are key, for they allow us to deduce the entire space and time dependence of G_B in this regime using a simple argument. The main point is that the long-distance and time correlators are controlled by the theory $\mathcal{L}_{FL}$, which is conformally invariant. Then we may use arguments essentially identical to those in Section 10.4.3 where we considered the high-T limit of the quantum Ising chain. The latter was controlled by the conformally invariant, $z = 1$, theory $\mathcal{L}_I$ in (10.28) at $\Delta = 0$. As we showed in Section 10.4.2, the $T > 0$ equal-time long-distance decay in (10.94) allowed us to deduce the complete spacetime-dependent correlation function in (10.95) and also the exact $T = 0$ correlator at the critical point in (10.91). Proceeding in precisely the same manner here, we may conclude that the $T = 0$ bosonic correlator obeys, for $\mu > 0$,

$$G_B(x, \tau) \sim \frac{1}{(x^2 + v_F^2 \tau^2)^{1/4}}, \tag{16.85}$$

where the normalization constant will be fixed shortly. Indeed (16.85) follows simply from (16.84) and the relativistic invariance of $\mathcal{L}_{FL}$. Therefore, as noted earlier, the bosonic superfluid correlations decay with a power law in the $\mu > 0$ ground state. At finite T, the analog of (10.95) is

$$G_B(x, \tau) = \left(\frac{mT}{2}\right)^{1/2} \frac{2^{-1/2} G_X(\infty)}{[\sin(\pi T(\tau + ix/v_F)) \sin(\pi T(\tau - ix/v_F))]^{1/4}}. \tag{16.86}$$

Note that this result obeys the reduced scaling form characteristic of the scaling dimensions of the theory $\mathcal{L}_{FL}$ in (16.30):

$$G_B(x, t) = \left(\frac{mT}{2}\right)^{1/2} \phi_X\left(\frac{Tx}{v_F}, Tt\right). \tag{16.87}$$

From this expression we can also explicitly relate the reduced scaling function ϕ_X to the global scaling function Φ_{G_B} (this is the scaling function of G_B defined as the bosonic analog of (16.20)) by the $\mu/T \to \infty$ of the latter:

$$\phi_X(\bar{x}, \bar{t}) = \lim_{y \to \infty} \Phi_{G_B}(2\sqrt{y}\bar{x}, \bar{t}, y). \tag{16.88}$$

The physical properties of these dynamical correlations are essentially identical in form to the dynamic responses discussed in Section 10.4.3, and particularly in Figs. 10.7 and 10.8, and so need not be described here. Correlations decay exponentially with a length $\xi \sim v_F/T$ and on a phase coherence time $\tau_\varphi \sim 1/T$. Both these scales are those expected in the "high-T" limit of a critical theory with $z = 1$. In the present case this is the theory $\mathcal{L}_{FL}$ characterizing the line of $\mu > 0$ critical points. Remember, though, that the present region is a low-T region of the global theory $\mathcal{L}_B$.

High-T limit, $k_B T \gg |\mu|$

Now we have from (16.78) and (16.79)

$$G_B(x, 0) = G_X(0)\left(\frac{mT}{2}\right)^{1/2} \exp\left(-F_X(0)(2mT)^{1/2}|x|\right) \text{ as } |x| \to \infty, \tag{16.89}$$

where $F_X(0) = \zeta(3/2)(1 - 1/\sqrt{8})/\sqrt{\pi} = 0.952781471\ldots$ and $G_X(0) = 0.86757\ldots$ are pure numbers. All scales are set by T, and the correlation length $\sim T^{-1/2}$, as expected from the $z = 2$ value at the $\mu = 0$ critical point. Notice also the similarity of this correlation length to that of the fermionic correlator in (16.42); only the numerical factors are different.

Asymptotics of dynamic correlation functions in this regime have been obtained by Korepin and collaborators [242, 278] by the quantum inverse scattering method. However, this is the one limiting regime where their approach appears indispensable, and an alternative derivation using the simpler physical arguments employed here does not exist; finding such a derivation remains an important open problem. (Korepin and collaborators [242, 278] also give the dynamic analogs of (16.77) containing the crossovers between the different finite-T regimes; the methods discussed here cannot give these either.) Their results are quite lengthy and are not reproduced here. We will just be satisfied by noting that, as expected by scaling arguments in the high-T regime of a continuum theory, time-dependent correlations decay exponentially on a phase coherence time of order $1/T$.

Summary

We summarize all of the structure in the dynamic correlations of $\mathcal{Z}_B$ in Fig. 16.6; note the similarity to (and some differences from) the corresponding figure for the Ising chain in Fig. 10.12. First, in all the three universal regions of Fig. 16.2, the short-time properties are essentially the same: a free nonrelativistic particle propagating quantum mechanically, without yet having felt the influence of any other particle. The interactions with other particles appear at longer times, and their consequences are rather different in the various regimes.

In the low-T regime for $\mu < 0$ ($T \ll |\mu|$), the concentration of other particles is exponentially small, and so the decoherence and spectral line broadening due to collisions are not felt until the very long time $\tau_\varphi \sim (1/T)e^{|\mu|/T}$.

In the opposing low-T regime for $\mu > 0$ ($T \ll \mu$) the behavior is rather different. Now the particles are dense and degenerate, and at times longer than $1/\mu$, the Pauli exclusion principle leads to the quantum coherence of a Fermi liquid ground state (more generally for large μ, a Tomonaga–Luttinger liquid; see Chapter 20). This state is described by the separate $z = 1$ theory $\mathcal{L}_{FL}$ in (16.29), and for a while the systems appear to be in the ground state of $\mathcal{L}_{FL}$. However, eventually thermal effects cause decoherence and relaxation

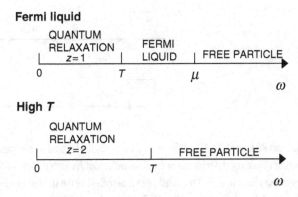

Fig. 16.6 Crossovers as a function of frequency for the boson model $\mathcal{Z}_B$ (in (16.1)) in $d = 1$ in the regimes of Fig. 16.2; this model is equivalent to $\mathcal{Z}_F$ (in (16.3)) in $d = 1$.

at a time $\tau_\varphi \sim 1/T$ and a length scale $\xi \sim 1/T$. This last crossover is entirely a property of $\mathcal{L}_{FL}$ and is characterized by its $z = 1$ critical exponents.

Finally, in the high-T regime, we have a completely different behavior. Now the value of μ is unimportant, and we may as well set $\mu = 0$. The crossover from the free particle behavior to relaxational dynamics happens at a time $\tau_\varphi \sim 1/T$ and a length scale $\xi \sim 1/\sqrt{T}$, which are characteristic of the $z = 2$ critical point at $\mu = 0$. The mean spacing between the particles is of order their de Broglie wavelength, and thermal and quantum effects are equally important.

16.4 The dilute spinful Fermi gas: the Feshbach resonance

This section turns to the case of the spinful Fermi gas with short-range interactions; as we noted in the introduction, this is a problem which has acquired renewed importance because of new experiments on ultracold fermionic atoms.

The partition function of the theory examined in this section was displayed in (16.4). The renormalization group properties of this theory in the zero density limit are *identical* to those of the dilute Bose gas considered in Section 16.3. The scaling dimensions of the couplings are the same, the scaling dimension of $\Psi_{F\sigma}$ is $d/2$ as for Ψ_B in (16.43), and the flow of the u is given by (16.47). Thus for $d < 2$, a spinful Fermi gas with repulsive interactions is described by the stable fixed point in (16.49).

However, for the case of a spinful Fermi gas, we can consider another regime of parameters which is of great experimental importance. We can also allow u to be attractive: unlike the Bose gas case, the $u < 0$ case is not immediately unstable, because the Pauli exclusion principle can stabilize a Fermi gas even with attractive interactions. Furthermore, at the same time we should also consider the physically important case with $d > 2$, when $\epsilon < 0$. The distinct nature of the RG flows predicted by (16.47) for the two signs of ϵ are shown in Fig. 16.7.

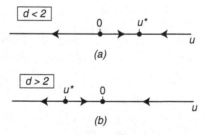

Fig. 16.7 The exact RG flow of (16.47). (*a*) For $d < 2$ ($\epsilon > 0$), the infrared stable fixed point at $u = u^* > 0$ describes quantum liquids of either bosons or fermions with repulsive interactions which are generically universal in the low-density limit. In $d = 1$ this fixed point is described by the spinless free Fermi gas ("Tonks" gas), for all statistics and spin of the constituent particles. (*b*) For $d > 2$ ($\epsilon < 0$) the infrared unstable fixed point at $u = u^* < 0$ describes the Feshbach resonance which obtains for the case of attractive interactions. The relevant perturbation ($u - u^*$) corresponds to detuning from the resonant interaction.

Note the *unstable* fixed point present for $d > 2$ and $u < 0$. Thus accessing the fixed point requires fine-tuning of the microscopic couplings. As discussed in [362, 363], this fixed point describes a Fermi gas at a *Feshbach resonance*, where the interaction between the fermions is universal. For $u < u^*$, the flow is to $u \to -\infty$: this corresponds to a strong attractive interaction between the fermions, which then bind into tightly bound pairs of bosons, which then Bose condense; this corresponds to the so-called "BEC" regime. On the other hand, for $u > u^*$, the flow is to $u \nearrow 0$, and the weakly interacting fermions then form the Bardeen–Cooper–Schrieffer (BCS) superconducting state.

Note that the fixed point at $u = u^*$ for $\mathcal{Z}_{Fs}$ has *two* relevant directions for $d > 2$. As in the other problems considered in this chapter, one corresponds to the chemical potential μ. The other corresponds to the deviation from the critical point $u - u^*$, and this (from (16.47)) has RG eigenvalue $-\epsilon = d - 2 > 0$. This perturbation corresponds to "detuning" from the Feshbach resonance, ν (not to be confused with the symbol for the correlation length exponent); we have $\nu \propto u - u^*$. Thus we have

$$\dim[\mu] = 2 \ , \quad \dim[\nu] = d - 2. \tag{16.90}$$

These two relevant perturbations have important consequences for the phase diagram, as we see shortly.

For now, let us understand the physics of the Feshbach resonance better. For this, it is useful to compute the two-body T matrix exactly by summing the graphs in Fig. 16.3, along with a direct interaction first order in u_0. The second-order term has already been evaluated for the bosonic case in (16.45) for zero external momentum and frequency, and has an identical value for the present fermionic case. Here, however, we want the off-shell T-matrix, for the case in which the incoming particles have momenta $k_{1,2}$, and frequencies $\omega_{1,2}$. Actually for the simple momentum-independent interaction u_0, the T matrix depends only upon the sums $k = k_1 + k_2$ and $\omega = \omega_1 + \omega_2$, and is independent of the final state of the particles, and the diagrams in Fig. 16.3 form a geometric series. In this manner we obtain

$$\frac{1}{T(k, i\omega)} = \frac{1}{u_0}$$

$$+ \int \frac{d\Omega}{2\pi} \int \frac{d^d p}{(2\pi)^d} \frac{1}{(-i(\Omega + \omega) + (p+k)^2/(2m))} \frac{1}{(i\Omega + p^2/(2m))}$$

$$= \frac{1}{u_0} + \int_0^\Lambda \frac{d^d p}{(2\pi)^d} \frac{m}{p^2} + \frac{\Gamma(1 - d/2)}{(4\pi)^{d/2}} m^{d/2} \left[-i\omega + \frac{k^2}{4m} \right]^{d/2-1}. \tag{16.91}$$

In $d = 3$, the s-wave scattering amplitude of the two particles, f_0, is related to the T matrix at zero center of mass momentum and frequency k^2/m by $f_0(k) = -mT(0, k^2/m)/(4\pi)$, and so we obtain

$$f_0(k) = \frac{1}{-1/a - ik}, \tag{16.92}$$

where the scattering length, a, is given by

$$\frac{1}{a} = \frac{4\pi}{mu_0} + \int_0^\Lambda \frac{d^3 p}{(2\pi)^3} \frac{4\pi}{p^2}. \tag{16.93}$$

For $u_0 < 0$, we see from (16.93) that there is a critical value of u_0 where the scattering length diverges and changes sign: this is the Feshbach resonance. We identify this critical value with the fixed point $u = u^*$ of the RG flow (16.47). It is conventional to identify the deviation from the Feshbach resonance by the detuning ν

$$\nu \equiv -\frac{1}{a}. \tag{16.94}$$

Note that $\nu \propto u - u^*$, as claimed earlier. For $\nu > 0$, we have weak attractive interactions, and the scattering length is negative. For $\nu < 0$, we have strong attractive interactions, and a positive scattering length. Importantly, for $\nu < 0$, there is a two-particle bound state, whose energy can be deduced from the pole of the scattering amplitude; recalling that the reduced mass in the center of mass frame is $m/2$, we obtain the bound state energy, E_b,

$$E_b = -\frac{\nu^2}{m}. \tag{16.95}$$

We can now draw the zero temperature phase diagram [362] of $\mathcal{Z}_{Fs}$ as a function of μ and ν, and the result is shown in Fig. 16.8.

For $\nu > 0$, there is no bound state, and so no fermions are present for $\mu < 0$. At $\mu = 0$, we have an onset of nonzero fermion density, just as in the other sections of this chapter. These fermions experience a weak attractive interaction, and so experience the Cooper instability once there is a finite density of fermions for $\mu > 0$. So the ground state for $\mu > 0$ is a paired Bardeen–Cooper–Schrieffer (BCS) superfluid, as indicated in Fig. 16.8. For small negative scattering lengths, the BCS state modifies the fermion state only near

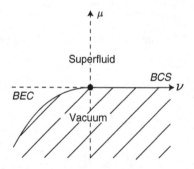

Fig. 16.8   Universal phase diagram at zero temperature for the spinful Fermi gas in $d = 3$ as a function of the chemical potential μ and the detuning ν. The vacuum state (shown hatched) has no particles. The position of the $\nu < 0$ phase boundary is determined by the energy of the two-fermion bound state in (16.95): $\mu = -\nu^2/(2m)$. The density of particles vanishes continuously at the second-order quantum phase transition boundary of the superfluid phase, which is indicated by the thin continuous line. The quantum multicritical point at $\mu = \nu = 0$ (denoted by the filled circle) controls all the universal physics of the dilute spinful Fermi gas near a Feshbach resonance. The universal properties of the critical line $\mu = 0$, $\nu > 0$ map onto the theory of Section 16.2, while those of the critical line $\mu = -\nu^2/(2m)$, $\nu < 0$ map onto the theory of Section 16.3. This implies that the $T > 0$ crossovers in Fig. 16.2 apply for $\nu > 0$ (the "Fermi liquid" region of Fig. 16.2 now has BCS superconductivity at an exponentially small T), while those of Fig. 16.4 apply for $\nu < 0$.

the Fermi level. Consequently as $\mu \searrow 0$ (specifically for $\mu < v^2/m$), we can neglect the pairing in computing the fermion density. We therefore conclude that the universal critical properties of the line $\mu = 0$, $v > 0$ map precisely on to two copies (for the spin degeneracy) of the noninteracting fermion model $\mathcal{Z}_F$ studied in Section 16.2. In particular the $T > 0$ properties for $v > 0$ map onto the crossovers in Fig. 16.2. The only change is that the BCS pairing instability appears below an exponentially small T in the "Fermi liquid" regime. However, the scaling functions for the density as a function of μ/T remain unchanged.

For $v < 0$, the situation changes dramatically. Because of the presence of the bound state (16.95), it pays to introduce fermions even for $\mu < 0$. The chemical potential for a fermion pair is 2μ, and so the threshold for having a nonzero density of paired fermions is $\mu = E_b/2$. This leads to the phase boundary shown in Fig. 16.8 at $\mu = -v^2/(2m)$. Just above the phase boundary, the density of fermion pairs is small, and so these can be treated as canonical bosons. Computations of the interactions between these bosons [362] show that they are repulsive. Therefore we map their dynamics onto those of the dilute Bose gas studied in Section 16.3. Thus the universal properties of the critical line $\mu = -v^2/(2m)$ are equivalent to those of $\mathcal{Z}_B$. Specifically, this means that the $T > 0$ properties across this critical line map onto those of Fig. 16.4.

Thus we reach the interesting conclusion that the Feshbach resonance at $\mu = v = 0$ is a multicritical point separating the density onset transitions of $\mathcal{Z}_F$ (Section 16.2) and $\mathcal{Z}_B$ (Section 16.3). This conclusion can be used to sketch the $T > 0$ extension of Fig. 16.8, on either side of the $v = 0$ line.

We now need a practical method of computing universal properties of $\mathcal{Z}_{Fs}$ near the $\mu = v = 0$ fixed point, including its crossovers into the regimes described by $\mathcal{Z}_F$ and $\mathcal{Z}_B$. The fixed point (16.47) of $\mathcal{Z}_{Fs}$ provides an expansion of the critical theory in the powers of $\epsilon = 2 - d$. However, observe from Fig. 16.7, the flow for $u < u^*$ is to $u \to -\infty$. The latter flow describes the crossover into the dilute Bose gas theory, $\mathcal{Z}_B$, and so this cannot be controlled by the $2 - d$ expansion. The following subsections propose two alternative analyses of the Feshbach resonant fixed point which address this difficulty.

16.4.1 The Fermi–Bose model

One successful approach is to promote the two-fermion bound state in (16.95) to a canonical boson field Ψ_B. This boson should also be able to mix with the scattering states of two fermions. We are therefore led to consider the following model

$$\mathcal{Z}_{FB} = \int \mathcal{D}\Psi_{F\uparrow}(x, \tau)\mathcal{D}\Psi_{F\downarrow}(x, \tau)\mathcal{D}\Psi_B(x, \tau) \exp\left(-\int d\tau d^d x \, \mathcal{L}_{FB}\right),$$

$$\mathcal{L}_{FB} = \Psi_{F\sigma}^* \frac{\partial \Psi_{F\sigma}}{\partial \tau} + \frac{1}{2m}|\nabla\Psi_{F\sigma}|^2 - \mu|\Psi_{F\sigma}|^2$$

$$+ \Psi_B^* \frac{\partial \Psi_B}{\partial \tau} + \frac{1}{4m}|\nabla\Psi_{F\sigma}|^2 + (\delta - 2\mu)|\Psi_B|^2$$

$$- \lambda_0\left(\Psi_B^* \Psi_{F\uparrow}\Psi_{F\downarrow} + \Psi_B \Psi_{F\downarrow}^* \Psi_{F\uparrow}^*\right). \tag{16.96}$$

Here we have taken the bosons to have mass $2m$, because that is the expected mass of the two-fermion bound state by Galilean invariance. We have omitted numerous possible quartic terms between the bosons and fermions above, and these turn out to be irrelevant in the analysis below.

The conserved U(1) charge for $\mathcal{Z}_{FB}$ is

$$Q = \Psi_{F\uparrow}^* \Psi_{F\uparrow} + \Psi_{F\downarrow}^* \Psi_{F\downarrow} + 2\Psi_B^* \Psi_B, \tag{16.97}$$

and so $\mathcal{Z}_{FB}$ is in the class of models being studied in this chapter. The factor of 2 in (16.97) accounts for the 2μ chemical potential for the bosons in (16.96). For μ sufficiently negative it is clear that $\mathcal{Z}_{FB}$ will have neither fermions nor bosons present, and so $\langle Q \rangle = 0$. Conversely for positive μ, we expect $\langle Q \rangle \neq 0$, indicating a transition as a function of increasing μ. Furthermore, for δ large and positive, the Q density will be primarily fermions, while for δ negative the Q density will be mainly bosons; thus we expect a Feshbach resonance at intermediate values of δ, which then plays the role of detuning parameter.

We have thus argued that the phase diagram of $\mathcal{Z}_{FB}$ as a function of μ and δ is qualitatively similar to that in Fig. 16.8, with a Feshbach resonant multicritical point near the center. The main claim of this section is that the universal properties of $\mathcal{Z}_{FB}$ and $\mathcal{Z}_{Fs}$ are *identical* near this multicritical point [362, 363]. Thus, in a strong sense, the theories $\mathcal{Z}_{FB}$ and $\mathcal{Z}_{Fs}$ are equivalent. Unlike the equivalence between $\mathcal{Z}_B$ and $\mathcal{Z}_F$, which held only in $d = 1$, the present equivalence applies for $d > 2$.

We establish the equivalence by an exact RG analysis of the zero density critical theory. We scale the spacetime coordinates and the fermion field as in (16.12), but allow an anomalous dimension η_b for the boson field relative to (16.43):

$$
\begin{aligned}
x' &= xe^{-\ell}, \\
\tau' &= \tau e^{-z\ell}, \\
\Psi_{F\sigma}' &= \Psi_{F\sigma} e^{d\ell/2}, \\
\Psi_B' &= \Psi_B e^{(d+\eta_b)\ell/2}, \\
\lambda_0' &= \lambda_0 e^{(4-d-\eta_b)\ell/2},
\end{aligned}
\tag{16.98}
$$

where, as before, we have $z = 2$. At tree level, the theory $\mathcal{Z}_{FB}$ with $\mu = \delta = 0$ is invariant under the transformations in (16.98) with $\eta_b = 0$. At this level, we see that the coupling λ_0 is relevant for $d < 4$, and so we have to consider the influence of λ_0. This also suggests, as in Chapter 4, that we may be able to obtain a controlled expansion in powers of $(4 - d)$.

Upon considering corrections in powers of λ_0 in the critical theory, it is not difficult to show that there is a non-trivial contribution from only a single Feynman diagram: this is the self-energy diagram for Ψ_B which is shown in Fig. 16.9. All other diagrams vanish in the zero-density theory, for reasons similar to those discussed for $\mathcal{Z}_B$ below (16.49). This diagram is closely related to the integrals in the T-matrix computation in (16.91), and leads to the following contribution to the boson self-energy Σ_B:

Fig. 16.9 Feynman diagram contributing to the RG. The dark triangle is the λ_0 vertex, the full line is the Ψ_B propagator, and the dashed line is the Ψ_F propagator.

$\Sigma_B(k, i\omega)$

$$
= \lambda_0^2 \int \frac{d\Omega}{2\pi} \int_{\Lambda e^{-\ell}}^{\Lambda} \frac{d^d p}{(2\pi)^d} \frac{1}{(-i(\Omega + \omega) + (p+k)^2/(2m))} \frac{1}{(i\Omega + p^2/(2m))}
$$

$$
= \lambda_0^2 \int_{\Lambda e^{-\ell}}^{\Lambda} \frac{d^d p}{(2\pi)^d} \frac{1}{(-i\omega + (p+k)^2/(2m) + p^2/(2m))}
$$

$$
= \lambda_0^2 \int_{\Lambda e^{-\ell}}^{\Lambda} \frac{d^d p}{(2\pi)^d} \frac{m}{p^2} - \lambda_0^2 \left(-i\omega + \frac{k^2}{4m} \left(2 - \frac{4}{d} \right) \right) \int_{\Lambda e^{-\ell}}^{\Lambda} \frac{d^d p}{(2\pi)^d} \frac{m^2}{p^4}. \tag{16.99}
$$

The first term is a constant that can be absorbed into a redefinition of δ. For the first time, we see above a special role for the spatial dimension $d = 4$, where the momentum integral is logarithmic. Our computations below turn out to be an expansion in powers of $(4 - d)$, and so we evaluate the numerical prefactors in (16.99) with $d = 4$. The result turns out to be correct to all orders in $(4 - d)$, but to see this explicitly we need to use a proper Galilean-invariant cutoff in a field-theoretic approach [362]. The simple momentum shell method being used here preserves Galilean invariance only in $d = 4$.

With the above reasoning, we see that the second term in the boson self-energy in (16.99) can be absorbed into a rescaling of the boson field under the RG. We therefore find a nonzero anomalous dimension

$$
\eta_b = \lambda^2, \tag{16.100}
$$

where we have absorbed phase space factors into the coupling λ by

$$
\lambda_0 = \frac{\Lambda^{2-d/2}}{m\sqrt{S_d}} \lambda. \tag{16.101}
$$

With this anomalous dimension, we use (16.98) to obtain the exact RG equation for λ:

$$
\frac{d\lambda}{d\ell} = \frac{(4-d)}{2} \lambda - \frac{\lambda^3}{2}. \tag{16.102}
$$

For $d < 4$, this flow has a stable fixed point at $\lambda = \lambda^* = \sqrt{(4-d)}$. The central claim of this subsection is that the theory $\mathcal{Z}_{FB}$ at this fixed point is identical to the theory $\mathcal{Z}_{Fs}$ at the fixed point $u = u^*$ for $2 < d < 4$.

Before we establish this claim, note that at the fixed point, we obtain the exact result for the anomalous dimension of the boson field

$$
\eta_b = 4 - d. \tag{16.103}
$$

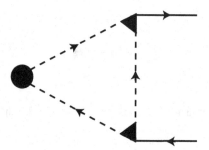

Fig. 16.10 Feynman diagram for the mixing between the renormalization of the $\Psi_F^\dagger \Psi_F$ and $\Psi_B^\dagger \Psi_B$ operators. The filled circle is the $\Psi_F^\dagger \Psi_F$ source. Other notation is as in Fig. 16.9.

Let us now consider the spectrum of relevant perturbations to the $\lambda = \lambda^*$ fixed point. As befits a Feshbach resonant fixed point, there are two relevant perturbations in $\mathcal{Z}_{FB}$, the detuning parameter δ and the chemical potential μ. Apart from the tree-level rescalings, at one loop we have the diagram shown in Fig. 16.10. This diagram has a $\Psi_{F\sigma}^\dagger \Psi_{F\sigma}$ source, and it renormalizes the coefficient of $\Phi^\dagger \Phi$: it evaluates to

$$
2\lambda_0^2 \int \frac{d\Omega}{2\pi} \int_{\Lambda e^{-\ell}}^{\Lambda} \frac{d^d p}{(2\pi)^d} \frac{1}{(-i\Omega + p^2/(2m))^2 (i\Omega + p^2/(2m))}
$$
$$
= 2\lambda_0^2 \int_{\Lambda e^{-\ell}}^{\Lambda} \frac{d^d p}{(2\pi)^d} \frac{m^2}{p^4}. \tag{16.104}
$$

Combining (16.104) with the tree-level rescalings, we obtain the RG flow equations

$$
\frac{d\mu}{d\ell} = 2\mu,
$$
$$
\frac{d}{d\ell}(\delta - 2\mu) = (2 - \eta_b)(\delta - 2\mu) - 2\lambda^2 \mu, \tag{16.105}
$$

where the last term arises from (16.104). With the value of η_b in (16.100), the second equation simplifies to

$$
\frac{d\delta}{d\ell} = (2 - \lambda^2)\delta. \tag{16.106}
$$

Thus we see that μ and δ are actually eigen-perturbations of the fixed point at $\lambda = \lambda^*$, and their scaling dimensions are

$$
\dim[\mu] = 2 \ , \ \ \dim[\delta] = d - 2. \tag{16.107}
$$

Note that these eigenvalues coincide with those of $\mathcal{Z}_{Fs}$ in (16.90), with δ identified as proportional to the detuning ν. This, along with the symmetries of Q conservation and Galilean invariance, establishes the equivalence of the fixed points of $\mathcal{Z}_{FB}$ and $\mathcal{Z}_{Fs}$.

The utility of the present $\mathcal{Z}_{FB}$ formulation is that it can provide a description of universal properties of the unitary Fermi gas in $d = 3$ via an expansion in $(4 - d)$. Further details of explicit computations can be found in [363].

16.4.2 Large-N expansion

We now return to the model $\mathcal{Z}_{Fs}$ in (16.4), and examine it in the limit of a large number of spin components [362, 525]. We also use the structure of the large-N perturbation theory to obtain exact results relating different experimental observables of the unitary Fermi gas.

The mechanics of the large-N expansion turn out to be quite similar to those in Chapter 11, although here we are dealing with fermionic fields. The basic idea is to endow the fermion with an additional flavor index $a = 1 \ldots N/2$, so that the fermion field is $\Psi_{F\sigma a}$, where we continue to have $\sigma = \uparrow, \downarrow$. Then, we write $\mathcal{Z}_{Fs}$ as

$$\mathcal{Z}_{Fs} = \int \mathcal{D}\Psi_{F\sigma a}(x, \tau) \exp\left(-\int_0^{1/T} d\tau \int d^d x \, \mathcal{L}_{Fs}\right),$$

$$\mathcal{L}_{Fs} = \Psi_{F\sigma a}^* \frac{\partial \Psi_{F\sigma a}}{\partial \tau} + \frac{1}{2m} |\nabla \Psi_{F\sigma a}|^2 - \mu |\Psi_{F\sigma a}|^2$$

$$+ \frac{2u_0}{N} \Psi_{F\uparrow a}^* \Psi_{F\downarrow a}^* \Psi_{F\downarrow b} \Psi_{F\uparrow b}, \tag{16.108}$$

where there is an implied sum over $a, b = 1 \ldots N/2$. The case of interest has $N = 2$, but we consider the limit of large even N, where the problem becomes tractable.

As written, there is an evident O($N/2$) symmetry in $\mathcal{Z}_{Fs}$ corresponding to rotations in flavor space. In addition, there is U(1) symmetry associated with Q conservation, and an SU(2) spin rotation symmetry. Actually, the spin and flavor symmetry combine to make the global symmetry U(1)$\times$Sp(N), but we do not make much use of this interesting observation.

The large-N expansion proceeds by decoupling the quartic term in (16.108) by a Hubbard–Stratonovich transformation (last seen in (9.32)). For this we introduce a complex bosonic field $\Psi_B(x, \tau)$ and write

$$\mathcal{Z}_{Fs} = \int \mathcal{D}\Psi_{F\sigma a}(x, \tau) \mathcal{D}\Psi_B(x, \tau) \exp\left(-\int_0^{1/T} d\tau \int d^d x \, \widetilde{\mathcal{L}}_{Fs}\right),$$

$$\widetilde{\mathcal{L}}_{Fs} = \Psi_{F\sigma a}^* \frac{\partial \Psi_{F\sigma a}}{\partial \tau} + \frac{1}{2m} |\nabla \Psi_{F\sigma a}|^2 - \mu |\Psi_{F\sigma a}|^2$$

$$+ \frac{N}{2|u_0|} |\Psi_B|^2 - \Psi_B \Psi_{F\uparrow a}^* \Psi_{F\downarrow a}^* - \Psi_B^* \Psi_{F\downarrow a} \Psi_{F\uparrow a}. \tag{16.109}$$

Here, and below, we assume $u_0 < 0$, which is necessary for being near the Feshbach resonance. Note that Ψ_B couples to the fermions just like the boson field in the Bose–Fermi model in (16.96), which is the reason for choosing this notation. If we perform the integral over Ψ_B in (16.109), we recover (16.108), as required. For the large-N expansion, we have to integrate over $\Psi_{F\sigma a}$ first and obtain an effective action for Ψ_B; this is the analog of the integration over $\mathbf{n}$ in Chapter 11. Because the action in (16.109) is Gaussian in the $\Psi_{F\sigma a}$, the integration over the fermion field involves evaluation of a functional determinant, and has the schematic form

$$\mathcal{Z}_{Fs} = \int \mathcal{D}\Psi_B(x, \tau) \exp\left(-N\mathcal{S}_{\text{eff}}[\Psi_B(x, \tau)]\right), \tag{16.110}$$

where S_{eff} is the logarithm of the fermion determinant of a single flavor. The key point is that the only N dependence is in the prefactor in (16.110), and so the theory of Ψ_B can controlled in powers of $1/N$, just as in Chapter 11.

We can expand S_{eff} in powers of Ψ_B: the pth term has a fermion loop with p external Ψ_B insertions. Details can be found in [362, 525]. Here, we note only the expansion to quadratic order at $\mu = \delta = T = 0$, in which case the coefficient is precisely the inverse of the fermion T matrix in (16.91):

$$S_{\text{eff}}[\Psi_B(x, \tau)] = -\frac{1}{2} \int \frac{d\omega}{2\pi} \frac{d^d k}{(2\pi)^d} \frac{1}{T(k, i\omega)} |\Psi_B(k, \omega)|^2 + \dots \qquad (16.111)$$

Given S_{eff}, we then have to find its saddle point with respect to Ψ_B. At $T = 0$, we find the optimal saddle point at a $\Psi_B \neq 0$ in the region of Fig. 16.8 with a nonzero density: this means that the ground state is always a superfluid of fermion pairs. The traditional expansion about this saddle point yields the $1/N$ expansion, and many experimental observables have been computed in this manner [362, 525, 526].

We conclude our discussion of the unitary Fermi gas by deriving an exact relationship between the total energy, E, and the momentum distribution function, $n(k)$, of the fermions [500, 501]. We do this using the structure of the large-N expansion. However, we drop the flavor index a below, and quote results directly for the physical case of $N = 2$. As usual, we define the momentum distribution function by

$$n(k) = \langle \Psi_{F\sigma}^{\dagger}(k, t) \Psi_{F\sigma}(k, t) \rangle, \qquad (16.112)$$

with no implied sum over the spin label σ. The Hamiltonian of the system in (16.108) is the sum of kinetic and interaction energies: the kinetic energy is clearly an integral over $n(k)$ and so we can write

$$\begin{aligned} E &= 2V \int \frac{d^d k}{(2\pi)^d} \frac{k^2}{2m} n(k) + u_0 V \langle \Psi_{F\uparrow}^{\dagger} \Psi_{F\downarrow}^{\dagger} \Psi_{F\downarrow} \Psi_{F\uparrow} \rangle \\ &= 2V \int \frac{d^d k}{(2\pi)^d} \frac{k^2}{2m} n(k) - u_0 \frac{\partial \ln \mathcal{Z}_{Fs}}{\partial u_0}, \end{aligned} \qquad (16.113)$$

where V is the system volume, and all the Ψ_F fields are at the same x and t. Now let us evaluate the u_0 derivative using the expression for $\mathcal{Z}_{Fs}$ in (16.109); this leads to

$$\frac{E}{V} = 2 \int \frac{d^d k}{(2\pi)^d} \frac{k^2}{2m} n(k) + \frac{1}{u_0} \langle \Psi_B^*(x, t) \Psi_B(x, t) \rangle. \qquad (16.114)$$

Now using the expression (16.93) relating u_0 to the scattering length a in $d = 3$, we can write this expression as

$$\frac{E}{V} = \frac{m}{4\pi a} \langle \Psi_B^* \Psi_B \rangle + 2 \int \frac{d^3 k}{(2\pi)^3} \frac{k^2}{2m} \left(n(k) - \frac{\langle \Psi_B^* \Psi_B \rangle m^2}{k^4} \right). \qquad (16.115)$$

This is the needed universal expression for the energy, expressed in terms of $n(k)$ and the scattering length, and independent of the short-distance structure of the interactions.

At this point, it is useful to introduce "Tan's constant" C, defined by [500, 501]

$$C = \lim_{k \to \infty} k^4 n(k). \qquad (16.116)$$

The requirement that the momentum integral in (16.115) is convergent in the ultraviolet implies that the limit in (16.116) exists, and further specifies its value

$$C = m^2 \langle \Psi_B^* \Psi_B \rangle. \tag{16.117}$$

We now note that the relationship $n(k) \to m^2 \langle \Psi_B^* \Psi_B \rangle / k^4$ at large k is also as expected from a scaling perspective. We saw in Section 16.4.1 that the fermion field Ψ_F does not acquire any anomalous dimensions, and has scaling dimension $d/2$. Consequently $n(k)$ has scaling dimension zero. Next, note that the operator $\Psi_B^* \Psi_B$ is conjugate to the detuning from the Feshbach critical point; from (16.107) the detuning has scaling dimension $d - 2$, and so $\Psi_B^* \Psi_B$ has scaling dimension $d + z - (d - 2) = 4$. Combining these scaling dimensions, we explain the k^{-4} dependence of $n(k)$.

It now remains to establish the claimed exact relationship in (16.117) as a general property of a spinful Fermi gas near unitarity. As a start, we can examine the large k limit of $n(k)$ in the BCS mean-field theory of the superfluid phase: the reader can easily verify that the textbook BCS expressions for $n(k)$ do indeed satisfy (16.117). However, the claim of [55, 479] is that (16.117) is exact beyond mean-field theory, and also holds in the non-superfluid states at nonzero temperatures. A general proof was given in [479], and relied on the operator product expansion (OPE) applied to the field theory (16.109). The OPE is a general method for describing the short-distance and time (or large momentum and frequency) behavior of field theories. Typically, in the Feynman graph expansion of a correlator, the large momentum behavior is dominated by terms in which the external momenta flow in only a few propagators, and the internal momentum integrals can be evaluated after factoring out these favored propagators. For the present situation, let us consider the $1/N$ correction to the fermion Green's function given by the diagram in Fig. 16.11. Representing the bare fermion and boson Green's functions by G_F and G_B, respectively, Fig. 16.11 evaluates to

$$G_F^2(k, \omega) \int \frac{d^d p}{(2\pi)^d} \frac{d\Omega}{2\pi} G_B(p, \Omega) G_F(-k + p, -\omega + \Omega). \tag{16.118}$$

Here G_B is the propagator of the boson action $\mathcal{S}_{\text{eff}}$ specified by (16.111). In the limit of large k and ω, the internal p and Ω integrals are dominated by p and Ω much smaller than k and ω; so we can approximate (16.118) by

$$G_F^2(k, \omega) G_F(-k, -\omega) \int \frac{d^d p}{(2\pi)^d} \frac{d\Omega}{2\pi} G_B(p, \Omega)$$
$$= G_F^2(k, \omega) G_F(-k, -\omega) \langle \Psi_B^* \Psi_B \rangle. \tag{16.119}$$

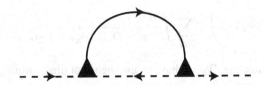

Fig. 16.11   Order $1/N$ correction to the fermion Green's function. Notation is as in Fig. 16.9.

This analysis can now be extended to all orders in $1/N$. Among these higher order contributions are terms which contribute self-energy corrections to the boson propagator G_B in (16.119): it is clear that these can be summed to replace the bare G_B in (16.119) by the exact G_B. Then the value of $\langle |\Psi_B|^2 \rangle$ in (16.119) also becomes the exact value. All remaining contributions can be shown [479] to fall off faster at large k and ω than the terms in (16.119). So (16.119) is the exact leading contribution to the fermion Green's function in the limit of large k and ω after replacing $\langle |\Psi_B|^2 \rangle$ by its exact value. We can now integrate (16.119) over ω to obtain $n(k)$ at large k. Actually the ω integral is precisely that in (16.104), which immediately yields the needed relation (16.117).

Similar analyses can be applied to determine the spectral functions of other observables [56, 92, 209, 448, 449, 479, 526].

Determination of the specific value of Tan's constant requires numerical computations in the $1/N$ expansion of (16.110). From the scaling properties of the Feshbach resonant fixed point in $d = 3$, we can deduce that the result obeys a scaling form similar to (16.20):

$$C = (2mT)^2 \Phi_C \left(\frac{\mu}{T}, \frac{\nu}{\sqrt{2mT}} \right), \tag{16.120}$$

where Φ_C is a dimensionless universal function of its dimensionless arguments; note that the arguments represent the axes of Fig. 16.8. The methods of [362, 525] can now be applied to (16.117) to obtain numerical results for Φ_C in the $1/N$ expansion. We illustrate this method here by determining C to leading order in the $1/N$ expansion at $\mu = \nu = 0$. For this, we need to generalize the action (16.111) for Ψ_B to $T > 0$ and general N. Using (16.91) we can modify (16.111) to

$$\mathcal{S}_{\text{eff}} = NT \sum_{\omega_n} \int \frac{d^3k}{8\pi^3} [D_0(k, \omega_n) + D_1(k, \omega_n)] |\Psi_B(k, \omega_n)|^2, \tag{16.121}$$

where D_0 is the $T = 0$ contribution, and D_1 is the correction at $T > 0$:

$$D_0(k, \omega_n) = \frac{m^{3/2}}{16\pi} \sqrt{-i\omega_n + \frac{k^2}{4m}},$$

$$D_1(k, \omega_n) = \frac{1}{2} \int \frac{d^3p}{8\pi^3} \frac{1}{(e^{p^2/(2mT)} + 1)} \frac{1}{(-i\omega + p^2/(2m) + (p+k)^2/(2m))}. \tag{16.122}$$

We now have to evaluate $\langle \Psi_B^* \Psi_B \rangle$ using the Gaussian action in (16.121). It is useful to do this by separating the D_0 contribution, which allows us to properly deal with the large frequency behavior. So we can write

$$\langle \Psi_B^* \Psi_B \rangle = \frac{1}{N} T \sum_{\omega_n} \int \frac{d^3k}{8\pi^3} \left[\frac{1}{D_0(k, \omega_n) + D_1(k, \omega_n)} - \frac{1}{D_0(k, \omega_n)} \right] + D_{00}. \tag{16.123}$$

In evaluating D_{00} we have to use the usual time-splitting method to ensure that the bosons are normal-ordered, and evaluate the frequency summation by analytically continuing to the real axis:

$$D_{00} = \frac{1}{N} \int \frac{d^3k}{8\pi^3} \lim_{\eta \to 0} T \sum_{\omega_n} \frac{e^{i\omega_n \eta}}{D_0(k, \omega_n)}$$

$$= \frac{16\pi}{Nm^{3/2}} \int \frac{d^3k}{8\pi^3} \int_{\frac{k^2}{4m}}^{\infty} \frac{d\Omega}{\pi} \frac{1}{(e^{\Omega/T} - 1)} \frac{1}{\sqrt{\Omega - k^2/(4m)}}$$

$$= \frac{8.37758}{N} T^2. \tag{16.124}$$

The frequency summation in (16.123) can be evaluated directly on the imaginary frequency axis: the series is convergent at large ω_n, and is easily evaluated by a direct numerical summation. Numerical evaulation of (16.123) now yields

$$C = (2mT)^2 \left(\frac{0.67987}{N} + \mathcal{O}(1/N^2) \right) \tag{16.125}$$

at $\mu = \nu = 0$.

16.5 Applications and extensions

Experiments on the loss of superfluidity of ^{4}He adsorbed in aerogel [98, 99, 402] provide a realization of the dilute Bose gas theory in the presence of a random external potential [148]. There are few analytic results on this random problem, although some detailed numerical studies have been undertaken [484, 535]; however, a reconciliation between theory and experiments has not yet occurred. The experiments have been carried out both in bulk ($d = 3$) and in films ($d = 2$).

Quantum antiferromagnets in the presence of an external magnetic field provide some of the best experimental realizations of the dilute Bose gas quantum critical point. We defer discussion of experiments on this case until Chapter 19 where the connection is explicitly discussed.

Corrections to the T_c of a dilute Bose gas in $d = 3$ beyond the result (16.69) have been studied numerically [183] and by a renormalization group method somewhat different from our analysis here [47]. It would be interesting to compare these results with those obtained here, after accounting for the fact that the critical point is at a nonzero $R_c \sim (TU)^{1/2}$ (see the discussion below (14.19)); this has not yet been done.

Parts II and III focus on quantum phase transitions in which the primary degrees of freedom are bosonic. However, many of the experimentally most interesting transitions involve fermions in a crucial way. We met the examples of the dilute Fermi gases in Chapter 16, where the quantum critical point itself was at zero fermion density. Consequently, the singularities in the fermion Green's function are at momentum $\vec{k} = 0$, the point at which the single-particle energy vanishes quadratically. This and the following chapter explore fermion systems at nonzero density; the systems have different types of singularity in momentum space. The present chapter explores fermion systems with fermionic excitations with a relativistic Dirac spectrum with an energy which vanishes linearly at isolated points in the Brillouin zone. We find that this leads to field theories which can be studied by methods similar to those developed earlier for bosonic systems. Chapter 18 turns to Fermi liquids, whose zero energy excitations lie along an extended Fermi surface in the Brillouin zone, and their transitions have numerous qualitative new features. The methods developed in the present chapter for Dirac fermions prove useful in our analysis of Fermi surfaces in Chapter 18.

Further motivation for the study of Dirac fermions is provided by numerous new experimental realizations. These include d-wave superconductors, graphene, and topological insulators. Here we present the theory in the context of d-wave superconductors, but the methods are easily extended to the other situations [217, 254].

We begin in Section 17.1 with a discussion of the origin of d-wave superconductivity in correlated electron systems, and the appearance of Dirac fermions in their spectrum. Then we consider two quantum phase transitions, both involving a simple Ising order parameter. The first in Section 17.2, with time-reversal symmetry breaking, leads to a relativistic quantum field theory closely related to the so-called Gross–Neveu model. The second model of Section 17.4 involves breaking of a lattice rotation symmetry, leading to "Ising-nematic" order. The theory for this model is not relativistically invariant: it is strongly coupled, but can be controlled by a $1/N$ expansion.

17.1 *d*-wave superconductivity and Dirac fermions

We begin with a review of the origin of d-wave superconductivity in correlated electron models with antiferromagnetic exchange interactions on the square lattice [344, 446]. We then show that the theory of low-energy electronic excitations can be expressed in terms of a continuum theory of Dirac fermions.

We apply the standard Bardeen–Cooper–Schrieffer (BCS) mean-field theory of superconductivity to the following Hamiltonian for electrons hopping between the sites of the
square lattice

$$H_{tJ} = \sum_k \varepsilon_k c_{ka}^\dagger c_{ka} + J_1 \sum_{\langle ij \rangle} \hat{\mathbf{S}}_i \cdot \hat{\mathbf{S}}_j, \tag{17.1}$$

where c_{ja} is the annihilation operator for an electron on site j with spin $a = \uparrow\downarrow$, c_{ka} is
its Fourier transform to momentum space, and ε_k is the dispersion of the electrons; we
can include first- and second-neighbor hopping to obtain $\varepsilon_k = -2t_1(\cos(k_x) + \cos(k_y)) -
2t_2(\cos(k_x + k_y) + \cos(k_x - k_y)) - \mu$, with μ the chemical potential (the Fermi surface
associated with this dispersion appears in Fig. 18.1). The J_1 term is the nearest-neighbor
antiferromagnetic exchange interaction with

$$\hat{S}_{j\alpha} = \frac{1}{2} c_{ja}^\dagger \sigma_{ab}^\alpha c_{jb} \tag{17.2}$$

and σ^α the Pauli matrices. We consider the consequences of the further neighbor exchange
interactions for the superconductor in Section 17.2 below.

It is now useful to re-express the exchange interaction in (17.1) in a different form, using
simple mathematical identities obeyed by the Pauli matrices:

$$\hat{\mathbf{S}}_i \cdot \hat{\mathbf{S}}_j = -\frac{1}{2}\left(\epsilon_{\alpha\gamma} c_{i\alpha}^\dagger c_{j\gamma}^\dagger\right)\left(\epsilon_{\beta\delta} c_{j\delta} c_{i\beta}\right) + \frac{1}{4}\left(c_{i\alpha}^\dagger c_{i\alpha}\right)\left(c_{j\beta}^\dagger c_{j\beta}\right), \tag{17.3}$$

where $\epsilon_{\alpha\beta}$ is the unit antisymmetric tensor. Note that the first term on the right-hand side
is of the form of an attractive interaction between singlet pairs of electrons: it is this attraction which leads to superconductivity. The second term is a nearest-neighbor repulsion
which, in mean-field theory, only renormalizes the chemical potential: we therefore neglect
it below. Inserting (17.3) into (17.1), and performing a BCS mean-field factorization of the
attractive interaction leads to the Bogoliubov Hamiltonian

$$H_{BCS} = \sum_k \varepsilon_k c_{ka}^\dagger c_{ka} - \frac{J_1}{2} \sum_{j,\mu=x,y} \Delta_\mu \left(c_{j\uparrow}^\dagger c_{j+\hat{\mu},\downarrow}^\dagger - c_{j\downarrow}^\dagger c_{j+\hat{\mu},\uparrow}^\dagger\right) + \text{H.c.} \tag{17.4}$$

Here Δ_x and Δ_y are mean-field variational parameters which have to be determined by
solution of the self-consistency equations

$$\Delta_\mu = -\left\langle c_{j\uparrow} c_{j+\hat{\mu},\downarrow} - c_{j\downarrow} c_{j+\hat{\mu},\uparrow}\right\rangle. \tag{17.5}$$

We can diagonalize H_{BCS} by a Bogoliubov transformation similar to that applied to (10.16),
and so obtain the ground state energy, and the spectrum of Bogoliubov fermion excitations
γ_{ka}. The ground state energy is

$$E_{BCS} = \frac{J_1}{2}\left(|\Delta_x|^2 + |\Delta_y|^2\right) - \int \frac{d^2k}{4\pi^2}\left[E_k - \varepsilon_k\right], \tag{17.6}$$

and this is expressed in terms of the excitation energy of the fermionic quasiparticles, γ_{ka},

$$E_k = \left[\varepsilon_k^2 + |J_1(\Delta_x \cos k_x + \Delta_y \cos k_y)|^2\right]^{1/2}. \tag{17.7}$$

It now remains to determine the parameters Δ_x and Δ_y. These are determined either by solving (17.5), or by minimizing (17.7); when generalized to $T > 0$, both yield the equation

$$\Delta_\mu = \int \frac{d^2k}{4\pi^2} \frac{(\Delta_x \cos k_x + \Delta_y \cos k_y)}{E_k} \tanh\left(\frac{E_k}{2T}\right) \cos k_\mu. \qquad (17.8)$$

For the band structure ε_k appropriate to the cuprates (specified below (17.1)), numerical analysis shows that the optimum choice is always $\Delta_x = -\Delta_y$. This is the $d_{x^2-y^2}$ pairing solution. Qualitatively, we can understand this by noting [463] that the energy in (17.6) is efficiently minimized if we can choose the gap function $\Delta_k \equiv \Delta_x \cos k_x + \Delta_y \cos k_y$ to be as large as possible near the region of the Brillouin zone where $|\varepsilon_k|$ is small, i.e. near the Fermi surface; for the cuprate Fermi surface (see Fig. 18.1), this happens for $\Delta_x = -\Delta_y$. We therefore set $\Delta_x = -\Delta_y = \Delta_{x^2-y^2}$ below, and the value of $\Delta_{x^2-y^2}$ is determined by the solution of (17.8).

Now we show that the above d-wave superconductor has quasiparticle excitations which are Dirac fermions. With the choice $\Delta_x = -\Delta_y$, the energy of the quasiparticles, E_k, vanishes at the four points $(\pm Q, \pm Q)$ at which $\varepsilon_k = 0$, shown at the left of Fig. 17.2; these points are determined by the crossing of the diagonals (where $\cos k_x = \cos k_y$) with the Fermi surface in Fig. 18.1 (where $\varepsilon_k = 0$). We are especially interested in the low-energy quasiparticles in the vicinity of these points, and so we perform a gradient expansion of H_{BCS} near each of them. We label the points $\vec{Q}_1 = (Q, Q)$, $\vec{Q}_2 = (-Q, Q)$, $\vec{Q}_3 = (-Q, -Q)$, $\vec{Q}_4 = (Q, -Q)$ and write

$$c_{ja} = f_{1a}(\vec{r}_j)e^{i\vec{Q}_1 \cdot \vec{r}_j} + f_{2a}(\vec{r}_j)e^{i\vec{Q}_2 \cdot \vec{r}_j} + f_{3a}(\vec{r}_j)e^{i\vec{Q}_3 \cdot \vec{r}_j} + f_{4a}(\vec{r}_j)e^{i\vec{Q}_4 \cdot \vec{r}_j}, \qquad (17.9)$$

while assuming the $f_{1-4,a}(\vec{r})$ are slowly varying functions of x. We also introduce the bispinors

$$\Psi_1 = \begin{pmatrix} f_{1\uparrow} \\ f_{3\downarrow}^\dagger \\ f_{1\downarrow} \\ -f_{3\uparrow}^\dagger \end{pmatrix}, \quad \Psi_2 = \begin{pmatrix} f_{2\uparrow} \\ f_{4\downarrow}^\dagger \\ f_{2\downarrow} \\ -f_{4\uparrow}^\dagger \end{pmatrix}, \qquad (17.10)$$

and then express H_{BCS} in terms of $\Psi_{1,2}$ while performing a spatial gradient expansion. This yields the following effective action for the fermionic quasiparticles:

$$S_\Psi = \int d\tau d^2r \left[\Psi_1^\dagger \left(\partial_\tau - i\frac{v_F}{\sqrt{2}}(\partial_x + \partial_y)\tau^z - i\frac{v_\Delta}{\sqrt{2}}(-\partial_x + \partial_y)\tau^x \right) \Psi_1 \right.$$
$$\left. + \Psi_2^\dagger \left(\partial_\tau - i\frac{v_F}{\sqrt{2}}(-\partial_x + \partial_y)\tau^z - i\frac{v_\Delta}{\sqrt{2}}(\partial_x + \partial_y)\tau^x \right) \Psi_2 \right], \qquad (17.11)$$

where the $\tau^{x,z}$ are 4×4 matrices which are block diagonal, the blocks consisting of 2×2 Pauli matrices. The velocities $v_{F,\Delta}$ are given by the conical structure of E_k near the Q_{1-4}:

we have $v_F = \left|\nabla_k \varepsilon_k|_{k=Q_a}\right|$ and $v_\Delta = |J_1 \Delta_{x^2-y^2} \sqrt{2} \sin(Q)|$. In this limit, the energy of the Ψ_1 fermionic excitations is

$$E_k = (v_F^2 (k_x + k_y)^2/2 + v_\Delta^2 (k_x - k_y)^2/2)^{1/2}, \tag{17.12}$$

(and similarly for Ψ_2), which is the spectrum of massless Dirac fermions.

We can now study the influence of short-range interactions between the Dirac fermions, using arguments similar to those in Section 16.2. The free Dirac theory in (17.11) is invariant under the scaling transformation

$$x' = x e^{-\ell}, \quad \tau' = \tau e^{-\ell}, \quad \Psi' = \Psi e^{\ell}. \tag{17.13}$$

Let us assume there is a contact four Fermi interaction, u; unlike (16.17), a term without spatial gradients is now allowed because the multiple indices on the Dirac fermion allow us to evade the exclusion principle. Simple power-counting shows that $\dim[u] = -1$, and so the interaction is irrelevant. Using the arguments similar to those around (14.37), we conclude therefore that the self-energy of the Dirac fermions scales as $\mathrm{Im}\,\Sigma(k = 0, \omega) \sim u^2 \omega^3$. This is a weak perturbation on the bare Green's function, and so the Dirac quasiparticles remain well defined.

Interesting departures from the free Dirac fermion behavior appear only upon considering the vicinity of quantum phase transitions which change the nature of the fermionic spectrum. We consider two such examples in the remainder of this chapter.

17.2 Time-reversal symmetry breaking

We have now shown that the low-energy spectrum of a BCS d-wave superconductor on a square lattice consists of two massless 4-component Dirac fermions (see (17.11)). We now consider a quantum phase transition involving the breaking of time-reversal symmetry within the superconducting state. The superconducting order itself will only be weakly affected by this transition, and so will be regarded as a spectator to the analysis below: the main role of the superconductivity is to provide the low-energy Dirac fermions, and they control the critical fluctuations. This is why similar theories apply to symmetry-breaking transitions in graphene [217, 254].

We consider a simple model of time-reversal symmetry breaking in which the pairing symmetry of the superconductor changes from $d_{x^2-y^2}$ to $d_{x^2-y^2} \pm i d_{xy}$. The choice of the phase between the two pairing components leads to a breaking of time-reversal symmetry.

The mean-field theory of this transition can be explored entirely within the context of BCS theory, as we review below. However, fluctuations about the BCS theory are strong, and lead to nontrivial critical behavior involving both the collective order parameter and the Bogoliubov fermions: this is probably the earliest known example [527, 528] of the failure of BCS theory in two (or higher) dimensions in a superconducting ground state.

We extend H_{tJ} in (17.1) so that BCS mean-field theory permits a region with d_{xy} super-conductivity. With a J_2 second-neighbor interaction, (17.1) is modified to:

$$\widetilde{H}_{tJ} = \sum_k \varepsilon_k c_{k\sigma}^\dagger c_{k\sigma} + J_1 \sum_{\langle ij \rangle} \mathbf{S}_i \cdot \mathbf{S}_j + J_2 \sum_{\text{nnn } ij} \mathbf{S}_i \cdot \mathbf{S}_j. \tag{17.14}$$

We follow the evolution of the ground state of $\widetilde{H}_{tJ}$ as a function of J_2/J_1. It is worth noting here that the model in (17.14) has been argued [463] to describe the pnictide superconductors, with ε_k modified to realize the band structure of these compounds.

The mean-field Hamiltonian is now modified from Eq. (17.4) to

$$\widetilde{H}_{BCS} = \sum_k \varepsilon_k c_{k\sigma}^\dagger c_{k\sigma} - \frac{J_1}{2} \sum_{j,\mu} \Delta_\mu \left(c_{j\uparrow}^\dagger c_{j+\hat{\mu},\downarrow}^\dagger - c_{j\downarrow}^\dagger c_{j+\hat{\mu},\uparrow}^\dagger \right) + \text{h.c.}$$

$$- \frac{J_2}{2} \sum_{j,\nu}' \Delta_\nu \left(c_{j\uparrow}^\dagger c_{j+\hat{\nu},\downarrow}^\dagger - c_{j\downarrow}^\dagger c_{j+\hat{\nu},\uparrow}^\dagger \right) + \text{h.c.}, \tag{17.15}$$

where the second summation over ν is along the diagonal neighbors $\hat{x} + \hat{y}$ and $-\hat{x} + \hat{y}$. We have now a total of four variational parameters: Δ_x, Δ_y, Δ_{x+y}, Δ_{-x+y}. For ε_k as in the cuprates, the optimum solution has $\Delta_x = -\Delta_y = \Delta_{x^2-y^2}$ as before, and d_{xy} pairing along the diagonals with $\Delta_{x+y} = -\Delta_{-x+y} = \Delta_{xy}$: these choices for the spatial structure of the pairing amplitudes (which determine the Cooper pair wavefunction) are summarized in Fig 17.1. For the pnictide band structure, the appropriate choices turn out to be $\Delta_x = \Delta_y$ and $\Delta_{x+y} = \Delta_{-x+y}$, which realizes a form of extended-s pairing [326, 463]. We do not consider the latter choice here, and focus on the $d_{x^2-y^2}$ and d_{xy} pairing shown in Fig. 17.1.

The values of the pairing amplitudes $\Delta_{x^2-y^2}$ and Δ_{xy} are to be determined by minimizing the ground state energy (generalizing (17.6))

$$E_{BCS} = J_1 |\Delta_{x^2-y^2}|^2 + J_2 |\Delta_{xy}|^2 - \int \frac{d^2k}{4\pi^2} \left[E_k - \varepsilon_k \right], \tag{17.16}$$

where the quasiparticle dispersion is now (generalizing (17.7))

$$E_k = \left[\varepsilon_k^2 + \left| J_1 \Delta_{x^2-y^2} (\cos k_x - \cos k_y) + 2 J_2 \Delta_{xy} \sin k_x \sin k_y \right|^2 \right]^{1/2}. \tag{17.17}$$

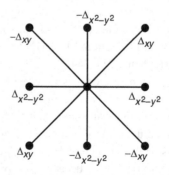

Fig. 17.1   Values of the pairing amplitudes, $-\langle c_{i\uparrow} c_{j\downarrow} - c_{i\downarrow} c_{j\uparrow} \rangle$ with i the central site, and j one of its eight near neighbors; for the band structure of the cuprates (Fig. 18.1).

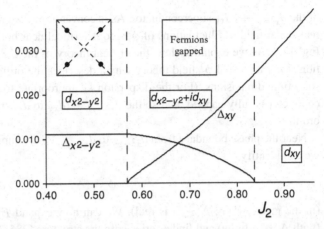

Fig. 17.2 BCS solution of the phenomenological Hamiltonian $\widetilde{H}_{tJ}$ in (17.14). Shown are the optimum values of the pairing amplitudes $|\Delta_{x^2-y^2}|$ and $|\Delta_{xy}|$ as a function of J_2 for $t_1 = 1$, $t_2 = -0.25$, $\mu = -1.25$, and J_1 fixed at $J_1 = 0.4$. The relative phase of the pairing amplitudes was always found to obey (17.18). The dashed lines denote locations of phase transitions between $d_{x^2-y^2}$, $d_{x^2-y^2} + id_{xy}$, and d_{xy} superconductors. The pairing amplitudes vanish linearly at the first transition corresponding to the exponent $\beta_{BCS} = 1$ in Eq. (17.21). The Brillouin zone location of the gapless Dirac points in the $d_{x^2-y^2}$ superconductor is indicated by filled circles. For the dispersion ε_k appropriate to the cuprates, the d_{xy} superconductor is fully gapped, and so the second transition is ordinary Ising.

Note that the energy depends upon the relative phase of $\Delta_{x^2-y^2}$ and Δ_{xy}: this phase is therefore an observable property of the ground state.

It is a simple matter to numerically carry out minimization of (17.17), and the results for a typical choice of parameters are shown in Fig. 17.2 as a function J_2/J_1. One of the two amplitudes $\Delta_{x^2-y^2}$ or Δ_{xy} is always nonzero and so the ground state is always superconducting. The transition from pure $d_{x^2-y^2}$ superconductivity to pure d_{xy} superconductivity occurs via an intermediate phase in which *both* order parameters are nonzero. Furthermore, in this regime, their relative phase is found to be pinned to $\pm\pi/2$, i.e.

$$\arg(\Delta_{xy}) = \arg(\Delta_{x^2-y^2}) \pm \pi/2. \tag{17.18}$$

The reason for this pinning can be seen intuitively from (17.17): only for these values of the relative phase does the equation $E_k = 0$ never have a solution. In other words, the gapless nodal quasiparticles of the $d_{x^2-y^2}$ superconductor acquire a finite energy gap when a secondary pairing with relative phase $\pm\pi/2$ develops. With a level repulsion picture, we can expect that gapping out the low-energy excitations should help lower the energy of the ground state. The intermediate phase obeying (17.18) is called a $d_{x^2-y^2} + id_{xy}$ superconductor.

The choice of the sign in (17.18) leads to an overall two-fold degeneracy in the choice of the wavefunction for the $d_{x^2-y^2} + id_{xy}$ superconductor. This choice is related to the breaking of time-reversal symmetry, and implies that the $d_{x^2-y^2} + id_{xy}$ phase is characterized by the nonzero expectation value of a Z_2 Ising order parameter; the expectation value of this order vanishes in the two phases (the $d_{x^2-y^2}$ and d_{xy} superconductors) on either side of the $d_{x^2-y^2} + i d_{xy}$ superconductor. As is conventional, we represent the Ising order by a real scalar field ϕ. Fluctuations of ϕ become critical near both of the phase boundaries in Fig. 17.2. As we explain below, the critical theory of the $d_{x^2-y^2}$ to $d_{x^2-y^2} + i d_{xy}$ transition is *not* the usual ϕ^4 field theory which describes the ordinary Ising transition in three spacetime dimensions. (For the dispersion ε_k appropriate to the cuprates, the d_{xy} superconductor is fully gapped, and so the $d_{x^2-y^2} + i d_{xy}$ to d_{xy} transition in Fig. 17.2 will be ordinary Ising.)

Near the phase boundary from $d_{x^2-y^2}$ to $d_{x^2-y^2} + i d_{xy}$ superconductivity it is clear that we can identify

$$\phi = i \Delta_{xy}, \tag{17.19}$$

(in the gauge where $\Delta_{x^2-y^2}$ is real). We can now expand E_{BCS} in (17.16) for small ϕ (with $\Delta_{x^2-y^2}$ finite) and find a series with the structure [285, 298]

$$E_{BCS} = E_0 + r\phi^2 + v|\phi|^3 + \dots, \tag{17.20}$$

where r, v are coefficients and the ellipses represent regular higher order terms in even powers of ϕ; r can have either sign, whereas v is always positive. Note the nonanalytic $|\phi|^3$ term that appears in the BCS theory – this arises from an infrared singularity in the integral in (17.16) over E_k at the four nodal points of the $d_{x^2-y^2}$ superconductor, and is a preliminary indication that the transition differs from that in the ordinary Ising model, and that the Dirac fermions play a central role. We can optimize ϕ by minimizing E_{BCS} in (17.20) – this shows that $\langle \phi \rangle = 0$ for $r > 0$, and $\langle \phi \rangle \neq 0$ for $r < 0$. So $r \sim (J_2/J_1)_c - J_2/J_1$ where $(J_2/J_1)_c$ is the first critical value in Fig. 17.2. Near this critical point, we find

$$\langle \phi \rangle \sim (-r)^\beta, \tag{17.21}$$

and the present BCS theory yields the exponent $\beta_{BCS} = 1$; this differs from the usual mean-field exponent $\beta_{MF} = 1/2$, and this is of course due to the nonanalytic $|\phi|^3$ term in (17.20).

The form of (17.20) and the unusual β exponent are tell-tale signatures that the time-reversal symmetry breaking transition is *not* described by the universality class of the quantum Ising model studied in Parts II and III. We need a new field theory, and an associated RG analysis, and this is described in the following section.

17.3   Field theory and RG analysis

We perform our analysis of the critical point by going back to the basic strategy of Chapters 3 and 4: identify the complete set of low-energy degrees of freedom, and write down the most general local action of these degrees of freedom, while being consistent with all global symmetries. Clearly, the Ising order parameter, ϕ, should remain as one of our degrees of freedom. However, we also have to keep the Dirac fermions alive, even though they are not directly related to the order parameter; they are responsible for the nonanalyticity in (17.20), and so we should not eliminate them from the low-energy theory.

Thus we need a local field theory expressed in terms of both ϕ and Ψ. Clearly, the action is still allowed to have the terms of the pure Ising model of Part II. So we include the terms in (2.11) for $N = 1$, which we write here as

$$S_\phi = \int d^d x d\tau \left[\frac{1}{2} \left((\partial_\tau \phi)^2 + c^2 (\nabla_x \phi)^2 + c^2 (\partial_y \phi)^2 + r\phi^2 \right) + \frac{u}{24} \phi^4 \right], \qquad (17.22)$$

where for the RG analysis we consider the general case of d spatial dimensions. Note that, unlike (17.20), we do not have any nonanalytic $|\phi|^3$ terms in the action: this is because we have not integrated out the low-energy Dirac fermions, and the terms in (17.22) are viewed as arising from high-energy fermions away from the nodal points.

For the Dirac fermions, we already have the low-energy excitations described by the field theory of massless Dirac fermions, S_Ψ in (17.11).

Finally, we need to couple the ϕ and $\Psi_{1,2}$ excitations. Their coupling is already contained in the last term in (17.15): expressing this in terms of the $\Psi_{1,2}$ fermions using (17.9) we obtain

$$S_{\Psi\phi} = \lambda \int d^d x d\tau \left[\phi \left(\Psi_1^\dagger \tau^y \Psi_1 - \Psi_2^\dagger \tau^y \Psi_2 \right) \right], \qquad (17.23)$$

where λ is a coupling constant.

Collecting all terms, we obtain the needed field theory for the time-reversal symmetry breaking transition from a $d_{x^2-y^2}$ superconductor to a $d_{x^2-y^2} + id_{xy}$ superconductor:

$$\mathcal{Z}_{did} = \int \mathcal{D}\phi \mathcal{D}\Psi_1 \mathcal{D}\Psi_2 \exp\left(-S_\Psi - S_\phi - S_{\Psi\phi} \right). \qquad (17.24)$$

It can now be checked that if we were to integrate out the $\Psi_{1,2}$ fermions for a spacetime independent ϕ using the above theory, we would indeed obtain a $|\phi|^3$ term in the effective potential for ϕ.

Note that the field theory defined above has a formal similarity to that proposed for the spinful Fermi gas near a Feshbach resonance in Section 16.4.1: both theories have bosonic and fermionic degrees of freedom coupled together in a trilinear term. In the previous case, the dispersion of the particles was quadratic in momentum, whereas here it is linear.

Let us now analyze $\mathcal{Z}_{did}$ using the momentum shell RG described in Chapter 4, and also used in Section 16.4.1. First, we rescale coordinates as in (4.1), (10.30), and (17.13):

$$x' = xe^{-\ell}; \quad \tau' = \tau e^{-z\ell}. \qquad (17.25)$$

At the mean-field level, the dynamic exponent $z = 1$ is appropriate for both the boson and fermion actions. At higher orders, we choose z to maintain a constant coefficient for the spatial derivative of the ϕ field; we can therefore choose units to set $c = 1$ from now on. The fermionic velocities v_F and v_Δ flow under RG, as we see below.

The corresponding renormalizations of the fields parallel those in (4.3), (10.30), and (17.13):

$$\phi' = \phi \, e^{(d-1+\eta_b)\ell/2}, \quad \Psi' = \Psi \, e^{(d+\eta_f)\ell/2}, \qquad (17.26)$$

and these are chosen to keep the temporal derivative terms of both the bosons and fermions invariant. Note that there are separate anomalous dimensions, η_b and η_f, for the bosons and fermions, and these become nonzero and important once we include loop corrections.

At the level of the mean-field action, let us now examine the flow of the quartic coupling, u, and the "Yukawa" coupling, λ. Ignoring the anomalous dimensions, which vanish at this level, and setting $z = 1$, we have

$$u' = u \, e^{(3-d)\ell}, \quad \lambda' = \lambda \, e^{(3-d)\ell/2}. \tag{17.27}$$

The change in u is as in (4.10). The key observation we can now make is that the coupling between the fermions and the bosons, λ, is relevant for $d < 3$. Thus we need to control the expansion in powers of λ using the RG, just as we did earlier for the expansion in powers of u. The relevancy of λ also explains why β_{BCS} was not equal to the mean-field value above.

To proceed further we need the one-loop RG equations, the analog of those in (4.24). Note that the scaling dimensions of both u and λ become small as d approaches 3: this implies that we are able to control the RG flow in an expansion in powers of $(3 - d)$, as was the case for (4.24).

There is one important difference from (4.24) here. We now find that the boson and fermion self-energies depend upon external momenta and frequencies already at the one-loop level. Thus we find nonvanishing contributions to the anomalous dimensions η_f and η_b even in our one-loop computation, along with nontrivial flows of the velocities c, v_F, and v_Δ. In addition to the boson graphs already considered in Fig. 4.1, we also have to consider the Feynman graphs in Fig. 17.3. The computation of the Feynman graphs is straightforward, but lengthy: the details are presented in [528]. The main subtlety arises from the non-Lorentz invariant structure of the propagators which leads to spacetime integrals which are not rotationally invariant. The rotational averages over spacetime momenta were performed directly in $d = 2$, and the radial momentum integrals were then performed for d close to 4.

An analysis of the RG equations so obtained showed [528] that on the critical surface ($r = 0$ to leading order), there was only one stable fixed point, and that this fixed point had relativistic invariance with $v_F = v_\Delta = c$ and $z = 1$. Thus even if the three

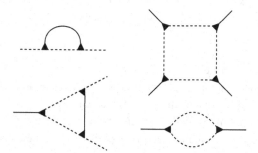

Fig. 17.3 Feynman diagrams which contribute to the one-loop RG, along with those in Fig. 4.1. The dark triangle is the λ vertex, the full line is the ϕ propagator, and the dashed line is the Ψ propagator.

velocities are distinct to begin with, they are attracted by renormalization to the same common value, leading to emergent relativistic invariance at the critical point. The Yukawa coupling in (17.23) can also be written in a relativistically invariant form, after transforming to a representation using the Dirac gamma matrices, which we are not using here.

Here, we will be satisfied by presenting the form of the RG flow equations assuming relativistic invariance with $v_F = v_\Delta = c = 1$ and $z = 1$. Then, evaluating the trace over the Dirac matrices in $d - 2$, and performing the momentum space integrals close to $d = 3$, we obtain the flow equations (which generalize (4.24)):

$$\frac{du}{d\ell} = \epsilon u - \frac{3}{2}u^2 - 8\lambda^2 u + 48\lambda^4,$$
$$\frac{d\lambda}{d\ell} = \frac{\epsilon}{2}\lambda - \frac{7}{2}\lambda^3, \tag{17.28}$$

where, as in (4.25), $\epsilon = 3 - d$. We have rescaled the coupling constants by $u \to u/S_4$ and $\lambda^2 \to \lambda^2/S_4$, where the phase space constant S_4 is defined below (4.24). This relativistically invariant theory has $z = 1$, and anomalous dimensions given by

$$\eta_b = 4\lambda^2, \quad \eta_f = \lambda^2/2. \tag{17.29}$$

We can now solve (17.28), and conclude that phase transition is described by the fixed point

$$\lambda^{*2} = \frac{\epsilon}{7}, \quad u^* = \frac{16\epsilon}{21}; \tag{17.30}$$

the values of the anomalous dimensions now follow from using these fixed point values in (17.29).

The same fixed point can also be accessed in the $1/N$ expansion [264], where N is the number of fermion species. It has also been studied in the particle physics literature [410], where the relativistic theory is known as the Higgs–Yukawa or Gross–Neveu model: a quantum Monte Carlo simulation of this model also exists [260], and provides probably the most accurate estimate of the exponents.

This nontrivial fixed point has strong implications for the correlations of the fermionic excitations of the superconductor. By analogy with the arguments made in Chapter 7 for the boson field ϕ, we conclude that the quantum critical fermion correlation function $G_1 = \langle \Psi_1 \Psi_1^\dagger \rangle$ obeys

$$G_1(k, \omega) = \frac{\omega + v_F k_x \tau^z + v_\Delta \tau^x}{(v_F^2 k_x^2 + v_\Delta^2 k_y^2 - \omega^2)^{(1-\eta_f)/2}}. \tag{17.31}$$

Away from the critical point, in the $d_{x^2-y^2}$ superconductor with $r > 0$, (17.31) holds with $\eta_f = 0$, and this is the BCS result, with sharp quasiparticle poles in the Green's function. At the critical point $s = s_c$, (17.31) holds with the fixed point values for the velocities (which satisfy $v_F = v_\Delta = c$) and with the anomalous dimension $\eta_f \neq 0$, and the spectral function resembles Fig. 7.2. This is clearly non-BCS behavior, and the fermionic quasiparticle pole in the spectral function has been replaced by a branch-cut representing the continuum of critical excitations.

The corrections to BCS theory extend also to correlations of the Ising order ϕ: its expectation value vanishes as (17.21) with the Monte Carlo estimate $\beta \approx 0.877$ [260]. The critical point correlators of ϕ have the anomalous dimension $\eta \approx 0.754$ [260], which is clearly different from the very small value of the exponent η at the unstable $\lambda = 0$ fixed point found in Part II. The value of β is related to η by the usual scaling law $\beta = (1+\eta)\nu/2$, with $\nu \approx 1.00$ the correlation length exponent (which also differs from the exponent ν of the Ising model of Part II).

17.4 Ising-nematic ordering

We now consider an Ising transition associated with "Ising-nematic" ordering in the d-wave superconductor. This is associated with a spontaneous reduction of the lattice symmetry of the Hamiltonian from "square" to "rectangular." Our study is motivated by experimental observations of such a symmetry breaking in the cuprate superconductors [23, 108, 221].

One of the main purposes of this section is to describe briefly a quantum critical point which is not relativistically invariant. The model in Section 17.2 had an order parameter which coupled to the Dirac fermions in a manner compatible with relativistic invariance, but this is not the case with most symmetry breaking possibilities.

The ingredients of Ising-nematic ordering are actually already present in our simple review of BCS theory in Section 17.1. In (17.4), we introduce two variational pairing amplitudes Δ_x and Δ_y. Subsequently, we assumed that minimization of the energy led to a solution with $d_{x^2-y^2}$ pairing symmetry with $\Delta_x = -\Delta_y = \Delta_{x^2-y^2}$. However, it is possible that upon including the full details of the microscopic interactions we are led to a minimum where the optimal solution also has a small amount of s-wave pairing. Then $|\Delta_x| \neq |\Delta_y|$, and we can expect all physical properties to have distinct dependencies on the x and y coordinates. Thus, one measure of the Ising-nematic order parameter is $|\Delta_x|^2 - |\Delta_y|^2$.

The derivation of the field theory for this transition follows closely our presentation in Section 17.2. We allow for small Ising-nematic ordering by introducing a scalar field ϕ and writing

$$\Delta_x = \Delta_{x^2-y^2} + \phi; \quad \Delta_y = -\Delta_{x^2-y^2} + \phi. \tag{17.32}$$

The evolution of the Dirac fermion spectrum under such a change is indicated in Fig. 17.4. Note that the gapless Dirac particles survive on both sides of the transition, and behave like essentially free fermions, as discussed at the end of Section 17.1. However, the gapless fermions reside at different locations in the Brillouin zone on the two sides of the transition. As we approach the quantum critical point, there are precursor fluctuations of the motion of the gapless points, and these lead to interesting effects on the fermionic spectra that we now briefly discuss.

A description of these fluctuations requires an effective action for ϕ and the Dirac fermions $\Psi_{1,2}$. The result is essentially identical to that in Section 17.2, apart from a change

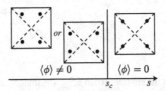

Fig. 17.4 Phase diagram of Ising-nematic ordering in a d-wave superconductor as a function of the coupling s in $\mathcal{S}_\phi$. The filled circles indicate the location of the gapless fermionic excitations in the Brillouin zone. The two choices for $s < s_c$ are selected by the sign of $\langle \phi \rangle$.

in the structure of the Yukawa coupling. Thus we obtain a theory $\mathcal{S}_\Psi + \mathcal{S}_\phi + \overline{\mathcal{S}}_{\Psi\phi}$, defined by (17.11) and (17.22), and where (17.23) is now replaced by

$$\overline{\mathcal{S}}_{\Psi\phi} = \lambda_I \int d^2r\,d\tau \left[\phi \left(\Psi_1^\dagger \tau^x \Psi_1 + \Psi_2^\dagger \tau^x \Psi_2 \right) \right]. \tag{17.33}$$

The seemingly innocuous change between (17.23) and (17.33), however, has strong consequences. This is linked to the fact that $\overline{\mathcal{S}}_{\Psi\phi}$ cannot be relativistically invariant even after all velocities are adjusted to equal. A weak-coupling renormalization group analysis in powers of the coupling λ_I was performed in $(3-d)$ dimensions in [527, 528], as described in Section 17.3, and led to flows to strong coupling with no accessible fixed point: thus no firm conclusions on the nature of the critical theory were drawn.

This problem remained unsolved until the recent works of [228, 266]. The main realization was that the critical point should not be described in an expansion in powers of the fermion–boson coupling λ_I, and the critical theory should be formulated at *fixed* λ_I. The approach described below can also be applied to the transition of Section 17.2, but we chose there to use the simpler perturbative analysis in the coupling λ which works only in that case. Also, the analysis described below parallels that for the Fermi gas near a Feshbach resonance in Section 16.4.2.

Consider the leading contribution to the self-energy of ϕ from the fermions Ψ: this is conveniently expressed as a contribution to the Gaussian effective action for ϕ after integrating out the Ψ:

$$\delta\mathcal{S}_\phi = \frac{N}{2} \int \frac{d^2k}{4\pi^2} \int \frac{d\omega}{2\pi} \Gamma_2(k, \omega) |\phi(k, \omega)|^2. \tag{17.34}$$

Here we have generalized to a model with N spin components ($N = 2$ is the physical case). The function Γ_2 is given by one fermion loop Feynman graphs with two insertions of the external ϕ vertices, as shown in Fig. 17.5. The explicit form of $\Gamma_2(k, \omega)$ for the Ising-nematic case can be written as

$$\Gamma_2(k, \omega) = \lambda_I^2 \left(\Pi_2(k_x, k_y, \omega) + \Pi_2(k_y, k_x, \omega) \right), \tag{17.35}$$

with the two terms representing the contributions of the Ψ_1 and Ψ_2 fermions, respectively. The one fermion loop diagram yields

Fig. 17.5 Feynman graph for the effective action $\delta \mathcal{S}_\phi$. The smooth lines are fermion propagators, while the wavy lines are ϕ insertions.

$$
\Pi_2(k_x, k_y, \omega) = \int \frac{d^2 p}{4\pi^2} \int \frac{d\Omega}{2\pi} \text{Tr}\Big[\left(-i\Omega + v_F p_x \tau^z + v_\Delta p_y \tau^x\right)^{-1}
$$
$$
\times \tau^x \left(-i(\Omega + \omega) + v_F(p_x + k_x)\tau^z + v_\Delta(p_y + k_y)\tau^x\right)^{-1} \tau^x \Big]
$$
$$
= \frac{1}{16 v_F v_\Delta} \frac{(\omega^2 + v_F^2 k_x^2)}{(\omega^2 + v_F^2 k_x^2 + v_\Delta^2 k_y^2)^{1/2}}. \tag{17.36}
$$

Here, we have omitted a constant which can be absorbed into the value of r in $\mathcal{S}_\phi$. Now the key observation is that the momentum and frequency dependence in $\delta \mathcal{S}_\phi$ implied by (17.36) is more important in the infrared than that in the bare action in (17.22): the above result for Γ_2 scales with one power of momentum/frequency, while that in (17.22) has two powers. Thus, for the critical theory, it seems reasonable simply to drop the spatial and temporal gradients in (17.22).

Working with the Gaussian action $\delta \mathcal{S}_\phi$ has immediate and important consequences for the RG analysis. At tree level, the scaling of Ψ remains the same as before, while that for ϕ becomes

$$
\phi' = \phi e^{d\ell/2}. \tag{17.37}
$$

Comparing with (17.26), we see that we have the anomalous exponent $\eta_b = 1$ at the outset. The corresponding rescaling of the fermion–boson coupling now becomes

$$
\lambda_I' = \lambda_I e^{(2-d)\ell/2}, \tag{17.38}
$$

replacing that in (17.27). We now observe that λ_I does not flow in $d = 2$. It therefore pays to work in a theory with a *fixed* λ_I. Furthermore, we find $u' = u e^{(1-d)\ell}$: so u is irrelevant, and we can set $u = 0$ from now on.

It is useful to interpret these ideas in an RG picture. We have decided to drop the bare gradient terms in $\mathcal{S}_\phi$ because they are not as important as the induced terms. Normally, such gradient terms are used to set the normalization of the field scale of ϕ. Now, we can use this new-found freedom to set the field scale by choosing $\lambda_I = 1$. In other words, we use the fermion–boson Yukawa coupling to set the field scale for ϕ, and work in a theory in which this coupling remains fixed at all scales. This will lead to a theory in which the boson anomalous dimension $\eta_b = 1$ at the start of the computation.

As shown in [228], higher order loop corrections can be computed in powers of $1/N$. These lead to further RG flows in the field scales and the coupling constants. Actually, after setting $\lambda_I = 1$ and $u = 0$, the only remaining couplings in the theory are the velocities v_F and v_Δ. And the only dimensionless coupling characterizing the theory is the velocity ratio v_Δ/v_F. The RG equation for this ratio was derived in [228], and showed that the flow at long scales was $v_\Delta/v_F \to 0$; the approach to zero was roughly with the inverse logarithm of the length or time scale. Thus the Ising-nematic critical point is characterized by a highly anisotropic velocity ratio, in strong contrast to the relativistic theory of Section 17.2. One important consequence of this anisotropic critical point was the presence of "arc-like" spectra for the Dirac fermions, see [266], where other experimental implications for the cuprate superconductors are also discussed. Associated singularities in the thermal conductivity have also been computed [158].

Fermi liquids, and their phase transitions

The Fermi liquid is perhaps the most familiar quantum many-body state of solid state physics; we met it briefly in Section 16.2.2. It is the generic state of fermions at nonzero density, and is found in all metals. Its basic characteristics can already be understood in a simple free fermion picture. Noninteracting fermions occupy the lowest energy single-particle states, consistent with the exclusion principle. This leads to the fundamental concept of the Fermi surface: a surface in momentum space separating the occupied and empty single fermion states. The lowest energy excitations then consist of quasiparticle excitations which are particle-like outside the Fermi surface, and hole-like inside the Fermi surface. Landau's Fermi liquid theory is a careful justification for the stability of this simple picture in the presence of interactions between fermions. Just as we found in Chapters 5 and 7 for the quantum Ising and rotor models, interaction corrections modify the wavefunction of the quasiparticle and so introduce a quasiparticle residue $\mathcal{A}$; however, they do not destabilize the integrity of the quasiparticle, as we review in Section 18.1.

The purpose of this chapter is to describe two paradigms of symmetry breaking quantum transitions in Fermi liquids. In the first class, studied in Section 18.2, the broken symmetry is related to the point-group symmetry of the crystal, while translational symmetry is preserved; consequently, the order parameter resides at zero wavevector. In the second class, studied in Section 18.3, the order parameter is at a finite wavevector, and so translational symmetry is also broken. We find that these transitions have distinct effects on the Fermi surface, and so lead to very different critical theories. We study both critical theories using a simple example in each class, both motivated by the physics of the cuprate superconductors. For the first class we consider the case of Ising-nematic ordering, while in the second class we consider the onset of spin density wave order.

Among our aims is to understand the possible breakdown of Landau's Fermi liquid theory in Fermi gases. The most prominent example of this breakdown is in spatial dimension $d = 1$, where we generically obtain (not necessarily near any quantum phase transitions) a different quantum state known as the Tomonaga–Luttinger liquid: this is described in Chapter 20. We meet examples of Fermi liquid breakdown in $d \geq 2$ in the present chapter: in the examples considered here, and unlike Chapter 20, the breakdown occurs only at quantum critical points assciated with symmetry breaking transitions.

A comprehensive theoretical treatment of symmetry breaking transitions in a Fermi liquid was given by Hertz [218], although many important points were anticipated in earlier work [43, 347, 348, 391]. We review this treatment here, adapted to our field-theoretic approach. A key step in Hertz's work is to completely integrate out the fermionic excitations near the Fermi surface, resulting in an effective action for the order parameter characterizing the symmetry breaking alone. Such an approach seems natural from the perspective of the classical phase transitions considered in Chapter 3, in which we need only pay attention to the low-energy fluctuations of the order parameter. However, here we also have the low-energy quasiparticles near the Fermi surface: they are not associated directly with the broken symmetry, but their existence is protected by the requirement of the presence of a Fermi surface. It seems dangerous to integrate them out, and it would be preferable to make them active participants in the critical theory. This is a subtle question which we address carefully in the present chapter. The main conclusion is that the Hertz strategy remains largely correct in $d \geq 3$, but that it fails badly in the important case of $d = 2$. This conclusion applies to both classes of symmetry breaking transitions in a Fermi liquid, with order parameters at zero and nonzero momentum.

After a presentation of the critical theories at zero temperature, we also briefly address the nature of the crossovers at nonzero temperatures. We limit this discussion to the case of the spin density wave transition using the framework of the Hertz theory in Section 18.3.3. These crossovers were computed by Millis [336] who pointed out the universal features and emphasized the basic similarities of the crossovers to those in the dilute Bose gas; related results, not using the perspective of quantum phase transitions, were also available in the earlier work of Moriya [204, 347–349], Ramakrishnan and Mishra [341, 390, 391], and others. Many results for finite-temperature crossovers near a magnetic quantum phase transition in Fermi liquids were also anticipated by Lonzarich and Taillefer [304, 305]. We have already studied the dilute Bose gas in Section 16.3.2, where we also noticed the similarity to the quantum Ising/rotor models above their upper-critical dimension as treated in Section 14.2.2. Here we are able to use the techniques developed in these earlier sections to arrive rapidly at the needed generalization. The study of the finite-temperature crossovers is restricted here to those above the Fermi liquid state and in the high-T regime.

18.1 Fermi liquid theory

Let us begin with a review of some basic ideas from the Fermi liquid theory of interacting fermions in d dimensions. We consider spin-1/2 fermions c_{ka} with momentum k and spin $a = \uparrow, \downarrow$ and dispersion ε_k. Thus the noninteracting fermions are described by the action

$$S_c = \int d\tau \int \frac{d^d k}{(2\pi)^d} c_{ka}^\dagger \left(\frac{\partial}{\partial \tau} + \varepsilon_k \right) c_{ka}. \tag{18.1}$$

As an example, it is useful to keep in mind the dispersion ε_k appropriate for the cuprate superconductors, which is shown in Fig. 18.1. The fermion Green's function under the free fermion action S_c has the simple form

$$G_0(k, \omega_n) = \frac{1}{-i\omega_n + \varepsilon_k}. \tag{18.2}$$

After analytically continuing to real frequencies, we observe that this Green's function has a pole at energy ε_k with residue 1. Thus there are quasiparticle excitations with residue $\mathcal{A} = 1$, much like those found in the strong- or weak-coupling expansions of the quantum

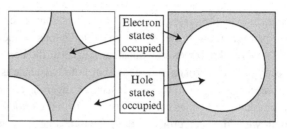

Two views of the Fermi surface of the cuprate superconductors (hole and electron doped). We use the dispersion ε_k specified below (17.1). The chemical potential is included in the dispersion ε_k, and so the Fermi surface is determined by $\varepsilon_k = 0$. The left panel has the momentum $k = (0, 0)$ (the "Γ point") in the center of the square Brillouin zone, while the right panel has the Γ point at the left edge. The momenta with both up and down electron states occupied are shaded gray.

Ising model in Chapter 5. However, unlike those excitations, these quasiparticles can have both positive and negative energies, as ε_k can have either sign; the Fermi surface is the locus of points where ε_k changes sign. The positive energy quasiparticles are electron-like, while those with negative energy are hole-like, i.e. they correspond to the absence of an electron. Note that the existence of negative energy quasiparticles is not an indication of the instability of the ground state. All true excitation energies are positive: the excitations are electron-like on one side of the Fermi surface, and hole-like on the other side. It is just convenient to combine the electron and hole quasiparticles within a single Green's function, by identifying hole-like quasiparticles with negative energy electron-quasiparticles.

We now wish to examine the stability of quasiparticles to interactions between them. In keeping with the strategy followed in this book, this should be preceded by an effective action for the low-energy quasiparticles. The latter is usually done by a gradient expansion, leading to an effective field theory. However, here we face a unique difficulty: there are zero-energy quasiparticles along a $d - 1$ dimensional Fermi surface identified by $\varepsilon_k = 0$. It would therefore seem that we should expand about all points on the Fermi surface. This is indeed the strategy followed in textbook treatments of Fermi liquid theory: we measure momenta, $k_\perp$, from the Fermi surface, choose a cutoff so that $|k_\perp| < \Lambda$, and then perform an RG which reduces the value of Λ [465]. This procedure is illustrated in Fig. 18.2. Formally, for each direction $\hat{n}$, we define the position of the Fermi surface by the wavevector $\vec{k}_F(\hat{n})$, so that $\hat{n} = \vec{k}_F(\hat{n})/|\vec{k}_F(\hat{n})|$. Then we identify wavevectors near the Fermi surface by

$$\vec{k} = \vec{k}_F(\hat{n}) + k_\perp \hat{n}. \tag{18.3}$$

Now we should expand in small momenta $k_\perp$. For this, we define the infinite set of fields $\psi_{\hat{n}a}(k_\perp)$, which are labeled by the spin a and the direction $\hat{n}$, related to the fermions c by

$$c_{\vec{k}a} = \frac{1}{\sqrt{S_F}} \psi_{\hat{n}a}(k_\perp), \tag{18.4}$$

where $\vec{k}$ and $k_\perp$ are related by (18.3), and S_F is the area of the Fermi surface. Note that this parameterization can be considered to be a generalization of (16.28), where we had only the two fields Ψ_L, Ψ_R, rather than the infinite number of fields labeled by the direction $\hat{n}$.

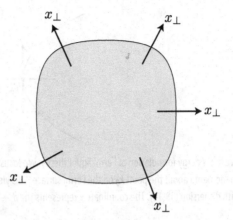

Fig. 18.2 Traditional low-energy limit of Fermi liquid theory. The Fermi surface has one-dimensional chiral fermions on every point, moving along the direction $x_\perp$. There fermions are present for momenta $|k_\perp| < \Lambda$; i.e. in a momentum shell of width 2Λ around the Fermi surface.

Inserting (18.4) into (18.1), expanding in $k_\perp$, and Fourier transforming to real space $x_\perp$, we obtain the low-energy theory

$$S_{\mathrm{FL}} = \int d\Omega_{\hat{n}} \int dx_\perp \psi_{\hat{n}a}^\dagger(x_\perp) \left(\frac{\partial}{\partial \tau} - i v_F(\hat{n}) \frac{\partial}{\partial x_\perp} \right) \psi_{\hat{n}a}(x_\perp), \qquad (18.5)$$

where the Fermi velocity is the energy gradient on the Fermi surface $v_F(\hat{n}) = |\nabla_k \varepsilon_{\vec{k}_F(\hat{n})}|$. For each $\hat{n}$, (18.5) describes fermions moving along the single dimension $x_\perp$ with the Fermi velocity: this is a one-dimensional chiral fermion, which we met in Section 16.2.2, and will meet again in (20.6); the "chiral" refers to the fact that the fermion only moves in the positive $x_\perp$ direction, and not the negative $x_\perp$ direction. In other words, the low-energy theory of the Fermi liquid is an infinite set of one-dimensional chiral fermions, one chiral fermion for each point on the Fermi surface.

Apart from the free Fermi term in (18.5), Landau's Fermi liquid theory also allows for contact interactions between chiral fermions along different directions [465]. These are labeled by the Landau parameters, and lead only to shifts in the quasiparticle energies which depend upon the densities of the other quasiparticles. Such shifts are important when computing the response of the Fermi liquid to external density or spin perturbations. However, the resulting fixed-point action of Fermi liquid theory does not offer a route to computing the decay of quasiparticles: the stability of the quasiparticles is implicitly assumed in the fixed-point theory. Our primary purpose here is to verify the stability of the quasiparticles, so that we are prepared for the breakdown of Fermi liquid theory at quantum critical points. So we refer the reader to the many textbook treatments of the traditional formulation of Landau's Fermi liquid theory, and turn to an alternative analysis below.

A shortcoming of the effective action (18.5) is that it includes only the dispersion of the fermions transverse to the Fermi surface. Thus, if we discretize the directions $\hat{n}$, and pick a given point on the Fermi surface, the Fermi surface is effectively *flat* at that point. We shortly see that the curvature of the Fermi surface is important in understanding the

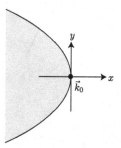

Alternative low-energy formulation of Fermi liquid theory. We focus on an extended patch of the Fermi surface, and expand in momenta about the point $\vec{k}_0$ on the Fermi surface. This yields a theory of d-dimensional fermions ψ in (18.7), with dispersion (18.14). The coordinate y represents the $d - 1$ dimensions parallel to the Fermi surface.

decay and breakdown of quasiparticles. Thus we have to take the continuum scaling limit in a manner which keeps the curvature of the Fermi surface fixed, and does not scale it to zero. For this, as shown in Fig. 18.3, we focus attention on a single arc of the Fermi surface in the vicinity of any chosen point $\vec{k}_0$. We show in Section 18.1.1 that the results are independent of the choice of $\vec{k}_0$ on the Fermi surface, but we defer that issue for now. Then we choose our cutoff Λ to scale towards the single point k_0 (the cutoff is defined more carefully below), rather than scaling to all points on the Fermi surface, as we did for (18.5).

With $\vec{k}_0$ chosen as in Fig. 18.3, let us now define our low-energy theory and scaling limit [333]. Unlike the one-dimensional chiral fermions which appeared in (18.5), we now use a d-dimensional fermion $\psi_a(x, y)$. Here x is the one-dimensional coordinate orthogonal to the Fermi surface, and $\vec{y}$ represents the $(d - 1)$-dimensional transverse coordinates. After Fourier transformations, this fermion is related to the underlying fermions c_{ka} simply by

$$\psi_a(k) = c_{\vec{k}_0+\vec{k},a}. \tag{18.6}$$

In other words, we only shift the origin of momentum space from $\vec{k} = 0$ to $\vec{k} = \vec{k}_0$. Inserting (18.6) in (18.1), and expanding the dispersion in the vicinity of $\vec{k}_0$ (contrast to the expansion away from all points on the Fermi surface in (18.5)), we obtain the low-energy theory

$$\mathcal{S}_0 = \int d\tau \int dx \int d^{d-1}y \; \psi_a^\dagger \left(\zeta \frac{\partial}{\partial \tau} - i v_F \frac{\partial}{\partial x} - \frac{\kappa}{2} \nabla_y^2 \right) \psi_a. \tag{18.7}$$

We have added a coefficient ζ to the temporal gradient term for future convenience: we are interested in $\zeta = 1$, but see later that in non-Fermi liquid states it is convenient to allow ζ to renormalize. Note the additional second-order gradients in y which were missing from (18.5): the coefficient κ is proportional to the curvature of the Fermi surface at $\vec{k}_0$. Also, as we have already noted, the fermion field in (18.7) is d-dimensional, while that in (18.5) in one-dimensional. One benefit of (18.7) is now immediately evident: it has zero energy excitations when

$$v_F k_x + \kappa \frac{k_y^2}{2} = 0, \tag{18.8}$$

and so (18.8) defines the position of the Fermi surface, which is then part of the low-energy theory including its curvature. Note that (18.7) now includes an extended portion of the Fermi surface; contrast that with (18.5), where the one-dimensional chiral fermion theory for each $\hat{n}$ describes only a single point on the Fermi surface.

The gradient terms in (18.7) define a natural momentum space cutoff, and associated scaling limit. We take such a limit at *fixed* ζ, v_F, and κ. Note that momenta in the x direction scale as the square of the momenta in the y direction, and so we can choose $v_F^2 k_x^2 + \kappa^2 k_y^4 < \Lambda^4$. As we reduce Λ, we scale towards the single point $\vec{k}_0$ on the Fermi surface, as we required above.

It is now a simple matter to apply the RG analysis to the fermion theory in (18.7): the analysis closely parallels that in Section 16.2. At fixed ζ, v_F, and κ, the action (18.7) is invariant under the following rescalings of spacetime:

$$x' = xe^{-2\ell}, \quad y' = ye^{-\ell}, \quad \tau' = \tau e^{-2\ell}. \tag{18.9}$$

Note that we have chosen the directions parallel to the Fermi surface as the ones defining the primary length scale, with $\dim[y] = -1$, and the transverse direction has $\dim[x] = -2$. The temporal direction rescaling implies that we have the dynamic exponent $z = 2$ when measured relative to the y spatial directions. The RG invariance of (18.7) also requires the field rescaling

$$\psi' = \psi e^{(d+1)\ell/2}. \tag{18.10}$$

We now have the tools needed to determine the role of fermion interactions. The simplest contact interaction has the form

$$S_1 = u_0 \int d\tau \int dx \int d^{d-1}y \ \Psi_a^\dagger \Psi_b^\dagger \Psi_b \Psi_a. \tag{18.11}$$

Applying the RG rescalings in (18.10), we find

$$u_0' = u_0 e^{(1-d)\ell}. \tag{18.12}$$

In other words, the interaction between the fermions u_0 is irrelevant in all dimension $d > 1$. This strongly suggests that the Fermi liquid picture of noninteracting fermions is indeed RG stable.

Let us understand the stability of Fermi liquid theory a bit better by computing corrections to the fermion Green's function in (18.2) using the methods of Section 7.2.1. Let us write the interaction corrected Green's function as

$$G(k, \omega) = \frac{1}{-\zeta\omega + \varepsilon_k - \Sigma(k, \omega)}, \tag{18.13}$$

where now

$$\varepsilon_k = v_F k_x + \kappa \frac{k_y^2}{2}. \tag{18.14}$$

To first order in u_0, the fermion self-energy is real (for real frequencies), and so only modifies the quasiparticle dispersion and residue, $\mathcal{A}$, but does not destabilize the existence of the quasiparticle pole.

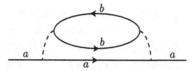

Fig. 18.4 Feynman diagram for the decay of quasiparticles at order u_0^2. The dashed line is the interaction u_0, and a, b are spin labels.

So let us move to second order in u_0. First, we use the analog of the RG argument used around (14.37). We are interested in the imaginary part of the self-energy, and let us assume for now at small ω

$$\text{Im}\Sigma(k = 0, \omega) \sim u_0^2 \omega^p. \tag{18.15}$$

As in (14.37), we determine p by scaling arguments. From (18.13) we know that $\dim[\Sigma] = z = 2$, and so conclude from matching dimensions in (18.15) that $p = d$. However, there is a subtlety here: scaling arguments only yield the power laws of singular corrections, and do not say anything about analytic backgrounds that may be allowed from the structure of the theory. Here, a term with $p = 2$ is permitted because $\text{Im}\Sigma$ is an even function of ω. So the proper conclusion is

$$p = \min(d, 2). \tag{18.16}$$

The above scaling argument is fine as it stands, but cannot substitute for the insight gained by an explicit computation. The Feynman diagram contributing to the quasiparticle decay at order u_0^2 is indicated in Fig. 18.4. We evaluate it in two stages. First we evaluate the fermion loop of the fermions with spin label b; this gives us the fermion polarizability

$$\Pi(q, \omega_n) = \int \frac{d^d k}{(2\pi)^d} \int \frac{d\epsilon_n}{2\pi} G_0(k + q, \epsilon_n + \omega_n) G_0(k, \epsilon_n), \tag{18.17}$$

which is the analog of the polarizability of the rotor model used extensively in Section 13.2. This enters the self-energy by

$$\Sigma(k, \epsilon_n) = u_0^2 \int \frac{d^d q}{(2\pi)^d} \int \frac{d\omega_n}{2\pi} \Pi(q, \omega_n) G_0(k + q, \epsilon_n + \omega_n), \tag{18.18}$$

which is the analog here of (13.41).

We first explicitly evaluate $\Pi(q, \omega_n)$. We are only interested in terms that are singular in q and ω_n, and drop regular contributions from regions of high momentum and frequency. In this case, it is permissible to reverse the conventional order of integrating over frequency first in (18.17), and to first integrate over k_x. It is a simple matter to perform the integration over k_x in using the method of residues to yield

$$\Pi(q, \omega_n) = \frac{1}{2v_F} \int \frac{d^{d-1}k_y}{(2\pi)^{d-1}} \int \frac{d\epsilon_n}{2\pi} \frac{\text{sgn}(\epsilon_n + \omega_n) - \text{sgn}(\epsilon_n)}{\left(\zeta\omega_n + iv_F q_x + i\kappa q_y^2/2 + i\kappa \vec{q}_y \cdot \vec{k}_y\right)}$$

$$= \frac{|\omega_n|}{2\pi v_F} \int \frac{d^{d-1}k_y}{(2\pi)^{d-1}} \frac{1}{\left(\zeta\omega_n + iv_F q_x + i\kappa q_y^2/2 + i\kappa \vec{q}_y \cdot \vec{k}_y\right)}. \tag{18.19}$$

We now integrate along the component of $\vec{k}_y$ parallel to the direction of $\vec{q}_y$ to obtain

$$
\begin{aligned}
\Pi(q, \omega_n) &= \frac{|\omega_n|}{2\pi v_{FK}|q_y|} \int \frac{d^{d-2}k_y}{(2\pi)^{d-2}} \\
&= \frac{|\omega_n|}{2\pi v_{FK}|q_y|} \Lambda^{d-2}.
\end{aligned}
\tag{18.20}
$$

Note that in $d = 2$ the last nonuniversal factor is not present, and the result for Π is universal with $\Lambda^{d-2} = 1$. Note also that ζ has dropped out of the result Π: this is important in our subsequent treatment of quantum critical points.

Now we insert (18.20) into (18.18). After evaluating the integral over q_x we obtain

$$
\begin{aligned}
\Sigma(k, \omega_n) &= i \frac{u_0^2}{2\pi v_F^2 \kappa} \int \frac{d^{d-1}q_y}{(2\pi)^{d-1}} \int \frac{d\epsilon_n}{2\pi} \frac{\mathrm{sgn}(\epsilon_n + \omega_n)|\epsilon_n|}{|q_y|} \\
&= i\, \mathrm{sgn}(\omega_n)\omega_n^2 \frac{u_0^2}{4\pi v_F^2 \kappa} \int \frac{d^{d-1}q_y}{(2\pi)^{d-1}} \frac{1}{|q_y|} \\
&= i\, \mathrm{sgn}(\omega_n)\omega_n^2 \frac{u_0^2}{4\pi v_F^2 \kappa} \Lambda^{d-2}. \qquad d > 2
\end{aligned}
\tag{18.21}
$$

Again, ζ has dropped out. This result is in perfect accord with the scaling arguments in (18.15) and (18.16).

Let us consider the important case $d = 2$. There is an infrared divergence in the q_y integral in (18.21) at small q_y. This is only cut off after we include a self-consistent damping of the quasiparticle propagators in the Feynman diagram of Fig. 18.4, rather than the bare propagators we have used above. After including this damping, we expect that (18.15) will be modified to

$$
\mathrm{Im}\Sigma(k, \omega) \sim u_0^2 \omega^2 \log\left(\frac{\Lambda}{u_0|\omega|}\right), \qquad d = 2
\tag{18.22}
$$

thus the scaling result is modified by a logarithm in $d = 2$.

With $\mathrm{Im}\Sigma \sim u_0^2 \omega^2$ (up to logarithms), we can now easily examine the fate of the quasiparticles from (18.13). The situation differs somewhat from what we found for the rotor model in Section 7.2.1. There we found that the quasiparticle pole remained infinitely sharp for a finite range of momenta, to all orders in the interactions. Here, from (18.13), we see that the quasiparticle pole is always broadened: the width of the quasiparticle peak is $\sim u_0^2 \epsilon_k^2$ for a quasiparticle with energy $\omega = \epsilon_k$. Thus the quasiparticle width vanishes as the square of the distance from the Fermi surface. Asymptotically close to the Fermi surface, the quasiparticle width is much smaller than the quasiparticle energy: this is sufficient to regard the quasiparticle as a sharp excitation, and confirm the validity of Landau's Fermi liquid theory.

An important and frequently used diagnostic of the stability of the quasiparticle is the discontinuity in the fermion momentum distribution function $n(k) = \langle c_{ka}^\dagger c_{ka} \rangle$. This can be computed from the real frequency Green's function $G(k, \omega)$ by (here we set $\zeta = 1$)

$$
n(k) = \int_{-\infty}^{0} \frac{d\omega}{(2\pi)} \mathrm{Im}G(k, \omega).
\tag{18.23}
$$

Assuming a pole in the Green's function of the form

$$G(k, \omega) = \frac{\mathcal{A}}{-\omega + \varepsilon_k + ic\omega^2} + \ldots, \tag{18.24}$$

we find a step discontinuity in the momentum distribution function at the Fermi surface

$$n(k) = \mathcal{A}\theta(-\varepsilon_k) + \ldots, \tag{18.25}$$

of strength $\mathcal{A}$.

18.1.1 Independence of choice of $\vec{k}_0$

Our theory of the Fermi liquid state is now contained in the action $\mathcal{S}_0 + \mathcal{S}_1$ defined by (18.7) and (18.11). It focused on an arc of the Fermi surface, as shown in Fig. 18.3, and then expanded in gradients about the point $\vec{k}_0$ on the Fermi surface. To complete our discussion, we now wish to show that the theory is *independent* of the choice of $\vec{k}_0$.

As shown in Fig. 18.5, we could equally well have defined the theory about the point $\vec{k}_0'$ on the Fermi surface. Consistency requires that the fermion Green's function at the point P should have the same value whether it is computed using the theory at $\vec{k}_0$ or at $\vec{k}_0'$. This section shows that this is indeed the case.

Note that such a consistency requirement is not present for the representation in terms of chiral one-dimensional fermions in Fig. 18.2. There, each point in the momentum space is associated only with a single one-dimensional theory. It is our use of a d-dimensional theory which induces our redundant description.

Let us choose our momentum space coordinates centered at $\vec{k}_0$, and let $\vec{k}_0' = (k_x, k_y)$ in this coordinate system. Because $\vec{k}_0'$ is on the Fermi surface, (18.8) is obeyed. Now let the point P in Fig. 18.5 have coordinates (p_x, p_y) relative to $\vec{k}_0$, and coordinates (p_x', p_y') relative to $\vec{k}_0'$. The latter are obtained from the old coordinates by a shift in origin followed by a rotation by an angle θ, where $\tan\theta = \kappa k_y/v_F$; this yields

$$\begin{aligned}
p_x' &= p_x - k_x + (\kappa/v_F)\vec{k}_y \cdot (\vec{p}_y - \vec{k}_y), \\
\vec{p}_y' &= \vec{p}_y - \vec{k}_y,
\end{aligned} \tag{18.26}$$

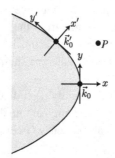

Fig. 18.5 The fermion correlator at the point P can be described either in terms of the field theory $\mathcal{S}_0 + \mathcal{S}_1$ at $\vec{k}_0$, or that at $\vec{k}_1$.

where we only keep terms to the needed accuracy of $\mathcal{O}(x, y^2)$. The equality of the physics in the two coordinate systems implies that the Green's function of the theory $\mathcal{S}_0 + \mathcal{S}_1$ must satisfy

$$G((p_x, p_y), \omega) = G((p'_x, p'_y), \omega), \tag{18.27}$$

for any momenta p, p' related by (18.26) and any k obeying (18.8). A simpler statement of the constraint follows from the easily verified identity

$$v_F p_x + \frac{\kappa}{2} p_y^2 = v_F p'_x + \frac{\kappa}{2} p_y'^2. \tag{18.28}$$

Thus, the Green's function should not depend separately on the momenta, but on the quasiparticle energy in (18.14) alone:

$$G((p_x, p_y), \omega) = G(\varepsilon_p, \omega). \tag{18.29}$$

We now prove (18.29) is true. The result relies upon the invariance of $\mathcal{S}_0 + \mathcal{S}_1$ on the following transformation

$$\psi(x, \vec{y}) \rightarrow \exp\left(-i\frac{v_F}{\kappa}\left(\vec{\theta} \cdot \vec{y} + \frac{\theta^2}{2}x\right)\right) \psi(x, \vec{y} + \vec{\theta} x), \tag{18.30}$$

where θ is an arbitrary $(d-1)$-dimensional vector; this invariance is easily verified by direct substitution in (18.7) and (18.11). The change in the arguments of ψ shows that this transformation corresponds to a local rotation of the Fermi surface, which effectively moves the point $\vec{k}_0$ to a neighboring point. Now taking the Fourier transform of (18.30), we immediately establish (18.27) and (18.29).

18.2 Ising-nematic ordering

Having established the stability of quasiparticles in the Fermi liquid, we turn to the first of the symmetry breaking transitions of this chapter. We consider one of the simplest order parameters at zero wavevector: the breaking of lattice rotation symmetry. In two dimensions, a simple choice is the change from "square" to "rectangular" symmetry considered in Section 17.4 for a d-wave superconductor. In higher dimensions, we consider the same symmetry breaking in the x-y plane embedded in the higher-dimensional space.

As in Section 17.4, we consider an Ising-nematic transition driven by strong interactions between the electrons. The symmetry breaking is characterized by the real scalar field ϕ, which is described as before by the action $\mathcal{S}_\phi$ in (17.22), reproduced here for completeness

$$\mathcal{S}_\phi = \int d^d x d\tau \left[\frac{1}{2}\left((\partial_\tau \phi)^2 + c^2(\partial_x \phi)^2 + c^2(\partial_y \phi)^2 + r\phi^2\right) + \frac{u}{24}\phi^4\right]. \tag{18.31}$$

The key difference from Section 17.4 is that the electrons are not in a d-wave superconducting state, but in a metallic Fermi liquid. Thus, there is a Fermi surface of fermions c_{ka},

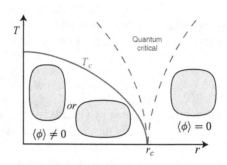

Phase diagram of Ising-nematic ordering in a metal as a function of the coupling s in $\mathcal{S}_\phi$ and temperature T. The Fermi surface for $r > 0$ is as in the overdoped region of the cuprates, with the shaded region indicating the occupied hole (or empty electron) states (compare Fig. 18.1). The choice between the two quadrapolar distortions of the Fermi surface is determined by the sign of $\langle\phi\rangle$. The line of $T > 0$ phase transitions at T_c is described by Onsager's solution of the classical two-dimensional Ising model. We are interested here in the quantum critical point at $r = r_c$, which controls the quantum-critical region.

described here by $\mathcal{S}_c$ in (18.1). Finally, we need to couple the Ising order ϕ to the fermions. This can be deduced by symmetry considerations, as in (17.33). A convenient choice is

$$\mathcal{S}_{c\phi} = \int d\tau \int \frac{d^dk\,d^dq}{(2\pi)^{2d}} d(k)\phi(q) c^\dagger_{k+q/2,a} c_{k-q/2,a}. \tag{18.32}$$

The momentum-dependent form factor, $d(k)$, can be any even parity function which changes sign under $x \leftrightarrow y$, as is required by the symmetry properties of ϕ; a simple choice is $d(k) \sim \cos k_x - \cos k_y$. The integral over q is over small momenta, while that over k extends over the entire Brillouin zone.

The theory for the nematic ordering transition is now described by $\mathcal{S}_c + \mathcal{S}_\phi + \mathcal{S}_{c\phi}$ in (18.1), (18.31), and (18.32), and forms the basis of the discussion in the remainder of this chapter. A schematic phase diagram as a function of the coupling s in $\mathcal{S}_\phi$ and temperature T is shown in Fig. 18.6. Note that there is a line of Ising phase transitions at $T = T_c$: this transition is in the same universality class as the classical two-dimensional Ising model. However, quantum effects and fermionic excitations are crucial at the $T = 0$ critical point at $r = r_c$ and its associated quantum critical region.

18.2.1 Hertz theory

As indicated in the introduction, Hertz's strategy is to integrate out all the fermionic excitations, and derive an effective action for the Ising order parameter ϕ. The same strategy was the starting point of the analysis in Section 17.4.

The integration is easily performed using our Fermi liquid theory results in Section 18.1. The important term is the fermion loop contribution to the ϕ^2 term in the effective action, and this is given by the Feynman diagram in Fig. 18.7. We can determine the structure of the fermion loop integral by taking the continuum limit of fermion theory in a Fermi surface patch about $\vec{k}_0$ as in (18.7), and then adding up the contributions of all the patches. For a given patch, the fermion loop contributes $d^2(k_0)\Pi(q, \omega)$, where Π is the fermion

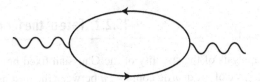

Fig. 18.7   Fermion loop contribution to the action of the order parameter ϕ. The wavy line is ϕ.

polarizability in (18.17). Using the result for Π in (18.20), averaging over different patches on the Fermi surface, and combining with the terms of the ϕ action in (18.31), we obtain the Hertz action for the order parameter at the Ising-nematic quantum critical point in d dimensions:

$$\mathcal{S}_H = \int \frac{d^d k}{(2\pi)^d} T \sum_{\omega_n} \frac{1}{2} \left[k^2 + \gamma \frac{|\omega_n|}{|k|} + r \right] |\phi(k, \omega_n)|^2$$
$$+ \frac{u}{24} \int d^d x \, d\tau \, \phi^4(x, \tau). \tag{18.33}$$

Compared to (18.31), the crucial new term is the one proportional to γ, which represents the nonlocal consequences of low-energy particle–hole excitations near the Fermi surface; the value of γ is determined from an average of the coefficient in (18.20) over the Fermi surface. In a system with a spherical Fermi surface, the $|k|$ in the denominator is simply $\sqrt{\vec{k}^2}$, arising from the average of (18.20) over different patches. However, without spherical symmetry, it is a more complex function which depends upon the details of the Fermi surface structure. Nevertheless, it retains the property of being an even function of k with scaling dimension 1, and that is all that we need below.

We are now ready to perform an RG analysis of $\mathcal{S}_H$, just as we did for (17.34) in the Dirac fermion case. As in Chapter 4, we begin with an analysis of the Gaussian part of $\mathcal{S}_H$, which is scale-invariant at $r = 0$, under the transformations:

$$x' = xe^{-\ell}, \quad \tau' = \tau e^{-z\ell}, \quad \phi' = \phi e^{(d+z-2)\ell/2}, \tag{18.34}$$

with dynamic critical exponent $z = 3$. This exponent can also be understood from the characteristic frequency scale $\omega \sim k^3$ emerging from a comparison of the first two terms in $\mathcal{S}_H$. With the transformations in (18.34), we see that the quartic coupling has scaling dimension

$$\dim[u] = 4 - d - z. \tag{18.35}$$

In other words, with $z = 3$, the Gaussian fixed point is stable for all $d > 1$. All computations in this chapter have implicitly assumed that $d > 1$, and the present approach is not sensible in $d = 1$ (which is discussed separately in Chapter 20). So the Gaussian theory appears to describe the quantum critical point for all values of d. This is one of Hertz's primary conclusions.

We argue below that the above conclusion is not correct. The Gaussian fixed point does, in fact, yield a proper description in $d = 3$. However, in the physically important case of $d = 2$, the Hertz approach fails.

18.2.2 Fate of the fermions

Our analysis of the stability of the Gaussian fixed point of the Hertz theory relied on the irrelevance of the quartic coupling u between the bosons. However, there are also fermionic quasiparticle excitations in the underlying theory, and it is important to apply the RG to these excitations too. In particular, they couple to ϕ via the "Yukawa" coupling of $\mathcal{S}_{c\phi}$ in (18.32), and to establish the stability of the Gaussian fixed point we need to examine the scaling dimension of $d(k)$ on the Fermi surface.

Answering this question requires a continuum theory for the fermionic sector also, and its coupling to the bosonic sector. Fortunately, we can now directly use the continuum limit presented in Section 18.1. We pick a Fermi surface patch centered at the momentum $\vec{k}_0$, and describe the low-energy fermionic quasiparticles by $\mathcal{S}_0$ in (18.7). Next, we apply the substitution (18.6) to (18.32), and obtain the Yukawa coupling

$$\mathcal{S}_{\psi\phi} = \lambda \int d\tau \int dx \int d^{d-1}y \, \phi \, \psi_a^\dagger \psi_a, \tag{18.36}$$

where $\lambda = d(k_0)$. Finally, we also need to distinguish the directions parallel and transverse to the Fermi surface in $\mathcal{S}_H$, and so write its Gaussian part as

$$\mathcal{S}_{HG} = \frac{1}{2} \int \frac{dk_x}{(2\pi)} \int \frac{d^{d-1}k_y}{(2\pi)^{d-1}} \int \frac{d\omega_n}{2\pi} \left[k_y^2 + \gamma \frac{|\omega_n|}{|k_y|} + r \right] |\phi(k, \omega_n)|^2. \tag{18.37}$$

We have dropped the k_x dependence in (18.18) because it is irrelevant compared to the k_y dependence under the rescaling (18.9).

Now our task is to determine the scaling dimension of λ under the theory $\mathcal{S}_0 + \mathcal{S}_{HG} + \mathcal{S}_{\psi\phi}$ defined in (18.7), (18.37), and (18.36).

We scale x and y as in (18.9). For the rescaling of time, we clearly need to focus on the critical excitations of the Ising-nematic order, which have dynamic exponent $z = 3$. Thus we choose

$$x' = xe^{-2\ell}, \quad y' = ye^{-\ell}, \quad \tau' = \tau e^{-3\ell}. \tag{18.38}$$

An immediate consequence is that the coefficient ζ of the temporal derivative term in $\mathcal{S}_0$ in (18.7) is no longer invariant under the RG: it has scaling dimension

$$\dim[\zeta] = -1, \tag{18.39}$$

and so scales to zero. Thus this temporal derivative is irrelevant at the Ising-nematic quantum critical point. It is important at this point that our derivation of $\mathcal{S}_H$ in (18.33) showed that all terms had a finite limit as $\zeta \to 0$: in fact, the damping coefficient γ was shown to be independent of ζ in (18.20). Thus setting $\zeta = 0$ seems safe now, although this conclusion has to be re-examined at higher orders.

With these choices for the spacetime rescalings, it is a simple matter to compute the rescalings of the fields from their respect Gaussian actions:

$$\dim[\psi] = (d+2)/2, \quad \dim[\phi] = (d+2)/2. \tag{18.40}$$

Finally, we obtain the needed renormalization of the fermion–boson coupling in (18.36)

$$\dim[\lambda] = \frac{(2-d)}{2}. \tag{18.41}$$

So, only for $d > 2$ is the Gaussian action stable to the presence of the cubic nonlinearity associated with the fermion–boson coupling. This is the primary conclusion of this subsection.

For $d > 2$, the above arguments suggest that we can estimate the fate of the fermionic quasiparticles in a perturbation theory in λ. Let us first try to guess the structure of the answer using scaling arguments similar to those used in (14.37) and (18.15); we expect

$$\text{Im}\Sigma \sim \lambda^2 \omega^p. \tag{18.42}$$

Matching scaling dimensions with $\dim[\Sigma] = 2$ (because $\dim[k_x] = \dim[k_y^2] = 2$), $\dim[\omega] = 3$ and (18.41), we obtain

$$p = \frac{d}{3}. \tag{18.43}$$

In $d = 3$, the integer value of p suggests that there should be additional logarithms, and we indeed see this in an explicit computation below. Examination of the quasiparticle spectral weight using (18.13), and as discussed below (18.22) shows that the quasiparticles are only marginally well defined with a width of the same order as the quasiparticle energy upon approaching the Fermi surface. Such states have been named "marginal Fermi liquids" [523], but the present argument shows this terminology is a misnomer in the RG sense: the coupling λ is irrelevant and not marginal. The RG argument also has a bonus in implying that higher orders in λ will only produce higher powers of ω in the self-energy; perturbation theory about the Gaussian fixed point directly yields the terms most important in the infrared already at low order in the expansion.

The situation is very different in $d = 2$. In this case $\text{Im}\Sigma \sim \omega^{2/3}$, and so it is now clear from (18.22) that the quasiparticle is no longer well defined, and we are dealing with a non-Fermi liquid. Applying (18.23), we find that the momentum distribution function does not have a step discontinuity on the Fermi surface; there is a weaker power-law singularity with

$$n(k) \sim \text{sgn}(-\varepsilon_k)|\varepsilon_k|^{1/3}. \tag{18.44}$$

Importantly, the scaling dimension of the boson–fermion coupling λ is 0, and so it is not clear whether perturbation theory in λ is reliable. By analogy with the similar situation in Section 17.4 for Dirac fermions, the implication is that the critical theory should be formulated at a fixed λ, and that the perturbative Hertz approach has broken down. We turn to a discussion of the needed critical field theory in Section 18.2.3.

However, before we turn to that crucial question, we need to verify the scaling estimate for the self-energy in (18.42) by an explicit computation. The needed contribution to the self-energy at order λ^2 is given by the Feynman diagram in Fig. 18.8, which evaluates to

$$\Sigma(k, \omega_n) = \lambda^2 \int \frac{d^d q}{(2\pi)^d} \int \frac{d\epsilon_n}{2\pi} \frac{1}{q_y^2 + \gamma |\epsilon_n|/|q_y|} G_0(k + q, \epsilon_n + \omega_n). \tag{18.45}$$

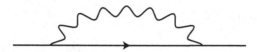

Fig. 18.8   Order λ^2 contribution to the fermion self-energy.

This can be evaluated by the same methods used for (18.18). Integrating over q_x, we find the analog of (18.21)

$$\Sigma(k, \omega_n) = i \frac{\lambda^2}{v_F} \int \frac{d^{d-1}q_y}{(2\pi)^{d-1}} \int \frac{d\epsilon_n}{2\pi} \frac{\text{sgn}(\epsilon_n + \omega_n)|q_y|}{|q_y|^3 + \gamma |\epsilon_n|}$$

$$= i \frac{\lambda^2}{\pi v_F \gamma} \text{sgn}(\omega_n) \int \frac{d^{d-1}q_y}{(2\pi)^{d-1}} |q_y| \ln\left(\frac{|q_y|^3 + \gamma |\omega_n|}{|q_y|^3}\right). \qquad (18.46)$$

Evaluation of the q_y integral yields a result which agrees with (18.42) and (18.43) in $d = 2$, and with the expected logarithmic corrections in $d = 3$. In the physically important case of $d = 2$, the q_y integral evaluates to

$$\Sigma(k, \omega_n) = \frac{\lambda^2}{\pi v_F \gamma^{1/3}\sqrt{3}} \text{sgn}(\omega_n)|\omega_n|^{2/3}, \quad d = 2 \qquad (18.47)$$

in agreement with (18.43).

18.2.3 Non-Fermi liquid criticality in $d = 2$

Section 18.2.2 established that a perturbative analysis in the fermion–boson coupling λ, in the spirit of the familiar "random-phase-approximation" (RPA) of many body physics, led to a valid theory of the Ising-nematic quantum critical point in $d = 3$. However, the RPA-like Hertz approach broke down in $d = 2$. Here we provide a field-theoretic description of the quantum criticality in $d = 2$, using the approach proposed in [333].

An important feature of the discussion in Section 18.2.2 was that the low-energy fermion modes at the Fermi surface point $\vec{k}_0$ coupled most strongly to ϕ fluctuations with momenta parallel to the Fermi surface. This is clear from the k_y dependence of $\mathcal{S}_{HG}$ in (18.37). Physically, this is because a fermion at $\vec{k}_0$ scattered by ϕ by momentum k tangent to the Fermi surface only changes its energy $\sim k^2$, while in all other directions its energy change $\sim k$. Consistent with this, if we compute the induced four-point ϕ vertex in the theory $\mathcal{S}_0 + \mathcal{S}_{HG} + \mathcal{S}_{\psi\phi}$, we find an enhancement dependent upon the ϕ momenta only if the momenta are parallel or anti-parallel. This suggests that all couplings between ϕ fluctuations with noncollinear momenta, such as, e.g., those induced by the u term in (18.33), are formally irrelevant, just as in the Hertz theory. This asymptotic decoupling indicates that we may treat noncollinear directions of ϕ in separate critical theories. Thus we end up with an infinite number of 2+1 dimensional field theories, labeled by the momentum direction of ϕ.

The reader will recall our discussion of an infinite number of 1+1 dimensional field theories of chiral fermions in our discussion of Fermi liquid theory associated with (18.5).

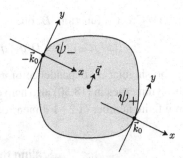

Fig. 18.9   A ϕ fluctuation at wavevector $\vec{q}$ couples most efficiently to the fermions $\psi_\pm$ near the Fermi surface points $\pm\vec{k}_0$.

A crucial difference here is that we have an infinite number of 2+1 dimensional field theories: this is needed because, as discussed above, the dominant scattering processes for the fermions are tangent to the Fermi surface. The present description necessarily induces a redundant description: however, by a simple generalization of the arguments in Section 18.1.1, we see that the redundant description is consistent.

So let us focus on a given direction $\vec{q}$ of the momentum carried by ϕ and derive the associated critical theory, as illustrated in Fig. 18.9. Such a ϕ field couples most strongly to fermions near *two* points on the Fermi surface: those at $\vec{k}_0$ and at $-\vec{k}_0$. We generalize (18.6) by introducing two fermionic fields $\psi_{\pm a}$ by

$$\psi_{+a}(\vec{k}) = c_{\vec{k}_0+\vec{k},a}, \quad \psi_{-a}(\vec{k}) = c_{-\vec{k}_0+\vec{k},a}. \tag{18.48}$$

We allow the spin index a to extend over the N values: the physical case is $N = 2$, but the large-N expansion provides a useful computational tool.

Now we expand all terms in $\mathcal{S}_c + \mathcal{S}_\phi + \mathcal{S}_{c\phi}$ (defined in (18.1), (18.31), and (18.32)) in spatial and temporal gradients. Using the coordinate system illustrated in Fig. 18.9, performing appropriate rescaling of coordinates, and dropping terms which can later be easily shown to be irrelevant, we obtain the 2+1 dimensional Lagrangian

$$\mathcal{L} = \psi_{+a}^\dagger \left(\zeta \frac{\partial}{\partial \tau} - i \frac{\partial}{\partial x} - \frac{\partial^2}{\partial y^2} \right) \psi_{+a} + \psi_{-a}^\dagger \left(\zeta \frac{\partial}{\partial \tau} + i \frac{\partial}{\partial x} - \frac{\partial^2}{\partial y^2} \right) \psi_{-a}$$

$$- \lambda \phi \left(\psi_{+a}^\dagger \psi_{+a} + \psi_{-a}^\dagger \psi_{-a} \right) + \frac{N}{2} (\partial_y \phi)^2 + \frac{Nr}{2} \phi^2. \tag{18.49}$$

Here ζ, λ, and r are coupling constants, with r the tuning parameter across the transition; we will see that all couplings apart from r can be scaled away or set equal to unity.

A first crucial property of $\mathcal{L}$ is that it continues to have fermion Green's functions with singularities on the original Fermi surface, as established in Section 18.1.1. This follows from generalizing the symmetry (18.30) to

$$\phi(x, y) \to \phi(x, y + \theta x), \quad \psi_s(x, y) \to e^{-is(\frac{\theta}{2}y + \frac{\theta^2}{4}x)} \psi_s(x, y + \theta x), \tag{18.50}$$

where θ is now one-dimensional and $s = \pm$. This is an emergent symmetry of $\mathcal{L}$ for arbitrary shapes of the Fermi surface. An immediate consequence of (18.50) is that (18.29) is obeyed by both $\psi_\pm$, so that all singularities slide without change along the Fermi surface.

Furthermore, the ϕ Green's function, D, obeys

$$\langle |\phi(q, \omega)|^2 \rangle \equiv D(q_x, q_y, \omega) = D(q_y, \omega), \tag{18.51}$$

indicating that no singular q_x dependence of ϕ is generated by the theory $\mathcal{L}$. As in Section 18.1.1, the symmetries in (18.50) also help establish the consistency of our description in terms of an infinite number of 2+1 dimensional field theories.

Scaling theory

We now generalize the scaling analysis of Section 18.2.2 to the continuum field theory $\mathcal{L}$. Because $\mathcal{L}$ is strongly coupled, we have to allow for anomalous dimensions at all stages.

As before we choose

$$\dim[y] = -1, \quad \dim[x] = -2. \tag{18.52}$$

The invariance in (18.29) implies that no anomalous dimension appears in the relative scaling of x and y. However, we do have to allow for an anomalous dimension in time, and so keep the rescaling of the temporal coordinate general:

$$\dim[\tau] = -z. \tag{18.53}$$

Note that the dynamic critical exponent z is defined relative to the spatial coordinate y tangent to the Fermi surface (other investigators sometimes define it relative to the coordinate x normal to the Fermi surface, leading to a difference by a factor of 2). We define the engineering dimensions of the fields so that coefficients of the y derivatives remain constant. Allowing for anomalous dimensions η_ϕ and η_ψ from loop effects, we have

$$\dim[\phi] = (1 + z + \eta_\phi)/2, \quad \dim[\psi] = (1 + z + \eta_\psi)/2. \tag{18.54}$$

Using these transformations, we can examine the scaling dimensions of the couplings in $\mathcal{L}$ at tree level

$$\dim[\zeta] = 2 - z - \eta_\psi, \quad \dim[\lambda] = (3 - z - \eta_\phi - 2\eta_\psi)/2. \tag{18.55}$$

We have seen from our low-order loop computations in Sections 18.2.1 and 18.2.2 that $z = 3$. Assuming that the anomalous dimensions η_ϕ and η_ψ are small, we see that the coupling ζ is strongly irrelevant. Thus we can send $\zeta \to 0$ in all our computations. However, we do not set $\zeta = 0$ at the outset, because the temporal derivative term is needed to define the proper analytic structure of the frequency loop integrals [292].

Also note that these estimates of the scaling dimensions imply $\dim[\lambda] \approx 0$. Thus the fermion and order parameter fluctuations remain strongly coupled at all scales in $d = 2$, as we anticipated in Section 18.2.2. Conversely, we can also say that the requirement of working in a theory with fixed λ implies that $z \approx 3$; this circumvents the appeal to loop computations for taking the $\zeta \to 0$ limit. With a near zero scaling dimension for λ, we cannot expand perturbatively in powers of λ. This feature was also found in Section 17.4.

Moving beyond tree-level considerations, we note that another Ward identity obeyed by the theory $\mathcal{L}$ allows us to fix the scaling dimension of ϕ exactly. This Ward identity is linked to the fact that ϕ appears in the Yukawa coupling like the x component of a gauge field coupled to the fermions [333]. The usual arguments associated with gauge invariance then imply that $\dim[\phi] = 2$ (the same as the scaling dimension of ∂_x), and that we can work in a theory in which the "gauge coupling" λ is set equal to unity at all scales. Note that with this scaling dimension, we have the exact relation

$$\eta_\phi = 3 - z. \qquad (18.56)$$

Note also that (18.54) now implies that $\dim[\lambda] = \eta_\psi$ at tree level, which is the same as the tree-level transformation of the spatial derivative terms. The latter terms have been set equal to unity by rescaling the fermion field, and so it is also consistent to set $\lambda = 1$ from now on.

We reach the remarkable conclusion that at the critical point $r = r_c$, $\mathcal{L}$ is independent of all coupling constants. The only parameter left is N, and we have no choice but to expand correlators in powers of $1/N$. The characterization of the critical behavior only requires computations of the exponents z and η_ψ, and associated scaling functions.

We can combine all the above results into scaling forms for the ϕ and Ψ Green's functions at the quantum critical point at $T = 0$. These are, respectively

$$D^{-1}(q_x, q_y, \omega) = q_y^{z-1} \mathcal{F}_D\left(\frac{\omega}{q_y^z}\right), \qquad (18.57)$$

$$G_+^{-1}(q_x, q_y, \omega) = (q_x + q_y^2)^{1-\eta_\psi/2} \mathcal{F}_G\left(\frac{\omega}{(q_x + q_y^2)^{z/2}}\right), \qquad (18.58)$$

where $\mathcal{F}_D$ and $\mathcal{F}_G$ are nontrivial scaling functions. Note that the second scaling form shows that the singularity of the fermion Green's function in momentum space is invariant along the Fermi surface, and depends only upon the distance from the Fermi surface.

We have come as far as possible by symmetry and scaling analyses alone on $\mathcal{L}$. Further results require specific computations of loop corrections, and these can only be carried out within the context of the $1/N$ expansion. At leading order, the $1/N$ expansion reproduces the results in Section 18.2.2. It is important to note that in the notation of the present section, the results of Section 18.2.2 for D and the fermion self-energy turn out to be *independent* of ζ. Although ζ appears at intermediate stages, it cancels out in the final result: the reader is urged to verify this crucial feature of the theory. Consequently there is no problem in taking the $\zeta \to 0$ limit. Higher order computations are involved, and raise numerous complicated issues we do not wish to enter into here: we refer the reader to [292, 333]. It was found that $z = 3$ was preserved up to three loops, but a nonzero value for η_ψ did appear at three-loop order.

18.3 Spin density wave order

We now turn to the second major class of symmetry breaking transitions of Fermi liquids: those involving an order parameter which is spatially modulated and so breaks translational symmetry. As a canonical example of such a transition we consider spin density wave

(SDW) ordering in the Hamiltonian (17.1) describing a single band on the square lattice, which was motivated by the physics of the cuprate superconductors. However, our methods and results are easily extended to other types of ordering with spatial modulations.

We have already met the analog of spin density wave ordering in insulating antiferromagnets. In that context it was referred to as Néel ordering, and characterized by the order parameter $\mathbf{n}$ of the O(3) quantum rotor model: we discussed this in Section 1.4.3 on spin–ladder models, and consider the square lattice in more detail in Section 19.3. In the continuum soft-spin limit, such ordering was described by the three-component real field ϕ_α whose fluctuations were controlled by the action $\mathcal{S}_\phi$ in (2.11). The field ϕ_α will remain the order parameter for SDW ordering in a metal being considered here, and its action $\mathcal{S}_\phi$ will be an important ingredient in our theory.

As in Section 18.2, apart from the order parameter, we also have to consider the fermionic excitations near the Fermi surface, and these are described as before by $\mathcal{S}_c$ in (18.1).

Finally we need to couple the c_{ka} fermions to the SDW order ϕ_α. The two-sublattice Néel order on the square lattice carries momentum $\vec{K} = (\pi, \pi)$, and we can consider a general $\vec{K}$ which leads to two-sublattice SDW ordering. Other values of $\vec{K}$ lead to complex order parameter fields (rather than the real case we considered), but we will not consider this relatively straightforward generalization. Translational invariance implies that ϕ_α scatters the fermions with momentum $\vec{K}$, and so the natural generalization of the fermion–boson coupling in (18.32) is

$$\mathcal{S}_{c\phi} = \int d\tau \int \frac{d^d k\, d^d q}{(2\pi)^{2d}} \phi_\alpha(q) \sigma_{ab}^\alpha c_{k+K+q,a}^\dagger c_{kb} + \text{c.c.}, \tag{18.59}$$

where q is a small momentum associated with a long-wavelength SDW fluctuation, while the integral over the momentum $\vec{k}$ extends over the entire Brillouin zone.

Our complete theory for the SDW transition is $\mathcal{S}_c + \mathcal{S}_\phi + \mathcal{S}_{c\phi}$ in (18.1), (2.11), and (18.59). This forms the basis of the discussion in the remainder of this section.

18.3.1 Mean-field theory

The theory has two phases: the ordinary Fermi liquid with $\langle \phi_\alpha \rangle = 0$ and Fermi surface as in Fig. 18.1, and the SDW state with $\langle \phi_\alpha \rangle \neq 0$. We describe here the configuration of the Fermi surface in the SDW state. We replace ϕ_α by its expectation value $\langle \phi_\alpha \rangle = (0, 0, \phi)$; by rotational symmetry we can take the SDW order in the z direction without loss of generality. Now $\mathcal{S}_c + \mathcal{S}_{c\phi}$ is a bilinear in the fermions and can be diagonalized to yield a fermion band structure. The fermion Hamiltonian takes the form of a 2×2 matrix coupling together c_{ka} and $c_{k+K,a}$. Diagonalizing this matrix leads to the single fermion energy eigenvalues

$$E_k = \frac{\varepsilon_k + \varepsilon_{k+K}}{2} \pm \left(\left(\frac{\varepsilon_k + \varepsilon_{k+K}}{2} \right)^2 + \phi^2 \right)^{1/2}. \tag{18.60}$$

We now have to occupy the lowest energy bands in this band structure, and deduce the configuration of the Fermi surface. Such a solution is illustrated in Fig 18.10 for the case

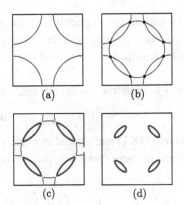

Fig. 18.10 The transformation of the Fermi surface of cuprates by SDW order [432]. (a) Fermi surface without SDW order, as in Fig. 18.1. (b) The original Fermi surface along with the Fermi surface shifted by wavevector (π, π). These intersect at the hot spots shown by the filled circles. (c) With the onset of a nonzero spin density wave order with $\langle \phi_\alpha \rangle \neq 0$, gaps open at the hot spots leading to electron (thin lines) and hole (thick lines) pockets. (d) With increasing $|\langle \phi_\alpha \rangle|$ the electron pockets shrink to zero for the hole-doped case, leaving only hole pockets. In the electron-doped case, the hole pockets shrink to zero, leaving only electron pockets (this is not shown). Finally, in the half-filled case, the electron and hole pockets shrink to zero simultaneously.

of $\vec{K} = (\pi, \pi)$ ordering in cuprate superconductors [432]. A key feature is that the original "large" Fermi surface splits into "small" electron and hole pockets upon the onset of SDW order: this is also a generic property of SDW ordering on other lattices.

18.3.2 Continuum theory

We now wish to develop a continuum theory for the quantum critical point for the onset of SDW order, which is accompanied by a change from a large Fermi surface to small pockets, as indicated in Fig. 18.10. The general strategy is similar to that in Section 18.2, although there are key differences which we highlight below.

An important and new concept here is that of a "hot" manifold on the Fermi surface. As in Section 18.2, we expect the boson–fermion coupling to be most efficient if it scatters fermions between nearly degenerate low-energy states. Let us pick a point $\vec{k}_1$ on the Fermi surface, where the fermion has zero energy: this is described by a d-dimensional vector upon which we impose one constraint to place it on the Fermi surface. Now the ϕ_α field will scatter this fermion to the point $\vec{k}_2 = \vec{k}_1 + K$. We require that $\vec{k}_2$ is also on the Fermi surface, to ensure that the final fermion state has zero energy: this places a second constraint on $\vec{k}_2$. The solution of these constraints yields a pair of $(d-2)$-dimensional manifolds specifying the allowed values of $\vec{k}_1$ and $\vec{k}_2$: these are the "hot" manifolds. In $d = 2$, they become the "hot spots" shown in Fig. 18.10 for the cuprate case.

We take the continuum limit by focusing on one generic point on the hot manifold, say $\vec{k}_1$. Then its partner, $\vec{k}_2 = \vec{k}_1 + \vec{K}$ will also be on the hot manifold. We focus on the patches of the Fermi surface near these points. There are several other pairs of patches in the Brillouin zone, as is clear from Fig. 18.10: these are described by parallel theories which we will not discuss explicitly. Within a given pair of Fermi surface patches, our results do not depend upon the specific choices of $\vec{k}_{1,2}$ on the hot manifold: this is evident from our continuum theory below, and does not require a proof which is the analog of Section 18.1.1.

Near $\vec{k}_1$ and $\vec{k}_2$, we define continuum fields $\psi_{1,2}$, as in (18.48)

$$\psi_{1a}(\vec{k}) = c_{\vec{k}_1 + \vec{k}, a}, \quad \psi_{2a}(\vec{k}) = c_{\vec{k}_2 + \vec{k}, a}. \tag{18.61}$$

We insert this into (18.1) and expand in powers of k. Unlike the situation in Sections 18.1 and 18.2, it turns out here that it is sufficient to keep terms only linear in k, and the analog of (18.7) is

$$S_\psi = \int d\tau \int d^d x \left[\psi_{1a}^\dagger \left(\zeta \frac{\partial}{\partial \tau} - i \vec{v}_1 \cdot \nabla_x \right) \psi_{1a} \right.$$
$$\left. + \psi_{2a}^\dagger \left(\zeta \frac{\partial}{\partial \tau} - i \vec{v}_2 \cdot \nabla_x \right) \psi_{2a} \right]. \tag{18.62}$$

Here $\vec{v}_1 = \nabla_k \varepsilon_k |_{k_1}$ is the Fermi velocity at $\vec{k}_1$, and similarly for $\vec{v}_2$. We have inserted factors of ζ in front of the temporal derivatives by analogy with Section 18.2.2, anticipating that the temporal derivatives ultimately become irrelevant near the critical point. In the $\psi_{1,2}$ formulation, the configurations of the Fermi surfaces and hot manifolds are shown in Fig. 18.11. The Fermi surface of the ψ_1 fermions is defined by $\vec{v}_1 \cdot \vec{k} = 0$, and the Fermi surface of the ψ_2 fermions is defined by $\vec{v}_2 \cdot \vec{k} = 0$. Finally, the hot manifold is defined by the $\vec{k}$ which satisfy *both* these conditions. We assume here and below that $\vec{v}_1$ and $\vec{v}_2$ are not collinear: the collinear case corresponds to the "nesting" of the Fermi surfaces, and we do not consider that here.

It is now a simple matter to take the continuum limit of the boson–fermion coupling in (18.59). Including the important terms in S_ϕ in (2.11), we obtain

$$S_{\psi\phi} = \int d^d x \int d\tau \left[\frac{1}{2}(\nabla_x \phi_\alpha)^2 + \frac{r}{2}\phi_\alpha^2 + \frac{u}{4}\left(\phi_\alpha^2\right)^2 \right.$$
$$\left. + \lambda \phi_\alpha \sigma_{ab}^\alpha \left(\psi_{1a}^\dagger \psi_{2b} + \psi_{2a}^\dagger \psi_{1b} \right) \right]. \tag{18.63}$$

Here we have omitted the temporal gradient term in ϕ_α because it later turns out to be irrelevant.

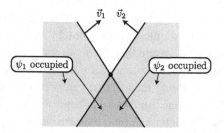

Fig. 18.11   Fermi surfaces of ψ_1 and ψ_2 fermions in the plane defined by the Fermi velocities $\vec{v}_1$ and $\vec{v}_2$. The Fermi surfaces are $(d-1)$-dimensional, and are indicated by the full lines. The $(d-2)$-dimensional "hot manifold" intersects this plane at the filled circle at the origin.

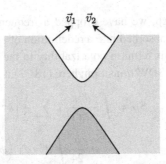

Fig. 18.12

Modification of the Fermi surfaces in Fig. 18.11 by SDW order with $\langle \phi_\alpha \rangle \neq 0$. The full lines are the Fermi surfaces, and the white, light shaded, and dark shaded regions denote momenta where 0, 1, and 2 of the bands in (18.60) are occupied. The upper and lower lines are boundaries of hole and electron pockets, respectively. There are eight instances of such Fermi surface configurations in Fig. 18.10c, centered on the eight hotspots.

The remaining analysis of this section works with the continuum theory of bosons and fermions defined by $\mathcal{S}_\psi + \mathcal{S}_{\psi\phi}$ defined in (18.62) and (18.63). Our steps in the following subsection closely parallel those for the Ising-nematic case in Section 18.2.

We begin by noting the Fermi surface change in the SDW phase of $\mathcal{S}_\psi + \mathcal{S}_{\psi\phi}$. Setting $\phi_\alpha = (0, 0, \phi)$, and diagonalizing the $\psi_{1,2}$ spectrum, it is a simple matter to show that the Fermi surfaces in Fig. 18.11 are modified to those in Fig. 18.12.

18.3.3 Hertz theory

As in Section 18.2.1, the Hertz theory for the SDW order ϕ_α is obtained by integrating out the $\psi_{1,2}$ fermions from $\mathcal{S}_\psi + \mathcal{S}_{\psi\phi}$ in (18.62) and (18.63).

Again, the most important contribution is the coefficient of the ϕ_α^2 term, which is given by the fermion polarizability in Fig. 18.7. Here the explicit expression for the polarizability maps from (18.17) to

$$\Pi(q, \omega_n) = \int \frac{d^d k}{(2\pi)^d} \int \frac{d\epsilon_n}{2\pi} \frac{1}{[-i\zeta(\epsilon_n + \omega_n) + \vec{v}_1 \cdot (\vec{k} + \vec{q})][-i\zeta\epsilon_n + \vec{v}_2 \cdot \vec{k}]}. \quad (18.64)$$

We define oblique coordinates $p_1 = \vec{v}_1 \cdot \vec{k}$ and $p_2 = \vec{v}_2 \cdot \vec{k}$. It is then clear that the integrand in (18.64) is independent of the $(d - 2)$ transverse momenta, whose integral yields an overall factor Λ^{d-2} (in $d = 2$ this factor is precisely 1). Also, by shifting the integral over k_1 we note that the integral is independent of q. So we have

$$\Pi(q, \omega_n) = \frac{\Lambda^{d-2}}{|\vec{v}_1 \times \vec{v}_2|} \int \frac{dp_1 dp_2 d\epsilon_n}{8\pi^3} \frac{1}{[-i\zeta(\epsilon_n + \omega_n) + p_1][-i\zeta\epsilon_n + p_2]}. \quad (18.65)$$

Next, we evaluate the frequency integral to obtain

$$\begin{aligned}
\Pi(q, \omega_n) &= \frac{\Lambda^{d-2}}{\zeta|\vec{v}_1 \times \vec{v}_2|} \int \frac{dp_1 dp_2}{4\pi^2} \frac{[\mathrm{sgn}(p_2) - \mathrm{sgn}(p_1)]}{-i\zeta\omega_n + p_1 - p_2} \\
&= -\frac{|\omega_n|\Lambda^{d-2}}{4\pi|\vec{v}_1 \times \vec{v}_2|}. \quad (18.66)
\end{aligned}$$

In the last step, we have dropped a frequency-independent, cutoff-dependent constant, which can be absorbed into a redefinition of r. Note also that the factor of ζ has cancelled.

Inserting this fermion polarizability in the effective action for ϕ_α, we obtain the Hertz action for the SDW transition; here (18.33) is replaced by

$$
\mathcal{S}_H = \int \frac{d^d k}{(2\pi)^d} T \sum_{\omega_n} \frac{1}{2} \left[k^2 + \gamma |\omega_n| + r \right] |\phi_\alpha(k, \omega_n)|^2
$$
$$
+ \frac{u}{24} \int d^d x \, d\tau \left(\phi_\alpha^2(x, \tau) \right)^2 . \tag{18.67}
$$

The main difference from (18.33) is that the $|\omega_n|/|k|$ has been replaced by a $|\omega_n|$: this is a direct consequence of the Fermi surface structure in Fig. 18.11, which leads to a density of states of particle–hole excitations which are linear in energy, and independent of momentum.

The subsequent analysis of the above Hertz action proceeds just as in (18.34) and (18.35), but now with dynamic exponent $z = 2$. Again, this exponent characterizes the frequency scale $\omega \sim k^2$ emerging from a comparison of the first two terms in (18.67). Now we see that the Gaussian fixed point of (18.67) is stable for $d > 2$.

Just as in Section 18.2, we see below that the Hertz approach is essentially correct in $d = 3$, but that it fails in the physically important case of $d = 2$, and not just by a marginal correction. The key to this is an examination of the fermion spectrum and the RG flow of the fermion-boson coupling, to which we turn in Section 18.3.4.

18.3.4 Fate of the fermions

We proceed just as in Section 18.2.2. With the scaling dimensions of space–time and ϕ as in (18.34) with $z = 2$, the action $\mathcal{S}_\psi$ in (18.62) implies the tree-level rescaling

$$
\dim[\psi] = (d+1)/2, \quad \dim[\zeta] = -1. \tag{18.68}
$$

Thus, just as in Sections 18.2.2 and 18.2.3, ζ is irrelevant, and can eventually be sent to 0. Finally, evaluating the scaling dimension of the λ term in $\mathcal{S}_{\psi\phi}$ in (18.63), we obtain the same result as in (18.41), namely

$$
\dim[\lambda] = \frac{(2-d)}{2}. \tag{18.69}
$$

The conclusions then are the same as in Section 18.2.2: a perturbative theory in λ, in the Hertz/RPA approach is valid in $d = 3$, but requires a new formulation in terms of a fixed λ theory in $d = 2$.

The remaining analysis in this subsection tracks that below (18.41) in Section 18.2.2.

The perturbative estimate for the fermion damping on the hot manifold is given by (18.42). Using the values here, $\dim[\Sigma] = 1$ and $\dim[\omega] = 2$, we obtain instead of (18.43) that

$$
p = \frac{d-1}{2}. \tag{18.70}
$$

In $d = 3$, this has the same value as in (18.43), and so the conclusions are the same.

In $d = 2$, we obtain non-Fermi liquid behavior at the hot spots with $\text{Im}\Sigma \sim \sqrt{\omega}$.

We verify this result by explicitly computing the value of the fermion damping from the graph in Fig. 18.8. At zero momentum for the ψ_1 fermion we have, instead of (18.45),

$$\Sigma_1(0, \omega_n) = \lambda^2 \int \frac{d^d q}{(2\pi)^d} \int \frac{d\epsilon_n}{2\pi} \frac{1}{\left[q^2 + \gamma |\epsilon_n|\right]\left[-i\zeta(\epsilon_n + \omega_n) + \vec{v}_2 \cdot \vec{q}\right]}. \quad (18.71)$$

We first perform the integral over the $\vec{q}$ direction parallel to $\vec{v}_2$, while ignoring the subdominant dependence on this momentum in the boson propagator. The dependence on ζ immediately disappears, and yields in place of (18.46),

$$\Sigma_1(0, \omega_n) = i\frac{\lambda^2}{|v_2|} \int \frac{d^{d-1}q}{(2\pi)^{d-1}} \int \frac{d\epsilon_n}{2\pi} \frac{\text{sgn}(\epsilon_n + \omega_n)}{|q|^2 + \gamma |\epsilon_n|}$$

$$= i\frac{\lambda^2}{\pi |v_2| \gamma} \text{sgn}(\omega_n) \int \frac{d^{d-1}q}{(2\pi)^{d-1}} \ln\left(\frac{|q|^2 + \gamma |\omega_n|}{|q|^2}\right). \quad (18.72)$$

Again, evaluation of the q integral yields results in agreement with the scaling estimate (18.70). Specifically, in $d = 2$, (18.72) evaluates to

$$\Sigma_1(0, \omega) = i\frac{\lambda^2}{\pi |v_2| \sqrt{\gamma}} \text{sgn}(\omega_n) \sqrt{|\omega_n|}, \quad d = 2 \quad (18.73)$$

as expected from (18.70).

18.3.5 Critical theory in $d = 2$

With the conclusion of Section 18.3.4 that the Hertz theory only applies for $d = 3$, let us turn to the strong-coupling dynamics in $d = 2$. Following Section 18.2.3, we need to understand the renormalization structure of the underlying theory of fermions and bosons, without an integration of the fermionic modes.

In the present case, the needed critical theory in $d = 2$ was already formulated in Section 18.3.2. It is defined by $\mathcal{S}_\psi + \mathcal{S}_{\psi\phi}$ in (18.62) and (18.63).

We formulate the RG with the same spacetime rescalings used in Section 18.3.3:

$$\dim[x] = -1, \quad \dim[\tau] = -2. \quad (18.74)$$

Note that we do not allow an anomalous dimension in the rescaling of τ. This does not mean that we necessarily have dynamic exponent $z = 2$. We allow the Fermi velocities $\vec{v}_1$ and $\vec{v}_2$ to flow under the RG, and a nontrivial z will arise from the nature of their flow to large scales. It is convenient to use such a formulation because the scaling of the fermion spectrum depends sensitively on the direction in momentum space, and the shape of the Fermi surface also evolves.

We do, however, have to allow for anomalous dimensions in the field rescalings, which become

$$\dim[\phi] = (2 + \eta_\phi)/2, \quad \dim[\psi] = (3 + \eta_\psi)/2. \quad (18.75)$$

Contributions to these anomalous dimensions do arise from loop fluctuation contributions.

Next, as in (18.55), we have the tree-level rescalings of the couplings associated with the fermions:

$$\text{dim}[\zeta] = -1 - \eta_\psi, \quad \text{dim}[\lambda] = -(\eta_\phi + 2\eta_\psi)/2. \qquad (18.76)$$

We reach the same conclusions from these results that we did in Section 18.2.3. We can safely assume that $\zeta = 0^+$, and use this value in loop computations. With the absence of the ζ term, we can choose the fermion field scale renormalization so that the theory maintains $\lambda = 1$. Thus, there is a strong fermion–boson coupling at all scales, and no independent renormalization group flow for λ.

The task before us is now clear, in principle. We have to evaluate higher loop diagrams and so determine the RG flow of the couplings $\vec{v}_1$, $\vec{v}_2$, r, and u, and the anomalous dimensions η_ϕ and η_ψ. Note that the boson damping coefficient γ in the Hertz action (18.67) does *not* appear as an independent coupling. In reality, it is a parameter in the boson spectral function, and its value is pinned to the underlying couplings via (18.66); it is reassuring that the value in (18.66) has no nonuniversal cutoff dependence in $d = 2$.

The evaluation of these higher loop diagrams is very involved, and the reader is referred to [1–3, 334] for further details. These papers describe the rather complex structure of the dynamic response of the fermions and the bosons near the quantum critical point. Here we focus on just one striking aspect: the shape of the Fermi surface. This is determined by the flow of the velocities near the critical point. Let us write the velocities as

$$\vec{v}_1 = (v_x, v_y), \quad \vec{v}_2 = (-v_x, v_y). \qquad (18.77)$$

Then to two-loop order, the RG flow of the velocity ratio is given by

$$\frac{d\alpha}{d\ell} = -\frac{12}{\pi n}\frac{\alpha^2}{\alpha^2 + 1}, \quad \alpha \equiv \frac{v_y}{v_x}, \qquad (18.78)$$

in a model with n pairs of hot spots (the Fermi surface in Fig. 18.10 has $n = 4$). Integrating (18.78), we observe that α scales logarithmically to zero with momentum scale (this is similar to the logarithmic flow of the velocity ratio found in Section 17.4 for a transition in a d-wave superconductor). We can use the distance from the hot spot to set the momentum scale. The location of the ψ_1 Fermi surface is given by $\vec{v}_1 \cdot \vec{k} = 0$, or $k_y = -v_x k_x/v_y = -k_x/\alpha$. Evaluating α at the scale k_x, we find the Fermi surfaces of the $\psi_{1,2}$ at

$$k_y = \pm \frac{12}{\pi n} k_x \log(1/|k_x|). \qquad (18.79)$$

Such Fermi surfaces are sketched in Fig. 18.13.

18.4 Nonzero temperature crossovers

We limit our discussion of the nonzero temperature crossovers to the Hertz theory for the spin density wave transition in (18.67).

Here it is useful to note that the only difference between $\mathcal{S}_H$ and the dilute Bose gas model analyzed in Section 16.3 is that $\mathcal{S}_H$ contains a $|\omega_n|$ frequency dependence in the

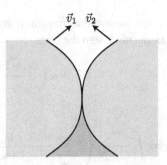

Fig. 18.13   Modification of the Fermi surfaces in Fig. 18.11 at the SDW quantum critical point. As in Figs. 18.11 and 18.12, the full lines are the Fermi surfaces, and the white, light shaded, and dark shaded regions denote momenta where 0, 1, and 2 of the bands in (18.60) are occupied. The equations of the Fermi surfaces are given in (18.79).

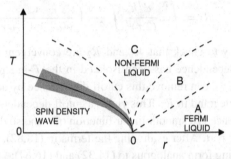

Fig. 18.14   Phase diagram of a Fermi liquid undergoing an instability to a spin density wave state for $2 \leq d < 4$. The regimes A, B, and C and their crossover boundaries are described in the text. Compare to Fig. 16.4 for the dilute Bose gas.

quadratic term, while the Bose gas had $-i\omega_n$. Consequently we can map the techniques and computations of the Bose gas theory to the present situation. The computation of the $T > 0$ crossovers is essentially identical to that in Section 16.3.2 and leads to the phase diagram shown in Fig. 18.14 [336], which is very similar to Fig. 16.4. We integrate out the $\omega_n \neq 0$ modes and obtain an effective action for the static modes, which takes the form $\mathcal{S}_{\phi,\mathrm{eff}}$ in (14.16) and is characterized by the couplings R and U. To leading order in u we have $U = u$, while for R we have (analogous to (14.17), (14.18), and (16.62)):

$$R = r + u\left(\frac{N+2}{6}\right)\int \frac{d^d k}{(2\pi)^d}\left(T\sum_{\omega_n \neq 0}\frac{1}{\gamma|\omega_n| + k^2 + r} + \frac{T}{k^2} - \int\frac{d\omega}{2\pi}\frac{1}{\gamma|\omega| + k^2}\right).$$

$$(18.80)$$

The next step is the mathematical one of evaluating the frequency and momentum sums and integrals in (18.80). The main subtlety, as in (14.31), is that while the result does depend upon a large momentum cutoff Λ, the divergent momentum integral can be separated out into a T-independent term. The remaining momentum integrals are convergent in the ultraviolet, and we can safely set $\Lambda \to \infty$ in them at the cost of ignoring some uninteresting and noncritical dependence on T. We show a few intermediate steps on how this separation is performed. The basic idea, as discussed below (11.47), is to subtract from each frequency summation the frequency integral of precisely the same quantity. In this manner we manipulate R into the form

$$R = r + u \left(\frac{N+2}{6} \right) [R_1 + R_2 + R_3], \tag{18.81}$$

with

$$R_1 = \int \frac{d^d k}{(2\pi)^d} \left(T \sum_{\omega_n} \frac{1}{\gamma |\omega_n| + k^2 + r} - \int \frac{d\omega}{2\pi} \frac{1}{\gamma |\omega| + k^2 + r} \right),$$

$$R_2 = -T \int \frac{d^d k}{(2\pi)^d} \left(\frac{1}{k^2 + r} - \frac{1}{k^2} \right),$$

$$R_3 = \int^\Lambda \frac{d^d k}{(2\pi)^d} \int \frac{d\omega}{2\pi} \left(\frac{1}{\gamma |\omega| + k^2 + r} - \frac{1}{\gamma |\omega| + k^2} \right). \tag{18.82}$$

It is easy to check that R_1 and R_2 are convergent at large momenta and that all of the cutoff dependence has been isolated in the T-independent term R_3. As discussed below (14.31), we can remove this cutoff dependence by adding and subtracting $r/(\gamma |\omega| + k^2)^2$ to the integrand in R_3. This yields a cutoff dependence term $\sim r \Lambda^{d-2}$. Note that this cutoff dependence is a smooth linear function of r and so does not affect the remaining universal singular part. After evaluating the terms in (18.82), we can write the final result for R in the scaling form analogous to (14.32) and (16.64):

$$R = r(1 - c_2 u \Lambda^{d-2}) + \frac{u}{\gamma} (\gamma T)^{d/2} \left(\frac{N+2}{6} \right) L \left(\frac{r}{\gamma T} \right), \tag{18.83}$$

where the universal scaling function $L(y)$ is given by

$$L(y) = \frac{1}{\pi} \int \frac{d^d k}{(2\pi)^d} \left[\ln \left(\frac{k^2}{2\pi} \right) - \psi \left(1 + \frac{k^2 + y}{2\pi} \right) + \frac{\pi + y}{k^2} \right], \tag{18.84}$$

where ψ is the digamma function. We point out the now familiar property of all such crossover functions: it is analytic at $y = r/\gamma T = 0$, reflecting the absence of any thermodynamic singularity at $r = 0$, $T > 0$ (see Fig. 18.14). From (18.84) it is easily seen that $L(y)$ is analytic for $y > -2\pi$; the singularity at $y = -2\pi$ is of no physical consequence as it is within the ordered phase.

Knowing the values of R and U, we can work out the predictions for physical observables. The expression for the order parameter correlations, correct for small u, is (compare (14.34) and (16.67))

$$\langle |\phi_\alpha(k, \omega)|^2 \rangle = \frac{1}{-i\omega + k^2 + \xi^{-2}}, \tag{18.85}$$

where from (14.19) and (14.5) we have (compare (16.68))

$$\xi^{-2} = R - Tu \left(\frac{N+2}{6} \right) \frac{2\Gamma((4-d)/2)}{(d-2)(4\pi)^{d/2}} R^{(d-2)/2}. \tag{18.86}$$

As in (16.70) we can compute the free energy density and obtain

$$\mathcal{F}(T, r) = \frac{TN}{2} \int \frac{d^d k}{(2\pi)^d} \left[\sum_{\omega_n} \ln(\gamma |\omega_n| + k^2 + r) + \ln\left(\frac{k^2 + \xi^{-2}}{k^2 + r}\right) \right], \quad (18.87)$$

where the numerator of the second logarithm is the contribution of the $\omega_n = 0$ modes, while the remainder come from the $\omega_n \neq 0$ modes. Note that, for $T > 0$, the expression (18.87) has no singularity at $r = 0$. This is as expected from the absence of a thermodynamic singularity in the middle of region C in Fig. 18.14. It is advantageous to subtract out the free energy of the system at the critical point $r = 0$, $T = 0$ from the above (this was simply 0 for the dilute Bose gas) and evaluate $\Delta\mathcal{F} \equiv \mathcal{F}(T, r) - \mathcal{F}(0, 0)$; for this we get

$$\Delta\mathcal{F} = \frac{TN}{2} \int^\Lambda \frac{d^d k}{(2\pi)^d} \left[-\frac{2}{\pi} \int_0^\infty \frac{d\Omega}{(e^{\Omega/T} - 1)} \tan^{-1}\left(\frac{\gamma\Omega}{k^2 + r}\right) \right.$$
$$\left. + \frac{k^2 \ln(k^2) - (k^2 + r) \ln(k^2 + r)}{\pi\gamma} + \ln\left(\frac{k^2 + \xi^{-2}}{k^2 + r}\right) \right] + \cdots, \quad (18.88)$$

where we have omitted background terms that are T independent and only depend upon positive integer powers of r. The momentum integral has a remaining cutoff dependence that cannot be removed and does affect the singular T and r dependence. This is a consequence of being above the upper-critical dimension.

We discuss the implications of the above results for the order parameter susceptibility; thermodynamic properties follow from results such as (18.88) and more explicit results are available in the literature [236, 432, 558]. In the low-T "Fermi liquid" region A in Fig. 18.14, defined by $T \ll r/\gamma$, the susceptibility is given by (18.85); by evaluating the large-y limit of (18.84) and inserting into (18.83) and (18.86) we get for the T dependence of the correlation length

$$\xi^{-2}(T) = \xi^{-2}(T = 0) + \frac{u\gamma}{r^{(4-d)/2}} \frac{(N+2)\Gamma((4-d)/2)}{36(4\pi)^{d/2}} T^2. \quad (18.89)$$

Note the characteristic T^2 dependence of a Fermi liquid. Conversely, in the high-T limit, $T \gg r/\gamma$, we take the $y \to 0$ limit of (18.84) and obtain the leading result

$$\xi^{-2}(T) = r + (\gamma T)^{d/2} \frac{(N+2)u}{6\gamma} L(0), \quad (18.90)$$

where $L(0)$ is a number. In region B of Fig. 18.14, $r \ll \gamma T \ll (\gamma r/u)^{2/d}$, the first term in (18.90) dominates, while in region C, $\gamma T \gg (\gamma |r|/u)^{2/d}$, the second T-dependent term is larger. So in the high-T region C we have $\xi \sim T^{-3/4}$, which does not agree with the naive scaling estimate $\xi \sim T^{-1/z}$. As we noted in Sections 14.2.2 and 16.3.2, this violation appears because of the presence of a dangerously irrelevant coupling u. Note also that if we insert (18.90) into (18.85), the resulting dynamic response function does not scale as a function of ω/T, and this is again because the present system is above its upper-critical dimension.

As noted earlier, the results above are analytic at $r = 0$ for $T > 0$, and so they apply also for $r < 0$. For this case the correlation length diverges at a critical value of r, and this determines the position of the phase boundary in Fig. 18.14 at $T = T_c(r)$, where to leading order in u

$$T_c(r) = \frac{1}{\gamma} \left[-\frac{6\gamma r}{(N+2)u} \right]^{2/d}. \tag{18.91}$$

Finally, we note the case of $d = 2$. For the Hertz action (18.67), this case corresponds to the upper-critical dimension, and can be analyzed like the corresponding situation for the $d = 3$ quantum Ising/rotor models in Part III: logarithmic terms arise from the flow of the coupling u to zero. However, as we noted in Section 18.3.4, the Hertz model breaks down in $d = 2$ as a representation of the underlying Fermi surface and its transformation under spin density wave order. The critical theory of Section 18.3.5 has to be analyzed at $T > 0$, a problem which we shall not address here.

18.5 Applications and extensions

Important applications of the spin density wave to Fermi liquid transition appear in studies of the heavy fermion compounds [25, 93, 94, 336, 514]. A case that has been intensively studied is $CeCu_{6-x}Au_x$ [408, 451, 492, 532], and there is also related work on $CeCu_{6-x}Ag_x$ [220]. The Cambridge group [182, 255, 300, 321] has examined a different series of Ce compounds ($CeNi_2Ge_2$, $CePd_2Si_2$, and $CeIn_3$) and these show similar transitions under pressure, but at stoichiometric compositions at which disorder is quite small; they have reported the existence of superconductivity near the antiferromagnet/Fermi liquid quantum critical point [321]. A comprehensive study of quantum transitions involving loss of antiferromagnetic order in metallic and insulating phases of V_2O_3 has also been performed [35]. A puzzling feature of present experiments in the Ce compounds and V_2O_3 is that while thermodynamic and transport properties are in rough agreement with the theory discussed in this chapter, the dynamic neutron scattering experiments show clear scaling of the response functions as a function ω/T (where ω is the measurement frequency). Such scaling was discussed at length in Part II but is only a property of quantum critical points below their upper-critical dimension; in contrast, the theories used to explain thermodynamic measurements are above their upper-critical dimension and do not predict scaling of response functions as a function of ω/T. Resolving this inconsistency is an important direction for future work. A range of theoretical work on quantum transitions between spin density waves and Fermi liquids is reported in [258, 342, 350, 375, 407]. An interesting perspective on open questions has been given by Coleman [91], who, following [441, 456, 472, 476, 477], has sketched a scenario for a strong-coupling critical point with anomalous exponents and ω/T scaling.

We refer the reader to a separate review article [302] for a comprehensive review of the experimental situation in this vast subject.

19 Heisenberg spins: ferromagnets and antiferromagnets

Part II of this book deals with the magnetically ordered and quantum paramagnetic phases of models of N-component quantum rotors. In Chapter 9 we showed how the $N = 2$ rotors could be mapped onto certain boson models in the vicinity of a phase transition between a Mott insulator and a superfluid. In this chapter we consider models of Heisenberg spins: these directly represent the spin fluctuations of physical electrons in insulators or other systems with an energy gap toward charged excitations (e.g. certain quantum Hall states). We describe the conditions under which certain models of Heisenberg spins reduce to $N = 3$ quantum rotor models, thus providing the long-promised physical motivation for studying the latter models; recall that a preview of this mapping has already appeared in Section 1.4.3. We also discuss the physical properties of Heisenberg spin models under conditions in which they do not map onto the rotor models of Part II.

We consider lattice models with the Hamiltonian

$$H_S = -\sum_{i,j} J_{ij} \hat{\mathbf{S}}_i \cdot \hat{\mathbf{S}}_j - \mathbf{H} \cdot \sum_i \hat{\mathbf{S}}_i. \tag{19.1}$$

Here the magnetic field $\mathbf{H}$ is precisely the same (with no overall scale factor) as that appearing in the rotor Hamiltonian (11.1): $\mathbf{H}$ couples to a conserved total spin (or for the rotors the total angular momentum), which, as we see, commutes with the rest of the Hamiltonian. The $\hat{\mathbf{S}}_i$ are Heisenberg spin operators whose basic properties were introduced in Section 1.4.3. They satisfy the commutation relations (1.29) on each site i and act on the $2S + 1$ states (1.30) of the spin S representation on each site. The J_{ij} are a set of translationally invariant exchange interactions between these sites.

We begin in Section 19.1 by showing how to set up a path integral for systems with states restricted in the manner (1.30) and (1.31) on each site. Then Section 19.2 considers the properties of *ferromagnets* in which all $J_{ij} > 0$, and the ground state is the fully polarized state with all spins parallel and the total spin takes its maximum possible value. The properties of antiferromagnets in which the ground state has negligible total spin are discussed in Section 19.3 – these are likely to arise when all $J_{ij} < 0$. Finally, Section 19.4 considers more complex situations with partial uniform polarization of the spins, which is accompanied by a certain "canted" order in dimensions $d > 1$.

19.1 Coherent state path integral

We described the coherent state path integral in general in Section 9.2, and then applied it to bosons in Section 9.2.1. An important feature of the path integral was the "Berry phase" term $b^\dagger db/d\tau$ in (9.32), which accounted for the kinematics of ordinary bosons and played an important role in the structure of the Mott insulating phases and the nature of their transitions to the superfluid. In this section we apply the same method to spin systems via "spin coherent" states. Many derivations of this path integral exist in the literature, but we follow here the approach used in [395], which has the advantage of explicitly maintaining spin rotation invariance. The reader is also referred to a collection of reprints [272] for further information on coherent states and their relationship to path integrals.

We deal in this section with a single Heisenberg spin and therefore drop the site index. In the formalism of Section 9.2, $\hat{\mathbf{S}}$ is the vector of spin operators of representation S. We construct the states $|\mathbf{N}\rangle$ explicitly below, where we choose $\mathbf{N}$ to be a unit vector with $\mathbf{N}^2 = 1$. Thus the coherent states are labeled by points on the unit sphere. With this definition, (9.21) is modified here to

$$\langle \mathbf{N}|\hat{\mathbf{S}}|\mathbf{N}\rangle = S\mathbf{N}. \tag{19.2}$$

The completeness relation (9.19) takes the form

$$\int \frac{d\mathbf{N}}{2\pi}|\mathbf{N}\rangle\langle \mathbf{N}| = 1 = \sum_{m=-S}^{S} |S,m\rangle\langle S,m|, \tag{19.3}$$

where the integral of $\mathbf{N}$ is over the unit sphere. The state $|\mathbf{N}\rangle$ is almost like a classical spin of length S pointing in the $\mathbf{N}$ direction; indeed, the spin-coherent states are the minimum uncertainty states localized as much in the $\mathbf{N}$ direction as the principles of quantum mechanics will allow, and in the large-S limit, $|\mathbf{N}\rangle$ reduces to a classical spin in the $\mathbf{N}$ direction.

Let us now explicitly construct the spin coherent states. For $\mathbf{N} = (0, 0, 1)$, the state $|\mathbf{N}\rangle$ is easy to determine; we have

$$|\mathbf{N} = (0, 0, 1)\rangle = |S, m = S\rangle \equiv |\Psi_0\rangle. \tag{19.4}$$

We have labeled this particular coherent state as a reference state $|\Psi_0\rangle$ as it is needed frequently in the following. It should be clear that for other values of $\mathbf{N}$ we can obtain $|\mathbf{N}\rangle$ simply by acting on $|\Psi_0\rangle$ by an operator that performs an $SU(2)$ rotation from the direction $(0, 0, 1)$ to the direction $\mathbf{N}$. In this manner we obtain the following explicit representation for the coherent state $|\mathbf{N}\rangle$:

$$|\mathbf{N}\rangle = \exp(z\hat{S}_+ - z^*\hat{S}_-)|\Psi_0\rangle, \tag{19.5}$$

where the complex number z is related to the vector $\mathbf{N}$. This relationship is simplest in spherical coordinates; if we parameterize $\mathbf{N}$ as

$$\mathbf{N} = (\sin\theta\cos\phi, \sin\theta\sin\phi, \cos\theta), \tag{19.6}$$

then

$$z = -\frac{\theta}{2}\exp(-i\phi). \tag{19.7}$$

We leave it as an exercise for the reader to verify that (19.5) satisfies (9.18), (9.19), and (9.20); this verification is aided by the knowledge that the value of the expression

$\exp(-i\mathbf{a}\cdot\hat{\mathbf{S}})\hat{\mathbf{S}}\exp(i\mathbf{a}\cdot\hat{\mathbf{S}})$, where $\mathbf{a}$ is some vector, is determined solely by the spin commutation relations (1.29) and can therefore be worked out by temporarily assuming that the $\hat{\mathbf{S}}$ are 1/2 times the Pauli matrices; the result, when expressed in terms of $\hat{\mathbf{S}}$, is valid for arbitrary S.

It is useful for our subsequent formulation to rewrite the above results in a somewhat different manner, making the $SU(2)$ symmetry more manifest. Define the 2×2 matrix of operators $\hat{S}$ by

$$\hat{S} = \begin{pmatrix} \hat{S}_z & \hat{S}_x - i\hat{S}_y \\ \hat{S}_x + i\hat{S}_y & -\hat{S}_z \end{pmatrix}. \tag{19.8}$$

Then (19.4) can be rewritten as

$$\langle\mathbf{N}|\hat{S}_{\alpha\beta}|\mathbf{N}\rangle = SW_{\alpha\beta}, \tag{19.9}$$

where the matrix W is

$$W = \begin{pmatrix} N_z & N_x - iN_y \\ N_x + iN_y & -N_z \end{pmatrix} \equiv \mathbf{N}\cdot\vec{\sigma}, \tag{19.10}$$

where $\vec{\sigma}$ are the Pauli matrices. So instead of labeling the coherent states with the unit vector $\mathbf{N}$, we could equally well use the traceless Hermitean matrix W. Furthermore, there is a simple relationship between W and the complex number z. In particular, if we use the spin-1/2 version of the operator in (19.5)

$$U = \exp\left[\begin{pmatrix} 0 & z \\ -z^* & 0 \end{pmatrix}\right], \tag{19.11}$$

(U is thus a 2×2 matrix), then we find

$$W = U\sigma_z U^\dagger. \tag{19.12}$$

Now let us apply these results to the path integral representation in (9.25) and (9.26). Clearly, the τ dependence of $\mathbf{N}(\tau)$ implies a τ-dependent $z(\tau)$ through (19.7). From (19.5) we have therefore

$$\frac{d}{d\tau}|\mathbf{N}(\tau)\rangle = \frac{d}{d\tau}\exp(z(\tau),\hat{S}_+ - z^*(\tau)\hat{S}_-)|\Psi_0\rangle. \tag{19.13}$$

Taking this derivative is, however, not so simple. Note that if an operator $\hat{O}$ does not commute with its derivative $d\hat{O}/d\tau$ then

$$\frac{d}{d\tau}\exp(\hat{O}) \neq \frac{d\hat{O}}{d\tau}\exp(\hat{O}). \tag{19.14}$$

The correct form of this result is in fact

$$\frac{d}{d\tau}\exp(\hat{O}) = \int_0^1 du\,\exp(\hat{O}(1-u))\frac{d\hat{O}}{d\tau}\exp(\hat{O}u), \tag{19.15}$$

where u is just a dummy integration variable. This result can be checked by expanding both sides in powers of $\hat{O}$ and verifying that they agree term by term. More constructively, a "hand-waving" derivation can be given as follows:

$$
\begin{aligned}
\frac{d}{d\tau}\exp(\hat{O}) &= \frac{d}{d\tau}\exp\left(\hat{O}\int_0^1 du\right)\\
&= \lim_{M\to\infty}\frac{d}{d\tau}\exp\left(\sum_{i=1}^M \hat{O}\Delta u_i\right)\quad\text{with }\Delta u_i = 1/M\\
&\approx \lim_{M\to\infty}\frac{d}{d\tau}\prod_{i=1}^M \exp(\hat{O}\Delta u_i)\\
&\approx \lim_{M\to\infty}\sum_{j=1}^M\prod_{i=1}^j \exp(\hat{O}\Delta u_i)\frac{d\hat{O}}{d\tau}\Delta u_j\prod_{i=j+1}^M \exp(\hat{O}\Delta u_i).
\end{aligned}
$$
(19.16)

Finally, taking the limit $M\to\infty$, we obtain the needed result (19.15). Now using (19.13) and (19.15) we find

$$
\begin{aligned}
\mathcal{S}_B &= \int_0^{1/T} d\tau\,\langle\mathbf{N}(\tau)|\frac{d}{d\tau}|\mathbf{N}(\tau)\rangle\\
&= \int_0^{1/T} d\tau\int_0^1 du\,\langle\mathbf{N}(\tau,u)|\left(\frac{\partial z}{\partial\tau}\hat{S}_+ - \frac{\partial z^*}{\partial\tau}\hat{S}_-\right)|\mathbf{N}(\tau,u)\rangle,
\end{aligned}
$$
(19.17)

where $\mathbf{N}(\tau,u)$ is defined by

$$
|\mathbf{N}(\tau,u)\rangle = \exp(u(z(\tau)\hat{S}_+ - z^*(\tau)\hat{S}_-))|\Psi_0\rangle.
$$
(19.18)

From this definition, three important properties of $\mathbf{N}(\tau,u)$ should be apparent:

$\mathbf{N}(\tau,u=1)\equiv\mathbf{N}(\tau)$,

$\mathbf{N}(\tau,u=0)=(0,0,1)$, and

$\mathbf{N}(\tau,u)$ moves with u along the great circle

between $\mathbf{N}(\tau,u=0)$ and $\mathbf{N}(\tau,u=1)$.

We can visualize the dependence on u by imagining a *string* connecting the physical value of $\mathbf{N}(\tau)=\mathbf{N}(\tau,u=1)$ to the North pole, along which u decreases to 0. Associated with each $\mathbf{N}(\tau,u)$ we can also define a u-dependent $W(\tau,u)$ as in (19.10); the analog of (19.19) is $W(\tau,u=1)\equiv W(\tau)$ and $W(\tau,u=1)=\sigma_z$. A simple explicit expression for $W(\tau,u)$ is also possible: we simply generalize (19.11) to

$$
U(\tau,u) = \exp\left[u\begin{pmatrix}0 & z\\ -z^* & 0\end{pmatrix}\right];
$$
(19.19)

then the relationship (19.12) gives us $W(\tau, u)$. Now we can use the expression (19.9) to rewrite (19.17) as

$$\mathcal{S}_B = S \int_0^{1/T} d\tau \int_0^1 du \left[\frac{\partial z}{\partial \tau} W_{21}(\tau, u) - \frac{\partial z^*}{\partial \tau} W_{12}(\tau, u) \right], \tag{19.20}$$

As everything is a periodic function of τ, we may freely integrate this expression by parts and obtain

$$\mathcal{S}_B = -S \int_0^{1/T} d\tau \int_0^1 du \, \mathrm{Tr} \left[\begin{pmatrix} 0 & z(\tau) \\ -z^*(\tau) & 0 \end{pmatrix} \partial_\tau W(\tau, u) \right], \tag{19.21}$$

where the trace is over the 2×2 matrix indices. The definitions (19.12) and (19.19) can be used to easily establish the identity

$$\begin{pmatrix} 0 & z(\tau) \\ -z^*(\tau) & 0 \end{pmatrix} = -\frac{1}{2} W(\tau, u) \frac{\partial W(\tau, u)}{\partial u}, \tag{19.22}$$

which when inserted into (19.21) yields the expression for $\mathcal{S}_B$ in one of its final forms

$$\mathcal{S}_B = \int_0^{1/T} d\tau \int_0^1 du \left[\frac{S}{2} \mathrm{Tr} \left(W(\tau, u) \frac{\partial W(\tau, u)}{\partial u} \frac{\partial W(\tau, u)}{\partial \tau} \right) \right]. \tag{19.23}$$

An expression for $\mathcal{S}_B$ solely in terms of $\mathbf{N}(\tau, u)$ can be obtained by substituting in (19.10); this yields the final expression for $\mathcal{S}_B$, which when inserted in (9.25) gives us the coherent state path integral for a spin:

$$\mathcal{S}_B = iS \int_0^{1/T} d\tau \int_0^1 du \, \mathbf{N} \cdot \left(\frac{\partial \mathbf{N}}{\partial u} \times \frac{\partial \mathbf{N}}{\partial \tau} \right). \tag{19.24}$$

This expression has a simple geometric interpretation. The function $\mathbf{N}(\tau, u)$ is a map from the rectangle $0 \leq \tau \leq 1/T, 0 \leq u \leq 1$ to the unit sphere. As $\mathbf{N}$ moves from $\mathbf{N}(\tau)$ to $\mathbf{N}(\tau + \Delta \tau)$ it drags along the string connecting it to the North pole represented by the u dependence of $\mathbf{N}(\tau, u)$ (recall (19.19)). It is easy to see that the contribution to $\mathcal{S}_B$ of this evolution is simply iS times the oriented area swept out by the string. The value of this area clearly depends upon the fact that the $u = 0$ end of the string was pinned at the North pole. This was a "gauge" choice, and by choosing the phases of the coherent states differently, we could have pinned the point $u = 0$ anywhere on the sphere. However, when we consider the complete integral over τ in (19.24), the boundary condition $\mathbf{N}(1/T) = \mathbf{N}(0)$ (required by the trace in (9.22)) shows that $\mathbf{N}(\tau)$ sweeps out a closed loop on the unit sphere. Then the total τ integral in (19.24) is the area contained within this loop and is independent of the choice of the location of the $u = 0$ point. Actually this last statement is not completely correct: the "inside" of a closed loop is not well defined and the location of the $u = 0$ point makes the oriented area uncertain modulo 4π (which is the total area of the unit sphere). Thus the net contribution of $e^{\mathcal{S}_B}$ is uncertain up to a factor of $e^{i4\pi S}$. For consistency, we can now demand that this arbitrary factor always equal unity, which, of course, leads to the familiar requirement that $2S$ be an integer.

19.2 Quantized ferromagnets

We turn to the lattice model H_S in (19.1) and consider the case of ferromagnetic interactions where all $J_{ij} > 0$. In this case, the state with all spins parallel $\prod_i |S, S\rangle_i$ is the exact ground state (see, e.g., [28]; we have assumed that the field $\mathbf{H}$ points along the spin quantization z-axis). The adjective "quantized" in the title refers to the fact that the magnetization density, M_0 (this is the magnitude of the expectation value of the total spin magnetization $\sum_i \hat{\mathbf{S}}_i$ divided by the system volume), is pinned at a simple value, which can be determined a priori, and which does not vary as the exchange constants J_{ij} are varied. In Section 19.4, we meet examples of quantized ferromagnets in which the magnetic moment is quantized, but not at a fully polarized value. Although fractional quantization is also possible, in every case twice the average total spin moment per unit cell is an integer. The discussion in this chapter applies to the low-energy properties of all such quantized ferromagnets, but we only explicitly refer to the fully polarized case.

Apart from their quantized moment, the characteristic property of quantized ferromagnets is that the only low-lying excitation that carries spin is a "spin wave" that arises from a slow rotation of the orientation of the ordered moment. Many readers may be familiar with the fact that the wave function of a single spin-wave excitation can also be written down exactly for a fully polarized, quantized ferromagnet. These well-known results also emerge below. The purpose of our discussion is twofold: (*i*) to obtain a continuum field theory of the low-lying excitations of the quantized ferromagnet, and to understand its behavior under a scaling transformation, and (*ii*) to use the continuum theory to enumerate systematically the parameters required to describe the low-T properties of such ferromagnets.

We begin by constructing the continuum field theory for the low-lying excitations above the fully polarized ferromagnetic ground state. It is reasonable to expect that these will consist of fluctuations in which the orientations of the spins vary slowly from site to site. We start with a functional integral such as (9.25) for the spin orientation $\mathbf{N}_i(\tau)$ on each site i and perform a gradient expansion by introducing the continuum field $\mathbf{N}(x, \tau)$. Keeping terms up to second spatial derivatives we obtain for the partition function $\mathcal{Z} = \mathrm{Tr}e^{-H_S/T}$ [248]:

$$
\mathcal{Z} = \int \mathcal{D}\mathbf{N}(x, \tau)\delta(\mathbf{N}^2 - 1) \exp\left(-\int_0^{1/T} d\tau \int d^d x \mathcal{L}_F\right),
$$
$$
\mathcal{L}_F = iM_0 \int_0^1 du \mathbf{N} \cdot \left(\frac{\partial \mathbf{N}}{\partial u} \times \frac{\partial \mathbf{N}}{\partial \tau}\right) - M_0 \mathbf{N} \cdot \mathbf{H} + \frac{\rho_s}{2}(\nabla \mathbf{N})^2, \tag{19.25}
$$

where $M_0 \equiv S/v$ is the magnetization density of the ground state, v is the volume per site, and ρ_s is the spin stiffness. We introduced the analogous stiffness for the rotor model in Section 11.2.3; here, the gradient expansion upon the partition function of H_S gives us

$$
\rho_s = \frac{S^2}{2v} \sum_m J_m x_{m1}^2, \tag{19.26}
$$

where the J_m are the set of exchange constants coupling a given site i to the other sites separated from i by $(x_{m1}, x_{m2}, \ldots, x_{md})$; the sum over m includes separate terms for $\vec{x}_m$ and

$-\vec{x}_m$. The continuum theory (19.25) should really be regarded as a convenient schematic representation of the quantum ferromagnet, and we will often need to go back to the underlying lattice model H_S to regulate short-distance singularities.

We consider the behavior of $\mathcal{L}_F$ under a rescaling transformation [399] at $T = 0$. The continuum theory is characterized by two dimensionful couplings M_0 and ρ_s, and despite the nonlinear constraint in (19.25), some special properties of the quantum theory make it possible to determine their exact renormalization group flow equations (this should be contrasted with the rotor theory (11.4) where no such exact results were available). First, we noted at the end of Section 19.1 that the single-spin Berry phase was uncertain up to an additive constant of $4\pi S$, and this imposed the requirement that S be integer or half-integer. Precisely the same argument applied to the Berry phase of the continuum ferromagnet (19.25) in a hypercubic box of volume L^d implies that $2M_0 L^d$ must be an integer (this is just a fancy way of saying that the continuum ferromagnet must model an integral number of spins). This integer cannot change under any scaling transformation, and as L transforms as a physical length, the invariance of $M_0 L^d$ leads to the exact flow equation

$$\frac{dM_0}{d\ell} = dM_0. \tag{19.27}$$

This equation describes the quantization of the average magnetic moment at its fully saturated value.

A closely related scaling equation holds for ρ_s, and this follows from the exactly known single spin-wave spectrum. To prepare for some future computations, we derive this by going back to the lattice Hamiltonian, H_S, and then taking the continuum limit of the resulting response functions. The most convenient formalism for computations is provided by the Dyson–Maleev transformation [124, 318] from the spin operators $\hat{\mathbf{S}}_i$ to Bose operators $\hat{b}_i$. Explicitly, the mapping is

$$\hat{S}_{+i} = \sqrt{2S}\hat{b}_i^\dagger,$$

$$\hat{S}_{-i} = \sqrt{2S}\left(\hat{b}_i - \frac{1}{2S}\hat{b}_i^\dagger\hat{b}_i\hat{b}_i\right),$$

$$\hat{S}_z = -S + \hat{b}_i^\dagger\hat{b}_i. \tag{19.28}$$

Along with the constraint $\hat{b}_i^\dagger\hat{b}_i \leq 2S$, this defines an exact mapping between the Hilbert space of the spin S spins ($2S + 1$ states per spin) and the bosons ($2S + 1$ possible boson occupation numbers); in practice, one does not even have to impose the constraint $\hat{b}_i^\dagger\hat{b}_i \leq 2S$, as all matrix elements out of the physical sector vanish. The reader can verify that the operators in (19.28) do indeed satisfy the commutation relations (1.29). The relations (19.28) do not satisfy the hermiticity requirement $\hat{S}_{+i} = (\hat{S}_{-i})^\dagger$, but this can be repaired by performing a similarity transformation on the space of spin states. (The reader should consult [17] for more information, as here we mainly use (19.28) as a black-box tool.) Inserting (19.28) into (19.1), and Fourier transforming to momentum space by defining $\hat{b}(\vec{k}) = \sqrt{v}\sum_i \hat{b}_i e^{-i\vec{k}\cdot\vec{x}}$ (these Bose operators then satisfy the canonical continuum commutation relations $[\hat{b}(\vec{k}), \hat{b}^\dagger(\vec{k}')] = (2\pi)^3 \delta^d(\vec{k} - \vec{k}'))$, the Hamiltonian becomes

$$H_S = \int \frac{d^d k}{(2\pi)^d} \{S[J(0) - J(\vec{k})] + H\} \hat{b}^\dagger(\vec{k}) \hat{b}(\vec{k})$$

$$+ \frac{v}{2} \int \prod_{i=1}^{4} \frac{d^d k_i}{(2\pi)^d} (2\pi)^d \delta^d(\vec{k}_1 + \vec{k}_2 - \vec{k}_3 - \vec{k}_4)[J(\vec{k}_1) - J(\vec{k}_1 - \vec{k}_4)]$$

$$\times \hat{b}^\dagger(\vec{k}_1) \hat{b}^\dagger(\vec{k}_2) \hat{b}(\vec{k}_3) \hat{b}(\vec{k}_4), \tag{19.29}$$

where all momentum integrals are over the first Brillouin zone of the lattice, and

$$J(\vec{k}) = \sum_m J_m e^{-i\vec{k}\cdot\vec{x}_m}. \tag{19.30}$$

This bosonic form for H_S can be analyzed by the methods developed in Chapter 16 for (16.1). The ground state is the vacuum, $|0\rangle$, with no $\hat{b}$ particles (the fully polarized ferromagnet), whereas the lowest excitations are single boson states, $\hat{b}^\dagger(\vec{k})|0\rangle$ ("spin waves"), which are exact eigenstates of H_S with energy $\varepsilon_{\vec{k}} = S(J(0) - J(\vec{k})) + H$. We have $\varepsilon_{\vec{k}} > 0$ for all $\vec{k}$, which indicates that the choice of the no boson state as the ground state is a consistent one. At $T = 0$, the one-particle propagator is given exactly by the free-particle propagator, as in (16.51), for there are no other particles present. Taking the small momentum limit of this propagator, and using the correspondence between the continuum fields

$$\hat{b}^\dagger(\vec{k}, \omega_n) = (M_0/2)^{1/2} N_+(-\vec{k}, -\omega_n), \tag{19.31}$$

which follows from our definitions above ($N_\pm = N_x \pm i N_y$), we obtain an exact result for a two-point correlator of (19.25):

$$\langle N_-(-\vec{k}, -\omega_n) N_+(\vec{k}, \omega_n)\rangle = \frac{2}{-i\omega_n M_0 + \rho_s k^2 + M_0 H}. \tag{19.32}$$

This represents the propagation of spin waves with the exact dispersion $\varepsilon_k = (\rho_s/M_0)k^2 + H$. The consistency of this dispersion with the scaling transformation requires $\dim[H] = z$ (as before in (11.29)) and the exact scaling equation

$$\frac{d\rho_s}{d\ell} = (d + z - 2)\rho_s. \tag{19.33}$$

Because the spin wave disperses quadratically with momentum at small k, it is convenient to choose $z = 2$ (other choices are also permissible, as physical observables will have compensating scale dependence arising from that of ρ_s).

The exact results (19.27), (19.32), and (19.33) are strongly reminiscent of the behavior of the Bose gas in Section 16.9. In both cases, the simplicity is due to the fluctuationless nature of the ground state and the exactly known single-particle excitations. For the case of the Bose gas we had an additional nonlinearity u, whose renormalization was determined by examining the two-particle scattering amplitude. In the present situation, the dimensionful parameters ρ_s and M_0 determine both the single-particle dispersion (19.32) and the strengths of the nonlinear couplings. It might therefore seem that the finite-T properties of

(19.25) must be given by universal functions of T and the *bare* couplings ρ_s and M_0, consistent with the requirements of scaling and engineering dimensional analysis. However, this is the case only if a short-distance cutoff scale (explicitly present in (19.29) but not in (19.25)) does not influence the low-energy properties. Such a scale might be required to cut off large momentum (ultraviolet) divergences of momentum integrals over virtual excitations. Motivated by the structure of the Bose gas problem in Section 16.3, we look for ultraviolet divergences in the two spin-wave scattering amplitude at $T = 0$ (we need not consider $T > 0$ explicitly as the finite-T corrections all involve Bose functions that fall off exponentially at large momentum). For the Bose gas problem we found ultraviolet divergences for $d \geq 2$, and this identified $d = 2$ as the upper-critical dimension below which the universality of the continuum theory was robust.

We compute the on-shell T matrix of two spin waves coming in with momenta $\vec{k}_1$ and $\vec{k}_2$ and scattering into spin waves with momenta $\vec{k}_1 + \vec{q}$ and $\vec{k}_2 - \vec{q}$. Conservation of energy requires

$$J(\vec{k}_1) + J(\vec{k}_2) = J(\vec{k}_1 + \vec{q}) + J(\vec{k}_2 - \vec{q}). \tag{19.34}$$

To zeroth order in $1/S$, the Hamiltonian (19.29) gives us the bare T-matrix element $v[J(\vec{k}_1 + \vec{q}) + J(\vec{k}_2 - \vec{q}) - J(\vec{k}_1 + \vec{q} - \vec{k}_2) - J(\vec{q})]$. The first order in $1/S$ correction to the T matrix is given by the first diagram in Fig 16.3, and by standard quantum mechanical perturbation theory [502], it evaluates to (this expression is the analog of (16.45))

$$\frac{v^2}{S} \int \frac{d^d q_1}{(2\pi)^d} [J(\vec{k}_1 + \vec{q}_1) + J(\vec{k}_2 - \vec{q}_1) - J(\vec{k}_1 + \vec{q}_1 - \vec{k}_2) - J(\vec{q}_1)]$$

$$\times \frac{[J(\vec{k}_1 + \vec{q}) + J(\vec{k}_2 - \vec{q}) - J(\vec{k}_1 + \vec{q} - \vec{k}_2 + \vec{q}_1) - J(\vec{q} - \vec{q}_1)]}{J(\vec{k}_1) + J(\vec{k}_2) - J(\vec{k}_1 + \vec{q}_1) - J(\vec{k}_2 - \vec{q}_1)}. \tag{19.35}$$

To understand the implications of this result for the continuum theory (19.25), we allow the external momenta $\vec{k}_1, \vec{k}_2, \vec{q}$ to become small, but for the moment we allow the internal momentum $\vec{q}_1$ to be large. Then there is a term from (19.35) that is quadratic in external momenta; however, this can be seen to vanish after use of the identity $\int d^d q_1 e^{-i\vec{q}_1 \cdot \vec{x}_m} = 0$ (valid because all the $\vec{x}_m \neq 0$). Clearly the lattice regularization is crucial in obtaining this result, and it is mainly this step which cannot be deduced from the continuum theory (19.25). The next term is *quartic* in external momenta, and it simplifies to

$$\frac{v^2}{S} \int \frac{d^d q_1}{(2\pi)^d} \frac{\left[\sum_m J_m e^{-i\vec{q}_1 \cdot \vec{x}_m} (\vec{k}_1 \cdot \vec{x}_m)(\vec{k}_2 \cdot \vec{x}_m)\right]^2}{\sum_m J_m (1 - e^{-i\vec{q}_1 \cdot \vec{x}_m})}. \tag{19.36}$$

We take the small-$\vec{q}_1$ limit of (19.36) and obtain the result for the correction to the two spin-wave T matrix [280] at low momenta:

$$\frac{4\rho_s}{M_0^3} (\vec{k}_1 \cdot \vec{k}_2)^2 \int \frac{d^d q_1}{(2\pi)^d} \frac{1}{q_1^2}; \tag{19.37}$$

this expression involves only couplings present in $\mathcal{L}_F$ in (19.25) and so could also have been obtained directly from the continuum quantum theory after ignoring ultraviolet divergences in terms lower order in the external momenta. The integral in (19.37) is dominated by the ultraviolet for $d > 2$ and so we have to return to the lattice expression (19.36). However, it is ultraviolet finite for $d < 2$, and the continuum theory is insensitive to lattice perturbations; the infrared divergence will of course be cut off by the external momenta, which have not been kept in the propagator in the above approximation. Thus, as in the case of the dilute Bose gas in Section 16.3, we see the emergence of $d = 2$ as a critical dimension.

It is very useful to interpret (19.37) in renormalization group sense. If we imagine we are integrating out virtual spin-wave fluctuations between momentum scales Λ and $\Lambda e^{-\ell}$ (Λ is a momentum cutoff), then these become the boundaries of the integration in (19.37), and the result generates a four-gradient term to $\mathcal{L}_F$. The generated term cannot be quadratic in $\mathbf{N}$, as that would modify the exactly known spin-wave dispersion. The simplest terms that modify *only* the two spin-wave scattering amplitude are quartic also in $\mathbf{N}$; by noting the momentum dependence on (19.37), using the low-momentum limit of the energy conservation equation (19.34), and imposing the restrictions of rotational invariance, a simple analysis shows that the generated term is [399]

$$\mathcal{L}_F \rightarrow \mathcal{L}_F + \lambda(\nabla_a N_\alpha \nabla_a N_\alpha \nabla_b N_\beta \nabla_b N_\beta - 2\nabla_a N_\alpha \nabla_b N_\alpha \nabla_a N_\beta \nabla_b N_\beta), \qquad (19.38)$$

where λ is a new coupling constant of the continuum theory. Converting from scattering amplitudes of b to $\mathbf{N}$ quanta using (19.31), (19.38), and (19.37) implies the flow equation

$$\frac{d\lambda}{d\ell} = (d - 2)\lambda + \frac{\rho_s}{M_0}. \qquad (19.39)$$

As with (16.47), this flow equation is believed to be exact. So for $d < 2$, λ is attracted to a universal critical value, and the parameters ρ_s and M_0 completely determine the low-energy physics of the continuum theory (19.25). However, λ becomes large at long distances for $d \geq 2$, and its bare value is important for it is responsible for temperature-dependent corrections to the magnetization computed by Dyson [124].

For $d < 2$ these considerations imply that we may write down universal scaling forms for the continuum ferromagnet (19.25). The usual scaling and dimensional considerations imply for the free energy density [399]

$$\mathcal{F} \asymp T M_0 \Phi_{fm} \left(\frac{\rho_s}{M_0^{(d-2)/d} T}, \frac{H}{T} \right), \qquad (19.40)$$

where Φ_{fm} is a universal function; corresponding results follow for observables that are derivatives of the free energy. Actually, our arguments for universality have really been made in an expansion in powers of $1/S$, and so the result (19.40) only holds as an asymptotic expansion in inverse powers of $\rho_s/(M_0^{(d-2)/d} T)$, and this is represented by the symbol $\asymp$. Indeed, (19.40) is expected to be true to all orders in $\rho_s/(M_0^{(d-2)/d} T)$, but this is not the same thing as being exactly true. Lattice effects become significant when $T \sim \rho_s/M_0^{(d-2)/d}$, for then the wavelength of the characteristic spin wave is of order $M_0^{1/d}$, which is of order a lattice spacing; these effects appear as essential singularities and destroy strict equality for (19.40). Some short-distance regularization at the scale $M_0^{1/d}$ is always required for any consistent theory of quantum ferromagnets [191]. Similar considerations apply for expansions in $1/N$ [27, 29, 504], and for ferromagnets with more complicated replica and supersymmetries [184, 185].

Finally, we briefly note that effective classical models for thermal fluctuations in ferromagnets can be derived for $T \ll \rho_s / M_0^{(d-2)/d}$, precisely as was done for the rotor models in Part II. In $d = 1$ we would get the effective theory (6.45) with $\xi = \rho_s / T$ [497], while in $d = 2$ we would obtain the model (13.8) [277] with a $\Lambda_{\overline{MS}}$ that can be computed from (19.29) by methods parallel to those in Section 13.1.1.

19.3 Antiferromagnets

This section considers models H_S in (19.1) with all $J_{ij} < 0$. Classically (i.e. in the limit $S \to \infty$), such models minimize their energies by making nearest-neighbor spins acquire an antiparallel orientation. On bipartite lattices (i.e. lattices that can be split into two equivalent sublattices so that all nearest neighbors of any site on one sublattice belong to the other sublattice) with nearest-neighbor interactions, the antiparallel constraint is easy to satisfy: the spins simply point in opposite directions on the two sublattices. Note that any pair of spins is either parallel or antiparallel, and so such an ordering is collinear. We begin in Section 19.3.1 by exclusively considering quantum antiferromagnets whose classical ground state is collinear. Such an ordering is expected to be present at least over short distances in the quantum case. Noncollinear ordering arises on nonbipartite lattices or even on bipartite lattices with further neighbor interactions. Such antiferromagnets are classically frustrated and possess ground states in which the spins are *coplanar* (as on the triangular lattice with nearest-neighbor interactions) or, in some rare cases, can even form structures that are three-dimensional in spin space. We consider the noncollinear cases in Section 19.3.4.

19.3.1 Collinear antiferromagnetism and the quantum nonlinear sigma model

For definiteness, we begin by considering antiferromagnets on a d-dimensional hypercubic lattice with only a nearest-neighbor exchange $J_{ij} = -J < 0$; other collinear antiferromagnets can be treated in a similar manner. In the classical limit of large S, as noted above, the ground state has spins oriented in opposite directions on the two sublattices: this is the so-called Néel-ordered state. For smaller S this orientation should survive at least over a few lattice spacings, suggesting that a continuum description of the quantum antiferromagnet may be possible [6, 7, 192]. We therefore begin by introducing a parameterization of the unit length spin field $\mathbf{N}_i(\tau)$ that captures this local ordering. We write

$$\mathbf{N}_i(x, \tau) = \lambda_i \mathbf{n}(x_i, \tau)) \sqrt{1 - (a^d/S)^2 \mathbf{L}^2(x_i, \tau)} + (a^d/S) \mathbf{L}(x_i, \tau), \qquad (19.41)$$

where λ_i equals ± 1 on the two sublattices and a is the lattice spacing. The fields $\mathbf{n}(x_i)$ and $\mathbf{L}(x_i)$ parameterize the staggered and uniform components of the Heisenberg spins. The prefactor of a^d/S has been associated with $\mathbf{L}$ so that the spatial integral of $\mathbf{L}$ over any region is precisely the total magnetization inside it. Both fields are assumed to be slowly varying on the scale of a lattice spacing. This is certainly true as $S \to \infty$, and it is hoped that this assumption remains valid down to $S = 1/2$. Consequently, we treat $\mathbf{n}(x, \tau)$ and $\mathbf{L}(x, \tau)$ as continuum quantum fields that can be expanded in spatial gradients over separations of order a. These continuum fields satisfy the constraints

$$\mathbf{n}^2 = 1, \qquad \mathbf{n} \cdot \mathbf{L} = 0, \tag{19.42}$$

which combined with (19.41) imply that $\mathbf{N}_i^2 = 1$ is obeyed. Further, spins on nearby sites are expected to be predominantly antiparallel, and so the uniform component $\mathbf{L}$ should be small; more precisely we have

$$\mathbf{L}^2 \ll S^2 a^{-2d}. \tag{19.43}$$

The field $\mathbf{n}(x, \tau)$ clearly plays the role of the order parameter associated with Néel ordering. Note that although $\mathbf{n}$ varies slowly on the scale of a lattice spacing, values of $\mathbf{n}$ on well-separated points can be considerably different, leaving open the possibility of a quantum paramagnetic phase with no magnetic long-range order. Magnetic Néel order requires that the time-average orientation of $\mathbf{n}(x, \tau)$ is correlated across the sample. Whether this happens is determined by the effective action for $\mathbf{n}$ fluctuations, which we now derive.

We insert the decomposition (19.41) for $\mathbf{N}_i$ into $H_S(S\mathbf{N}_i(\tau))$ and expand the result in gradients, and in powers of $\mathbf{L}$. This yields

$$H_S = \int d^d x \left[\frac{JS^2 a^{2-d}}{2} (\nabla_x \mathbf{n})^2 + 2d J a^d \mathbf{L}^2 - \mathbf{H} \cdot \mathbf{L} \right]$$

$$\equiv \frac{1}{2} \int d^d x \left[\frac{Nc}{g} (\nabla_x \mathbf{n})^2 + \frac{cg}{N} \mathbf{L}^2 - 2\mathbf{H} \cdot \mathbf{L} \right]. \tag{19.44}$$

In the second equation we have introduced the couplings $c = 2\sqrt{d} J S a$ and $g = (N/S) 2\sqrt{d} a^{d-1}$. The notation is informative and anticipates our eventual mapping of the present model to the rotor models in (2.12) and (11.4). In the present case $N = 3$, but we introduced a general factor of N for notational consistency with Part II. If we had used a different form for H_S with modified short-range exchange interactions, the continuum limit of H would have been the same but with new values of g and c.

To complete the expression for the coherent state path integral of the antiferromagnet in the continuum limit, we also need the expression for $\mathcal{S}_B$ in terms of $\mathbf{n}$, $\mathbf{L}$. We insert (19.41) into (19.24) and retain terms up to linear order in $\mathbf{L}$. This yields

$$\mathcal{S}_B = \mathcal{S}_B' + i \int d^d x \int_0^{1/T} d\tau \int_0^1 du \left[\mathbf{n} \cdot \left(\frac{\partial \mathbf{n}}{\partial u} \times \frac{\partial \mathbf{L}}{\partial \tau} \right) \right.$$

$$\left. + \mathbf{n} \cdot \left(\frac{\partial \mathbf{L}}{\partial u} \times \frac{\partial \mathbf{n}}{\partial \tau} \right) + \mathbf{L} \cdot \left(\frac{\partial \mathbf{n}}{\partial u} \times \frac{\partial \mathbf{n}}{\partial \tau} \right) \right], \tag{19.45}$$

where

$$\mathcal{S}_B' = iS \sum_i \lambda_i \int_0^{1/T} d\tau \int_0^1 du \, \mathbf{n}(x_i) \cdot \left(\frac{\partial \mathbf{n}(x_i)}{\partial u} \times \frac{\partial \mathbf{n}(x_i)}{\partial \tau} \right). \tag{19.46}$$

The evaluation of S'_B in the continuum limit is a rather subtle matter, as the leading λ_i in (19.46) shows that it is the sum of terms that oscillate in sign on the two sublattices. The naive assumption would be that these oscillating terms just cancel out, and therefore $S'_B = 0$ in the continuum limit. For some purposes this assumption is in fact adequate, but there are a number of important cases where S'_B is nonvanishing and is crucial for a complete understanding of the physics. We postpone a careful evaluation of S'_B to the following subsections where we consider its consequences in $d = 1$ and $d = 2$ separately. Let us first simplify the other terms in (19.45) a little further.

We use the fact that the vectors $\mathbf{L}$, $\partial \mathbf{n}/\partial \tau$, and $\partial \mathbf{n}/\partial u$ are all perpendicular to $\mathbf{n}$; hence, they lie in a plane and have a vanishing triple product:

$$\mathbf{L} \cdot \left(\frac{\partial \mathbf{n}}{\partial u} \times \frac{\partial \mathbf{n}}{\partial \tau} \right) = 0. \tag{19.47}$$

Using (19.47) in (19.45) we find

$$S_B = S'_B + i \int d^d x \int_0^{1/T} d\tau \int_0^1 du \left[\frac{\partial}{\partial \tau} \left(\mathbf{n} \cdot \left(\frac{\partial \mathbf{n}}{\partial u} \times \mathbf{L} \right) \right) \right.$$

$$\left. + \frac{\partial}{\partial u} \left(\mathbf{n} \cdot \left(\mathbf{L} \times \frac{\partial \mathbf{n}}{\partial \tau} \right) \right) \right]. \tag{19.48}$$

The total τ derivative yields 0 after using the periodicity of the fields in τ, while the total u derivative yields a surface contribution at $u = 1$. This gives finally

$$S_B = S'_B - i \int d^d x \int_0^{1/T} d\tau \mathbf{L} \cdot \left(\mathbf{n} \times \frac{\partial \mathbf{n}}{\partial \tau} \right). \tag{19.49}$$

Putting together (19.44) and (19.49) in (9.25) we obtain the following path integral for the partition function of the antiferromagnet:

$$\mathcal{Z} = \int \mathcal{D}\mathbf{n} \mathcal{D}\mathbf{L} \delta(\mathbf{n}^2 - 1) \delta(\mathbf{L} \cdot \mathbf{n}) \exp(-S'_B - S'_n),$$

$$S'_n = \frac{1}{2} \int_0^{1/T} d\tau \int d^d x \left[\frac{Nc}{g} (\nabla_x \mathbf{n})^2 + \frac{cg}{N} \mathbf{L}^2 - 2i\mathbf{L} \cdot \left(\mathbf{n} \times \frac{\partial \mathbf{n}}{\partial \tau} - i\mathbf{H} \right) \right]. \tag{19.50}$$

The functional integral over $\mathbf{L}$ can be carried out explicitly (after imposing the constraint $\mathbf{L} \cdot \mathbf{n} = 0$, e.g. by adding a term $w(\mathbf{L} \cdot \mathbf{n})^2$ to the Hamiltonian and taking the limit $w \to \infty$ after carrying out the integral) and we obtain the final result of this section [6, 7, 192]:

$$\mathcal{Z} = \int \mathcal{D}\mathbf{n} \delta(\mathbf{n}^2 - 1) \exp(-S'_B - S_n),$$

$$S_n = \frac{N}{2cg} \int_0^{1/T} d\tau \int d^d x \, [c^2(\nabla_x \mathbf{n})^2 + (\partial_\tau \mathbf{n} - i\mathbf{H} \times \mathbf{n})^2]. \tag{19.51}$$

Note that S_n is identical to the rotor model action studied in (11.4). However, before we can carry over all the results of Part II here, we have to examine the consequences of S'_B, and this is done separately in the following two subsections in dimensions $d = 1$ and $d = 2$, respectively.

19.3.2 Collinear antiferromagnetism in $d = 1$

It is simpler to evaluate S'_B in $d = 1$ by a geometric argument, rather than by working directly with the formal expression (19.46). We have already argued below (19.24) that the contribution of each site i in (19.46) equals $\lambda_i S$ times the area on the unit sphere contained inside the close loop defined by the periodic time evolution of $\mathbf{n}(x_i, \tau)$. We define this area to equal $\mathcal{A}_i$. Let us examine the contribution of two neighboring sites, i and $i + 1$, to S'_B. The weight λ_i will have opposite signs on these sites, and so the net contribution will be the difference of the areas. We can further assume that the order parameter field $\mathbf{n}(x_i)$ only varies slightly between i and $i + 1$. Under these conditions, and using the definition of an area element on the sphere, we have (after defining $\Delta\mathbf{n}(x_i) = \mathbf{n}(x_{i+1}) - \mathbf{n}(x_i)$)

$$\mathcal{A}_{i+1} - \mathcal{A}_i \approx \int_0^{1/T} d\tau\, \mathbf{n}(x_i) \cdot \left(\Delta\mathbf{n}(x_i) \times \frac{\partial \mathbf{n}(x_i)}{\partial\tau} \right)$$

$$\approx a \int d\tau\, \mathbf{n}(x_i) \cdot \left(\frac{\partial \mathbf{n}(x_i)}{\partial x_i} \times \frac{\partial \mathbf{n}(x_i)}{\partial\tau} \right). \tag{19.52}$$

The summation in (19.46) can be carried out over pairs of sites. All terms are of the same sign and therefore the summation can be easily converted into an integral. In this manner we obtain our final result for S'_B in $d = 1$ [6, 7, 192]:

$$S'_B = i \frac{\theta}{4\pi} \int dx \int_0^{1/T} d\tau\, \mathbf{n} \cdot \left(\frac{\partial \mathbf{n}}{\partial x} \times \frac{\partial \mathbf{n}}{\partial\tau} \right), \tag{19.53}$$

where $\theta = 2\pi S$.

Some comments and/or cautions about the derivation leading up to (19.53) are in order. The arbitrary way in which the sites in (19.52) were paired suggests that the answer is sensitive to the boundary conditions, and depends upon whether there are an even or odd number of sites in the system. There are indeed interesting boundary effects in the physics of antiferromagnetic spin chains [13, 14, 188], but we will not discuss them here. The overall sign of the answer in (19.53) also depends upon the sign of λ_i, but, as we see shortly, the physics does not depend upon the sign of θ. Finally, the result (19.53) can also be derived by analytic computations from (19.46). We can write the oscillating sum as half the spatial integral of the spatial derivative of the contribution of each site (by the same arguments leading to (19.52)). Then using the fact that the triple product of $\partial\mathbf{n}/\partial x$, $\partial\mathbf{n}/\partial\tau$, and $\partial\mathbf{n}/\partial u$ must vanish we can obtain (19.53) using manipulations similar to those leading to (19.49).

In its present form, S'_B is the so-called topological θ-term, familiar in particle theory literature. The coefficient of θ in (19.53) computes a simple topological invariant, which, for periodic boundary conditions in space, is always an integer. If we consider the field configuration $\mathbf{n}(x, \tau)$ as a map from two-dimensional spacetime, with periodic boundary conditions, to the surface of a unit sphere, then the topological invariant is simply the number of times spacetime has been wrapped around the sphere. It is useful to visualize the simplest configuration of $\mathbf{n}(x, \tau)$ corresponding to the topological invariant of unity. Let the unit sphere be placed on an elastic sheet, representing spacetime. Now fold up the sheet to cover the sphere once. The orientation of **n** at (x, τ) is given by the point on the sphere adjacent to the point (x, τ) on the sheet. Such a spacetime configuration represents a tunneling event: deep in the past, or far in the future, **n** points to the North pole; however, at some time, in a certain compact region of space, the **n** orientation tunnels all the way to the vicinity of the South pole and back; configurations with larger topological invariants can be similarly interpreted. The result (19.17) implies that each such tunneling event yields a factor of $e^{i\theta} = (-1)^{2S}$ to the path integral for the partition function. This is the only consequence of the S'_B term. Of course, the terms in S_n give the usual positive weights (in imaginary time) also present for the rotor model. Note that as θ is always an integral multiple of π, the sign of θ does not change the value of $e^{i\theta}$.

We are now able to state our principal conclusions, first reached by Haldane. For integer S, the phase factor with topologically nontrivial tunneling events is simply unity, and the theory reduces to the rotor model action S_n, which has been studied in some detail in Chapters 11 and 12. For half-integer S, however, there are clearly substantial differences. But the present formulation of the theory in (19.51) is not a particularly convenient way of exploring the physics; it does, however, tell us that the low-energy properties of all the half-integer cases are the same. We explore the $S = 1/2$ case in Chapter 20 by alternative methods.

We anticipate these results by sketching the renormalization group flows for the dimensionless coupling g for the cases $\theta = 0$ and $\theta = \pi$ in Fig. 19.1. For the case of integer S, where $\theta = 0$, the flow just represents (12.8): all values of g flow eventually to strong coupling, and as we saw in Chapter 12, there is always an energy gap above the ground state. For the case $\theta = \pi$, the perturbative flow at small g is the same as before, as it is independent of θ. However, more sophisticated considerations [8, 12, 133, 554] discussed in Chapter 20 show that there is a fixed point at $g = g_c$, of order unity, that attracts all couplings with $g < g_c$. We also see that the ground state is then a so-called Tomonaga–Luttinger liquid and has gapless, linearly dispersing excitations. For $g > g_c$ (and $\theta = \pi$) the flow is again to strong coupling, and the ground state will be seen to be a "spin-Peierls" state with an energy gap to all excitations (such a state is described below for $d = 2$).

We conclude by reviewing more explicitly the implications of the results of Chapters 11 and 12 for antiferromagnetic chains of integer spins. The mapping between correlation

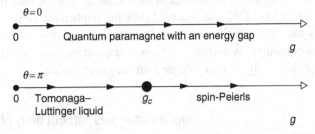

Fig. 19.1 Renormalization group flows for the dimensionless coupling g in (19.51) for $d = 1$ with S'_B given by (19.53). For $\theta = 0$, the flow is given by (12.8), and there is always an energy gap above the ground state. For $\theta = \pi$, there is a fixed point $g = g_c$, and near it the flow is $dg/d\ell \propto (g - g_c)^2$.

functions of the two theories is provided by (19.41). From this, we see that the correlator χ_u defined in (12.1) also specifies the fluctuations of the magnetization of the spin chain. At wavevector k this is a correlation function of the $\hat{\mathbf{S}}_i$ spins near the wavevector $q = k$. Further, the correlations of the order parameter $\mathbf{n}$, given by χ in (11.2), at wavevector k map onto correlations of $\hat{\mathbf{S}}_i$ at wavevector $q = k + Q$, where $Q = \pi/a$ is the ordering wavevector of the classical antiferromagnetic chain; all of the results for the rotor correlation functions in Chapter 12 can therefore be applied to integer spin antiferromagnets. We saw in Chapter 12 that the $d = 1$, $N = 3$ quantum rotor model always had a gap. The same is therefore true of integer spin antiferromagnetic chains – this is the so-called Haldane gap (we see in the following chapter that half-integer spin chains can be gapless). The $T = 0$ spectrum of the integer spin antiferromagnets is qualitatively the same as that discussed in the strong coupling expansion in Section 6.1: the lowest excited states are a triplet of $S = 1$ particles with infinite lifetime; for the spin chain, this particle appears as a pole in the $\hat{\mathbf{S}}$–$\hat{\mathbf{S}}$ correlation function, which has its minimum at $q = \pi/a$. Higher excited states consist of multiparticle continua of this triplet of particles.

19.3.3 Collinear antiferromagnetism in $d = 2$

We consider the properties of the theory (19.51) on the $d = 2$ square lattice.

This requires evaluation of the oscillating sum in $\mathcal{S}'_B$ in (19.46). Using techniques very similar to those used in $d = 1$, it is not difficult to establish an important result: $\mathcal{S}'_B$ vanishes for all smooth spacetime configurations of $\mathbf{n}(x, \tau)$. Simply evaluate (19.46) row by row on the square lattice. The sum on each row is precisely the same as that carried out in $d = 1$ and equals (19.53) on each row, up to an overall sign. Moreover, because of the structure of the sublattices, this overall sign oscillates as we move from row to row. Now, note that the arguments in Section 19.3.2 imply that the contribution of each row is quantized in integer multiples of θ. If, as we are assuming, $\mathbf{n}(x, \tau)$ is smoothly varying, the contribution of the rows must also change smoothly as we move from row to row. This is only compatible with the quantization if each row yields precisely the same integer. Hence their oscillating sum appearing in $\mathcal{S}'_B$ vanishes.

However, this is not the end of the story. There are important *singular* configurations of $\mathbf{n}(x, \tau)$ that do yield a nonvanishing contribution to $\mathcal{S}'_B$. We postpone discussion of the important consequences of these contributions to later in this section. We first ignore the effects of $\mathcal{S}'_B$ and discuss below the implication of the results of Parts II and III for square lattice antiferromagnets. We see later that most conclusions reached in this manner are qualitatively correct away from the quantum phase transition, but that neglect of $\mathcal{S}'_B$ is justified only for even integer S on the square lattice.

Antiferromagnets without Berry phases

The properties of the $N = 3$, $d = 2$ quantum rotor model were first discussed using the large-N expansion in Chapter 11, and then in some more detail in Chapters 13, 14, and 15. The most significant feature of these results was the existence of a quantum phase transition at a critical value $g = g_c$, separating a magnetically ordered ground state from a quantum paramagnetic ground state.

The magnetically ordered state of the rotor model corresponds to a "Néel" ground state of the antiferromagnet. This is a state in which the spin-rotation invariance of the Hamiltonian (19.1) is broken because of a nonzero, expectation value of the spin operator, which takes opposite signs on the two sublattices. From (19.41) we see that

$$\langle \hat{\mathbf{S}}_i \rangle \propto \lambda_i S \langle \mathbf{n}(x_i) \rangle = S N_0 \mathbf{e}_z, \tag{19.54}$$

where $\mathbf{e}_z$ is a unit vector pointing in the $\mathbf{e}_z$ direction (say) of spin space. Note that there was no state with such a broken symmetry in $d = 1$. The missing proportionality constant in (19.54) depends upon microscopic details and is not of any importance. In Part II we expressed physical properties of the rotor model on the ordered side in terms of N_0; these can be applied unchanged to the antiferromagnet simply by replacing N_0 by the actual expectation value of $\lambda_i \langle \hat{\mathbf{S}}_i \rangle$. As in $d = 1$, correlators of $\mathbf{L}$ at wavevector $\vec{k}$ map onto correlators of $\hat{\mathbf{S}}$ at $\vec{q} = \vec{k}$, whereas correlators of $\mathbf{n}$ at $\vec{k}$ map onto $\vec{q} = \vec{k} + \vec{Q}$, with $\vec{Q} = (\pi/a, \pi/a)$ the ordering wavevector. As was the case for the rotor model, the broken rotational invariance is restored at any nonzero temperature, and the antiferromagnet instead acquires an exponentially large correlation length given by (13.10) and (13.20). In these results, we take for the value of ρ_s the actual $T = 0$ spin stiffness of the quantum antiferromagnet. The nonzero temperature static and dynamic correlations are described by (13.1), with the function Φ_- as described in Chapter 13.

Numerical studies of square lattice antiferromagnets with nearest-neighbor antiferromagnets have shown fairly conclusively that the ground state has Néel order for all values of S including $S = 1/2$ [230, 401]. Thus it appears that all such antiferromagnets map onto the rotor model with $g < g_c$. For $S = 1/2$ it has been argued [85, 86] that the value of g is sufficiently close to g_c so that the universal crossover between the low- and high-T limits of the continuum rotor field theory shown in Fig. 11.2 can be observed with increasing temperature. For larger S, the antiferromagnets appear to go directly from the universal low-T region on the ordered side of Fig. 11.2 to a nonuniversal lattice high-T region [128].

Clearly, it would also be physically interesting to find collinear antiferromagnets that map onto rotor models with $g > g_c$ and therefore do not have Néel order in their ground state. A convenient choice, studied extensively in the literature, has been the square lattice antiferromagnet with first- and second-neighbor antiferromagnetic exchanges, labeled J_1 and J_2, respectively. The classical limit of this model has collinear Néel order for all J_2/J_1, and so the quantum fluctuations should continue to be described by (19.51). Numerical and series expansion studies [80, 84, 101, 137, 161, 162, 296, 346, 383, 454, 455] for $S = 1/2$ have shown that this model loses order (19.54) around $J_2/J_1 = J_c \approx 0.4$. Hence we can identify the point $J_2/J_1 = J_c$ with the quantum critical point $g = g_c$ of the rotor model. The quantum paramagnetic state of the rotor model should therefore yield the characteristics of the antiferromagnet with J_2/J_1 just above J_c: spin rotation invariance is restored, and there is a gap to all excitations. Nonzero temperature properties are described by (13.3) with Δ_+ the actual energy gap of the antiferromagnet.

One important property of the quantum paramagnetic state of the rotor model deserves special mention, as it has crucial implications for the corresponding antiferromagnet. Recall that the excited states of the rotor model were described in terms of an N-fold degenerate quasiparticle and its multiparticle continua. This leads to the spectrum shown in Fig. 7.1 and discussed in the strong-coupling expansion of Section 6.1. There is an infinitely sharp delta function in $\mathrm{Im}\,\chi(k, \omega)$ at the position of the quasiparticle energy $\omega = \varepsilon_k$. For $N = 3$, this is clearly a quasiparticle with total angular momentum $S = 1$; so the dominant excitation of this phase of quantum antiferromagnet is an $S = 1$ particle with its energy minimum at $\vec{q} = \vec{Q}$, and this leads to a delta function in the dynamic spin susceptibility at wavevectors near $\vec{Q}$. Note that this $S = 1$ particle exists for all values of the spin S of the individual spins of the underlying antiferromagnet. This gapped $S = 1$ excitation should also be contrasted with the spin-wave excitations of the ordered Néel state, which are gapless and twofold degenerate, and do not carry definite total spin (although they are eigenstates of total $\hat{S}_z$, with eigenvalues ± 1 for a Néel state polarized in the z direction).

Berry phases and valence bond solid order

We now turn to a careful evaluation of $\mathcal{S}'_B$, and a discussion of its consequences.

We consider the case of the square lattice with nearest-neighbor exchanges and possible further-neighbor exchanges that do not destroy the collinear, two sublattice ordering of the classical Néel state. We have already argued above that $\mathcal{S}'_B$ vanishes for smooth spacetime configurations of $\mathbf{n}(x, \tau)$. We should therefore consider singular configurations, and for the case of a three-component vector order parameter, the only topologically stable possibility is the so-called hedgehog singularity [193]. This is a singularity occurring at a point in spacetime and corresponds to a tunneling event in which the *Skyrmion number*, Y, of a given time slice of $\mathbf{n}(x, t)$ changes. The latter is defined by the spatial integral

$$Y(\tau) = \frac{1}{4\pi} \int d^2 x\, \mathbf{n} \cdot \left(\frac{\partial \mathbf{n}}{\partial x_1} \times \frac{\partial \mathbf{n}}{\partial x_2} \right). \tag{19.55}$$

Comparing (19.55) to the topological θ term in $d = 1$ of (19.53) we see that the two expressions are identical except that we now have an integral over space only, whereas earlier we had a spacetime integral. By the same arguments as made below (19.53), Y is an integer for periodic boundary conditions in space. Let us describe a hedgehog tunneling event in which Y changes from 1 to 0, in a pictorial language used by Haldane [193]. As below (19.53), we can represent a configuration with $Y = 1$ as an elastic sheet (now representing space, rather than spacetime) wrapped on a sphere. In reality, the spins lie on a lattice, and so the elastic sheet has a fine square mesh on it. Now imagine a tunneling event in which one square on the mesh expands and allows the sphere to pass through; the resulting configuration will have its Y changed to 0. It remains to evaluate the summation in (19.46) for the evolution of $\mathbf{n}(x, \tau)$ just described. Actually, we cannot consider hedgehog tunneling events singly, as then the periodic boundary conditions in τ, required for a meaningful evaluation of (19.46), will not be satisfied. We therefore consider a sequence of events at well-separated times, centered at the midpoints of plaquettes labeled a, and involving a change in Skyrmion number ΔY_a such that $\sum_a \Delta Y_a = 0$. These events are to be considered as saddle points in the evaluation of the coherent state path integral of the lattice antiferromagnet. The configuration of $\mathbf{n}(x, \tau)$ at the saddle point minimizes the action, and, provided the hedgehogs are well separated, it can reasonably be expected to have fourfold rotational symmetry about the plaquette a around which the tunneling occurs. As at the beginning of Section 19.3.2, let us write $\mathcal{S}'_B$ as

$$\mathcal{S}'_B = S \sum_i \lambda_i \mathcal{A}_i, \tag{19.56}$$

where $\mathcal{A}_i$ is the contribution of site i. Now we can evaluate $\mathcal{A}_i$ by following the area swept out on the unit sphere by each site on the elastic sheet during the tunneling event. From this it is simple to see the following important intermediate result: the lattice configuration of $\mathcal{A}_i$ has a vortex of strength $4\pi \Delta Y_a$ around plaquette a. As the sum in (19.56) cannot change from smooth changes in the lattice configuration of $\mathcal{A}_i$, we need only take a representative configuration that has the proper vortex singularities; for instance, we can take

$$\mathcal{A}_i = 2 \sum_a \Delta Y_a \arctan\left(\frac{x_{i1} - X_{a1}}{x_{i2} - X_{a2}}\right), \tag{19.57}$$

where $x_{i1,2}$ are the components of the lattice points x_i, and X_a is the position of the center of plaquette a. We have to insert (19.57) into (19.56) and evaluate the sum over i. This is a mathematical step, and the details are given by Haldane [193]. It is not difficult to see that the result takes the form

$$\mathcal{S}'_B = i\pi S \sum_a \Delta Y_a \zeta_a. \tag{19.58}$$

The values of ζ_a depend upon the coordinates of plaquette a; a number of choices for these values are possible, but $e^{-\mathcal{S}'_B}$ remains the same provided $\sum_a \Delta Y_a = 0$. A particular choice is $\zeta_a = 0, 1, 2, 3$ if the coordinates X_a are (even, even), (even, odd), (odd, odd), (odd, even).

Now to evaluate the partition function, we have to sum over all possible hedgehog events, while including the phase factors arising from $e^{\mathcal{S}'_B}$ with each such event. Read and Sachdev [396, 397] showed how such a summation could be carried out systematically in a certain large-N expansion. Describing this here would take us too far afield, and we refer the reader to a review article [429] for fairly explicit details. The hedgehog events are completely suppressed by the action arising from $\mathcal{S}_n$ for $g < g_c$, and therefore they have no significant consequence for the Néel phase. In contrast, for $g > g_c$, these events proliferate, and it has been shown in the quoted papers how the Berry phases in (19.58) necessarily lead to a spontaneously broken lattice symmetry unless S is an even integer. For S an even integer, (19.58) is always an integral multiple of $2\pi i$, and so $\mathcal{S}_B$ has no effect – the properties in this case are therefore the same as the rotor model, and there is no broken lattice symmetry [13].

Let us now describe the ordering associated with the broken lattice symmetry. The ordered state was originally referred to as "spin-Peierls," but it is now universally called a "valence bond solid" (VBS). Consider the quantity

$$P_{ij} = \langle \hat{\mathbf{S}}_i \cdot \hat{\mathbf{S}}_j \rangle. \tag{19.59}$$

Note that P_{ij} is a scalar under spin rotations, and so a nonzero value does not break a spin rotation symmetry. The Hamiltonian H_S in (19.1) is also invariant under a group of lattice

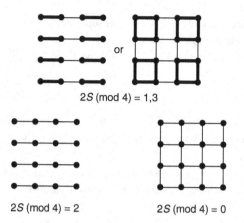

$2S \pmod 4 = 1,3$

$2S \pmod 4 = 2$ $2S \pmod 4 = 0$

Quantum paramagnetic ground states of the square lattice antiferromagnet as a function of $2S \pmod 4$. The values of P_{ij} on the nearest-neighbor links are schematically indicated by the different kinds of line on the links; those on thick lines are larger than those on thin lines, and weakest are on the empty links.

symmetries (involving lattice rotation, reflection, and translations), and the values of the P_{ij} for all pairs sites i, j should, in general, also respect these symmetries. A VBS state is one in which the values of P_{ij} break a lattice symmetry; this broken symmetry will be observable experimentally in lattice distortions whose pattern will reflect that in P_{ij}. This distortion arises from the coupling between the spin exchange energy and phonon displacements, which have not been included in the Hamiltonians we are considering here. For the case of a square lattice, the simplest possible patterns of VBS order are shown in Fig. 19.2. For $S = 1/2$, like values of P_{ij} line up in columns or plaquettes, which clearly break symmetry of rotation by 90 degrees about each lattice point; the ground state is fourfold degenerate, and a similar VBS ordering is expected for all half-integral S. If it was possible to obtain a quantum paramagnet for $S = 1$ (or other odd integer S) by a continuous transition from a Néel state, then it is predicted to have a twofold degenerate ground state, with the P_{ij} on the horizontal bonds differing from those on the vertical bonds (see Fig. 19.2). Finally, only for even integer S is the paramagnetic state nondegenerate and breaks no lattice symmetry [13, 14]. Related results exist for quantum paramagnetic states accessed from other collinear states on the square or other lattices. In all cases there are special values of S for which the quantum paramagnet is nondegenerate and has no VBS order; these special values extend to all values of S only for lattices with small symmetry groups.

One concern about the above arguments for the ubiquity of VBS order in collinear $S = 1/2$ antiferromagnets is that it relies on a semiclassical large-S limit to evaluate the Berry phase, and this could break down at small S. This issue has been addressed by studies designed to directly study $S = 1/2$ quantum antiferromagnets either by phenomenological [271, 406] or large-N approaches [395]. Neighboring spins are assumed to form singlet bonds in pairs, and then the low-lying, spin-singlet excitations arise from resonance between different arrangements of the bonds (the "resonating valence bond" picture [22, 38]). From both approaches, the *quantum dimer model* [406] appears as an effective Hamiltonian for the low-energy spin-singlet states. This latter model can be studied quite reliably by a series of duality transformations [151, 397, 434, 556] and an "instanton" gas model emerges that is, quite remarkably, equivalent to the hedgehog gas model obtained above from a semiclassical perspective. In particular, each instanton has a Berry phase that is given precisely by (19.58). In this context, the phases in (19.58) are a consequence of the constraint that each $S = 1/2$ spin can form a valence bond with exactly one of its neighbors, whereas here we obtained (19.58) from a very different coherent state path integral. The identity of these two distinct approaches reinforces our confidence in the correctness of (19.58), and in the presence of VBS order for $S = 1/2$, which follows quite robustly [397] from it. The quantum dimer model has also been examined in exact diagonalization studies, and again the evidence for VBS order is quite convincing [297]. A review of the current status of VBS ordering in two-dimensional ordering may be found in [430].

With the appearance of VBS order, do our conclusions on the nature of the spin excitation spectrum in the paramagnet without Berry phases discussed earlier still apply? The answer is yes: the lowest spinful excitation remains an $S = 1$ quasiparticle which contributes a delta function to the spectral density of the spin correlation. However, higher energy excitations, and the nature of the approach to the quantum critical point are significantly affected by the Berry phases. We defer further discussion of these subtle issues to Section 19.3.5, after we have considered noncollinear antiferromagnets.

19.3.4 Noncollinear antiferromagnetism in $d = 2$: deconfined spinons and visons

We turn to consideration of quantum antiferromagnets that have more complicated ordered magnetic states than those described so far. We consider models (19.1) on nonbipartite lattices or with further-neighbor interactions so that simple collinear states are not likely to be the ground states. Throughout, we only consider states that do not have a macroscopic magnetic moment (i.e. the expectation value of $\sum_i \hat{\mathbf{S}}_i$ in any low-lying state is not of the order of the number of sites in the system). Such states are expected to be preferred in models with all $J_{ij} < 0$. Also, we only consider the case of $d = 2$ here, as $d = 1$ antiferromagnets are better treated by the methods of the following chapter.

The simplest, and most thoroughly studied, example of a noncollinear antiferromagnet is the triangular lattice with a nearest-neighbor antiferromagnetic exchange. In the limit $S \to \infty$, the classical ground state is easy to work out. It is characterized by the expectation value

$$\langle \hat{\mathbf{S}}_i \rangle = S(\mathbf{n}_1 \cos(\vec{Q} \cdot \vec{x}_i) + \mathbf{n}_2 \sin(\vec{Q} \cdot \vec{x}_i)), \tag{19.60}$$

where the ordering wavevector $\vec{Q} = (4\pi/a)(1/3, 1/\sqrt{3})$ on a triangular lattice with $(a, 0, 0)$ one of the vectors connecting nearest-neighbor lattice sites, and $\mathbf{n}_{1,2}$ are arbitrary vectors in spin space satisfying

$$\mathbf{n}_1^2 = \mathbf{n}_2^2 = 1; \quad \mathbf{n}_1 \cdot \mathbf{n}_2 = 0. \tag{19.61}$$

These constraints define two orthogonal unit vectors, and each such pair defines a different classical ground state. This is a key difference from the collinear states in Section 19.3.3, where only a single unit vector was sufficient to characterize the ground state, as in (19.54).

Magnetically ordered ground state on the triangular lattice. The spins have been taken to lie in the plane of the triangular lattice, but this need not generally be the case.

Alternatively stated, the order parameter characterizing the broken symmetry in the classical ground state is a *pair* of orthogonal vectors [120, 196]. One possible ground state is shown in Fig. 19.3, for the case where $\mathbf{n}_1$, $\mathbf{n}_2$ lie in the plane of the lattice. Other antiferromagnets with *coplanar* ordering in their classical ground states can be treated in an essentially identical manner. Another important example studied in the literature is the square lattice antiferromagnet with first-, second-, and third-neighbor exchanges (the $J_1 - J_2 - J_3$ model). For a range of parameters this model has an incommensurate spiral ground state. Such an ordering is described, as in (19.60), but the wavevector $\vec{Q}$ is no longer pinned at a precise value and varies continuously as the values of exchange constants are changed. As we move from site to site in the direction $\vec{Q}$ the spin orientation rotates by some irrational angle in the plane defined by $\mathbf{n}_1$ and $\mathbf{n}_2$. Finally, antiferromagnets in which the spin arrangement is not even coplanar but is genuinely three-dimensional can be treated using similar methods, but these will not be considered here.

Instead of working with vectors $\mathbf{n}_1$, $\mathbf{n}_2$ that satisfy the constraints (19.61), it is convenient to introduce an alternative parameterization of the space of ground states. It takes six real numbers to specify the two vectors $\mathbf{n}_1$, $\mathbf{n}_2$, and the three constraints (19.61) reduce the degrees of freedom to three. We can use these three real numbers to introduce two complex numbers z_1, z_2 subject to the single constraint

$$|z_1|^2 + |z_2|^2 = 1. \tag{19.62}$$

We relate these numbers to $\mathbf{n}_1$, $\mathbf{n}_2$ by [26, 87]

$$n_{2\alpha} + i n_{1\alpha} = \sum_{a,b,c=1}^{2} \varepsilon_{ac} z_c \sigma_{ab}^{\alpha} z_b, \tag{19.63}$$

where $\alpha = x, y, z$, σ^{α} are the Pauli matrices, and ε_{ab} is the second-rank antisymmetric tensor $\varepsilon_{12} = -\varepsilon_{21} = 1$, $\varepsilon_{11} = \varepsilon_{22} = 0$. The reader can check that the parameterization

(19.63) for $\mathbf{n}_{1,2}$ automatically satisfies (19.61) provided the single constraint (19.62) holds. So we have succeeded in reducing the number of constraints from three to one. However, the mapping from $z_{1,2}$ to $\mathbf{n}_{1,2}$ is not one-to-one but two-to-one; the twofold redundancy is apparent from (19.63) as z_a and $-z_a$ correspond to precisely the same $\mathbf{n}_{1,2}$, and therefore the same spin configuration; this redundancy is crucial to our subsequent considerations. To describe it further, let us decompose z_a into its real and imaginary parts

$$z_1 = m_1 + im_2; \qquad z_2 = m_3 + im_4. \tag{19.64}$$

Then the order parameter becomes a four-component, real vector m_ρ ($\rho = 1, 2, 3, 4$) and (19.62) translates into the constraint that this vector has unit length (of course, there is no reason the effective action for m_ρ should be invariant under O(4) rotations in this space since the underlying symmetry is always O(3)). The identity of z_a and $-z_a$ means that m_ρ is a *headless* vector, much like a nematic liquid crystal, which is described by a headless three-vector.

We can proceed to examine the quantum fluctuations about the above classical states by precisely the same strategy as that followed in Section 19.3.3. We allow $\mathbf{n}_{1,2}$, and therefore z_a, to be slowly varying functions of spacetime. We also introduce a slowly varying uniform magnetization field $\mathbf{L}(x, t)$ such that the spatial integral over $\mathbf{L}$ equals precisely the total magnetization. Then, following (19.41), we parameterize

$$\mathbf{N}(i, \tau) = (\mathbf{n}_1(x_i, \tau) \cos(\vec{Q} \cdot \vec{x}_i) + \mathbf{n}_2(x_i, \tau) \sin(\vec{Q} \cdot \vec{x}_i))$$
$$\times \sqrt{1 - v^2 \mathbf{L}^2(x_i, \tau)} + v\mathbf{L}(x_i, \tau), \tag{19.65}$$

where v is the volume per site. This is to be inserted in the coherent state path integral of H_S in (19.1) and the result expanded in gradients. Finally, the uniform magnetization variable $\mathbf{L}$ is to be integrated out as below (19.49). The steps are similar to those in Section 19.3.3 and will not be explicitly carried out. Rather, let us try to anticipate the form of the answer on general symmetry grounds.

We list the constraints that must be obeyed by the final effective action:

(*i*) We must clearly require invariance under spin rotations. Such rotations are realized by the global $SU(2)$ transformation

$$\begin{pmatrix} z_1 \\ z_2 \end{pmatrix} \rightarrow U \begin{pmatrix} z_1 \\ z_2 \end{pmatrix} \equiv \begin{pmatrix} \alpha & \beta \\ -\beta^* & \alpha^* \end{pmatrix} \begin{pmatrix} z_1 \\ z_2 \end{pmatrix}, \tag{19.66}$$

where α and β are complex numbers satisfying $|\alpha|^2 + |\beta|^2 = 1$. Applying this transformation to (19.63), we see that it performs the rotation $n_{1,2\alpha} \rightarrow R_{\alpha\beta} n_{1,2\beta}$ where

$$U^\dagger \sigma^\alpha U = R_{\alpha\beta} \sigma^\beta. \tag{19.67}$$

(*ii*) Next, we consider the consequences of lattice translations. Any spatial configuration of $\mathbf{n}_{1,2}(x, \tau)$ should have its energy unchanged under translation by a lattice vector $\vec{y}$. By combining (19.63) with (19.65) we see that such a translation is realized by a simple overall phase change of the z:

$$z_a \rightarrow e^{-i\vec{Q} \cdot \vec{y}/2} z_a. \tag{19.68}$$

Note that this transformation is not a special case of (19.66), which was restricted to unitary matrices with unit determinant. For the case of the triangular lattices, (19.68) requires that the action be invariant under multiplication of z_a by the cube roots of unity. For incommensurate spiral states, by different choices of $\vec{y}$ we see that (19.68) requires invariance under multiplication of z_a by an arbitrary U(1) phase factor.

(*iii*) Time reversal inverts all the spins, and so maps $\mathbf{n}_{1,2} \rightarrow -\mathbf{n}_{1,2}$. This is realized by (19.63) by [228]

$$z_a \rightarrow i z_a. \tag{19.69}$$

(*iv*) Finally, let us recall the twofold redundancy in the mapping from z_a to the $\mathbf{n}_{1,2}$ discussed below (19.63). The change in sign of z_a can vary from point to point in space-time with no consequence for the $\mathbf{n}_{1,2}$. Therefore, we require invariance under the discrete Z_2 gauge transformation

$$z_a(x, \tau) \rightarrow \eta(x, \tau) z_a(x, \tau), \tag{19.70}$$

where $\eta(x, \tau) = \pm 1$ but can otherwise vary arbitrarily. In the naive continuum limit, the gauge nature of the transformation (19.70) does not impose any additional constraints beyond those arising from a constant η. However, the theory has to be regularized at short scales, and the Z_2 gauge symmetry does impose additional constraints on any effective lattice action. Moreover, the invariance (19.70) will also play a crucial role in the nature of the possible topological defects.

Let us write down the simplest action consistent with the above constraints in the naive continuum limit. Up to second order in spatial gradients, there are only two independent terms: $|\nabla z_a|^2$ and $|z_a^* \nabla z_a|^2$ (a third possibility, $|\varepsilon_{ab} z_a \nabla z_b|^2$, satisfies a simple linear relation with these two). Similar considerations apply to the terms with temporal gradients: here a term with one temporal gradient, $z_a^* \partial_\tau z_a$, is forbidden by time-reversal invariance. We are therefore led to the following effective action for the z_a, which plays the role of $\mathcal{S}_n$ in Section 19.3.3:

$$\mathcal{S}_z = \int d^2x \, d\tau \sum_{\mu=\vec{x},\tau} \frac{1}{g_\mu} \left[|\partial_\mu z_a|^2 + \gamma_\mu |z_a^* \nabla z_a|^2 \right], \tag{19.71}$$

where g_x, g_τ, γ_x, and γ_τ are coupling constants. In addition, as in Section 19.3.3, there could be Berry phases, associated with singular configurations of the z_a. These have to be analyzed on a lattice-by-lattice basis [543].

As in the case of our analysis of the collinear antiferromagnets, let us discuss the physics implied by the action (19.71), ignoring possible Berry phase effects for now. For small g_μ, we clearly have $\langle z_a \rangle \neq 0$, and this leads to a coplanar antiferromagnetic order as in Fig. 19.3. With increasing g_μ there is a transition to a paramagnetic phase with $\langle z_a \rangle \neq 0$, and we are interested both in the nature of this phase and of the quantum phase transition. There are at least two distinct possibilities: one leads to a "Z_2 spin liquid" phase with "topological order" and fractionalized excitations, and the other to a phase similar to the VBS state similar to that discussed in Section 19.3.3. We focus on the first possibility here, and briefly discuss the second possibility at the end of this section.

The Z_2 spin liquid emerged first in a direct large-N study [398,415,435] of the quantum antiferromagnet (19.1) on frustrated lattices, and related results emerge from studies of the continuum theory S_z in an expansion in the inverse of the number of z_a components, or in an expansion in $(d-1)$ [31,32,87,89]. The physical properties of both the magnetically ordered and paramagnetic phases can be rapidly understood by considering the case $\gamma_\mu = 0$ in (19.71), although this special value will not modify the general form of the following results. For $\gamma_\mu = 0$, we insert (19.64) into (19.71), and see straightforwardly that the action S_z is symmetric under O(4) rotations of the m_ρ, becoming precisely equivalent to the $N = 4$ case of the quantum rotor model S_n studied intensively in Parts II and III. Indeed, at the quantum critical point, it can be shown that the γ_μ are irrelevant perturbations, and so the global O(4) symmetry is asymptotically realized in the generic case. The properties of S_z therefore follow directly from the results of Part II. The magnetically ordered phase has $3 = 4 - 1$ linearly dispersing spin-wave excitations, and magnetic order disappears at any nonzero temperature. The quantum paramagnetic phase has an energy gap, Δ_+, and the excitations are built out of the Fock space of a fourfold degenerate particle.

Despite the mapping above to Part II, a crucial distinction exists in the physical interpretation of the structure of the quantum paramagnet, which is now more properly referred to as a "spin liquid." Its particle excitations are the bosonic quanta of the z_a field, and the transformation (19.66) under spin rotations makes it clear that these bosons carry spin $S = 1/2$. (This accounts for a twofold degeneracy of the particle states; an additional factor of two comes from accounting for the particle and antiparticle states.) This should be contrasted with the $S = 1$ particle that was found in the quantum paramagnet with collinear correlations in Section 19.3.3. These $S = 1/2$ bosonic particles are labeled "spinons." We can view the $S = 1$ particle as the bound state of two $S = 1/2$ particles, and therefore a quantum transition from a quantum paramagnet with collinear correlations to one with noncollinear correlations can be viewed as one of the deconfinement of spinons; a simple theory for such a transition has been discussed in [398,415,435]. Here let us discuss an important physical property of a spin liquid with deconfined spinons: we compute the dynamic susceptibility at the noncollinear ordering wavevector, defined by

$$\chi(k, i\omega_n)\delta_{\alpha\beta} = \frac{v}{M} \sum_{i,j} \int_0^{1/T} d\tau \langle \hat{S}_{i\alpha}(i\tau)\hat{S}_{j\beta}(0)\rangle e^{-i((\vec{k}+\vec{Q})\cdot(\vec{x}_i-\vec{x}_j)-\omega_n\tau)}. \qquad (19.72)$$

Using (19.63) and (19.65) we see that (ignoring the contribution of **L**, which will only renormalize a prefactor that can absorbed into a redefinition of the quasiparticle amplitude $\mathcal{A}$):

$$\chi(k, i\omega_n) = \frac{S^2}{6} \sum_{a,b=1}^{2} \int d^2x \int_0^{1/T} \langle z_a(x, i\tau)z_b(x, i\tau)z_a^*(0,0)z_b^*(0,0)\rangle. \qquad (19.73)$$

So χ is given by the propagator of two spinons, rather than the single-particle propagator that appeared in (11.2). As discussed above, the z quanta of the quantum paramagnet have a quasiparticle pole at $T = 0$ as in (10.82) or (11.18); the contribution of this pole leads to the expression

$$\chi(k, \omega_n) = \mathcal{A}^2 S^2 \Pi(k, \omega_n), \qquad (19.74)$$

where the two-particle propagator Π was discussed in (13.42). At $T = 0$, taking the imaginary part of (13.46) we obtain

$$\text{Im}\chi(k, \omega) = \frac{\mathcal{A}^2 S^2}{8c^2} \frac{\text{sgn}(\omega)}{\sqrt{\omega^2 - c^2 k^2}} \theta\left(|\omega| - (c^2 k^2 + 4\Delta_+^2)^{1/2}\right), \qquad (19.75)$$

where θ is the unit step function. Hence there is no pole in $\chi(k, \omega)$ as there was for the case of a quantum paramagnet with collinear spin correlations; rather there is a branch cut at frequencies greater than $(c^2 k^2 + 4\Delta_+^2)^{1/2}$, which corresponds to the threshold for the creation of a *pair* of spinons. This branch cut is a characteristic property of the deconfinement of spinons in a spin liquid.

Note also that the spinon changes sign under the gauge transformation in (19.70): in other words, the spinon carries an electric Z_2 gauge charge. In addition to the spinon, the Z_2 spin liquid has another gapped particle excitation which carries a "magnetic" Z_2 gauge charge, and which is crucial for an understanding of its "topological order" [283, 398, 435]. This excitation is a point-like vortex whose stability is ensured by the structure of the order parameter in the noncollinear antiferromagnet. It is now referred to as a vison [459]. The vison is best visualized in terms of the headless vector m_ρ, as is illustrated in Fig. 19.4. As one circles the core of the vison, m_ρ rotates by 180 degrees about a fixed axis orthogonal to m_ρ. So upon returning to the original point, m_ρ has now turned into $-m_\rho$, but this is acceptable as the overall sign of m_ρ is not significant (in mathematical terms, the order parameter m_ρ belongs to the space S_3/Z_2, and the vison is associated with its first homotopy group Z_2).

Note that the vison involves twist in the antiferromagnetic order parameter around it. Consequently, in the magnetically ordered phase, when the spin stiffness is nonzero, the energy of two visons grows logarithmically with the separation between them. However, in the spin liquid, once the stiffness is nonzero, the visons are free to separate, and become stable, deconfined, point-like excitations which carry Z_2 magnetic flux. Figure 19.4 makes it clear that a spinon picks a change in sign after circumnavigating a vison (and vice versa): this labels them as relative semions.

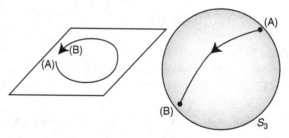

Fig. 19.4 A vison. On the left we show a circular path in real space. On the right is the space of magnetically ordered states represented by the complex spinor $(z_\uparrow, z_\downarrow)$ up to an overall sign. As we traverse the real-space circle, the path in order parameter space connects polar opposite points on S_3 (A and B), which are physically indistinguishable. A key point is that this vison excitation can be defined even in a state in which magnetic order is lost: the path on the right will fluctuate all over the sphere in quantum imaginary time, as will the location of the points A and B, but A and B remain polar opposites.

The topological order of the Z_2 spin liquid becomes clear when we consider the degeneracy of the ground state on a torus, in the limit of infinite torus size. We can pierce each hole of the torus by a vison with negligible energy cost, because the only contour carrying Z_2 flux must encircle the large hole of the torus. This leads to a fourfold degeneracy, characteristic of the Z_2 spin liquid.

Thus a complete description of the Z_2 spin liquid requires an effective theory of the dynamics of visons and spinons. Such theories have been considered in the literature [246, 459, 543], but will not be described here. They account for the Berry phase terms, and also lead to theories of transitions to confining phases with VBS order.

We conclude by mentioning the second possibility associated with the "quantum disordered" phase, after coplanar magnetic order has been lost. In (19.71) we used spinor variables z_a to describe the quantum fluctuations of the local magnetic order. However, we could equally well have used the two vector variables $\mathbf{n}_1$ and $\mathbf{n}_2$, and developed an effective action for their quantum fluctuations: the constraints (19.61) can be implemented in a soft-spin theory by appropriate nonlinearities. Without Berry phases, such models have been studied extensively [66], and lead to a continuous transition to a paramagnetic phase in which the $\mathbf{n}_1$ and $\mathbf{n}_2$ form a degenerate pair of spin $S = 1$ gapped quasiparticles. Berry phases are likely to lead to VBS order in such a confining paramagnetic, but this effect has not been specifically described.

19.3.5 Deconfined criticality

We now return to the collinear antiferromagnets of Section 19.3.1, and address the question of the phase transitions between the Néel and VBS states. Note that both phases break a global symmetry of the Hamiltonian (spin rotation and lattice rotations, respectively), and two symmetries are unrelated to each other. Under such circumstances, the conventional classical theory of phase transitions implies that such phases must be separated by a first-order phase transition.

For the present quantum case, the fact that one of the orderings (VBS) was driven by the Berry phases carried by "hedgehog" defects of the other (Néel) suggested that the situation might be different and speculations for the existence of a continuous phase transition were made in early work [86]. A specific theory for a continuous Néel transition has been presented in [351, 458, 462], and is outlined below.

An important input into the theory is the fact that density of free hedgehogs vanishes as the critical point is approached. These hedgehogs are responsible for the VBS ordering in the paramagnetic phase, but their density is computed to vanish with a power of the inverse correlation length in [353]. Given the fact that the hedgehogs are ultimately responsible for the confinement of spinons in the paramagnet, [351, 458, 462] argue that the critical theory should be expressed in terms of the spinon variables. Thus we should map the action (19.51) expressed in terms of the Néel vector $\mathbf{n}$ to one expressed in spinor variables: this has been done in a lattice model in [434]. Unlike the case of the coplanar antiferromagnet in Section 19.3.4, we only have a single Néel vector $\mathbf{n}$, and the analog of (19.63) is

$$n_\alpha = w_a^* \sigma_{ab}^\alpha w_b. \tag{19.76}$$

We are now using the symbol w_a to represent the spinon, unlike z_a as in Section 19.3.4; these spinon fields are distinct, and in particular, they have distinct transformations under time-reversal, as we see below.

Given the detailed arguments in [351, 458, 462], the construction of the Néel–VBS critical theory becomes simply a matter of writing down the simplest continuum theory for the w_a consistent with symmetries of the underlying Hamiltonian. The phase with $\langle w_a \rangle \neq 0$ corresponds to the Néel phase, as in Section 19.3.4. However, unlike Section 19.3.4, the region with $\langle w_a \rangle = 0$ is *not* a spin liquid with deconfined spinons, *except* at the quantum critical point. Away from the critical point in the paramagnet, the proliferation of hedge-hogs confines the w_a into the $S = 1$ quanta of $\mathbf{n}$, and the O(3) sigma model description of Section 19.3.3.

So it only remains to deduce the continuum theory of the critical spin liquid of the spinons w_a, using considerations of symmetry. Let us list the symmetry operations, which parallel those considered in Section 19.3.4:

(*i*) Invariance under spin rotations is just as in (19.66).

(*ii*) A lattice operation which interchanges the two sublattices maps $\mathbf{n} \to -\mathbf{n}$, and (19.76) therefore implies

$$w_a \to \varepsilon_{ab} w_b^*. \tag{19.77}$$

(*iii*) Time reversal also sends $\mathbf{n} \to -\mathbf{n}$, and so obeys (19.77). Note that (19.77) maps onto (19.69) if we set

$$z_a = e^{i\alpha}(w_a - i\varepsilon_{ab} w_b^*)/\sqrt{2}, \tag{19.78}$$

with α arbitrary, which is the connection obtained in specific microscopic models [228, 435].

(*iv*) Finally, we note that the Z_2 gauge invariance of (19.70) is now enlarged to a U(1) gauge invariance. Representation (19.76) is invariant under a gauge transformation by an arbitrary phase θ

$$w_a(x, \tau) \to e^{i\theta(x,\tau)} w_a(x, \tau). \tag{19.79}$$

Unlike (19.70), this continuous gauge transformation plays a crucial role in the allowed terms in the continuum theory. In particular, spatial and temporal gradients can only preserve gauge invariance with the help of a U(1) gauge field, analogous to that familiar from quantum electrodynamics. A microscopic derivation [396, 434] shows that a U(1) gauge field, A_μ, does emerge at low energies as a distinct degree of freedom. So our continuum theory will involve both w_a and A_μ.

The constraints above lead to the following relativistic theory, known as the CP^1 model, for the deconfined critical point:

$$\mathcal{S}_{zA} = \int d^2x d\tau \sum_{\mu=\vec{x},\tau} \frac{1}{g_\mu} |(\partial_\mu - iA_\mu) w_a|^2. \tag{19.80}$$

Compared to (19.71), we see the emergence of a U(1) gauge field: the "photons" of this gauge field are physical, gapless, spinless excitations which can, in principle, be detected

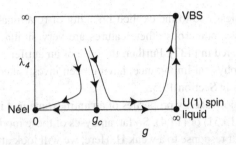

Schematic renormalization group flows for the $S = 1/2$ square lattice quantum antiferromagnet. Here λ_4 is a measure of the density of free hedgehogs (a hedgehog fugacity), g represents the couplings g_μ in (19.80). The theory $\mathcal{S}_{zA}$ in (19.80) describes only the line $\lambda_4 = 0$: it is therefore a theory for the transition between the Néel state and a U(1) spin liquid with a gapless "photon." However, the lattice antiferromagnet always has a nonzero bare value of the hedgehog fugacity λ_4. The λ_4 perturbation is irrelevant at the $g = g_c$ critical point of $\mathcal{S}_{zA}$: this critical point therefore also described the transition in the lattice antiferromagnet. However, the $g \to \infty$ U(1) spin liquid fixed point is *unstable* to λ_4, and the paramagnet is therefore a gapped VBS state.

in experiments. As a function of increasing g_μ, the theory $\mathcal{S}_{zA}$ exhibits a transition from the Néel state with $\langle w_a \rangle \neq 0$, to a U(1) spin liquid paramagnet with $\langle w_a \rangle = 0$. The w_a quanta are spin $S = 1/2$ excitations in this phase which interact via a 2+1 dimensional electromagnetic force. After including the effects of the "hedgehog" defects (which have been excluded from (19.80)), away from the critical point this spin liquid is unstable to confinement of the spinons and the appearance of VBS order, as we have argued above. This subtle state of affairs is summarized in Fig. 19.5. The photon excitation A_μ is also gapped near the VBS fixed point: thus there is an additional spinless gapped excitation in the VBS fixed point, in addition to the spin $S = 1$ **n** quanta discussed in Section 19.3.3.

19.4 Partial polarization and canted states

This section will interpolate between the ferromagnetic states studied in Section 19.2, with maximum uniform spin polarization in their ground states, and the antiferromagnets of Section 19.3, which had a thermodynamically negligible spin polarization. One way to do this would be to examine the ground states of models H_S in (19.1) at **H** = **0**, but with a set of J_{ij} that can take both signs. Models of this type were examined in [437], and it was argued that they could be described by a ferromagnetic extension of the rotor models studied in Part II. The properties of such models are quite intricate, and we refer the reader to the original paper for further details. Here, we look at a closely related model whose properties are significantly simpler to delineate. We begin with an antiferromagnet with all $J_{ij} < 0$ and attempt to force in a macroscopic moment by placing it in a strong uniform field **H**. Thus the uniform magnetization does not arise spontaneously from ferromagnetic exchange interactions but instead is induced by an external field. This causes important differences in the nature of certain spin-wave excitations, which are no longer required to be gapless due to the explicit breaking of rotational invariance in the Hamiltonian. Nevertheless, numerous other features are very similar to the far more complicated models considered in [437]. Further, the case of an antiferromagnet in a strong uniform field is of direct physical importance, having been investigated in several recent experiments, as we discuss in Section 19.5.

The low-energy properties of an antiferromagnet in a field $\mathbf{H}$ are described by the action $\mathcal{S}_n$ in (19.51) or (11.4). So far, analyses of these models have been restricted to $\mathbf{H} = \mathbf{0}$, and to linear response to a weak $\mathbf{H}$. Here, we will look at the full nonlinear response to a strong $\mathbf{H}$. It should be noted here that, in $d = 1$, closely related results can also be obtained by the bosonization technique of Chapter 20 [370], while making no reference to the rotor model, but we do not follow such an approach here.

We prefer to begin our analysis by placing the continuum model $\mathcal{S}_n$ on a lattice at some short-distance scale and working with the discrete lattice Hamiltonian. This is the inverse of the mapping carried out in Chapter 11, and we therefore obtain the rotor model Hamiltonian H_R in (11.1):

$$H_R = \frac{Jg}{2} \sum_i \hat{\mathbf{L}}_i^2 - J \sum_{\langle ij \rangle} \hat{\mathbf{n}}_i \cdot \hat{\mathbf{n}}_j - \mathbf{H} \cdot \sum_i \hat{\mathbf{L}}_i . \tag{19.81}$$

The lattice sites in this rotor Hamiltonian are not to be identified with the lattice sites of H_S in (19.1); rather each rotor is an effective degree of freedom for a cluster of an even number of spins in the original model. Each such cluster will have a spin-singlet ground state for $\mathbf{H} = \mathbf{0}$, as does the on-site Hamiltonian for each rotor in (19.81) (see (6.48)). The rotor also has an infinite tower of states with increasing angular momentum in (6.48); in contrast a cluster of p Heisenberg spins with spin S can have a maximum total angular momentum pS. Although this difference has some significant consequences for the topology of the phase diagram, it leaves many essential features unaltered; we comment on this issue later.

We proceed to understanding the properties of H_R in the remainder of this section. The analysis is quite similar to that discussed for the boson Hubbard model in Chapter 9, and the results bear some similarity to those in [256]; indeed, we find that the phase diagram of H_R is quite similar to that of H_B in (9.4), and the universality classes of the quantum phase transitions reduce either to the models studied in Part II, or to those in Chapter 16. This similarity is not surprising at one level: the model H_R in the presence of a nonzero $\mathbf{H}$ only has a global U(1) symmetry corresponding to rotations about an axis parallel to the field (rotations about all other axes are not allowed by the nonzero $\mathbf{H}$), and the model H_B also has only a U(1) symmetry. (In the models considered in [437], uniform moments appear spontaneously owing to ferromagnetic exchange in a model with full O(3) symmetry, and this reasoning does not hold; however, the similarity to H_B persists, with many (but not all) quantum critical points belonging to the same universality classes as those of H_B.)

Most of the physics of H_R already becomes apparent in a mean-field theory similar to that in Section 9.1. As in (9.7), we make a mean-field ansatz for H_{MF} as the sum of single-site Hamiltonians with initially arbitrary variational parameters:

$$H_{MF} = \sum_i \left(\frac{Jg}{2} \hat{\mathbf{L}}_i^2 - \mathbf{H} \cdot \sum_i \hat{\mathbf{L}}_i - \mathbf{N} \cdot \hat{\mathbf{n}}_i \right) . \tag{19.82}$$

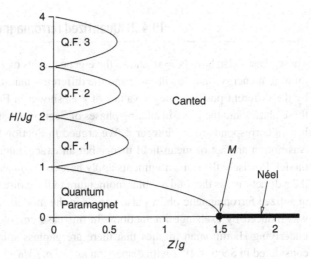

Fig. 19.6 Mean-field phase diagram of H_R (in (19.81)), the O(3) quantum rotor model in a field **H**. The notation Q.F. ℓ refers to a quantized ferromagnet with $\langle \hat{L}_z \rangle = \ell$. Compare with the phase diagram of the boson Hubbard model in Fig. 9.1. In the latter case, there is no special meaning to the vertical coordinate $= 0$, and the vertical axis is unbounded below. The positions of the phase boundaries follow from (19.84). The multicritical point M is precisely the critical point of the O(3) quantum rotor model studied in Part II.

Here the **N** are a set of three variational parameters that represent the effects of the exchange J with nearest neighbors in mean-field theory; they play a role similar to that of the complex number Ψ_B in Section 9.1. For simplicity we have assumed that the **N** are site independent and are therefore excluding the possibility of states with spatial structure. It is not difficult to extend our analysis to allow for broken translational symmetries in H_R.

Now the analysis proceeds as in Section 9.1: determine the ground state wavefunction of H_{MF}, and optimize the expectation value of H_R in this wavefunction toward variations in **N**. This has been done numerically and leads to the phase diagram in Fig. 19.6; we discuss the properties of each of the phases in turn and then consider the nature of the transitions between them.

19.4.1 Quantum paramagnet

The optimum value of the variational parameter is **N** $= 0$. For this value, H_{MF} is exactly diagonalizable: the eigenstates are simply the rotor eigenstates $|\ell, m\rangle$ of (6.2) and have eigenvalues $Jg\ell(\ell + 1)/2 - Hm$. The quantum paramagnet appears when parameters are such that the minimum energy state has $\ell = m = 0$. This happens for small H/J and large g. This quantum paramagnet is precisely the corresponding state of the rotor model studied in Part II. The field **H** couples only to the total spin, which is identically zero in the spin-singlet ground state; as a result the wavefunction and all equal-time correlations are unaffected by a nonzero **H**. The energy of the spin-triplet particle excitations does change, as was shown in (6.4), but their wavefunctions also remain unaffected.

19.4.2 Quantized ferromagnets

These phases also have $\mathbf{N} = 0$, and so the eigenenergies of H_{MF} are those listed above. The minimum energy state has $m = \ell$, and the different quantized ferromagnets are identified by the different positive integer values of ℓ as shown in Fig. 19.6. The analogy between these phases and the Mott-insulating phases of Section 9.1 should be clear: the boson number n_0 corresponds to the integer ℓ. We argued in Section 9.1 that the quantization of n_0 was not an artifact of mean-field theory but an exact statement about the full interacting model. Precisely the same arguments apply here to $\langle \hat{L}_z \rangle$ (we are assuming $\mathbf{H}$ is oriented in the z direction), as the total angular momentum in the z direction commutes with H_R. Such quantized ferromagnetic phases also appear in the models of [437] where ferromagnetism was induced by exchange interactions. In this case complete rotational symmetry of the underlying Hamiltonian implies that there are gapless spin-wave excitations of the type considered in Section 19.2 with dispersion $\varepsilon_k = (\rho_s/M_0)k^2$. In the present model, H_R, the spin-wave modes acquire a gap from the external field, and we have $\varepsilon_k = (\rho_s/M_0)k^2 + H$. In these respects these quantized ferromagnets are identical to the fully polarized ferromagnets of Section 19.2; we simply have to set M_0 equal to the actual quantized value of the ground state magnetization density.

Let us also note some aspects of the interpretation of these quantized ferromagnet phases for underlying spin models like H_S. We noted above that each rotor was an effective degree of freedom for an even number, p, of Heisenberg spins. Such a cluster has maximum spin pS, and so the quantized ferromagnets with $\ell > pS$ clearly cannot exist and are artifacts of the mapping to the rotor model, which introduced an infinite tower of states on each site. Also, for some antiferromagnets, making clusters of p spins may involve reducing the symmetry of the underlying lattice. In this case the quantized ferromagnets with $0 < \ell < pS$ necessarily involve a spontaneously broken translational symmetry. Each spin has an average fractional moment of ℓ/p and this can be quantized only if p spins spontaneously group together and carry a total moment ℓ together. This spontaneously broken symmetry affects the critical theory of the transition out of the quantized phase, but we will not discuss this further here. Finally, the rotor with $\ell = p$ is a fully polarized ferromagnet that can exist without any broken translational symmetry.

It should also be noted that very similar considerations apply for the case of p odd; then we have to work with rotors that carry half-integral angular momenta [437, 467].

19.4.3 Canted and Néel states

These states both have $\mathbf{N} \neq 0$ and are thus the analogs of the superfluid state of the boson model of Section 9.1. The Néel state occurs precisely at $\mathbf{H} = 0$, and the full rotational invariance of the Hamiltonian then implies that the direction of $\mathbf{N}$ is immaterial. The *canted* state occurs at nonzero $\mathbf{H}$. If we write $\mathbf{H} = H\mathbf{e}_z$, the numerical optimization of the mean-field Hamiltonian (19.82) shows that the vector $\mathbf{N}$ prefers to lie in the x–y plane; the direction within the plane is immaterial, reflecting the U(1) symmetry of the problem. This orientation of the Néel order parameter in a plane perpendicular to an applied uniform field is quite generic, and the reasons for it become more evident in Section 19.4.4 below. We choose $N_x \neq 0$ and $N_y = 0$. The resulting canted state is characterized by the nonzero expectation values

$$\langle \hat{n}_x \rangle = N_x/(JZ) \neq 0, \qquad \langle \hat{L}_z \rangle \neq 0, \qquad (19.83)$$

and all other components of $\hat{\mathbf{n}}$ and $\hat{\mathbf{L}}$ have vanishing expectation values. The first relation in (19.83) should be compared with (9.9) for its origin is the same. Both nonzero expectation values in (19.83) vary continuously as a function of J, g, or H, and nothing is pinned at a quantized value; because there is a nonzero, continuously varying ferromagnetic moment in the canted phase, this is an example of an *unquantized ferromagnet*. The results (19.83) also make the origin of the term "canted" clear (as illustrated within the canted region of Fig. 19.8). In terms of the underlying Heisenberg spins, a nonzero $\langle \hat{n}_x \rangle$ implies antiferromagnetic ordering within the x direction in spin space, while a nonzero $\langle \hat{L}_z \rangle$ implies a uniform ferromagnetic moment in the z direction.

We show a plot of the H dependence of the $T = 0$ magnetization $\langle \hat{L}_z \rangle$ in Fig. 19.7. Note that there are plateaus in the magnetization while the system is in the quantum paramagnetic or quantized ferromagnetic phases. In between these phases is the canted phase, or the unquantized ferromagnet, in which the magnetization continuously interpolates between the quantized values.

The excitation structure of the canted phase is easy to work out. We simply follow the same procedure as that used for the Néel state in Section 6.2. Examining equations of the motion of small fluctuations about the ordered state one finds a gapless spin-wave excitation with energy $\varepsilon_k \sim k$ corresponding to rotations of the $\hat{\mathbf{n}}$ in the x–y plane. For the case where the canted state appears in a model with full O(3) symmetry, there is an additional gapless mode with dispersion $\varepsilon_k \sim k^2$ [437].

The mean-field boundary between the canted/Néel states and the quantized ferromagnet/quantum paramagnet can be computed analytically, using the same analysis leading up

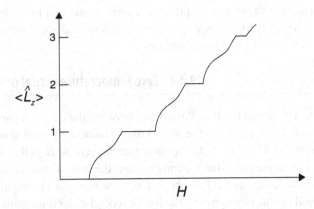

Fig. 19.7 The magnetization, $\langle \hat{L}_z \rangle$, as a function of the field $\mathbf{H}$ for the rotor model (19.81). It is assumed that the value of Z/g in Fig. 19.6 is small enough that a vertical line will intersect the Q.F. ℓ phases for $\ell \leq 3$. The magnetization is initially pinned at 0 when the system is in the quantum paramagnet and is subsequently pinned at ℓ in the Q.F. phases. The magnetization interpolates between these plateaus in the canted or "unquantized ferromagnetic" phase.

to (9.14) for the boson model. We expand the ground state energy of the quantized ferromagnet/quantum paramagnet in powers of N_x and demand that the coefficient of the N_x^2 vanish. This leads to the analog of the condition $r = 0$ with the expressions (9.15), (9.16); in the present situation we find the condition

$$\frac{g}{Z} = \frac{\ell + 1}{(2\ell + 3)(\ell + 1 - H/Jg)} - \frac{\ell}{(2\ell + 1)(\ell - H/Jg)} + \frac{1}{(2\ell + 1)(2\ell + 3)(\ell + 1 + H/Jg)},$$ (19.84)

for the instability of the quantized ferromagnet/quantum paramagnet with $\langle \hat{L}_z \rangle = \ell$ (the denominators in (19.84) are always positive over the range of applicability for a given value of ℓ). Simple application of (19.84) led to Fig. 19.6.

An important feature of the above results deserves special mention. Note that the only phase with a continually varying uniform magnetic moment (an unquantized ferromagnet) is the canted phase. This phase has a broken symmetry in the x–y plane and an associated gapless mode. This result is believed [437] to be a general principle: phases with continuously varying values of a ferromagnetic moment must have gapless spin modes in addition to the usual ferromagnetic spin waves that are present for the case of a spontaneously generated moment; moreover, unlike the spin waves, these gapless modes do not acquire a gap in the presence of a uniform field $\mathbf{H}$. In $d \geq 2$, for the rotor models considered here, the gapless modes are associated with the broken symmetry leading to canted order in such phases. In $d = 1$, the analysis in Chapter 16 shows that the order in the x–y plane becomes quasi-long range but the gapless mode survives.

(For completeness, we also note here another physical example of an unquantized ferromagnet: the Stoner ferromagnet [494] of an interacting Fermi gas, in which there are two Fermi surfaces, one each for up and down spins, with unequal Fermi wavevectors $k_{F\uparrow} \neq k_{F\downarrow}$. The values of $k_{F\uparrow}$ and $k_{F\downarrow}$ can vary continuously as the interaction strength is varied (provided they are both nonzero), and so can the mean magnetic moment. Consistent with the general principle above, in addition to the ferromagnetic spin waves, this system has low-energy spin-flip excitations at finite wavevectors involving particle–hole pairs near the two Fermi surfaces.)

19.4.4 Zero temperature critical properties

It is clear that the $\mathbf{H} = \mathbf{0}$ transition between the quantum paramagnet and the Néel state is precisely the same as the $N = 3$ model intensively studied in Part III; this critical point is denoted M in Fig. 19.6. We show that the generic $\mathbf{H} \neq \mathbf{0}$ transition between the quantized ferromagnet/quantum paramagnet and the canted state is in the universality class of the dilute Bose gas field theory in (16.1), which was thoroughly studied in Chapter 16. We will do this by examining the line of second-order transitions coming into the point M; the remaining portions of the phase boundary can be analyzed in a similar manner. It should also be noted that there are also special particle–hole symmetric points at the tips of the lobes surrounding the quantized ferromagnet phases where the $z = 1$ theory of Part II applies, just as was the case for the boson Hubbard model in Sections 9.1 and 9.3.

The promised result is most easily established by using the "soft-spin" theory of the point M studied in Chapter 14. In the presence of a field $\mathbf{H} = H\mathbf{e}_z$ the generalization of the $N = 3$ version of (14.2) is

$$S_\phi = \int d^d x \int_0^{1/T} d\tau \left\{ \frac{1}{2}\left[(\partial_\tau \phi_x - iH\phi_y)^2 + (\partial_\tau \phi_y + iH\phi_x)^2 \right. \right.$$
$$\left. \left. + (\partial_\tau \phi_z)^2 + c^2(\nabla_x \vec{\phi})^2 + r\phi_\alpha^2 \right] + \frac{u_0}{4!}\left(\phi_\alpha^2 \right)^2 \right\}. \qquad (19.85)$$

The uniform magnetic moment density is given by

$$\frac{1}{v}\langle \hat{L}_z \rangle = -\frac{\partial \mathcal{F}}{\partial H}, \qquad (19.86)$$

where v is the volume per rotor, and $\mathcal{F}$ is the free energy density associated with the action S_ϕ.

Let us first discuss the mean-field properties of S_ϕ, obtained by minimizing the action, while ignoring all spatial and time dependence of ϕ_α; this reproduces the structure in the vicinity of the point M in Fig. 19.6 obtained earlier using the mean-field Hamiltonian (19.82). Note that the components ϕ_x, ϕ_y have a quadratic term with coefficient $r - H^2$, while ϕ_z has the usual coefficient r; so ordering is preferred in the x–y plane, and this was the reason for the choice in the orientation of the $\mathbf{N}$ vector in Section 19.4.3. For $r - H^2 > 0$, the ground state has $\langle \phi_\alpha \rangle = 0$ and is therefore in the quantum paramagnetic phase. For $r - H^2 < 0$, the ground state has $\langle \phi_\alpha \rangle \neq 0$ and lies in the x–y plane. This is the C phase and the fields have the expectation values

$$\phi_\alpha = \left(\left(\frac{6(H^2 - r)}{u_0} \right)^{1/2}, 0, 0 \right), \qquad \frac{1}{v}\langle \hat{L}_z \rangle = \frac{6H(H^2 - r)}{u_0}, \qquad (19.87)$$

or any rotation of ϕ_α in the x–y plane. Note that $\langle \hat{L}_z \rangle$ vanishes for $H = 0$, and therefore the line $r < 0$, $H = 0$ is the Néel state. The resulting mean-field phase diagram is shown in Fig. 19.8 and is identical to the vicinity of the point M in Fig. 19.6. Let us focus on the vicinity of the generic transition between the quantum paramagnet and the canted phase: this corresponds to the regime $|r - H^2| \ll |r|$. In this region we can neglect ϕ_z fluctuations and focus only on the $\phi_x + i\phi_y$ that is undergoing Bose condensation. Further, the second-order time derivative in S_ϕ can be dropped as the low-energy properties are dominated by the more relevant first-order time derivative that appears by expanding the first two terms in S_ϕ. Making these approximations, and defining

$$\Psi = \frac{\phi_x + i\phi_y}{\sqrt{H}}, \qquad (19.88)$$

we see that S_ϕ reduces to

$$S_\Psi = \int d^2 x \int_0^{1/T} d\tau \left[\Psi^* \frac{\partial \Psi}{\partial \tau} + \frac{c^2}{2H}|\nabla_x \Psi|^2 + \frac{(r - H^2)}{2H}|\Psi|^2 + \frac{u_0}{24H^2}|\Psi|^4 \right].$$
$$(19.89)$$

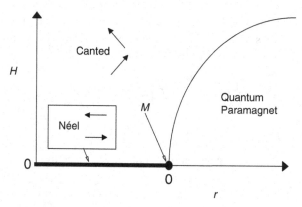

Mean-field phase diagram of S_ϕ (in (19.85)) at $T = 0$. The arrows denote the relative orientation of the spins in the corresponding phases of double layer systems, which map onto the rotor model as discussed in Section 1.4.3; the field **H** is assumed to point toward the top of the page. The multicritical point M is the $N = 3$ case of the quantum critical point studied in Part II. Note that the vicinity of M is similar to that in Fig. 19.6.

This is precisely the theory (16.1), establishing the claim made at the beginning of this subsection.

19.5 Applications and extensions

A survey of numerical and experimental studies of two-dimensional quantum antiferromagnets may be found in [430].

The analyses of Section 19.4 should make it clear that the dilute Bose gas quantum critical point of Chapter 16 describes the closing of a spin gap of an antiferromagnet by a strong external magnetic field [9, 10, 82, 438, 452, 482, 513]. This critical point has been intensively studied in a spin-ladder organic compound, $Cu_2(C_5H_{12}N_2)_2Cl_4$ [71–73, 130, 198]. The onset of magnetization plateaus at a finite field (as in Fig. 19.7) is also described by the same quantum critical point, and such plateaus have been observed recently in experiments on one-dimensional spin chains [356, 471].

A novel realization of the $d = 2$ continuum quantum ferromagnets of Section 19.2 is provided by magnetization studies of single-layer quantum Hall systems at filling factor $\nu = 1$ [135, 261, 262, 480]. These are electronic systems with a gap toward charged excitations and a strong ferromagnetic exchange between the electronic spins. As a result, the low-lying spin excitations are well described by the continuum theory (19.25). The magnetization of this system for different T and H has been measured in NMR [37] and optical [15, 319] experiments, and the results have been interpreted by computations on (19.25) [27, 213, 399, 504].

Also of interest are studies of double-layer quantum Hall systems, when two single-layer systems in a ferromagnetic quantum Hall state with a charge gap are brought close to each other [373, 374, 380, 445]. There is an antiferromagnetic exchange pairing between the layers [555], which suggests that we may consider the two layers to be similar to the two sublattices of an antiferromagnet and that there is an effective rotor model description of the spin excitations. Indeed, it has been argued [111, 112] that the system maps precisely onto the model studied in Section 19.4. Detailed light scattering studies have mapped out the phase diagram of the system [374], and the results are consistent with Figs. 19.6 and 19.8. Specific quantitative predictions for quantum critical behavior have been made in [111, 112, 324, 426, 509], and these and dynamical results like those in Section 14.3 could be tested in future experiments.

Spin chains: bosonization

This chapter has two central aims. The first is to describe a particular class of $S = 1/2$ antiferromagnets in $d = 1$ and to understand their properties in the context of the general discussion of antiferromagnets in Chapter 19. The second is to introduce the technical tool of bosonization and to illustrate its utility in the solutions of the models noted. The powerful bosonization method has been used extensively in recent years to understand a wide variety of systems in one dimension. We do not attempt to survey this vast literature here but refer the reader to a number of available reviews; a description of some important current topics appears in articles by Schulz [453] and Affleck [8]. However, most of the basic ideas and general principles make an appearance in our treatment here. The author benefited from unpublished Trieste lecture notes of T. Giamarchi in preparing this chapter.

The antiferromagnetic chain we study [190] has the Hamiltonian

$$H_{12} = J_1 \sum_i \left(\hat{\sigma}_i^x \hat{\sigma}_{i+1}^x + \hat{\sigma}_i^y \hat{\sigma}_{i+1}^y + \lambda \hat{\sigma}_i^z \hat{\sigma}_{i+1}^z \right)$$

$$+ J_2 \sum_i \left(\hat{\sigma}_i^x \hat{\sigma}_{i+2}^x + \hat{\sigma}_i^y \hat{\sigma}_{i+2}^y + \hat{\sigma}_i^z \hat{\sigma}_{i+2}^z \right), \tag{20.1}$$

where the $\hat{\sigma}_i^\alpha$ are Pauli matrices representing an $S = 1/2$ spin at site i, and the subscript 12 on H indicates the presence of first- and second-neighbor interactions. For $\lambda = 1$ this reduces to the $S = 1/2$ Heisenberg Hamiltonian H_S in (19.1), with first ($J_1 > 0$) and second ($J_2 > 0$) neighbor exchange in $d = 1$, which was studied in the continuum limit of the coherent state path integral in Chapter 19. We have introduced the anisotropy parameter λ to make contact with the quantum XX chain studied in Sections 16.1 and 16.3.3 by rather different methods; for $\lambda = 0$ and $J_2 = 0$, (20.1) reduces to the XX Hamiltonian in (16.6). We use these latter methods here and show how they can be combined with bosonization to examine the more general model (20.1). Recall that the XX chain had a global U(1) symmetry and an associated conserved charge $Q = (1/2) \sum_i \hat{\sigma}_i^z$. This U(1) symmetry is also present in H_{12} for general λ. Only the point $\lambda = 1$ has the full Heisenberg $SU(2)$ symmetry.

We begin by re-examining the XX model in Section 20.1, and then we re-obtain the results of Section 16.3.3 by introducing the bosonization method. The same method will be used to describe the phases and low-T properties of H_{12} in Section 20.2. Finally, in Section 20.3 we rectify an omission from Part II, examining the $N = 2$, $d = 1$ quantum rotor model and showing how it can be understood by a simple adaptation of the methods introduced in this chapter.

20.1 The *XX* chain revisited: bosonization

This section examines the model H_{12} at $\lambda = 0$ and $J_2 = 0$. Then, as noted earlier, H_{12} reduces to the Hamiltonian H_{XX} in (16.6). For antiferromagnetic exchange, $J_1 > 0$, we obtain H_{XX} with a coupling $w < 0$; it is somewhat inconvenient to work with a $w < 0$, but we can map to $w > 0$ by changing the signs of the $\hat{\sigma}^x\hat{\sigma}^x$ and $\hat{\sigma}^y\hat{\sigma}^y$ terms in H_{12} by rotating every second spin by 180 degrees about the z-axis. We have used the Jordan–Wigner transformation to map H_{XX} onto a model of free spinless fermions, and with the staggered rotation, the transformation (10.8, 10.9) becomes

$$\hat{\sigma}_i^z = 1 - 2c_i^\dagger c_i,$$
$$\hat{\sigma}_i^+ = (-1)^i \prod_{j<i}\left(1 - 2c_j^\dagger c_j\right)c_i,$$
$$\hat{\sigma}_i^- = (-1)^i \prod_{j<i}\left(1 - 2c_j^\dagger c_j\right)c_i^\dagger. \tag{20.2}$$

Inserting (20.2) into (20.1) we find

$$H_{12} = -\sum_i \Big[2J_1\left(c_{i+1}^\dagger c_i + c_i^\dagger c_{i+1}\right) + J_1\lambda\left(2c_i^\dagger c_i - 1\right)\left(2c_{i+1}^\dagger c_{i+1} - 1\right)$$
$$+ J_2\left(2c_i^\dagger c_i - 1\right)\left(2c_{i+2}^\dagger c_{i+2} - 1\right)$$
$$- 2J_2\left(c_i^\dagger\left(2c_{i+1}^\dagger c_{i+1} - 1\right)c_{i+2} + c_{i+2}^\dagger\left(2c_{i+1}^\dagger c_{i+1} - 1\right)c_i\right)\Big]. \tag{20.3}$$

So for $J_2 = 0$, $\lambda = 0$ we see that H_{12} reduces to the free fermion form (16.7) of H_{XX} with $w = 2J_1$ and $\mu = 0$.

The spin correlations of H_{XX} were examined in Section 16.3.3, and we were then especially interested in the quantum phase transition that occurred when the density of fermions in the ground state went from being pinned at zero to a nonzero value. As discussed in Section 16.1 and Fig. 16.1, this transition occurred at $\mu = 0$. Here we are interested in the case $\mu = 2w$, when the density of fermions is nonzero and large; by Fig. 16.1 the fermion band is exactly half filled. So we are well away from the quantum critical point of interest in Chapter 16 and are solely interested in the finite ground state fermion density region. This places us exclusively within the Tomonaga–Luttinger liquid region of Section 16.3.3. (This is the region labeled "Fermi liquid" in Fig. 16.2, which applies to general d.) We gave a complete derivation of the asymptotic form of the $T > 0$, equal-time correlators of the Tomonaga–Luttinger liquid region in Section 16.3.3, and we then deduced the ground state correlators in (16.85, 16.86) by appealing to a mapping based on conformal invariance. The analysis there was specialized to the case $\mu > 0$, with $|\mu| \ll w$, but precisely the same methods also work for $\mu = 2w$. Using the same steps as those leading up to (16.86) (or to (10.95) for the quantum Ising chain), we can obtain the following $T = 0$ correlators of H_{12} at $\lambda = 0$, $J_2 = 0$ [327, 420]:

$$\langle\hat{\sigma}_i^x\hat{\sigma}_{i+n}^x\rangle = \langle\hat{\sigma}_i^y\hat{\sigma}_{i+n}^y\rangle = (-1)^n\frac{8(G_I(0))^2}{(2\pi n)^{1/2}} \quad \text{as } n \to \infty, \tag{20.4}$$

where the numerical constant $G_I(0)$ was defined in (10.51) and its value was quoted above (10.94). The leading $(-1)^n$ prefactor is as expected from the staggered spin correlations in an antiferromagnet; technically it arises from the staggered rotation of the spins in (20.2). We can also directly use the first mapping in (20.2) to obtain correlators of $\hat{\sigma}^z$ quite simply:

$$\langle \hat{\sigma}_i^z \hat{\sigma}_{i+n}^z \rangle = \delta_{n,0} + (1 - \delta_{n,0}) \left(-\frac{2}{\pi^2 n^2} + \frac{2\cos(\pi n)}{\pi^2 n^2} \right). \tag{20.5}$$

This section obtains the power-law decays in (20.4) and (20.5) by the bosonization method [308, 309, 493, 505]. However, this approach abandons attempts to keep track of most of the prefactors (only the prefactor of the nonoscillating $1/n^2$ decay of the conserved z component of the spin in (20.5) is obtained exactly). This "sloppiness" is compensated by the important advantage that the method applies for nonzero λ and J_2. Further, the validity of the conformal mapping between $T > 0$ and $T = 0$ correlators noted above is explicitly demonstrated.

We begin by taking the continuum limit of H_{12} in (20.3) at $J_2 = \lambda = 0$ in precisely the same manner as discussed in Section 16.2.2 for the Fermi liquid region of Fig. 16.2. With lattice spacing a, we introduce the continuum Fermi field $\Psi_F(x, \tau)$ as in (10.23) and then parameterize it in terms of left (Ψ_L) and right (Ψ_R) moving excitations in the vicinity of the Fermi points as in (16.28). The fermion band is half filled, and so in this case $k_F = \pi/a$. The fields $\Psi_{L,R}$ are described by the simple Hamiltonian

$$H_{FL} = -iv_F \int dx \left(\Psi_R^\dagger \frac{\partial \Psi_R}{\partial x} - \Psi_L^\dagger \frac{\partial \Psi_L}{\partial x} \right), \tag{20.6}$$

which corresponds to the Lagrangean $\mathcal{L}_{FL}$ in (16.29); the Fermi velocity is given by $v_F = 4J_1 a$.

We examine $\mathcal{L}_{FL}$ a bit more carefully and show, somewhat surprisingly, that it can also be interpreted as a theory of free relativistic bosons. The mapping can be rather precisely demonstrated by placing $\mathcal{L}_{FL}$ on a system of finite length L. We choose to place antiperiodic boundary conditions on the Fermi fields, $\Psi_{L,R}(x + L) = -\Psi_{L,R}(x)$; this arbitrary choice will not affect the thermodynamic limit $L \to \infty$, which is ultimately all we are interested in. We can expand $\Psi_{L,R}$ in Fourier modes

$$\Psi_R(x) = \frac{1}{\sqrt{L}} \sum_{n=-\infty}^{\infty} \Psi_{Rn} e^{i(2n-1)\pi x/L}, \tag{20.7}$$

and similarly for Ψ_L. The Fourier components obey canonical Fermi commutation relations $\{\Psi_{Rn}, \Psi_{Rn'}^\dagger\} = \delta_{nn'}$ and are described by the simple Hamiltonian

$$\tilde{H}_R = \frac{\pi v_F}{L} \sum_{n=-\infty}^{\infty} (2n-1) \Psi_{Rn}^\dagger \Psi_{Rn} - E_0, \tag{20.8}$$

where the tilde in $\tilde{H}_R$ has been introduced to prevent confusion with the rotor Hamiltonian (11.1), and E_0 is an arbitrary constant setting the zero of energy, which we adjust to make the ground state energy of H_R exactly equal to 0; very similar manipulations apply to the left-movers Ψ_L. The ground state of H_R has all fermion states with

$n > 0$ empty, while those with $n \leq 0$ are occupied. We also define the total fermion number ("charge"), Q_R, of any state by the expression

$$Q_R = \sum_n : \Psi_{Rn}^\dagger \Psi_{Rn} : .$$ (20.9)

The colons are the so-called normal-ordering symbol – they simply indicate that the operator enclosed between them should include a c-number subtraction of its expectation value in the ground state of $\tilde{H}_R$, which of course ensures that $Q_R = 0$ in the ground state. Note that Q_R commutes with $\tilde{H}_R$ and so we need only consider states with definite Q_R, which allows us to treat Q_R as simply an integer. The partition function, Z_R, of $\tilde{H}_R$ at a temperature T is then easily computed to be

$$Z_R = \prod_{n=1}^{\infty} (1 + q^{2n-1})^2,$$ (20.10)

where

$$q \equiv e^{-\pi v_F / TL}.$$ (20.11)

The square in (20.10) arises from the precisely equal contributions from the states with n and $-n + 1$ in (20.8) after the ground state energy E_0 has been subtracted out.

We can provide an entirely different interpretation of the partition function Z_R. Instead of thinking in terms of occupation numbers of individual fermion states, let us focus on particle–hole excitations. We create a particle–hole excitation of "momentum" $n > 0$ above any fermion state by taking a fermion in an occupied state n' and moving it to the unoccupied fermion state $n' + n$. Clearly the energy change in such a transformation is $2n\pi v_F / L$ and is independent of the value of n'. This independence of n' is a crucial property and is largely responsible for the results that follow. It is a consequence of the linear fermion dispersion in (16.28), and of being in $d = 1$. We interpret the creation of such a particle–hole excitation as being equivalent to the occupation of a state with energy $2n\pi v_F / L$ created by the canonical boson operator $b_{Rn}^\dagger$. We can place an arbitrary number of bosons in this state, and we now show how this is compatible with the multiplicity of the particle–hole excitations that can be created in the fermionic language.

The key observation is that there is a precise one-to-one mapping between the fermionic labeling of the states and those specified by the bosons creating particle–hole excitations. Take any fermion state, $|F\rangle$, with an arbitrary set of fermion occupation numbers and charge Q_R. We uniquely associate this state with a set of particle–hole excitations above a particular fermion state we label $|Q_R\rangle$; this is the state with the lowest possible energy in the sector of states with charge Q_R, that is, $|Q_R\rangle$ has all fermion states with $n \leq Q_R$ occupied and all others unoccupied. The energy of $|Q_R\rangle$ is

$$\frac{\pi v_F}{L} \sum_{n=1}^{|Q_R|} (2n - 1) = \frac{\pi v_F Q_R^2}{L}.$$ (20.12)

To obtain the arbitrary fermion state, $|F\rangle$, with charge Q_R, first take the fermion in the "topmost" occupied state in $|Q_R\rangle$, (i.e. the state with $n = Q_R$) and move it to the topmost occupied state in $|F\rangle$ (see Fig. 20.1). Perform the same operation on the fermion in

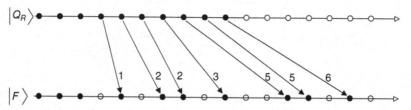

Fig. 20.1 Sequence of particle–hole excitations (bosons b_{Rn}) by which one can obtain an arbitrary fermion state $|F\rangle$ from the state $|Q_R\rangle$, which is the lowest energy state with charge Q_R. The filled (open) circles represent occupied (unoccupied) fermion states with energies that increase in units of $2\pi v_F/L$ to the right. The arrows represent bosonic excitations, b_{Rn}, with the integer representing the value of n. Note that the bosons act in descending order in energy upon the descending sequence of occupied states in $|Q_R\rangle$.

$n = Q_R - 1$ by moving it to the next lowest occupied state in $|F\rangle$. Finally, repeat until the state $|F\rangle$ is obtained. This order of occupying the boson particle–hole excitations ensures that the $b_{Rn}^{\dagger}$ act in descending order in n. Such an ordering allows one easily to show that the mapping is invertible and one-to-one. Given any set of occupied boson states, $\{n\}$, and a charge Q_R, we start with the state $|Q_R\rangle$ and act on it with the set of Bose operators in the same descending order; their ordering ensures that it is always possible to create such particle–hole excitations from the fermionic state, and one is never removing a fermion from an unoccupied state or adding it to an occupied state. The gist of these simple arguments is that the states of the many-fermion Hamiltonian $\tilde{H}_R$ in (20.8) are in one-to-one correspondence with the many-boson Hamiltonian

$$\tilde{H}'_R = \frac{\pi v_F Q_R^2}{L} + \frac{2\pi v_F}{L} \sum_{n=1}^{\infty} n b_{Rn}^{\dagger} b_{Rn}, \tag{20.13}$$

where Q_R can take an arbitrary integer value. It is straightforward to compute the partition function of $\tilde{H}'_R$ and we find

$$Z'_R = \left[\prod_{n=1}^{\infty} \frac{1}{(1 - q^{2n})} \right] \left[\sum_{Q_R=-\infty}^{\infty} q^{Q_R^2} \right]. \tag{20.14}$$

Our pictorial arguments above prove that we must have $Z_R = Z'_R$. That this is the case is an identity from the theory of elliptic functions. (The reader is invited to verify that the expressions (20.10) and (20.14) generate identical power series expansions in q.)

The above gives an appealing picture of bosonization at the level of states and energy levels, but we want to extend it to include operators. To this end, we consider the operator $\rho_R(x)$ representing the normal-ordered fermion density:

$$\rho_R(x) =: \Psi_R^{\dagger}(x)\Psi_R(x) := \frac{Q_R}{L} + \frac{1}{L} \sum_{n \neq 0} \rho_{Rn} e^{i2n\pi x/L}, \tag{20.15}$$

where the last step is a Fourier expansion of $\rho_R(x)$; the zero wavevector component is Q_R/L, while nonzero wavevector terms have coefficient ρ_{Rn}. The commutation relations of the ρ_{Rn} are central to our subsequent considerations and require careful evaluation;

we have

$$[\rho_{Rn}, \rho_{R-n'}] = \sum_{n_1, n_2} \left[\Psi^\dagger_{Rn_1} \Psi_{Rn_1+n}, \Psi^\dagger_{Rn_2} \Psi_{Rn_2-n'} \right]$$

$$= \sum_{n_2} \left(\Psi^\dagger_{Rn_2-n} \Psi_{Rn_2-n'} - \Psi^\dagger_{Rn_2} \Psi_{Rn_2+n-n'} \right). \tag{20.16}$$

It may appear that a simple change of variables in the summation over the second term in (20.16) ($n_2 \to n_2 + n$) shows that it equals the first, and so the combined expression vanishes. However, this is incorrect because it is dangerous to change variables in expressions that involve the summation over all integer values of n_2 and are therefore individually divergent; rather, we should first decide upon a physically motivated large-momentum cut-off that will make each term finite and then perform the subtraction. We know that the linear spectrum in (20.8) holds only for a limited range of momenta, and for sufficiently large $|n|$, lattice corrections to the dispersion become important. However, in the low-energy limit of interest here, the high fermionic states at such momenta are rarely, if ever, excited from their ground state configurations. We can use this fact to our advantage by explicitly sub-tracting the ground state expectation value ("normal-order") from every fermionic bilinear we consider; the fluctuations are then practically zero for the high-energy states in both the linear spectrum model (20.8) and the actual physical systems, and only the low-energy states, where (20.8) is actually a good model, matter. After such normal-ordering, the sum-mation over both terms in (20.15) is well defined and we are free to change the summation variable. As a result, the normal-ordered terms then do indeed cancel, and the expression (20.16) reduces to

$$[\rho_{Rn}, \rho_{R-n'}] = \delta_{nn'} \sum_{n_2} \left(\langle \Psi^\dagger_{Rn_2-n} \Psi_{Rn_2-n} \rangle - \langle \Psi^\dagger_{Rn_2} \Psi_{Rn_2} \rangle \right)$$

$$= \delta_{nn'} n. \tag{20.17}$$

This key result shows that the only nonzero commutator is between ρ_{Rn} and ρ_{R-n} and that it is simply the number n. By a suitable rescaling of the ρ_{Rn} it should be evident that we can associate them with canonical bosonic creation and annihilation operators. We will not do this explicitly but simply work directly with the ρ_{Rn} as a set of operators obeying the defining commutation relation (20.17), without making explicit reference to the fermionic relation (20.15). We assert that the Hamiltonians $\tilde{H}_R$, $\tilde{H}'_R$ are equivalent to

$$\tilde{H}''_R = \frac{\pi v_F Q_R^2}{L} + \frac{2\pi v_F}{L} \sum_{n=1}^{\infty} \rho_{R-n} \rho_{Rn}. \tag{20.18}$$

This assertion is simple to prove. First, it is clear from the commutation relations (20.17) that the eigenvalues and degeneracies of (20.18) are the same as those of (20.13). (The individual states are however not the same; there is a complicated linear relation between them, which is not difficult to reconstruct from our definitions of the operators ρ_{Rn} and b_{Rn}.) Second, the definition (20.18) and the commutation relations (20.17) imply that

$$\left[\tilde{H}''_R, \rho_{R-n} \right] = \frac{2\pi v_F n}{L} \rho_{R-n}. \tag{20.19}$$

Precisely the same commutation relation follows from the fermionic form (20.8) and the definition (20.15).

We have completed a significant part of the bosonization program. We have the "bosonic" Hamiltonian in (20.18) in terms of the operators ρ_{Rn}, which obey (20.17), and we also have the simple explicit relation (20.15) to the fermionic fields. Before proceeding further, we introduce some notation that will allow us to recast the results obtained so far in a compact, local, and physically transparent notation. We combine the operators ρ_{Rn} and ρ_{Ln} (the Fourier components of the left-moving fermions Ψ_L) into two local fields $\phi(x)$ and $\theta(x)$, defined by

$$\phi(x) = -\phi_0 + \frac{\pi Q x}{L} - \frac{i}{2} \sum_{n \neq 0} \frac{e^{i2n\pi x/L}}{n} [\rho_{Rn} + \rho_{Ln}],$$

$$\theta(x) = -\theta_0 + \frac{\pi J x}{L} - \frac{i}{2} \sum_{n \neq 0} \frac{e^{i2n\pi x/L}}{n} [\rho_{Rn} - \rho_{Ln}],$$

where $Q = Q_R + Q_L$ is the total charge, $J = Q_R - Q_L$, and ϕ_0 and θ_0 are a pair of angular variables that are canonically conjugate to J and Q, respectively; that is, the only nonvanishing commutation relations between the operators on the right-hand sides of (20.20) and (20.17) are $[\phi_0, J] = i$ and $[\theta_0, Q] = i$. Our objective in introducing these is to produce a number of simple and elegant results. First, using (20.20), and the commutators just noted, we have

$$[\nabla\phi(x), \theta(y)] = [\nabla\theta(x), \phi(y)] = i\pi\delta(x - y). \tag{20.20}$$

Second, (20.18) can now be written in the compact, local form

$$\tilde{H}''_R + \tilde{H}''_L = \frac{v_F}{2\pi} \int_0^L dx \left[\frac{1}{K}(\nabla\phi)^2 + K(\nabla\theta)^2 \right], \tag{20.21}$$

where the dimensionless coupling K has been introduced for future convenience; in the present situation $K = 1$, but we see later that moving away from H_{XX} to more general H_{12} leads to other values of K. The expressions (20.21) and (20.20) can be taken as defining relations, and we could have derived all the properties of the ρ_{Rn}, ρ_{Ln}, θ_0, ϕ_0 as consequences of the mode expansions (20.20), which follow after imposition of the periodic boundary conditions

$$\phi(x + L) = \phi(x) + \pi Q, \qquad \theta(x + L) = \theta(x) + \pi J. \tag{20.22}$$

These conditions show that $\phi(x)$ and $\theta(x)$ are to be interpreted as angular variables. Our final version of the bosonic form of $\tilde{H}_R + \tilde{H}_L$ in (20.8) is contained in (20.20), (20.21), and (20.22), and the two formulations are logically exactly equivalent. The Hilbert space splits into sectors defined by the integers $Q = Q_R + Q_L$, $J = Q_R - Q_L$ (and so $(-1)^Q = (-1)^J$), which measure the total charge of the left- and right-moving fermions. All fluctuations in each sector are defined by the fluctuations of the local angular bosonic fields $\phi(x)$ and $\theta(x)$, or equivalently by the fermionic fields $\Psi_R(x)$ and $\Psi_L(x)$.

We are going to make extensive use of the fields $\phi(x)$, $\theta(x)$ in the following, and so their physical interpretation would be useful. First, the field ϕ has nothing to do with the

O(N) order parameter ϕ_α used in other chapters of this book. Both notations are standard, and the context should prevent confusion. The meaning of ϕ follows from the derivative of (20.20), which with (20.15) gives

$$\nabla\phi(x) = \pi\rho(x) \equiv \pi(\rho_R(x) + \rho_L(x)). \tag{20.23}$$

So the gradient of ϕ measures the total density of particles, and $\phi(x)$ increases by π each time x passes through a particle. The expression (20.23) also shows that we can interpret $\phi(x)$ as the displacement of the particle at position x from a reference state in which the particles are equally spaced as in a crystal; that is, $\phi(x)$ is something like a phonon displacement operator whose divergence is equal to the local change in density. Turning to $\theta(x)$, one interpretation follows from (20.20), which shows that $\Pi_\phi(x) \equiv -\nabla\theta(x)/\pi$ is the canonically conjugate momentum variable to the field $\phi(x)$. In other words, Π_ϕ^2 is the kinetic energy associated with the "phonon" displacement $\phi(x)$. Using this interpretation, we can easily apply the methods of Chapter 3 to obtain the Lagrangean form of (20.21):

$$\mathcal{S}_{TL} = \frac{1}{2\pi K v_F} \int dx d\tau \big[(\partial_\tau\phi)^2 + v_F^2(\nabla\phi)^2 \big], \tag{20.24}$$

where the subscript TL represents Tomonaga–Luttinger. This is just the action of a free, massless, relativistic scalar field. Conversely, we also have a "dual" formulation of $\mathcal{S}_{TL}$ in which we interpret $\theta(x)$ as the fundamental degree of freedom and $\Pi_\theta \equiv -\nabla\phi/\pi$ as its canonically conjugate momentum; then we obtain the same action but with $K \to 1/K$

$$\mathcal{S}_{TL} = \frac{K}{2\pi v_F} \int dx d\tau \big[(\partial_\tau\theta)^2 + v_F^2(\nabla\theta)^2 \big], \tag{20.25}$$

for $\tilde{H}_R + \tilde{H}_L$. In this approach a direct physical interpretation of $\theta(x)$ is lacking; we see below that we can interpret it as an angular variable corresponding to the O(2) order-parameter correlations associated with the antiferromagnet H_{XX} in (16.6). In particular we find $\hat{\sigma}_+ \sim (\hat{\sigma}_x + i\hat{\sigma}_y) \sim (\phi_1 + i\phi_2) \sim (n_1 + in_2) \sim e^{i\theta}$ (here we have used the notation of Part II, where $\vec{\phi}$ and $\mathbf{n}$ represent an O(2) order parameter). Thus a slowly varying θ corresponds to ordering in the x–y plane in the original antiferromagnet. Also, if, as in Section 16.1, we interpret the $S = 1/2$ antiferromagnet as a hard-core Bose gas, then $e^{i\theta}$ is the superfluid order parameter. Another important property of θ is obtained by taking the gradient of (20.20), and we obtain the analog of (20.23):

$$\nabla\theta(x) = \pi(\rho_R(x) - \rho_L(x)); \tag{20.26}$$

hence gradients of θ measure the difference in density of right- and left-moving particles.

Let us also note a representation of the functional integrals in (20.24) and (20.25) which integrates over both θ and ϕ in a democratic manner. Such a representation is an integral over phase space, rather than configuration space, and has the action

$$\mathcal{S}_{TL} = \frac{v_F}{2\pi} \int dx d\tau \left[\frac{(\nabla\phi)^2}{K} + K(\nabla\theta)^2 \right] + i\pi \int dx d\tau \nabla\theta \partial_\tau\phi. \tag{20.27}$$

The final links in the bosonization procedure are expressions for the fermionic fields $\Psi_{R,L}(x)$ in terms of $\phi(x)$ and $\theta(x)$. The details of such a representation depend upon microscopic features of the particular model under consideration. These have been worked out explicitly for a fermion Hamiltonian known as the Luttinger model [189]; here we are considering H_{XX}, and for this more general case we are satisfied by an operator correspondence that gets the correct long-distance behavior but abandons attempts to get prefactors like those in (20.4) correct (for recent progress in computing prefactors for H_{12} at $J_2 = 0$ see [11, 306]). With this limited aim, the basic result can be obtained by some simple general arguments. First, note that if we annihilate a particle at the position x, from (20.23) the value of $\phi(y)$ at all $y < x$ has to be shifted by π. Such a shift is produced by the exponential of the canonically conjugate momentum operator Π_ϕ:

$$\exp\left(i\pi \int_{-\infty}^x \Pi_\phi(y)dy\right) = \exp\left(-i\theta(x)\right). \tag{20.28}$$

However, it is not sufficient to merely create a particle. We are creating a fermion, and the fermionic antisymmetry of the wavefunction can be accounted for if we pick up a minus sign for every particle to the left of x, that is, with a Jordan–Wigner-like factor

$$\exp\left(im\pi \int_{-\infty}^x \Psi_F^\dagger(y)\Psi_F(y)dy\right) = \exp\left(imk_Fx + im\phi(x)\right), \tag{20.29}$$

where m is any odd integer, and $\Psi_F^\dagger \Psi_F$ measures the *total* density of fermions (see (10.23)), including the contributions well away from the Fermi points. In the second expression in (20.29), the term proportional to k_F represents the density in the ground state, while $\phi(x)$ is the integral of the density fluctuation above that. Combining the arguments leading to (20.28) and (20.29) we can assert the basic operator correspondence [213–215]

$$\Psi_F(x) = \sum_{m \text{ odd}} A_m e^{imk_Fx + im\phi(x) - i\theta(x)}, \tag{20.30}$$

where the A_m are a series of unknown constants, which depend upon microscopic details. We see shortly that the leading contribution to (20.30) comes from the terms with $m = \pm 1$, and the remaining terms are subdominant at long distances. Comparison with (16.28) shows clearly that we may make the operator identifications for the right- and left-moving continuum Fermion fields

$$\Psi_R \sim e^{-i\theta + i\phi}, \qquad \Psi_L \sim e^{-i\theta - i\phi}. \tag{20.31}$$

The other terms in (20.30) arise when these basic fermionic excitations are combined with particle–hole excitations at wavevectors that are integer multiples of $2k_F$.

Similar arguments, and the above expressions, can also be applied to spin operators $\hat{\sigma}^\alpha$ via the relations (20.2). For $\hat{\sigma}^+$ the arguments are as above except that the "string" factor in (20.2) exactly compensates for the change in sign discussed above. We have then

$$\hat{\sigma}_j^+ = (-1)^j \sum_{m \text{ even}} B_m e^{imk_Fx_j + im\phi(x_j) - i\theta(x_j)}, \tag{20.32}$$

for some unknown B_m. The most important term in this expansion is $m = 0$, so that

$$\hat{\sigma}_j^+ \sim (-1)^j e^{i\theta}, \tag{20.33}$$

establishing our earlier claim of $e^{i\theta}$ as the order parameter for x–y spin correlations. Finally, $\hat{\sigma}^z$ is related to the fermion density, the slowly varying component of which can be reconstructed from (20.23), while additional contributions come from evaluating $\Psi_F^\dagger(x)\Psi_F(x)$ using (20.30):

$$\frac{\hat{\sigma}_j^z}{a} = -\frac{2}{\pi}\nabla\phi(x_j) + \sum_{m\neq 0,\text{even}} C_m e^{imk_F x_j + im\phi(x_j)}, \tag{20.34}$$

for some C_m, with a the lattice spacing. Note that the coefficient of the slowly varying term (which does not oscillate at a multiple of the wavevector k_F) is precisely determined; this is ultimately related to the fact that $\sum_j \hat{\sigma}_j^z = -2Q$ commutes with the Hamiltonian.

We have completed our derivation of the bosonization technology and are ready to apply it to obtain new results. The basic result is the equivalency of the fermionic Hamiltonian (20.8), and its left-moving partner, to the bosonic theory defined by (20.20), (20.21), and (20.22). Also key are the operator correspondences in (20.23), (20.30), (20.32), and (20.34).

We turn to the evaluation of the correlators of the spin operators, (20.32) and (20.34), under the theory (20.21), (20.24), or (20.25). These can be obtained by use of the basic identity

$$\langle e^{i\mathcal{O}} \rangle = e^{-\langle \mathcal{O}^2 \rangle/2}, \tag{20.35}$$

where $\mathcal{O}$ is an arbitrary linear combination of ϕ and θ fields at different spacetime points; this identity is a simple consequence of the free-field (Gaussian) nature of (20.21). In particular, all results can be reconstructed by combining (20.35) with repeated application of some elementary correlators. The first of these is the two-point correlator of ϕ:

$$\frac{1}{2}\langle(\phi(x,\tau) - \phi(0,0))^2\rangle = \pi v_F K \int \frac{dk}{2\pi} T \sum_{\omega_n} \frac{1 - e^{i(kx-\omega_n\tau)}}{\omega_n^2 + v_F k^2}$$
$$= \frac{K}{4} \ln\left[\frac{\cosh(2\pi T x/v_F) - \cos(2\pi T\tau)}{(2\pi T/v_F\Lambda)^2}\right], \tag{20.36}$$

where Λ is a large-momentum cutoff. Similarly, we have for θ, the correlator

$$\frac{1}{2}\langle(\theta(x,\tau) - \theta(0,0))^2\rangle = \frac{1}{4K} \ln\left[\frac{\cosh(2\pi T x/v_F) - \cos(2\pi T\tau)}{(2\pi T/v_F\Lambda)^2}\right]. \tag{20.37}$$

To obtain the θ, ϕ correlator we use the relation $\Pi_\phi = -\nabla\theta/\pi$ and the equation of motion $i\Pi_\phi = \partial_\tau\phi/(\pi v_F K)$ that follows from the Hamiltonian (20.21); then by an integration and differentiation of (20.36) we can obtain

$$\langle\theta(x,\tau)\phi(0,0)\rangle = \frac{i}{2}\arctan\left[\frac{\tan(\pi T\tau)}{\tanh(\pi T x/v_F)}\right]. \tag{20.38}$$

This expression can also be obtained directly from (20.27).

Applying (20.35) and (20.37) to (20.33) we get

$$(-1)^j\langle\hat{\sigma}_j^+(\tau)\hat{\sigma}_0^-(0)\rangle \sim \left[\frac{T^2}{\sin(\pi T(\tau + ix_j/v_F))\sin(\pi T(\tau - ix_j/v_F))}\right]^{\frac{1}{4K}}. \tag{20.39}$$

At the value $K = 1$ for H_{XX}, this agrees precisely with the result claimed earlier in (16.76) and (16.86). In this previous case we had obtained the $T > 0$ crossover functions by appealing to the mapping (10.47) between $T = 0$ and $T > 0$ correlators, which was claimed to be a consequence of the conformal invariance of the low-energy theory. Here we have shown that the low-energy theory is given by (20.24) or (20.25) and that its $T = 0$ and $T > 0$ correlators are indeed related by (10.47). Very similar arguments can also be advanced by a bosonization analysis of the quantum Ising chain to establish (10.47) for the model of Chapter 10. At $T = 0$, (20.39) gives an equal-time correlator that decays as $1/\sqrt{x}$, which is in agreement with the exact result (20.4). We can also obtain the form of the subleading terms by considering the correlators of the complete expression (20.32). At $T = 0$ we have the following structure in the asymptotic expansion of the equal-time correlators:

$$(-1)^j \langle \hat{\sigma}_j^+ \hat{\sigma}_0^- \rangle = \sum_{m=0}^{\infty} \frac{\tilde{B}_m}{(x_j)^{2m^2 K + 1/(2K)}} \cos(2mk_F x_j), \tag{20.40}$$

for some unknown coefficients $\tilde{B}_m$. Note, as claimed earlier, the $m > 0$ terms all decay faster than the dominant $m = 0$ term.

Precisely the same methods can be applied to the correlators of $\hat{\sigma}^z$. From (20.34), the analog of the expansion (20.40) for the $T = 0$ equal-time correlator is

$$\frac{1}{a^2} \langle \hat{\sigma}_j^z \hat{\sigma}_0^z \rangle = -\frac{2K}{\pi^2 x^2} + \sum_{m=1}^{\infty} \frac{\tilde{C}_m}{(x_j)^{2m^2 K}} \cos(2mk_F x_j), \tag{20.41}$$

for some unknown $\tilde{C}_m$. Note that the leading, nonoscillating, term agrees precisely with the first term in (20.5). For the special case of H_{XX}, we get $K = 1$, and the oscillating terms in (20.41) are in agreement with that in (20.5) for the special values $\tilde{C}_1 = 2/\pi^2$ and $\tilde{C}_{m>1} = 0$. The subleading terms in (20.41) do not appear for this special free fermion point, but there is no reason for them to vanish in the general case, which is considered in the following section.

Finally, we can also consider H_{12} in (20.3) to be a generic Hamiltonian of interacting spinless fermions in one dimension. Generally, as discussed in Chapter 18, such interacting fermion models, in greater than one dimension, are expected to be Fermi liquids, with a discontinuity in the momentum distribution function, $n(k)$ of the fermions at the Fermi wavevector; see (18.25). We now have the tools to compute $n(k)$ in one dimension. Using (20.30), and the results above for the correlation functions under (20.27) we obtain

$$\langle \Psi_F^\dagger(x) \Psi_F(0) \rangle \sim \frac{\sin(k_F |x|)}{|x|^{(K+1/K)/2}}. \tag{20.42}$$

Taking the Fourier transform of this, we conclude that $n(k)$ does indeed have a singularity at $k = k_F$, but that this singularity is not generally a step discontinuity:

$$n(k) \sim -\text{sgn}(k - k_F)|k - k_F|^{(K+1/K)/2 - 1}. \tag{20.43}$$

For $K = 1$, (20.43) does indeed have a step discontinuity. However, as demonstrated in Section 20.2, the full theory of interacting fermions in H_{12} does have phases which are adequately described by (20.27) at low energies but with $K \neq 1$. We refer to such a phase in Section 20.2 as a Tomonaga–Luttinger liquid. Thus, interpreted as a phase of interacting fermions, a Tomonaga–Luttinger liquid is a non-Fermi liquid: its momentum distribution function has a singularity at the Fermi surface, but the singularity is not a step discontinuity, and is instead given by (20.43). Note the similarity of this singularity to the non-Fermi liquids in two and higher dimensions noted in (18.44).

20.2 Phases of H_{12}

We are ready to address the properties of the Hamiltonian H_{12} in (20.1) for the case of general J_1, J_2, and λ [190]. We use exactly the same bosonization procedure developed in Section 20.2 for H_{XX} but apply it to the more interacting fermion Hamiltonian in (20.3). The first step, as in Section 20.1, is to focus on the low-energy degrees of freedom, which consist of fermionic excitations near the wavevectors $\pm k_F$. This is facilitated by taking the continuum limit of (20.3) by inserting the parameterization (10.23) and (16.28). Before doing this it is important to "normal-order" the terms in (20.3); in other words, we first perform a Hartree–Fock factorization to obtain the suitably renormalized one-particle Hamiltonian. In this manner we obtain the following continuum limit of H_{12}:

$$H_{12} = H_{FL} + H_a + H_b, \tag{20.44}$$

where H_{FL} was considered earlier in (20.6), and H_a and H_b are the two new terms arising from nonzero λ and J_2. The first of these has the form

$$H_a = 8(J_1\lambda + 2J_2)a \int dx \, [(\rho_R + \rho_L)(\rho_R + \rho_L)], \tag{20.45}$$

where ρ_R was defined in (20.15), and similarly for ρ_L; this term involves interactions in which left- or right-moving fermions scatter off each other while exchanging small momenta near the respective Fermi points. The second term, H_b, is more subtle. Its appearance relies on the special value of $k_F = \pi/2a$ demanded by a half-filled fermion band [190]. For this value of k_F, two right-moving fermions at k_F have a total momentum $2k_F = \pi/a$, which differs from the total momentum of two left-moving fermions $(-2k_F = -\pi/a)$, by a reciprocal lattice vector, $2\pi/a$. Hence it is possible to have an "umklapp" scattering event between these, as in

$$H_b = 4(J_1\lambda - 6J_2) \int dx \left[\Psi_R^\dagger \nabla\Psi_R^\dagger \Psi_L \nabla\Psi_L + \Psi_L^\dagger \nabla\Psi_L^\dagger \Psi_R \nabla\Psi_R\right]. \tag{20.46}$$

Note that this is the only instance in this chapter where the precise value of k_F has been important – all other expressions apply for general k_F and have been written as such.

We proceed to bosonize H_a and H_b using the prescriptions of Section 20.1. The case of H_a is straightforward: we use (20.23) to write H_a as

$$H_a = \frac{8(J_1\lambda + 2J_2)}{\pi^2} \int dx (\nabla\phi)^2. \tag{20.47}$$

This can be easily absorbed into the bosonized version of H_{FL} in (20.21) by a redefinition of v_F and K. In this way we have shown that the Hamiltonian $H_{FL} + H_{12}$ is equivalent to (20.21), but with the parameters

$$v_F \approx 4a \left[J_1 \left(J_1 + \frac{4(J_1\lambda + 2J_2)}{\pi} \right) \right]^{1/2},$$

$$K \approx \left[1 + \frac{4(J_1\lambda + 2J_2)}{J_1\pi} \right]^{-1/2}. \tag{20.48}$$

The values of the parameters only hold for small λ and J_2; however, the general result of a renormalization of v_F and K, but with no other change, is expected to hold more generally. Note that $K \neq 1$, but the results in Section 20.1 were quoted for general K and can now be used.

The consequences of H_b are slightly less trivial. We insert the expansions (20.30) into H_b and generate a number of terms; the most important of these arises from simply using the leading terms in (20.31), which yields

$$H_b = -v \int dx \cos(4\phi(x)) + \cdots, \tag{20.49}$$

where $v \sim (J_1\lambda - 6J_2)$. This is an important interaction modifying the simple Gaussian action in (20.21). The final bosonized version of H_{12} is then given by the action

$$\mathcal{S}_{SG} = \int dx d\tau \left[\frac{1}{2\pi K v_F} \left((\partial_\tau \phi)^2 + v_F^2 (\nabla\phi)^2 \right) - v\cos(4\phi) \right]. \tag{20.50}$$

This action represents the so-called sine–Gordon model and its properties are examined in the following subsection.

For now, let us note the physical implication of the $\cos(4\phi)$ term and some related issues. Recall from the commutation relations (20.20) that $\nabla\phi$ is canonically conjugate to the x–y order represented by the angular variable θ (see the relation above (20.25) that $\Pi_\theta = -\nabla\phi/\pi$). Then we can write the $\cos(4\phi)$ term as

$$\exp\left(-4\pi i \int_{-\infty}^{x} \Pi_\theta(y) dy \right) + h.c. \tag{20.51}$$

In this form it is clear that this operator translates $\theta \to \theta + 4\pi$ for all $y < x$. But this is the same as inducing a 4π *vortex* in the angular order parameter θ. Thus the effect of the $\cos(4\phi)$ term is to allow for 4π vortex tunneling events between different winding number sectors of the angular variable θ representing spin ordering in the x–y plane. This interpretation is also consistent with (20.26) and (20.46). In the latter equation we see that H_b turns two left-moving particles into two right-moving particles, and so by the former equation there must be a step of 4π in θ at the point this happens.

It is interesting that there is no 2π vortex event allowed above in H_{12}. We see shortly that the absence of such single vortices, and the presence only of double vortices, has some important consequences. The single 2π vortices are certainly permitted on general topological grounds, but to induce them requires modifying H_{12}. One possibility is a *staggered* exchange interaction

$$H_{12} \rightarrow H_{12} + J_3 \sum_i (-1)^i \vec{\hat{\sigma}}_i \cdot \vec{\hat{\sigma}}_{i+1}. \tag{20.52}$$

To obtain the bosonized version of this additional term, examine the structure of $\hat{\sigma}_i^+ \hat{\sigma}_{i+1}^-$ under the mapping (20.32); the staggering of the exchange means that we have to pick up the coefficient of $(-1)^i = e^{i2k_F x_i}$; this gives us the term $\int dx \sin(2\phi)$. The same term also arises from the corresponding mapping using (20.34) of the $\hat{\sigma}_i^z \hat{\sigma}_{i+1}^z$ term. So we have the operator correspondence

$$(-1)^i \vec{\hat{\sigma}}_i \cdot \vec{\hat{\sigma}}_{i+1} \sim \sin(2\phi). \tag{20.53}$$

A second possibility is a staggered field in the z direction; by a very similar argument from (20.34) we obtain the operator correspondence

$$(-1)^i \hat{\sigma}_i^z \sim \cos(2\phi). \tag{20.54}$$

The arguments in the previous paragraph show that adding either of the $\sin(2\phi)$ or $\cos(2\phi)$ terms to $\mathcal{S}_{SG}$ allows 2π vortex tunneling events. It is also interesting to note the fermionic form of these 2π tunneling events. By reversing the bosonization mapping, it is simple to see that (20.53) and (20.54) correspond to single-fermion scattering terms that turn left into right movers and vice versa and change total momentum by $2k_F$. In contrast, the original scattering term in (20.46) scattered two particles and changed momentum by $4k_F$.

20.2.1 Sine–Gordon model

We discuss some important properties of the sine–Gordon field theory $\mathcal{S}_{SG}$ in (20.50) as a function of the dimensionless coupling K and the dimensionful parameter v. The velocity v_F simply sets the relative scales of time and space but does not otherwise modify physical properties.

We have already obtained results for $\mathcal{S}_{SG}$ along the line $v = 0$. The model is a free, gapless, Gaussian field theory characterized by the following $T = 0$ equal-time correlators

$$\left\langle e^{ip\theta(x)} e^{-ip'\theta(0)} \right\rangle \sim \delta_{pp'}/x^{p^2/2K}, \qquad \left\langle e^{ip\phi(x)} e^{-ip'\phi(0)} \right\rangle \sim \delta_{pp'}/x^{p^2 K/2}; \tag{20.55}$$

for $p = p'$ these results follow directly from (20.35)–(20.37), while for $p \neq p'$ application of (20.35) leads to an infrared divergent integral in the exponent, and so the correlator vanishes. Note that these correlators are both power laws, indicating that the theory is scale invariant along the line $v = 0$ (indeed it is conformally invariant). From (20.55) we see that this is a line of critical points along which the exponents vary continuously as a function of the dimensionless parameter K. The technology of renormalization group scale transformations can therefore be applied freely at any point along this line. We can talk of scaling dimensions of operators, and the results (20.55) show that

$$\dim[e^{ip\theta}] = \frac{p^2}{4K}, \qquad \dim[e^{ip\phi}] = \frac{p^2 K}{4}. \tag{20.56}$$

Also, the relativistically invariant structure of the derivative terms in $\mathcal{S}_{SG}$ makes it clear that the dynamic exponent $z = 1$. Using this, and the scaling dimensions (20.56) for $p = 4$, we immediately obtain the scaling dimension $\dim[v] = 2 - 4K$ along the $v = 0$ line. This can be written as a renormalization group flow equation under the rescaling $\Lambda \to \Lambda e^{\ell}$:

$$\frac{dv}{d\ell} = (2 - 4K)v. \qquad (20.57)$$

So the critical fixed line $v = 0$ is stable for $K < 1/2$. However, this flow equation is not the complete story, especially when K approaches $1/2$. For $|K - 1/2| \sim |v|$ we see that the term on the right-hand side is not linear in the small parameter v, but quadratic. To be consistent, then, we also have to consider other terms of order v^2 that might arise in the flow equations. As we see below, there is a renormalization of K that appears at this order.

The flow equations at order v^2 are generated using an approach similar to that used in Section 12.1 for the $N \geq 3$ rotor model in $d = 1$. As in (12.5), we decompose the field $\phi(x, \tau)$ into a background slowly varying component $\phi_<(x, \tau)$ and a rapidly varying component $\phi_>(x, \tau)$, which is out to order v^2:

$$\phi(x, \tau) = \phi_<(x, \tau) + \phi_>(x, \tau), \qquad (20.58)$$

where $\phi_<$ has spatial Fourier components at momenta smaller than $\Lambda e^{-\ell}$, while $\phi_>$ has components between $\Lambda e^{-\ell}$ and Λ. Inserting (20.58) into (20.50), to linear order in v we generate the following effective coupling for $\phi_<$:

$$v \int d^2X \, \langle \cos(4\phi_<(X) + 4\phi_>(X)) \rangle_0$$

$$= v \int d^2X \cos(4\phi_<(X)) \left\langle e^{i4\phi_>(X)} \right\rangle_0$$

$$= v \int d^2X \cos(4\phi_<(X)) e^{-8\langle \phi_>^2 \rangle_0}$$

$$\approx v \left(1 - 4K\frac{d\Lambda}{\Lambda}\right) \int d^2X \cos(4\phi_<(X)), \qquad (20.59)$$

where $X \equiv (x, \tau)$ is a spacetime coordinate, the subscript 0 indicates an average with respect to the free $v = 0$ Gaussian action of $\phi_>$, and $d\Lambda = \Lambda(1 - e^{-\ell})$. When combined with a rescaling of coordinates $X \to Xe^{-\ell}$ to restore the cutoff to its original value, it is clear that (20.59) leads to the flow equation (20.57). The same procedure applies to quadratic order in v. As the algebra is a bit cumbersome, we only schematically indicate the steps. We generate terms such as

$$v^2 \int d^2X d^2Y \cos(4\phi_<(X) \pm 4\phi_<(Y)) \exp\left(\mp 16\langle \phi_>(X)\phi_>(Y) \rangle_0\right)$$

$$= v^2 \int d^2X d^2Y \cos(4\phi_<(X) \pm 4\phi_<(Y)) \exp\left(\mp f(X - Y)d\Lambda\right), \qquad (20.60)$$

where $f(X - Y)$ is some regularization-dependent function that decays on spatial scale $\sim \Lambda^{-1}$. For this last reason we may expand the other terms in (20.60) in powers of $X - Y$. The terms with the $+$ sign then generate a $\cos(8\phi)$ interaction; we ignore this term as the analog of the arguments used to obtain (20.57) show that it is strongly irrelevant for $K \sim 1/2$. The terms with the $-$ sign generate gradients on $\phi_<$ and therefore lead to a

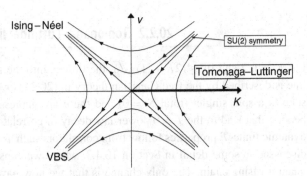

Fig. 20.2   Renormalization group flow trajectories and phase diagram of the sine–Gordon model $\mathcal{S}_{SG}$ in (20.50), as obtained
from (20.57) and (20.61). The origin is at $K = 1/2, v = 0$. The attractive fixed line $v = 0, K \geq 1/2$ controls the
Tomonaga–Luttinger liquid phase, which is described in Section 20.2.2. The points flowing off to $v \to -\infty$ are in
the VBS phase described in Section 20.2.3. Finally, the points flowing to $v \to \infty$ are in an Ising–Néel state discussed
in Section 20.2.4. The separatrices between these regions are $v = \pm(2K-1)/\sqrt{\delta}$. The line $v = (2K-1)\sqrt{\delta}$
corresponds to the $SU(2)$ symmetric H_{12} with $\lambda = 1$; different points on this line are accessed by varying J_2/J_1.

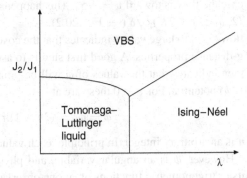

Fig. 20.3   Phase diagram of H_{12}, (20.1) deduced from the flows in Fig. 20.2 by [190]. The vertical line $\lambda = 1$ has $SU(2)$
symmetry and maps onto the line $v = (2K-1)/\sqrt{\delta}$ in Fig. 20.2. The multicritical point where all three phases meet
is the point $v = 0, K = 1/2$ in Fig. 20.2.

renormalization of K. In this manner we obtain the flow equation

$$\frac{dK}{d\ell} = -\delta v^2, \qquad (20.61)$$

where δ is a positive, regularization-dependent constant (it also depends upon K, but we
can ignore this by setting $K = 1/2$ in δ at this order).

A nearly complete understanding of the properties of $\mathcal{S}_{SG}$ follows from an analysis
of (20.57) and (20.61). The flow trajectories are shown in Fig. 20.2: they lie along the
hyperbolae $4\delta v^2 - (2 - 4K)^2 = \text{constant}$. There are three distinct possibilities for the
ultimate long-distance fate of the couplings, leading to three separate phases of $\mathcal{S}_{SG}$. We
consider each of these phases in the following subsections, followed by a discussion of the
critical lines and points between them. We also show the implications of the properties of
$\mathcal{S}_{SG}$ for a phase diagram of H_{12} in Fig. 20.3, with some needed justification to follow in
the subsections below.

20.2.2 Tomonaga–Luttinger liquid

For $K \geq 1/2$ and $|v| \leq (2K - 1)/\sqrt{\delta}$, the flow is into the fixed line $v = 0$, $K \geq 1/2$. This line is described by the free Gaussian theory in (20.21) or (20.24) or (20.25). The ground state is a spin singlet (total $S_z = 0$) and there are gapless excitations with a linear dispersion that lead to the $T = 0$ power-law decay of correlators in (20.40) and (20.41). The dynamic finite-T properties follow from correlators such as (20.39), whose properties were discussed in some detail in Section 10.4.3, where we considered the critical point of the quantum Ising chain. The only change is that we now have a general exponent K (compare (20.39) with (10.95)), but this does not make a qualitative change to the physical discussion; only some quantitative factors change, and these can be easily computed for arbitrary K.

20.2.3 Valence bond solid order

In this case the flow is toward $v = -\infty$. This happens for all $K \leq 1/2$ and $v < 0$, and for $K > 1/2$, $v < (1 - 2K)/\sqrt{\delta}$ (see Fig. 20.2).

The flow of $|v|$ to large values indicates that the $\cos(4\phi)$ term in $\mathcal{S}_{SG}$, (20.50), dominates the long-distance properties. A good first step is to assume that this is the dominant term, which then indicates that the values of ϕ will be pinned predominantly at the minima of the $\cos(4\phi)$ potential. For $v < 0$ these are at

$$\phi = \phi_n = (2n + 1)\pi/4, \tag{20.62}$$

where n is an arbitrary integer. In principle, each value of n labels a different ground state of $\mathcal{S}_{SG}$. However, ϕ is an angular variable, and physical observables depend only upon gradients or trigonometric functions of ϕ; one observable that can distinguish between the different ϕ_n is the staggered bond exchange energy in (20.53):

$$\langle (-1)^i \vec{\sigma}_i \cdot \vec{\sigma}_{i+1} \rangle \sim \sin(2\phi_n) = (-1)^n. \tag{20.63}$$

Thus there are only two distinct ground states, corresponding to even or odd values of n. There is a spontaneously broken translational symmetry in both of these states owing to the appearance of a staggering in the bond exchange energy. This is known as valence bond solid (VBS) ordering, as discussed for the $d = 2$ case in Section 19.3.3; the VBS is also referred to as a "spin-Peierls" state. A schematic of these VBS states is shown in Fig. 20.4. We emphasize that this ordering appears spontaneously in H_{12} and is not induced by a staggering of the exchange constants as in (20.52); the latter requires an explicit $\sin(2\phi)$ term in the action, which we have not included.

Fig. 20.4   The two VBS ground states of H_{12}. The thick lines represent larger values of $\langle \vec{\sigma}_i \cdot \vec{\sigma}_{i+1} \rangle$, while the unmarked near-neighbor pairs have smaller values.

We consider the excitations above either of the ground states. From the framework of the sine–Gordon theory it appears natural to parameterize $\phi(X) = \phi_n + \tilde{\phi}(X)$ and to expand the action in powers of $\tilde{\phi}$. At quadratic order the curvature at the minimum of the $\cos(4\phi)$ potential gives rise to a $\tilde{\phi}^2$ mass term, and so we can expect that there is a gap and the lowest-lying excitation is a massive $\tilde{\phi}$ particle. This expectation turns out to be incorrect, and there is an alternative massive excitation with a lower energy. For reasons we shall not fully discuss here, the most important excitation turns out to be a "soliton" (the reader can consult the book by Rajaraman [389] for further details). This is a topological excitation consisting of a localized lump at which ϕ interpolates between the two ground states. Thus, for example, we have $\phi(x \to \infty) = \phi_n$ and $\phi(x \to -\infty) = \phi_{n-1}$, and $\phi(x)$ moves between the two limits in the immediate vicinity of some point $x = x_0$. The disturbance around x_0 can move, and this constitutes a quantum particle of mass Δ/v_F^2. This solitonic particle has a lower energy than the $\tilde{\phi}$ particle for $K > 1/8$, and so, in keeping with the general notation in this book, we have used the symbol Δ for the energy gap of the VBS state (the action $\mathcal{S}_{SG}$ is relativistically invariant and so the energy–momentum dispersion of the solitonic particle is $\varepsilon_k = (\Delta^2 + v_F^2 k^2)^{1/2}$). The $\tilde{\phi}$ particle can be considered as a soliton/antisoliton bound state, and it is found to be stable toward decay into a pair of widely separated soliton and antisoliton particles only for $K < 1/4$. In any case, the low-temperature properties are dominated by those of a dilute gas of solitons and antisolitons for all $K > 1/8$.

It is also useful to have an interpretation of the soliton in terms of the underlying spin Hamiltonian H_{12} [468]. Note that each soliton involves a change $\Delta\phi = \pm\pi/2$. By the relation (20.23) between gradients of ϕ and the charge density, we see that each soliton carries a charge $Q = \pm 1/2$. This is to be contrasted with the charge $Q = \pm 1$ carried by the underlying Jordan–Wigner fermion Ψ_F. Of course this charge is also equal to the total spin S^z, and so the soliton is an $S_z = \pm 1/2$ particle – a spinon, in the terminology of Section 19.3.4. This suggests the simple pictorial representation shown in Fig. 20.5: the domain between the two VBS states requires a shift in the singlet bonds by one site, leading to a free $S_z = 1/2$ spin at the boundary.

Note that the $\cos(4\phi)$ in the action, representing tunneling only by $\pm 4\pi$ vortices, was crucial for the existence of free spinons. If we had an explicit staggering of the exchange constants, as in (20.52), the resulting action would allow $\pm 2\pi$ vortices with a corresponding $\cos(2\phi)$ term in the action, and a solitonic analysis similar to the one above would show that excitations were particles with *integer* spin. This confinement of spinons is also easy to understand from the pictorial representation in Fig. 20.5, as the explicit staggering would lead to an energy cost proportional to the length of the "wrong" domain between two spinons.

We turn to the low-temperature static and dynamic properties of this VBS phase. As already noted, these are dominated by a dilute gas of $S_z = \pm 1/2$ particles. The latter system can be analyzed using a method essentially identical to that employed in Section 12.2

Fig. 20.5   A $Q = 1/2$ spinon excitation interpolating between the two VBS ground states of Fig. 20.4.

for the low-temperature properties of the $d = 1$, O(3) quantum rotor model. In the latter case, we had particles with $S_z = 1, 0, -1$; this is one of the main substantive differences, and the presence here of particles with $S_z = 1/2, -1/2$ only leads to simple changes in various numerical prefactors, for the physical properties of the transport of magnetization density are identical to those discussed in Section 12.2. In particular, the spinon collisions are described by the low-momentum S matrix in (12.13), with the m_1, m_2, m'_1, m'_2 now taking the values $\pm 1/2$. (The arguments for this key property are the same as those presented below (12.13).) A second important difference is that the spin structure factor is not given by a single-particle propagator as in (12.28)–(12.30); instead we have to consider a convolution of two single-particle propagators, as in (19.72)–(19.75).

An explicit demonstration of the existence of the $S_z = \pm 1/2$ spinons in this phase can be given at the special value $K = 1/4$. This relies on a commonly used trick of "refermionization" of sine–Gordon-like field theories in $d = 1$, and this appears to be a convenient occasion to introduce it. Consider the fermionic fields

$$\psi_R \sim e^{-i\theta/2 + i2\phi}, \qquad \psi_L \sim e^{-i\theta/2 - i2\phi}. \tag{20.64}$$

Note that $\theta/2$ and 2ϕ obey the same commutation relations as those in (20.20), and so by working backwards through the arguments leading to (20.31), we see that $\psi_{R,L}$ are indeed fermionic operators annihilating particles with a linear dispersion. By the same arguments as those leading to (20.21) we may conclude that

$$-iv_F \int dx \left(\psi_R^\dagger \frac{\partial \psi_R}{\partial x} - \psi_L^\dagger \frac{\partial \psi_L}{\partial x} \right) = \frac{v_F}{2\pi} \int dx \left[4(\nabla\phi)^2 + \frac{1}{4}(\nabla\theta)^2 \right]. \tag{20.65}$$

However, this is precisely the Hamiltonian corresponding to the gradient terms in $\mathcal{S}_{SG}$ at $K = 1/4$. Furthermore, it is easy to see from (20.64) that the cosine term in $\mathcal{S}_{SG}$ can be obtained by *bilinear* combinations of the $\psi_{L,R}$. So we have the remarkable result that, at $K = 1/4$, $\mathcal{S}_{SG}$ is equivalent to the free fermion Hamiltonian

$$\int dx \left(-iv_F \psi_R^\dagger \frac{\partial \psi_R}{\partial x} + iv_F \psi_L^\dagger \frac{\partial \psi_L}{\partial x} + \frac{\Delta}{v_F} \left(\psi_R^\dagger \psi_L + \psi_L^\dagger \psi_R \right) \right), \tag{20.66}$$

where $\Delta \sim v$ multiplies a term arising from $\cos(4\phi)$ in $\mathcal{S}_{SG}$. However, (20.66) describes a free massive Dirac particle in $d = 1$. Also note that an identity analogous to (20.23) is

$$\frac{1}{2}(: \psi_R^\dagger \psi_R : + : \psi_L^\dagger \psi_L :) = \frac{1}{\pi} \nabla\phi; \tag{20.67}$$

the leading $1/2$ shows that the Dirac particle/antiparticles carry charges $\pm 1/2$ and identifies them as the spinons.

An important caution about the discussion above at $K = 1/4$ is in order. While the free Dirac particle mapping gives an appropriate picture of the elementary excitations above the ground state, its naive extension to $T > 0$ properties is quite misleading. In particular, if the spinons were really free, their two-particle S matrix for the collision in Fig. 12.2 would take the form

$$S^{m_1, m_2}_{m'_1, m'_2} = (-1)\delta_{m_1 m_2}\delta_{m'_1 m'_2}; \tag{20.68}$$

here we have included the (-1) arising from the exchange of two fermions explicitly in the S matrix. Comparing this with (12.13), we see a crucial difference in the structure of the spin indices: the spins are now "passing through" the collision, rather than "bouncing off." In fact (20.68) is never the appropriate result for any realistic condensed-matter system, and (12.13) always applies at low momenta. The important point is that it is not possible to ignore additional "irrelevant" terms not explicitly included in $\mathcal{S}_{SG}$. When these terms are carried through the refermionization above, they invariably lead to some four-fermion scattering terms; such terms are always important in the scattering of massive particles in $d = 1$, as discussed below (12.13), and lead to the "super-universal" S matrix in (12.13).

20.2.4 Néel order

Now the flow is toward $v = +\infty$. This happens for all $K \leq 1/2$ and $v > 0$, and for $K > 1/2$, $v > (2K - 1)/\sqrt{\delta}$ (see Fig. 20.2).

The reasoning then closely parallels that in Section 20.2.3 for $v \to -\infty$. The important minima of the $\cos(4\phi)$ potential are at

$$\phi = \tilde{\phi}_n = n\pi/2. \tag{20.69}$$

The physical properties of these minima are distinguished by the expectation value

$$\langle (-1)^i \hat{\sigma}_i^z \rangle \sim \cos(2\tilde{\phi}_n) = (-1)^n. \tag{20.70}$$

Thus there is a spontaneously broken symmetry characterized by a staggered expectation value in the z component of the spins. This is a Néel state with an Ising symmetry; it is to be contrasted with the Néel state in Section 19.3.3 in which the staggered moment could point in any direction in spin space. Here the anisotropy in the Hamiltonian picks out the z direction as a preferred one, and there is only a twofold degeneracy in the resulting Ising/Néel ground state. (Note that a fully isotropic Néel state is not possible in $d = 1$, as was indicated in Section 19.3.2, and discussed further below in Section 20.2.5.)

Apart from the shift in the minima of the cosine potential from (20.62) to (20.69) (and the resulting difference in the physical interpretation of the broken symmetry of the ground state), there is essentially no difference in the analysis of the fluctuations here from that in Section 20.2.3; indeed, we can map $v \to -v$ in $\mathcal{S}_{SG}$ by the shift $\phi \to \phi + \pi/4$. For $K > 1/8$ the lowest lying excitations are massive $S_z = \pm 1/2$ spinons, which interpolate between the two Ising/Néel ground states. Their collisions are described at low momenta by (12.13), and the low-T properties are as in Section 12.2 with the modifications noted above in Section 20.2.3.

20.2.5 Models with $SU(2)$ (Heisenberg) symmetry

Here we focus on the special point $\lambda = 1$ in H_{12}, where the Hamiltonian has full $SU(2)$ symmetry. We have argued in Sections 19.3.1 and 19.3.2 that this model should also be described by the $d = 1$, O(3) nonlinear sigma model (19.51) with an additional topological term (19.53) at $\theta = \pi$ (this θ represents the coefficient of the topological term and should not be confused with the angular bosonization field θ used elsewhere in this chapter).

The latter model is characterized by a single dimensionless coupling g (apart from the momentum cutoff Λ), and we can answer the following important question: what trajectory in the v–K phase diagram of S_{SG} in Fig. 20.2 does the $d = 1$, O(3) nonlinear sigma model at $\theta = \pi$ map onto as a function of g?

A first guess would be to simply set $\lambda = 1$ in the values of the couplings in (20.48) and in the value of v below (20.49). However, these results hold for small λ and J_2 and are not acceptable for $\lambda = 1$. A strategy that works is the following: let us focus on the Tomonaga–Luttinger phase of Section 20.2.2 and ask if there is any trajectory within it which corresponds to $\lambda = 1$. If there was such a trajectory, then $SU(2)$ symmetry demands that the $\hat{\sigma}^z \hat{\sigma}^z$ and $\hat{\sigma}^+ \hat{\sigma}^-$ correlators should decay with the same exponent. We compare the expansions in (20.40) and (20.41) and notice that their leading terms coincide only at $K = 1/2$ (the first subleading term also coincides at this value of K). So one point with $SU(2)$ symmetry in Fig. 20.2 is the very symmetrical point in the center $v = 0$, $K = 1/2$. Now if the renormalization group respects the underlying symmetry of the Hamiltonian, points flowing into and away from $v = 0$ and $K = 1/2$ could also be $SU(2)$ symmetric. By examining the trends in (20.48), and in the value of v below (20.49), we are then led to assert the following important result:

$$\text{the line } v = (2K - 1)/\sqrt{\delta} \text{ has } SU(2) \text{ symmetry,} \tag{20.71}$$

and therefore it corresponds to $\lambda = 1$ in H_{12}; we access different points on this line by varying J_2/J_1, and increasing J_2/J_1 corresponds to decreasing v and K. This line also maps onto the O(3) nonlinear sigma model at $\theta = \pi$, and increasing g also corresponds to decreasing v and K. The renormalization group flow along this line is easily deduced from either (20.57) or (20.61), and we have

$$\frac{dv}{d\ell} = -2\sqrt{\delta} v^2. \tag{20.72}$$

This flow has a fixed point at $v = 0$, which corresponds to some critical value of $J_2/J_1 = J_{2c}$ or $g = g_c$, as in Fig. 19.1. The O(3) nonlinear sigma model also has an additional unstable fixed point at $g = 0$, but that is inaccessible in the present sine–Gordon theory. This fixed point corresponds to the classical limit $S \to \infty$ (as $g \sim 1/S$) and so it is not surprising that it does not appear in an analysis set up explicitly for $S = 1/2$. Presumably, the $g = 0$ fixed point is present somewhere in the large K, v region of Fig. 20.2.

All points with $v > 0$ ($J_2/J_1 < J_{2c}$ or $g < g_c$) flow into $v = 0$ (Figs. 19.1 and 20.2). For these values the ground state is a Tomonaga–Luttinger liquid with correlations given by (20.40) and (20.41) at $K = 1/2$. The flow into the fixed point is logarithmically slow ($v(\ell) \sim 1/\ell$ for large ℓ), and this leads to logarithmic corrections to the correlators in a manner rather similar to the $d = 3$ quantum rotor model examined in Chapter 14. This critical state at $v = 0$, $K = 1/2$ is the closest a spin model in $d = 1$ can get to achieving long-range Néel order – the equal-time order parameter correlations decay as $1/x$. Without the topological term in the nonlinear sigma model, the correlations decay even faster (exponentially) as discussed in Chapters 11 and 12.

Points with $v < 0$ ($J_2/J_1 > J_{2c}$ or $g > g_c$) flow away to large negative values of v. This puts us in the gapped VBS phase already discussed in Section 20.2.3. Additional support for this identification comes from an interesting exact result of Majumdar and Ghosh [315,468]. They noted that at the special $SU(2)$ symmetric point, $\lambda = 1$, $J_2 = J_1/2$, it is possible to write down the exact wavefunction of the ground state of H_{12}. It can be checked that the following simple ansatz consisting of a product of pairs of singlet bonds is an exact eigenstate of H_{12}:

$$\ldots \mathcal{B}_{12}\mathcal{B}_{34}\mathcal{B}_{56}\mathcal{B}_{78} \ldots, \tag{20.73}$$

where $\mathcal{B}_{ij} = (|\uparrow\rangle_i|\downarrow\rangle_j - |\uparrow\rangle_j|\downarrow\rangle_i)/\sqrt{2}$; this state is degenerate with its symmetry-related partner

$$\ldots \mathcal{B}_{23}\mathcal{B}_{45}\mathcal{B}_{67}\mathcal{B}_{89} \ldots \tag{20.74}$$

Arguments proving that these are also the ground states are given by Majumdar and Ghosh. It should be clear that these are precisely the VBS states sketched in Fig. 20.4. (We also note that there are some interesting generalizations of the Majumdar–Ghosh construction of exact ground states to antiferromagnets on the square lattice [52, 469] (one of which has found a recent experimental realization [257, 343]), but these are for cases where the Hamiltonian does not have the full square lattice symmetry.)

We can also use the flow equation (20.72) to deduce how the energy gap vanishes, or the VBS order disappears, as $v \nearrow 0$ (or $g \searrow g_c$ or $J_2/J_1 \searrow J_{2c}$). The runaway flow for $v < 0$ from the $v = 0$ fixed point in (20.72) has precisely the same structure as the flow in (12.8) for the $d = 1$, O(3) rotor model. Using precisely the same arguments as those presented in Section 12.1 we may conclude here that the energy gap $\Delta \sim \exp(-1/(2\sqrt{\delta}|v|))$ for small $|v|$. Also, from (20.56), the VBS order parameter in (20.63) has scaling dimension $\dim[\sin(2\phi)] = 2^2 K/2 = 1$; so its expectation value vanishes as $\langle \sin(2\phi) \rangle \sim \Delta \sim \exp(-1/(2\sqrt{\delta}|v|))$.

20.2.6 Critical properties near phase boundaries

There are three phase boundaries in Fig. 20.3 and we consider properties in their vicinity in turn. The multicritical point where all three phases meet will not be considered since this point lies on the $SU(2)$ symmetric line $\lambda = 1$ and has therefore already been described in Section 20.2.5.

We first consider the transition from the Tomonaga–Luttinger liquid to the Néel phase. We cross the phase boundary by moving the initial values of v and K in Fig. 20.2 across the separatrix $v = (2K - 1)/\sqrt{\delta}$. Note that the last point within the Tomonaga–Luttinger liquid is on the separatrix, which was asserted earlier to have $\lambda = 1$ and $SU(2)$ symmetric correlations. To understand the growth of the Néel order parameter, we have to examine the flows from an initial point just across the separatrix, that is, from the point $v = (2K - 1 + \epsilon)/\sqrt{\delta}$ for small ϵ. To facilitate integration of the flow equations (20.57) and (20.61) we change variables to

$$y_{1,2} = \sqrt{\delta}v \mp (2K - 1). \tag{20.75}$$

Then (20.57) and (20.61) become

$$\frac{dy_1}{d\ell} = y_1(y_1 + y_2),$$

$$\frac{dy_2}{d\ell} = -y_2(y_1 + y_2).$$

It is clear from these equations that one integral is simply $y_1 y_2 = C$, where C is a constant determined by the initial conditions; the first equation is then easily integrated to give

$$\tan^{-1} \frac{y_1(\ell)}{\sqrt{C}} - \tan^{-1} \frac{y_1(0)}{\sqrt{C}} = \sqrt{C}\ell. \tag{20.76}$$

By the usual scaling argument, the characteristic energy gap, Δ, in the Néel phase is of order $e^{-\ell^*}$, where ℓ^* is the value of ℓ over which y_1 grows from an initial value of order $\epsilon \ll 1$ to a value of order unity. From the initial conditions, we expect the constant C to be of order ϵ also, and so let us choose $C = \epsilon$; then a straightforward analysis of (20.76) gives us

$$\Delta \sim \exp\left(-\frac{\pi}{2\sqrt{\epsilon}}\right). \tag{20.77}$$

This singularity, and the flow analysis above, are characteristic of a "Kosterlitz–Thouless" transition, which occurs in a variety of physical situations in both classical and quantum systems (the reader may find more details in the book by Itzykson and Drouffe [244]). Also note the difference between this singularity and that found for the $SU(2)$ case in Section 20.2.5 – there was no square root within the exponential in the latter case. By arguments similar to those presented in Section 20.2.5, we may also conclude here that the order parameter grows as $\sim \Delta$.

The transition between the Tomonaga–Luttinger liquid and the VBS phase is essentially identical to the above case, and little needs to be said. The energy gap in the VBS phase obeys (20.77) near the phase boundary, and the VBS order parameter vanishes as $\sim \Delta$. We note that the terminus of the Tomonaga–Luttinger liquid again has $K = 1/2$, $SU(2)$ symmetric exponents because the flow is again into the $v = 0$, $K = 1/2$ point; this happens even though the underlying model has $\lambda < 1$ (see Figs. 20.2 and 20.3).

Finally, let us consider the phase boundary between the VBS and Néel phases. This coincides with the line $K < 1/2$, $v = 0$ in Fig. 20.2. Along this line correlations of both order parameters decay with a power law determined by their common scaling dimension (from (20.53), (20.54), and (20.56)) $\dim[\sin(2\phi)] = \dim[\cos(2\phi)] = K$, that is, equal-time correlators decay as x^{-2K}. For nonzero v an energy gap appears, and its magnitude is determined by the relevant flow away from the $v = 0$ line in (20.57). This flow equation tells us that $1/v = \dim[v] = (2 - 4K)$, and as $z = 1$, the energy gap Δ behaves as

$$\Delta \sim |v|^{1/(2-4K)}. \tag{20.78}$$

The scaling dimensions of the order parameters above show that they vanish as Δ^K on either side of the phase boundary.

One interesting feature of this last phase boundary deserves further comment. Note that we have distinct broken symmetries on either side of the transition, characterized by very different order parameters, VBS and Néel. If we had attempted to construct a generic Landau-like mean-field theory for such distinct order parameters, we would have concluded that the two phases would not be separated by a second-order transition: a first-order line or coexistence between the two phases is generic. Nevertheless, we have found here a second-order transition across a line with continuously varying exponents. This is clearly a consequence of the strong quantum fluctuations in a low-dimensional system, and mean-field theory is not a suitable guide for the expected behavior. Recall also that a generic second-order phase boundary between Néel and VBS phases has also been proposed in certain collinear antiferromagnets in $d = 2$, as discussed in Section 19.3.5.

20.3 O(2) rotor model in $d = 1$

In Part II we examined the quantum/Ising rotor models in all spatial dimensions d and for all values of the number of rotor components, N. Only one case was omitted, as noted in Chapter 12: $d = 1$ and $N = 2$. For completeness, we discuss this case here, as only a simple extension of the methods already introduced is necessary.

We consider a chain of O(2) quantum rotors (defined in Section 6.3 and (6.35)) with the Hamiltonian

$$H_R = \frac{Jg}{2} \sum_i \hat{L}_i^2 - J \sum_i \hat{\mathbf{n}}_i \cdot \hat{\mathbf{n}}_{i+1}, \tag{20.79}$$

where $\hat{\mathbf{n}}_i$ are two-component unit vectors, there is only a single generator of O(2) rotations $\hat{L}_i$ on each site, and these operators obey the on-site commutation relations (6.34).

Let us parameterize

$$\mathbf{n}_i = (\cos\theta_i, \sin\theta_i), \tag{20.80}$$

and take the naive continuum limit of (20.79). This can be done using the methods discussed in Chapter 3; we obtain a continuum $d = 1$ quantum field theory for θ that has precisely the same action as $\mathcal{S}_{TL}$ in (20.25) but with the couplings

$$K \approx \frac{\pi}{\sqrt{8}}, \qquad v_F \approx \sqrt{g}Ja. \tag{20.81}$$

Under this action, equal-time correlators, from (20.55), decay as

$$\langle \hat{\mathbf{n}}_i \cdot \hat{\mathbf{n}}_j \rangle \sim \frac{1}{|x_i - x_j|^{1/(2K)}}. \tag{20.82}$$

However, this is clearly not the complete story. This naive continuum limit has explicitly prevented the introduction of vortices in the angular θ field. These are tunneling events in which the spatial winding number

$$\frac{1}{2\pi} \int dx \, \nabla\theta \tag{20.83}$$

changes between integer values. Such vortices can be conveniently introduced in the dual ϕ field formulation, as discussed below (20.51). In the present situation elementary 2π vortices are certainly allowed by the lattice Hamiltonian (20.79), and so by the arguments just before and after (20.51) we obtain the dual action

$$\tilde{\mathcal{S}}_{SG} = \int dx d\tau \left[\frac{1}{2\pi K v_F} \left((\partial_\tau \phi)^2 + v_F^2 (\nabla \phi)^2 \right) - \tilde{v} \cos(2\phi) \right]. \tag{20.84}$$

The most important difference from $\mathcal{S}_{SG}$ in (20.50) is that we have a $\cos(2\phi)$ rather than a $\cos(4\phi)$ term. Much of the analysis of $\mathcal{S}_{SG}$ in Section 20.2.1 can now be applied. The renormalization group equations (20.57) and (20.61) are modified to

$$\frac{d\tilde{v}}{d\ell} = (2 - K)\tilde{v},$$

$$\frac{dK}{d\ell} = -\tilde{\delta}\tilde{v}^2.$$

This leads to a renormalization group flow diagram as in Fig. 20.2, but in the vicinity of the point $K = 2$, $v = 0$ (instead of $K = 1/2$, $v = 0$). The model H_R therefore has a Kosterlitz–Thouless transition from a gapless phase with correlations decaying as (20.82) to a gapped phase (the gap increases as in (20.77)), and equal-time correlations decay exponentially as in (1.26). The exponent K takes the value $K = 2$ at this critical point. This is the most important difference from the corresponding transition in H_{12} where we had $K = 1/2$. As a result, the critical order parameter correlators decay as $1/x^{1/4}$. Also, the excitations in the gapped phase carry charges $Q = \pm 1$. This is a consequence of the transition being driven by single 2π vortices.

20.4 Applications and extensions

There is a great deal of experimental and theoretical work on $S = 1/2$ spin chains, and a complete survey will not be attempted here. For a discussion mainly of neutron scattering experiments see the review articles by Cowley [95] and Broholm [63]. Nuclear magnetic resonance experiments have also been important in measuring thermodynamic and low-frequency spin relaxation properties; a discussion of these may be found in [126, 416, 452, 486, 487, 499]. The VBS phase of Section 20.2.3 has an experimental realization in the intensively studied compound $CuGeO_3$, although the coupling between the spins and the phonon excitations [97] almost certainly has to be considered for a complete understanding of the experiments; a neutron scattering analysis may be found in [16], and a discussion of some theoretical issues in [30, 181].

Magnetic ordering transitions of disordered systems

This chapter has been co-authored with T. Senthil, and adapted from the Ph.D. thesis of T. Senthil, submitted to Yale University (1997, unpublished).

The last two chapters of this book move beyond the study of regular Hamiltonians that have the full translational symmetry of an underlying crystalline lattice and consider the physically important case of disordered systems described by Hamiltonians with couplings that vary from point to point in space. By the standards of the regular systems we have already discussed, the quantum phase transitions of disordered systems are very poorly understood, and only a few well-established results are available. A large amount of theoretical effort has been expended toward unraveling the complicated phenomena that occur, and they remain active topics of current research. The aims of our discussion here are therefore rather limited: we highlight some important features that are qualitatively different from those of nondisordered systems, make general remarks about insights that can be drawn from our understanding of the finite-T crossovers in Part II, and discuss the properties of some simple solvable models.

In keeping with the general strategy of this book, we introduce some basic concepts by studying the effects of disorder on the magnetic ordering transitions of quantum Ising/rotor models studied in Part II; we also make some remarks in Section 21.4 on the effects of disorder on the ordering transitions of Fermi liquids considered in Chapter 18. Models with much stronger disorder and frustrating interactions that have new phases not found in ordered systems are considered in Chapter 22.

Almost all of this chapter considers the following disordered Hamiltonians: for the case $N = 1$, we generalize (10.1) to

$$H_{Id} = -\sum_i g_i \hat{\sigma}_i^x - \sum_{\langle ij \rangle} J_{ij} \hat{\sigma}_i^z \hat{\sigma}_j^z, \qquad (21.1)$$

while for $N \geq 2$, we have the disordered version of (11.1):

$$H_{Rd} = \frac{1}{2} \sum_i g_i \hat{\mathbf{L}}_i^2 - \sum_{\langle ij \rangle} J_{ij} \hat{\mathbf{n}}_i \cdot \hat{\mathbf{n}}_j, \qquad (21.2)$$

where $\langle ij \rangle$ represents the sum over nearest neighbors on the sites, i, of a regular lattice, and the couplings $g_i \geq 0$, $J_{ij} \geq 0$ are random functions of position (note that g_i has the dimensions of energy, unlike the dimensionless g in (10.1) and (11.1), and the nondisordered case obtains with $g_i = gJ$ and $J_{ij} = J$). The restriction that the couplings all be nonnegative has an important simplifying consequence: there is no frustration in the exchange terms in (21.1) and (21.2), and so for small enough g_i, there is a magnetically ordered ground state, characterized by the same order parameter used for the nonrandom case. In the present case we define

$$N_0 = \overline{\langle \hat{\sigma}_i^z \rangle}, \qquad T = 0, \tag{21.3}$$

where the overbar denotes an average over different disorder configurations, and the generalization to $N \geq 2$ is obvious. For a specific realization of the disorder, the value of $\langle \hat{\sigma}_i^z \rangle$ in the magnetically ordered ground state varies from point to point due to the microscopic disorder, but there is an average uniform component, which is measured by N_0. This average can be computed by summing $\langle \hat{\sigma}_i^z \rangle$ over all sites i for a specific realization of the disorder, or by performing the disorder average as in (21.3) – the result is expected to be the same. Now, as we raise the value of all the g_i (say, by increasing their mean, while keeping their variance fixed), we expect a phase transition at a critical value of $\bar{g} = \langle g_i \rangle$ to a quantum paramagnet with $N_0 = 0$; for sufficiently large g_i, the strong-coupling methods of Sections 5.2 and 6.1 apply and show that the ground state must be a quantum paramagnet. It is this transition from a magnetically ordered state to a quantum paramagnet that forms the basis of most of our discussion of quantum phase transitions in disordered systems in this chapter.

We begin in Section 21.1 by discussing a general stability criterion that must be satisfied by a quantum critical point in any disordered system. This leads to the requirement that the correlation length exponent ν satisfy $\nu \geq 2/d$. Further general considerations appear in Section 21.2 where we discuss the low-energy spectrum of the phases away from the critical point. The presence of disorder introduces the so-called Griffiths–McCoy singularities. A first analysis of the models H_{Id} and H_{Rd} appears in Section 21.3 using the field-theoretic methods of Chapter 14. Two solvable cases of H_{Id} are considered next: models near the percolation transition in Section 21.5 and models in $d = 1$ in Section 21.6. Some concluding remarks then appear in Section 21.7.

21.1 Stability of quantum critical points in disordered systems

Because a random system is intrinsically inhomogeneous, it is not a priori clear that it can display a sharp second-order phase transition at a specific average coupling $\bar{g} = \bar{g}_c$ (say) at which the response functions become singular. After all, the couplings vary from point to point, and there will always be localized regions that are well away from the critical point, even though the average coupling is critical; consistency requires that such localized regions do not occur often enough. The restrictions this places on the classical critical point were first considered by Harris [200], who actually looked at the simpler question of whether the classical critical point of the nonrandom system was stable toward the introduction of a small amount of disorder. However, the restrictions that emerge apply also to quantum critical points of random systems, as was discussed by Chayes *et al.* [81] who also presented a rigorous argument.

We will be satisfied here presenting a simple heuristic argument, along the lines of Harris [200]. Let us tune the transition by varying the value of $\bar{g}$. Focus now on any region of size L; we can define a local critical point $\bar{g}_{c,r}$ at which this region will cross over from a magnetically ordered to a quantum paramagnetic state. The value of $\bar{g}_{c,r}$ will not necessarily equal the global value $\bar{g}_c$. We can expect that local random fluctuations will cause a deviation of order $L^{-d/2}$; this follows from the central limit theorem-like argument that the variance of order $\mathcal{N} = L^d$ independent random numbers (the local values of g_c) is of order $\sqrt{\mathcal{N}}$. Such a deviation is significant if it starts becoming of order $|\bar{g} - \bar{g}_c|$. This will happen at length scales shorter than $L = L_r \sim |\bar{g} - \bar{g}_c|^{-2/d}$. Now if L_r is shorter than the correlation length ξ, then by the time we renormalize out to the scale $\xi \sim |\bar{g} - \bar{g}|^{-\nu}$, the system has unambiguously decided what its critical point is, and local random fluctuations have been smoothed out. So we now have our stability requirement $L \ll \xi$, or

$$|\bar{g} - \bar{g}_c|^{-2/d} \ll |\bar{g} - \bar{g}_c|^{-\nu}. \tag{21.4}$$

Consistency of (21.4) leads to the main result of this section:

$$\nu \geq \frac{2}{d}, \tag{21.5}$$

an inequality that must be satisfied by all quantum critical points of disordered systems.

In our discussion above we considered the consequences of fluctuations in the local position of the critical point. In a field-theoretic language, we can induce such a fluctuation by perturbing the action with a random coupling multiplying the operator that tunes the system across the transition. More generally, consider the case where the randomness couples to some local operator $O(x, \tau)$, which has a scaling dimension $\dim[O] = \zeta_O$. This means that the effective action for the system will have an additional term

$$\int d^d x \int d\tau r(x) O(x, \tau), \tag{21.6}$$

where $r(x)$ is a fixed random function of space only. We assume that the spatial correlations in $r(x)$ are short-ranged (i.e. $r(x)$ and $r(x')$ are considered as independent random variables for moderate values of $|x - x'|$). In contrast, note that as $r(x)$ is time independent, there is an infinite correlation "length" along the imaginary time direction. It is this long-range correlation which makes the effects of randomness particularly severe in quantum systems. Now consider averaging over the disorder using replicas (this method is discussed briefly in Section 21.3). This generates a term $\delta^2 \int d^d x d\tau_1 d\tau_2 \sum_{ab} O_a(x, \tau_1) O_b(x, \tau_2)$, where δ^2 is the variance of r, and a, b are replica indices. The scaling of δ^2 is given by power counting to be $\dim[\delta'^2] = d + 2z - 2\zeta_O$. This type of randomness is therefore relevant if $d + 2z - 2\zeta_O > 0$. For the case of the energy density, the scaling dimension of the associated coupling constant is $1/\nu$, and so the dimension of the energy operator is $\zeta_O = d + z - 1/\nu$; the criterion for its relevance then becomes $\nu < 2/d$, as expected. Conversely, such random fluctuations are perturbatively irrelevant if $\nu > 2/d$.

21.2 Griffiths–McCoy singularities

In addition to the singularities in the spectrum at the quantum critical point, all disordered systems have additional "Griffiths–McCoy" (GM) singularities [180, 328, 329] that affect the *phases* on either side of the critical point (related singularities are also present in the statics and dynamics of classical spin systems [116, 117, 393]). The physics behind their appearance is quite different from those of the critical singularities, and the complicated interplay of these two distinct phenomena is at the heart of the difficulty of analyzing quantum phase transitions in disordered systems. One possibility is that GM singularities of the phases are quite weak and are simply idle spectators that are decoupled from the critical singularities – they are then not part of the universal scaling functions describing the crossover between the phases. At the other end of the possibilities, the GM and critical singularities could be tightly coupled, with no sharp distinction between the two – the GM singularities then become the critical singularities as one approaches the critical point. In any case, theoretical analyses cannot deal with one without considering the other, and unraveling the two is often quite difficult.

The central idea becomes clear by considering a specific case: the $N = 1$ model H_{Id} (see (21.1)). We are interested in the nature of the low-energy spectrum ($\omega \to 0$) in the quantum paramagnetic phase ($\bar{g} > \bar{g}_c$) not too far from the critical point; this is controlled by the GM singularities. (Notice the order of limits ($\omega \to 0$ followed by $\bar{g} \to \bar{g}_c$) characterizing these singularities; the opposite order of limits ($\bar{g} \to \bar{g}_c$ first and then $\omega \to 0$) lead to the critical singularities.) In the nondisordered case there was an energy gap $\Delta_+ \sim (g - g_c)^{z\nu}$ and so all spectral densities vanished for $\omega < \Delta_+$. We now argue, following [140, 142, 187, 400, 404, 503, 552] that there is no such gap for the disordered system, and there is always a nonzero spectral density at arbitrarily low energies. Due to the randomness, there would, in general, be a nonzero probability that any given bond is stronger than the critical bond strength at which the system orders as a whole. This would happen in an entire, compact region of linear size L with probability $P(L) \sim \exp(-cL^d)$, where c is a constant determined by the microscopic couplings, width of the random distribution, etc. Such regions constitute clusters of spins that are coupled strongly enough that if they were infinite in size, they would order.

Consider any such cluster of size L. For large L, all the spins in the cluster behave coherently in space, and it is legitimate to treat the cluster as a single giant spin in the presence of some effective transverse field g_L. Thus the cluster has two low-lying energy levels with an energy difference $2g_L$ well-separated from other higher energy levels. This effective field, and hence the splitting between the two levels, goes to zero as $L \to \infty$, thus giving rise to broken symmetry and "long-range" order within the cluster. For finite L, g_L can be estimated in perturbation theory as the ratio of the transverse field to the bond strength. To zeroth order of perturbation theory, there is no transverse field, and the cluster has two degenerate ground states (all spins up or down) and other excited states separated by a large energy (of order the bond strength). A nonzero transverse field breaks this ground state degeneracy, and there remains instead a doublet with a nonzero but small splitting. It is clear that this effect will appear only in a large order of the perturbation theory (= number of spins in the cluster) $\sim L^d$. Thus the splitting and hence g_L are exponentially small in L^d: $g_L \sim \bar{g} \exp(-c_1 L^d)$.

Now, we assume that different clusters in the system may be treated independently of one another. Consider the density of low-energy excitations as measured by the disorder average of the imaginary part of the local dynamic susceptibility, $\text{Im}\chi_L(\omega) = \int d^d k/(2\pi)^d \text{Im}\chi(k, \omega)$, with χ defined in (7.7):

$$\overline{\text{Im}\chi_L(\omega) = \sum_\alpha |\langle\alpha|\hat{\sigma}_i^z|0\rangle|^2 \delta(\omega - (E_\alpha - E_0))}, \qquad (21.7)$$

where $|\alpha\rangle$ refer to eigenstates of the system with energy E_α, and $|0\rangle$ is the ground state. For low ω in the paramagnetic phase, the only contribution will be from the rare clusters discussed above. Thus

$$\text{Im}\chi_L(\omega) \sim \int dL\, P(L)\delta(\omega - 2g_L), \qquad (21.8)$$

$$\sim \int dL\, e^{-cL^d}\delta\left(\omega - he^{-c_1 L^d}\right), \qquad (21.9)$$

$$\sim \frac{\omega^{d/\tilde{z}-1}}{(\ln(1/\omega))^{d/(d-1)}}, \qquad (21.10)$$

where $\tilde{z} = c_1 d/c$. Therefore we have the striking result that the paramagnetic phase is gapless with a singular power-law (up to logarithmic corrections) density of states at low energies. The power depends upon the nonuniversal exponent $\tilde{z}$ and could in principle even lead to a divergent density of states at zero energy. This power-law singularity leads to singularities in the thermodynamic properties of the system at low temperature. We have chosen the informative notation $\tilde{z}$ for the exponent, as it plays the role of the dynamic exponent for the GM singularities. The spectral density has units of density per unit energy, or (length)$^{-d}$/frequency, or the "scaling" dimension $d - \tilde{z}$ (the quotes emphasize that it is really not appropriate to think of the GM singularities as reflecting some underlying scale invariance).

The value of the exponent $\tilde{z}$ varies continuously with $\bar{g}$, and its limiting value as $\bar{g} \searrow \bar{g}_c$ is of some interest. However, it is important not to confuse this limiting value with the true critical exponent z of the critical singularities at $g = g_c$. For this we have, by (11.41),

$$\text{Im}\chi_L(\omega) \sim \omega^{(d-2+\eta)/z}. \qquad (21.11)$$

The values of $\lim_{\bar{g}\searrow\bar{g}_c} \tilde{z}$ and z are obtained from $\chi_L''(\omega)$ by different orders of limits and could, in principle, be distinct. There is numerical evidence in some simulations [403] that these two quantities coincide for H_{Id}, but there is no clear physical understanding why this should be so.

We emphasize that the GM singularities arise due to the presence of statistically rare clusters that are anomalously strongly coupled, and hence they are unique features of the disordered system. The effect becomes weaker with increasing dimension, ultimately vanishing in the limit of infinite dimension. Increasing the range of the interactions also weakens the effect – for infinite range interactions, there are no singularities. Finally, the effect is strongest for the $N = 1$ model with discrete symmetry. We turn below to the $N \geq 2$ cases and find much weaker singularities.

The analysis of the $N \geq 2$ case also focuses on the contribution of rare regions of size L that are almost ordered. We found above that such regions had a gap of order $\exp(-c_1 L^d)$ for $N = 1$, and we now need the corresponding result for $N \geq 2$. For this, we first offer an alternative interpretation of the magnitude of the gap for $N = 1$: we can model the time evolution of the correlated region of size L as a *one*-dimensional classical Ising chain, as is clear from the arguments in Section 5.5; this chain has an "exchange" of order L^d, and then the results (5.37) and (5.49) lead to the correct exponentially small gap. The same interpretation also works for $N \geq 2$; we again have an "exchange" of order L^d, but now, by (6.23) and (6.32), the gap is inversely proportional to the exchange (i.e. it now takes a much larger value of order L^{-d}). This larger gap indicates that the correlated region changes its orientation far more frequently and is less important for the low-energy physics. Inserting this gap into the analog of (21.10), we get

$$\chi_L''(\omega) \sim \int dL \, e^{-cL^d} \delta(\omega - c_1/L^d), \qquad (21.12)$$

$$\sim \exp(-cc_1/\omega), \qquad (21.13)$$

which is only a very weak essential singularity. It appears unlikely that such a weak GM effect will play an important role in the fluctuations at the quantum critical point.

It should be mentioned here that the above analysis of models with a continuous $O(N)$ symmetry is special to the rotor models and does not apply to random versions of the Heisenberg spin systems of Chapter 19. The GM of singularities of the latter are quite strong and have been considered in [46, 142].

21.3 Perturbative field-theoretic analysis

In this section, we attempt to analyze H_{Id} and H_{Rd} for the case of weak disorder, by extending the nondisordered system field-theoretic analysis of Chapter 14. For $N = 1$, the very strong GM singularities, and the dominance of rare regions suggests that the field-theoretic approach below is unlikely to be valid near the critical point. For $N \geq 2$, the weak GM singularities indicate that the field-theoretic approach is plausible.

A first question to ask is whether the nondisordered (or "pure") fixed point is stable against disorder. The arguments of Section 21.1 show that this will be the case if $\nu_{\text{pure}} > 2/d$. For $N = 1$ we know that $\nu_{\text{pure}} = 1/2$ for $d \geq 3$, and $\nu_{\text{pure}} \approx .632$ for $d = 2$ and $\nu_{\text{pure}} = 1$ for $d = 1$; thus weak randomness is relevant for all dimensions $d < 4$. A similar result holds for higher N. So for $d > 4$, sufficiently weak disorder should not change the critical properties from those of the pure system. For $d < 4$, we might hope that a renormalization group analysis would allow us to access a stable fixed point, at least for small $4 - d$. Such an analysis requires a disordered version of the pure system field theory $\mathcal{S}_\phi$ in (14.2). This is clearly realized by simply allowing all the coupling constants to become random functions of the spatial coordinate x. However, as could be expected from the arguments above, the most important spatial dependence is that of the parameter r, which controls the position of the critical point; we therefore consider the disordered action

$$S_{\phi d} = \int d^d x \int d\tau \left\{ \frac{1}{2} \left[(\partial_\tau \phi_\alpha)^2 + c^2 (\nabla_x \phi_\alpha)^2 + (r_0 + r(x)) \phi_\alpha^2(x) \right] + \frac{u}{4!} \left(\phi_\alpha^2(x) \right)^2 \right\}, \tag{21.14}$$

with $r(x)$ a random function of position with probability distribution $P[r(x)] \sim \exp(- \int d^d x\, r^2(x)/(2\delta^2))$.

While it is possible to work directly with $S_{\phi d}$, the subsequent analysis is made simpler by making an explicit average over disorder using the replica method. We do not discuss this method in any detail here but refer the reader to introductory discussions in the literature (e.g. the book by Fischer and Hertz [138]).

We are interested here in average correlators of the random system defined by

$$\overline{\int \mathcal{D}\phi\, e^{-S_{\phi d}}\, O / Z_d}, \tag{21.15}$$

where O is any observable, and note that the average over disorder must include the disorder-dependent partition function, $Z_d = \int \mathcal{D}\phi\, e^{-S_{\phi d}}$, in the denominator. To overcome this technical difficulty, we introduce n replicas of the field, $\phi_{\alpha a}$ ($a = 1, \ldots, n$ is the replica index). Then if the operator O involves only the field with $n = 1$, the integral over the remaining replicas gives a contribution Z_d^{n-1} in the functional integral over $S_{\phi d}$. Now note that in the limit $n \to 0$, this yields precisely the factor Z_d^{-1} appearing in (21.15). So the prescription of the replica method is to compute correlators with n arbitrary, and then we take the peculiar step of analytically continuing to a system with $n = 0$ field components. The advantage is that this allows us to average over the disorder in $e^{-S_{\phi d}}$ at an early stage.

Introducing n replicas of (21.14) and then averaging over $r(x)$, we obtain the following translationally invariant action of the field $\phi_{\alpha a}$ ($\alpha = 1, \ldots, N, a = 1, \ldots, n$):

$$S_{\phi d} = \int d^d x \int d\tau \sum_a \left\{ \frac{1}{2} \left[(\partial_\tau \phi_{\alpha a})^2 + c^2 (\nabla_x \phi_{\alpha a})^2 + r_0 \phi_{\alpha a}^2 \right] + \frac{u}{4!} \left(\phi_{\alpha a}^2 \right)^2 \right\}$$

$$- \frac{\delta^2}{2} \int d^d x \int d\tau d\tau' \sum_{a,b} \phi_{\alpha a}^2(x, \tau) \phi_{\beta b}^2(x, \tau'), \tag{21.16}$$

where all summations over replica indices are explicitly noted. The renormalization group analysis of this action can be carried out by standard methods – we simply treat n as an arbitrary integer, and only take the $n \to 0$ limit after the scaling equations are obtained. We perturb the theory in powers of the nonlinearities u and δ^2. First, simple power counting at zeroth order gives us the flow equations:

$$\frac{dr_0}{dl} = 2r_0,$$

$$\frac{du}{dl} = (3 - d)u,$$

$$\frac{d\delta^2}{dl} = (4 - d)\delta^2. \tag{21.17}$$

Thus δ^2 becomes relevant below four dimensions, as expected from the arguments at the beginning of this section. Note that the interaction strength u, however, remains irrelevant down to $d = 3$. At next order, these flow equations get modified to

$$\frac{dr_0}{dl} = 2r_0 + c_1 u - c_2 \delta^2, \tag{21.18}$$

$$\frac{du}{dl} = (3 - d)u - c_3 u^2 + c_4 u \delta^2, \tag{21.19}$$

$$\frac{d\delta^2}{dl} = (4 - d)\delta^2 + c_5 \delta^4 - c_6 u \delta^2, \tag{21.20}$$

where the c_i are all positive constants. These equations do not allow for a fixed point for small $4 - d$; instead δ^2 has runaway flows, suggesting a fundamental instability in the perturbation theory. This is a disappointing result, and we are unable to obtain any reliable information about the quantum critical point by this approach. Analysis of this problem by the large-N expansion [268] also fails, again because of runaway flows for the strength of the randomness. Thus the fixed-point theory presumably has a strong amount of randomness. At the level of the noninteracting theory, one expects that the lowest energy modes will be strongly localized. Physically, it is clear then that we cannot ignore the effects of interactions: condensation into a localized state leads to enhancement in interaction effects. It is necessary to include both disorder and interactions in a fundamental way.

An alternative approach was taken by Dorogovstev [122] and Boyanovsky and Cardy [54]. They extended (21.16) to a quantum field theory in d space and ϵ_τ time dimensions; formally this amounts to replacing $\int d\tau$ by $\int d^{\epsilon_\tau}\tau$ and using the standard field-theoretic methods of dimensional continuation. The quantum critical point of course corresponds to $\epsilon_\tau = 1$, but these authors suggested making an expansion in small ϵ_τ. The validity of such a procedure is not a priori clear because (21.16) represents the quantum mechanics of a Hamiltonian system only for $\epsilon_\tau = 1$, and it is also clear that a small ϵ_τ suppresses the GM singularities. Simple power counting shows that the equations for r_0 and δ in (21.17) remain unchanged, while that for u gets modified to

$$\frac{du}{d\ell} = (4 - \epsilon_\tau - d)u. \tag{21.21}$$

Now note that for small ϵ_τ, both u and δ become relevant about the $u = \delta = 0$ fixed point near $d = 4$. This allows interactions to control the instabilities due to disorder, and it raises the possibility that a stable fixed point may be found. This was indeed shown to be the case in [54], where it was found that a fixed point with nonzero disorder and interactions in a double expansion in ϵ_τ and $(4 - d)$ exhibited conventional dynamic scaling with exponents

$$z = 1 + \frac{(4 - N)(4 - d) + (2N + 4)\epsilon_\tau}{16(N - 1)},$$

$$v = \frac{1}{2} + \frac{(13N + 20)(4 - d) + 4(6N + 11)\epsilon_\tau}{32(N - 1)}, \tag{21.22}$$

at lowest order for $N > 1$. It would be useful to examine the GM singularities of the paramagnetic phase in this approach and to compute the value of $\tilde{z}$. This intriguing possibility for future work could lead to further insight into the validity of the ϵ_τ expansion.

21.4 Metallic systems

We now turn to the random case of the transitions of metallic systems considered in Chapter 18. We focus our attention here on the random version of the spin density wave transition in Section 18.3. It is assumed here that disorder does not remove the paramagnetic metallic phase itself, as may happen due to localization in low d (though, even for this case, there can still be a magnetic phase transition). Also, we ignore the complexities associated with the validity of the Hertz action which were investigated in Section 18.3.5.

We reverse the order of our discussion of the insulating quantum Ising/rotor models, and first consider the analog of the field-theoretic analysis of Section 21.3. Then, the central difference from the Ising/rotor models, as in the pure case, is that the frequency-dependent ω^2 term in the propagator for the order parameter gets replaced by a $|\omega|$ term as in (18.67). In this manner, the replicated field theory (21.16) generalizes to

$$\begin{aligned}
\mathcal{S}_{Hd} = \int \frac{d^d k}{(2\pi)^d} T \sum_{\omega_n} \frac{1}{2} \left[k^2 + |\omega_n| + r \right] |\phi_{\alpha a}(k, \omega_n)|^2 \\
+ \frac{u}{4!} \int d^d x d\tau \left(\phi_{\alpha a}^2(x, \tau) \right)^2 \\
- \frac{\delta^2}{2} \int d^d x \int d\tau d\tau' \sum_{a,b} \phi_{\alpha a}^2(x, \tau) \phi_{\beta b}^2(x, \tau').
\end{aligned} \tag{21.23}$$

This theory has been analyzed in a double expansion in $(4 - d)$ and ϵ_τ in [270].

Let us now turn to a discussion of the rare region arguments of Section 21.2, and consider Griffiths–McCoy effects in random metallic systems. As was pointed out in important recent work by T. Vojta and collaborators [225, 226, 529], the GM effects are actually much stronger for metallic systems than they are for insulators.

For the case of the insulating Ising/rotor models, Section 21.2 focused attention on rare nearly ordered regions of size L, and computed the quantum dynamics of this single region on its own. By considering the quantum tunneling between the degenerate ordered states, it was found that the order parameter correlations decayed exponentially in a time τ given by

$$\tau \sim \begin{cases} \exp(c_1 L^d), & N = 1 \\ L^d, & N \geq 2 \end{cases} \quad \text{insulators.} \tag{21.24}$$

Let us now consider the quantum dynamics of a similar ordered region in a metallic system. Then we can represent it by a path integral over the order parameter $\phi_\alpha(\tau)$ using an action obtained by localizing (21.23) to a single site, and Fourier transforming the $|\omega|$ to imaginary time dependence $\sim 1/\tau^2$; this yields

$$\frac{S_V}{L^d} = -\int d\tau \int d\tau' \frac{\phi_\alpha(\tau)\phi_\alpha(\tau')}{(\tau - \tau')^2} + u \int d\tau (\phi_\alpha^2(\tau) - m^2)^2; \tag{21.25}$$

here m is the mean order parameter in the cluster. Thus the quantum tunneling of a single cluster maps onto the one-dimensional classical statistical mechanical model defined by (21.25): this is a classical N component spin system with inverse square ferromagnetic interactions. Such spin systems have been thoroughly studied, and they have the following correlation "times" at large, but finite L:

$$\tau \sim \begin{cases} \infty, & N = 1 \\ \exp(c_2 L^d), & N \geq 2 \end{cases} \quad \text{metals.} \tag{21.26}$$

For $N = 1$, the infinite correlation time for $N = 1$ in (21.26) implies the presence of long-range order. In the quantum context, this means that a cluster of large, but finite, size L has its quantum tunneling suppressed by the dissipation from the low-energy metallic excitations. Thus at $T = 0$, even a rare cluster can acquire an ordered moment on its own. Vojta [529] examined the implication of this result for the thermodynamic system, using probabilistic arguments similar to those in Section 21.2. He concluded that such effects led to a *smeared* transition. Thus in the presence of metallic dissipation, there is no quantum phase transition as a function of the couplings, and there are always rare regions that contribute a nonzero ordered moment.

For $N \geq 2$, the key observation is that the correlation time in metals in (21.26) has a similar functional form to that for insulators at $N = 1$ in (21.24). We can immediately conclude that the GM singularities in the $N \geq 2$ metallic systems are the same as those for the insulating random quantum Ising model, which were computed in Section 21.2. Hoyos *et al.* [226] argued that this correspondence extended also to the quantum critical point: the $N \geq 2$ metallic systems do have a sharp quantum phase transition, and its universality class is the same as that of the insulating random quantum Ising model. Detailed numerical studies [113, 114] in the large-N limit have provided strong numerical evidence in support of this rather surprising conclusion.

The above conclusions reinforce the importance of the studies of the random quantum Ising model. So far, we have studied their GM singularities in Section 21.2, and examined a field-theoretic approach to their quantum critical point in Section 21.3. As we indicated earlier, the very strong GM singularities, and the dominance of rare regions suggests that the field-theoretic approach is unlikely to be valid near the critical point. The studies in the following two sections present evidence that this is indeed the case.

We focus on two simpler insulating models, which are amenable to an essentially exact analysis. Both models are restricted to the Ising case $N = 1$ and have very strong GM singularities. We are able to explicitly follow their evolution upon the approach to the critical point: we find that in these cases the GM singularities in fact become the critical singularities, and the resulting dynamic scaling is quite different from the one suggested above in (21.22) by the small ϵ_τ expansion above. Interpretations and attempts at a synthesis follow in the final section.

21.5 Quantum Ising models near the percolation transition

We consider here a special limiting case of the quantum Ising model H_{Id} in (21.1). Consider the following probability distribution of the exchange interactions:

$$J_{ij} = \begin{cases} 0 & \text{with probability } p, \\ J & \text{with probability } 1 - p, \end{cases} \tag{21.27}$$

and let us choose, for simplicity, all the transverse fields $g_i = g$ site independent (the results discussed below can be shown to hold also for a random distribution of g_i). Hence two neighboring sites either interact with an exchange J (such sites are "connected") or they have no direct coupling. Sets of mutually connected sites form clusters, and much is known about the geometry of such clusters in d spatial dimensions. (This is the geometrical "percolation" problem, and we quickly review some needed results for percolation theory in Section 21.5.1.) This sharp separation of sites into sets of disconnected clusters is an important simplifying feature and allows us to obtain a number of exact properties of the quantum critical point for general d. This simplification clearly relies upon the fact that J_{ij} becomes precisely zero with probability p. After our review of percolation in Section 21.5.1, we consider the classical Ising model (with $g = 0$) at nonzero T in Section 21.5.2 and finally consider the nonzero g case in Section 21.5.3.

21.5.1 Percolation theory

Removing bonds on a lattice with probability p (see (21.27)) yields the statistical problem of the geometry of connected clusters on the diluted lattice. This has been reviewed in the book by Stauffer and Aharony [488], and we will quote some needed results. There is a critical p_c, such that for $p > p_c$ there are (in the thermodynamic limit) only connected clusters of a finite size, while for $p < p_c$ there is a thermodynamically large connected cluster. Right at $p = p_c$, there are a large number of clusters with a broad distribution of sizes. These clusters are known to have a fractal structure. Though no cluster is thermodynamically large (i.e. the ratio of the number of sites in any cluster to the total number of sites in the system tends to zero in the thermodynamic limit), there is an infinite connected cluster with a fractal dimension $d_f < d$. An important fact about the critical percolating cluster is that it consists of arbitrarily long one-dimensional segments, which are crucial for its connectedness. Breaking these segments splits the cluster into two disjoint units.

For $p > p_c$ there is a finite probability that any given site belongs to the infinite cluster; this probability vanishes as $p \searrow p_c$ with the power law $\sim (p_c - p)^{\beta_p}$. We can also consider the probability $P(N, p)$ that any site belongs to a finite, large cluster of N sites; for p close to p_c this satisfies the scaling form

$$P(N, p) \sim N'^{-\tau} G(N/\xi^{d_f}), \tag{21.28}$$

where $\xi \sim |p - p_c|^{-\nu_p}$ is a characteristic finite cluster size, which diverges at $p = p_c$; τ, d_f, and ν_p are universal critical exponents; and G is a universal scaling function.

The exponents and scaling functions have been computed either exactly or numerically in $d = 2$ and 3, and we simply treat them here as known quantities. For some of our later results, we also need the limiting form of the function $G(y)$; it approaches 1 for $y \ll 1$, while for $y \gg 1$,

$$G(y, p > p_c) \sim y^{-\theta+\tau} e^{-c_+ y},$$
$$G(y, p < p_c) \sim y^{-\theta'+\tau} e^{-c_- y^{1-1/d}}, \tag{21.29}$$

where θ and θ' are additional known exponents.

Finally, we also need information on the correlation between pairs of sites. For $p \geq p_c$ the probability that two sites belong to the same cluster decays for large x as $\sim x^{-d+2-\eta_p} F(x/\xi)$, where η_p is another exponent ($2\beta_p = (d - 2 + \eta_p)\nu_p$) and F a scaling function.

21.5.2 Classical dilute Ising models

We warm up with a discussion of the properties of the classical Ising model, where $g = 0$ and T is nonzero; its phase diagram is shown in Fig. 21.1. At $p = 0$, as T is increased, there is a phase transition from an ordered state to a disordered one (see Fig. 21.1). On the other axis, when $T = 0$ there is a percolation transition at $p = p_c$; this transition coincides with loss of magnetic long-range order, as there is no infinite cluster and hence no spontaneous magnetization for $p > p_c$. The boundary of critical temperatures $T = T_c(p)$ approaches zero at $p = p_c$ as $T_c \sim \ln(1/(p_c - p))$. These results can be understood in the following way. As we mentioned earlier, the critical percolating cluster consists of a number of arbitrarily large one-dimensional segments. These segments are the weakest links in the cluster; correlations in the cluster will be destroyed if they are destroyed along the segments. From the low-temperature behavior of the classical Ising chain in Section 5.5, we know that any finite T will destroy correlations in a large segment over a length scale ξ_T that is exponentially large in $1/T$. Now consider the infinite cluster at $p < p_c$. This resembles the critical clusters at scales $\ll \xi$, where $\xi \sim (p_c - p)^{-\nu_p}$ is the percolation correlation length. At larger scales, there is a crossover to the geometry of a d-dimensional

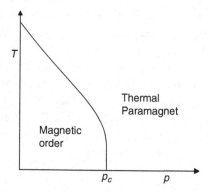

Fig. 21.1 Phase diagram of the classical dilute Ising model at finite temperature. The dilution probability is p. The phase boundary goes to zero as $p \to p_c$ as $T_c \sim 1/\ln(1/(p_c - p))$.

lattice. Thus thermal effects will destroy correlations in this cluster when $\xi_T \sim \xi$, which leads to $T_c \sim \ln(1/(p_c - p))$.

21.5.3 Quantum dilute Ising models

We are ready to consider the $T = 0$ properties of the quantum Ising model for $g \neq 0$. Its phase diagram as a function of g and p is shown in Fig. 21.2. At $p = 0$, as g is increased, there is a $T = 0$ transition from a magnetically ordered ground state to a quantum paramagnetic state, which is in the universality class of the models of Part II. On the other axis, when $g = 0$, so that the system is classical, there is a percolation transition at $p = p_c$. For $p < p_c$, for small enough g, the system retains long-range order. This is ultimately destroyed for some $g > g_c(p)$, with $g_c(p)$ expected to be a monotonically decreasing function of p. In contrast, if $p > p_c$, there is no long-range order for any g.

We now argue, as first noted by Harris [124, 201, 491], that the phase boundary of Fig. 21.2 remains vertical at $p = p_c$ for a finite range of $g \leq g_M$; we then show that a number of properties of the quantum phase transition across this vertical phase boundary can be computed exactly, as shown in [461]. The system in fact remains critical along the line $p = p_c$, $g < g_M$; to see this, note that, although there is no thermodynamically large connected cluster at p_c, there remains an infinite connected cluster with a fractal dimension smaller than d. The spins on this cluster align together at $g = 0$. A small but nonzero g is not sufficient to destroy this order on the critical cluster. That this is true can be seen by the following argument: the critical cluster is definitely more strongly connected than a one-dimensional chain of Ising spins. Even in $d = 1$ where fluctuation effects are strongest, a small g preserves long-range order in the Ising spins, as we know from Chapter 10. Thus a small g will certainly preserve the order in the critical cluster. Note that the effects of quantum fluctuations are thus quite different from the effects of thermal fluctuations discussed in the previous section. The root of this difference lies in the observation that whereas any

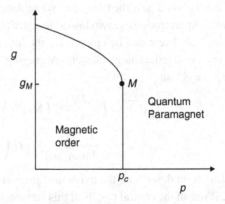

Fig. 21.2   Phase diagram of the dilute Ising model in a transverse field (g) at $T = 0$. The dilution probability is p. The multicritical point M is at $p = p_c$, $g = g_M$. The quantum transition along the vertical phase boundary ($g < g_M$, $p = p_c$) is controlled by the classical percolation fixed point at $p = p_c$, $g = 0$; quantum effects (due to a nonzero g) are dangerously irrelevant and lead to activated dynamic scaling near the $g < g_M$, $p = p_c$ line.

amount of thermal fluctuations destroys the order in the $d = 1$ Ising chain, it takes a finite strength of quantum fluctuations to do so. In fact, two spins on any sufficiently large finite cluster remain strongly correlated with each other in space for small g. (Of course, for a finite cluster there will be no long-range correlation in time.) The critical cluster eventually loses order when g reaches g_M.

Let us consider the equal-time two-point spin correlation function $C(x, 0)$ (see (10.2)). Spins at points 0 and x are correlated only if they belong to the same cluster; however, as argued above for $g < g_M$, once two spins are on the same cluster, they have an essentially perfect correlation (normalized to unity) even if they are very far apart. So the disorder-averaged $\overline{C(x, 0)}$ will be simply proportional to the probability that the two spins are on the same cluster; by the results of Section 21.5.1 we can then conclude at $p = p_c$ and for $h < h_M$, $\overline{C(x, 0)} \sim x^{-d+2-\eta_p}$. Therefore this line is critical with exponents given by that of ordinary percolation.

We can also compute a variety of static, dynamic, and thermodynamic properties across the $p = p_c$, $g < g_M$ critical line.

First, we describe some static properties. By precisely the same arguments as those above for $p = p_c$, we can conclude that for $p \geq p_c$, $\overline{C(x, 0)} \sim x^{-d+2-\eta_p} F(x/\xi)$; so the off-critical exponents and crossover functions are also those of percolation. For $p < p_c$, the spontaneous magnetization, N_0, is simply proportional to the probability that a given site lies on the infinite cluster, and so $N_0 \sim (p_c - p)^{\beta_p}$.

Now consider dynamic correlations. We compute the low-energy part of the contribution to $\chi_L''(\omega)$ by a cluster of N sites. The mean χ_L can then be computed by an average over $P(N, p)$. The energy levels of a cluster of N sites can be described for $g \ll J$ as follows: the two lowest levels correspond to the states of a single effective Ising spin with magnetic moment $\sim N$ in an effective transverse field $g_{\text{eff},N}$. For large N, $g_{\text{eff},N}$ can be estimated in Nth-order perturbation theory to be $\tilde{g} \exp(-cN)$, as discussed in Section 21.2. The quantities $\tilde{g}$ and c are of order h and $\ln(J/g)$, respectively, but their precise values depend on the particular cluster being considered. As the distributions of $\tilde{g}$ and c are not expected to become very broad near the transition, we replace them by their typical values g_0 and c_0, respectively. Apart from these two lowest levels, there are other levels separated from these by energies $\sim J$. These can be ignored for the low-energy physics, and for small $\omega \ll g$, we only need to consider large clusters. Averaging over all sites using $P(N, p)$ as written in (21.28), we obtain

$$\overline{\chi_L''(\omega)} \sim \int \frac{dN}{N^{\tau-1}} G\left(N/\xi^{d_f}\right) \delta\left(\omega - g_0 e^{-c_0 N}\right)$$

$$\sim \frac{1}{\omega (\ln(g_0/\omega))^{\tau-1}} G\left(\frac{\ln(g_0/\omega)}{c_0 \xi^{d_f}}\right). \tag{21.30}$$

This scaling form describing the dynamical properties across the vertical transition line in Fig. 21.2 is one of the central results of this section, and the reader should pause to consider its implications. Its most striking feature is that the characteristic length ξ scales as a power of the *logarithm* of the frequency ω; this is known as *activated dynamic scaling* and should be contrasted with the conventional behavior (considered in Section 21.3) where $\omega \sim \xi^{-z}$. The exponent z is effectively infinite if the dynamic scaling is activated. In the present case, the critical point $p = p_c$ contains clusters of all sizes, and as we have already seen, the characteristic excitation energy of a cluster of size L scales as $\exp(-cL^{d_f})$, which indicates the origin of the activated scaling.

The explicit results for the function G in (21.29) allow us to study the low-energy spectrum across the transition. For $p > p_c$ we get

$$\overline{\chi_L''(\omega)} \sim \frac{\omega^{d/\tilde{z}-1}}{(\ln(g_0/\omega))^{\theta-1}}, \tag{21.31}$$

which, apart from logarithms, is of the form (21.10) discussed earlier as a consequence of GM singularities. The dynamic "exponent" $\tilde{z}$ can be explicitly computed, and we find

$$\tilde{z} \sim \xi^{d_f}; \tag{21.32}$$

that is, $\tilde{z}$ *diverges* as we approach the quantum critical point. So the value of $\lim_{p \searrow p_c} \tilde{z}$ coincides with the activated dynamic scaling value of $z = \infty$. Precisely at $p = p_c$, the conventional dynamic scaling result (21.11) is replaced here by

$$\overline{\chi_L''(\omega)} \sim \frac{1}{\omega \, (\ln(g_0/\omega))^{\tau-1}}. \tag{21.33}$$

Finally, on the ordered side with $p < p_c$, the presence of the infinite cluster (and the associated long-range order) gives rise to a delta function at $\omega = 0$; for $\omega \neq 0$, $\chi_L''(\omega)$ is still determined by contributions from the finite clusters, and we find

$$\chi_L''(\omega \neq 0) \sim (1/\omega)(\ln(g_0/\omega))^{1-\theta'} \exp\left(-\kappa (\ln(g_0/\omega))^{1-1/d}\right), \tag{21.34}$$

with $\kappa \sim \xi^{-d_f(1-1/d)}$. Again the system is gapless, reflecting the GM singularities of the ordered phase.

Now we turn to the thermodynamic properties. The magnetization in response to a uniform external applied magnetic field h coupling to $\hat{\sigma}^z$ can be calculated similarly by an average over the response of clusters of size N. For small $h \ll g$, only large clusters contribute, and the magnetization per site is that of an Ising spin of magnetic moment N in a transverse field $g_{\text{eff},N}$; it is therefore given by

$$M_N(h) = \frac{Nh}{((Nh)^2 + g_{\text{eff},N}^2)^{1/2}}. \tag{21.35}$$

Thus the total magnetization per site (after subtracting the regular contribution of the infinite cluster for $p < p_c$) is

$$M(h) - M(h = 0) \sim \int dN \frac{1}{N^{\tau-1}} G\left(N/\xi^{d_f}\right) M_N(h). \tag{21.36}$$

The singular part therefore has the scaling form

$$M_{\text{sing}}(h) \sim \frac{1}{(\ln(g_0/h))^{\tau-2}} \Phi_M\left(c \frac{\ln(g_0/h)}{\xi^{d_f}}\right), \tag{21.37}$$

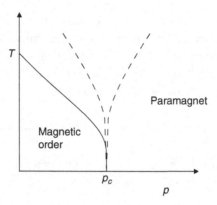

Fig. 21.3 Finite-temperature phase diagram for $g < g_M$. The dashed lines ($T \sim 1/\ln(1/|p - p_c|)$) represent crossovers from the high-T regime, characterized by spin fluctuations on the critical infinite cluster, to the low-T regimes. The solid line for $p < p_c$ is the phase transition ($T = T_c \sim 1/\ln(1/(p_c - p))$) where long-range order is destroyed.

with c a nonuniversal constant and $\Phi_M(y)$ a universal function, which is related to $G(y)$ by $\Phi_M(y) = \int_1^\infty w^{1-\tau} dw G(wy)$). Thus the consequence of activated scaling is that a power of the logarithm of the field scales as the correlation length. Using our earlier results for G, we can conclude that for $p > p_c$, $M_{\text{sing}}(h)$ rises as a power of h, with a continuously varying exponent that approaches 0 as $p \searrow p_c$, and so the linear susceptibility diverges over a whole region. At $p = p_c$ the magnetization is a power of $\ln(1/h)$. On the ordered side, $p < p_c$, $dM/dh \sim 1/h$ with weak corrections; thus the linear susceptibility diverges in the ordered side as well.

What about the finite-T properties of this quantum Ising model? For the *classical* dilute Ising model at $p = p_c$, the correlation length at finite T behaves as $\exp(\text{constant}/T)$. This is essentially due to the presence of one-dimensional segments in the critical percolating clusters. For the quantum problem for $g < g_M$, these one-dimensional segments would give rise to a thermal correlation length (ξ_T) with a similar exponential dependence on $1/T$ and a prefactor that is a power law in T; this is the behavior in the "high-T" region of Fig. 21.3. Away from the critical point, the crossovers are as shown in Fig. 21.3. The low-T behavior appears when $\xi \sim \xi_T$, or $T \sim \ln^{-1}(1/(|p - p_c|))$. On the paramagnetic side, the low-T system is described well as a collection of rigid Ising clusters with effective transverse fields and a size distribution as before; this leads, for instance, to a linear susceptibility $\chi_T \sim T^{-1+\kappa}$ (up to log corrections) with $\kappa \sim \xi^{-d_f}$. On the ordered side, there is a finite-temperature phase transition; as in the classical case, as $p \nearrow p_c$, the transition temperature falls to zero as $T_c \sim \ln^{-1}(1/(p_c - p))$. Finally, it would be interesting to understand the real-time dynamics at nonzero T, along the lines of our analysis in Part II. This remains an open problem of considerable experimental interest.

In summary, we have presented a simple example of a random quantum transition in dimensions $d > 1$ that exhibits activated dynamics scaling with $\ln(1/\text{energy scale}) \sim \xi^{d_f}$. There were Griffiths–McCoy regions on either side of the transition, with a singular density of states and a diverging susceptibility. Theoretically, an important feature of this transition is that it was controlled by a classical, static, percolation fixed point at $g = 0$, $p = p_c$, with dynamic, quantum fluctuations being "dangerously irrelevant." To see this, consider again the calculation of the field-dependent magnetization at the critical point. At the fixed point with $g = 0$, all the spins in the system align for any strength of the external field and the magnetization per spin would be 1 for any (positive) value of the field. This is, however, not correct. Quantum fluctuations prevent spins belonging to small clusters from contributing anything to the magnetization. Spins belonging to large clusters, however, contribute an amount of order 1. The crossover occurs for a cluster size $N_h \sim \ln(1/h)$. The leading scaling result is obtained by aligning all clusters with size bigger than N_h. Similarly, for the dynamics, the spin autocorrelation function is clearly just 1 at all times at the fixed point as it is classical. (Hence, we may describe such fluctuationless fixed points as static.) Again this is not correct, and we need to include the irrelevant quantum fluctuations to get the results we presented earlier.

21.6 The disordered quantum Ising chain

This section examines H_{Id} in (21.1) in dimension $d = 1$. In this case, as was shown by Fisher [140, 142], we find that for a *general* distribution of the couplings g_i, $J_{i,i+1}$ the quantum phase transition exhibits activated dynamic scaling very similar to that introduced in Section 21.5 for models on percolating clusters. This result is established using a renormalization group analysis of the entire probability distributions of the g_i and $J_{i,i+1}$, and it relies on the fact that these probability distributions become extremely broad at low-energy scales. So if we focus on the response at a given energy scale, ω, all couplings of nearby sites are either much smaller or much larger than ω: this suggests that we can set all the small couplings to zero and tightly couple the spins into clusters with the large couplings. This clustering now appears quite similar to the percolation model of Section 21.5 and explains the appearance of activated dynamic scaling in the present situation.

We begin by setting up the renormalization group analysis that will establish the above claims. We assume that the distribution of couplings is broad to begin with. Subsequent analysis shows this assumption to be self-consistent, and the resulting renormalized distributions have a large basin of attraction. The basic idea behind the procedure, first used by Dasgupta, Ma, and Hu [110, 313] in their study of the random antiferromagnetic spin chain, is to successively decimate the strongest coupling

$$\Omega \equiv \max\{g_i, J_{i,i+1}\}, \tag{21.38}$$

in the chain and get an effective Hamiltonian for the low-energy degrees of freedom. Consider the case when the maximum coupling is a field, say g_i. We first solve for the part of the Hamiltonian involving g_i: the ground state is the symmetric combination $|+\rangle_i = (|\uparrow\rangle_i + |\downarrow\rangle_i)/\sqrt{2}$, while the excited state $|-\rangle_i = (|\uparrow\rangle_i - |\downarrow\rangle_i)/\sqrt{2}$ is an energy $2g_i$ higher (see Section 5.2). As g_i is the largest energy around, it is legitimate to project the remaining Hamiltonian into the space with the state at i constrained to be $|+\rangle_i$, thereby eliminating the site i. This can be done in a simple perturbation theory in $1/g_i$, and to lowest nontrivial order the result is a new effective bond between the sites $i - 1$ and $i + 1$ of strength

$$J = \frac{J_{i-1,i} J_{i,i+1}}{g_i}. \tag{21.39}$$

We now have a new random quantum Ising chain with one fewer sites and one fewer bonds.

However, if the maximum coupling is a bond, say $J_{i,i+1}$, we first solve for the part of the Hamiltonian involving $J_{i,i+1}$. This is just the exchange interaction between the spins at sites i and $i + 1$: its ground state is doubly degenerate, $| \uparrow \rangle_i | \uparrow \rangle_{i+1}$ or $| \downarrow \rangle_i | \downarrow \rangle_{i+1}$, and the two excited states with the spins oriented in opposite directions are energy $2J_{i,i+1}$ higher (see Section 5.3). Clearly, we may think of the two degenerate ground states as corresponding to the two states of a single effective Ising degree of freedom with a magnetic moment equal to the sum of the moments of the individual spins. For large $J_{i,i+1}$, it is legitimate to project the remaining Hamiltonian into the space with the spins at i and $i + 1$ constrained to be in the same state. Again, we do this in second-order perturbation theory. The result is that the two sites i and $i + 1$ are replaced by a single Ising spin with an effective transverse field of strength

$$g = \frac{g_i g_{i+1}}{J_{i,i+1}}. \tag{21.40}$$

To this order of perturbation theory, the interaction of this effective site with the neighboring spins remains unmodified. We now again have a random quantum Ising chain with one fewer sites and one fewer bonds.

This decimation procedure is the basic renormalization group transformation. The strategy is to iterate this transformation till the maximum remaining coupling is of the order the energy at which we wish to probe the system. Note that no correlations are introduced between any of the couplings by this procedure. Thus the different bonds and fields continue to be independent random variables, though with probability distributions that are renormalized.

It is convenient to convert these recursion relations into flow equations for the distributions. From the form of the recursion relations, it is clear that it is natural to work in terms of the logarithmic variables. We therefore define

$$\Gamma = \ln(\Omega_I / \Omega),$$
$$\zeta = \ln(\Omega/J) \geq 0,$$
$$\beta = \ln(\Omega/g) \geq 0, \tag{21.41}$$

where Ω_I is the maximum coupling in the initial distributions, and Ω is the maximum at any given stage of the renormalization group. We denote the normalized probability distribution for the exchange constants by $P(\zeta; \Gamma)$ (satisfying $\int_0^\infty d\zeta \, P(\zeta; \Gamma) = 1$), and similarly the probability distributions for the transverse field by $R(\beta; \Gamma)$. As we reduce the high-energy cutoff Ω, notice that Γ becomes larger. The ultimate low energy is therefore controlled by the limit $\Gamma \to \infty$, and we are interested in the forms of the distributions P and R in this limit.

This paragraph contains a few intermediate steps showing how to transcribe the transformations (21.39) and (21.40) into partial differential equations for $P(\zeta; \Gamma)$ and $P(\beta; \Gamma)$; readers not interested in the details can move on to the next paragraph. Let $\mathcal{N}(\Gamma)$ be the total number of clusters at scale Γ, $\mathcal{N}_B(\zeta, \Gamma)$ be the total number of bonds of strength ζ at this scale, and $\mathcal{N}_S(\beta, \Gamma)$ be the total number of sites with transverse field of strength β. Then, by definition

$$P(\zeta; \Gamma) = \frac{\mathcal{N}_B(\zeta, \Gamma)}{\mathcal{N}(\Gamma)}, \qquad R(\beta; \Gamma) = \frac{\mathcal{N}_S(\beta, \Gamma)}{\mathcal{N}(\Gamma)}. \tag{21.42}$$

Now perform the basic renormalization group transformation by increasing Γ by an infinitesimal amount $d\Gamma$. This involves eliminating bonds with $\zeta \approx 0$ and sites with $\beta \approx 0$. In terms of Ω_I instead of Ω we have $\zeta = \ln(\Omega_I/J) - \Gamma$, and so when Γ is changed, ζ changes by $d\zeta = -d\Gamma$, and similarly for β. Therefore all bonds and sites with $0 < \zeta$, $\beta < d\Gamma$ are eliminated, which implies

$$\mathcal{N}(\Gamma + d\Gamma) = \mathcal{N}(\Gamma) - d\Gamma \left[\mathcal{N}_B(0, \Gamma) + \mathcal{N}_S(0, \Gamma)\right]. \tag{21.43}$$

Now consider the changes in $\mathcal{N}_B(\zeta, \Gamma)$. The transformation (21.39) will remove two bonds and add a new one. This leads to

$$\mathcal{N}_B(\zeta, \Gamma + d\Gamma) = \mathcal{N}_B(\zeta - d\zeta, \Gamma) + d\Gamma \int d\zeta_1 d\zeta_2 \mathcal{N}_S(0, \Gamma) P(\zeta_1; \Gamma) P(\zeta_2; \Gamma)$$
$$\times \left[\delta(\zeta - \zeta_1 - \zeta_2) - \delta(\zeta - \zeta_1) - \delta(\zeta - \zeta_2)\right]; \tag{21.44}$$

the first term within the square brackets represents the new bond that has been created and thus increases the probability (the delta function multiplying it is the logarithmic version of (21.39)), while the next two terms represent the two eliminated bonds. A very similar result holds for $\mathcal{N}_S$ from the transformation (21.40).

By combining (21.42)–(21.44) we obtain the required differential equations for the probability distributions:

$$\frac{\partial P(\zeta; \Gamma)}{\partial \Gamma} = \frac{\partial P(\zeta; \Gamma)}{\partial \zeta} + P(\zeta; \Gamma)(P(0; \Gamma) - R(0; \Gamma))$$
$$+ R(0; \Gamma) \int d\zeta_1 d\zeta_2 P(\zeta_1; \Gamma) P(\zeta_2; \Gamma) \delta(\zeta - \zeta_1 - \zeta_2),$$

$$\frac{\partial R(\beta; \Gamma)}{\partial \Gamma} = \frac{\partial R(\beta; \Gamma)}{\partial \beta} + R(\beta; \Gamma)(R(0; \Gamma) - P(0; \Gamma))$$
$$+ P(0; \Gamma) \int d\beta_1 d\beta_2 R(\beta_1; \Gamma) R(\beta_2; \Gamma) \delta(\beta - \beta_1 - \beta_2). \tag{21.45}$$

We are now faced with the following applied mathematics problem: given two initial arbitrary distributions $P(\zeta; \Gamma)$ and $R(\beta; \Gamma)$, evolve them with increasing Γ under (21.45); is it possible to make any general statements about possible universal forms of these distributions in the limit $\Gamma \to \infty$? This problem was solved by Fisher [142] through some rather intricate, but in principle straightforward, mathematical analysis. We are not interested in the details of this analysis here but simply assert the main results, which are then not difficult to verify a posteriori.

It was found that, for almost all initial conditions, the ultimate flow is toward one of two classes of probability distributions. In the first, most exchange constants are larger than all of the transverse fields, and this clearly represents a system that will then acquire magnetic long-range order in its ground state, as in Section 5.3. Conversely, in the second, most transverse fields are larger than all of the exchange constants, and this corresponds to a system with a quantum paramagnetic ground state, as in Section 5.2. It is of interest to first examine the critical point between these two classes of solution, in which case the two distributions P and R turn out to have precisely the same form (this a reflection of the "self-duality" of the quantum Ising chain [142]). Indeed, by setting $P = R$, it can be shown that in the limit $\Gamma \to \infty$ essentially all solutions of (21.45) are attracted to a unique fixed-point distribution; this distribution takes the scaling form

$$P(\zeta; \Gamma) = \frac{1}{\Gamma} \mathcal{P}\left(\frac{\zeta}{\Gamma}\right),$$

$$R(\beta; \Gamma) = \frac{1}{\Gamma} \mathcal{R}\left(\frac{\beta}{\Gamma}\right), \tag{21.46}$$

and the scaling functions take the simple explicit form

$$\mathcal{P}(y) = \mathcal{R}(y) = e^{-y}. \tag{21.47}$$

The reader is invited to verify that (21.46) and (21.47) constitute exact solutions of (21.45). Thus, even in terms of the logarithmic variables ζ and β, the distributions become extremely broad at low energies (the width of the distribution is $\sim \Gamma$, which rises indefinitely as we go to lower energies). This broad distribution justifies a posteriori the second-order perturbation theory used to obtain (21.39) and (21.40). If we choose the biggest transverse field g_i (say), it is overwhelmingly likely that the exchange couplings $J_{i-1,i}$ and $J_{i,i+1}$ to the neighboring sites will be much smaller. This also suggests that the results obtained by the flow equations (21.45) are asymptotically exact. In terms of the original physical couplings J and g, the fixed-point results (21.46) and (21.47) correspond to the distribution

$$P(J) \sim \frac{1}{J^{1-1/\Gamma}}, \tag{21.48}$$

and similarly for g. Note that the power in the denominator approaches 1 as Γ approaches ∞. Thus the distribution is highly singular at the origin – in fact for large enough Γ, the expectation value of $1/J$ will be divergent. It is this extreme broadness of the distribution that enables us to obtain physical properties of the system with the critical distribution through simple calculations, as we see shortly.

Let us consider perturbations of this critical solution. Linearizing the flow equations in the vicinity of the fixed point yields, as expected, a single relevant perturbation whose strength we parameterize by the coupling r; thus, as in Chapter 14, r represents the deviation from the critical point, with $r > 0$ putting the system in the quantum paramagnet. Fisher [140, 142] was also able to find a complete solution of (21.45) valid in the limit $\Gamma \to \infty$, $|r| \to 0$, but with $\Gamma|r|$ arbitrary. These solutions are expressed in scaling forms that generalize (21.47):

$$P(\zeta; \Gamma) = \frac{1}{\Gamma} \mathcal{P}\left(\frac{\zeta}{\Gamma}, r\Gamma\right),$$

$$R(\beta; \Gamma) = \frac{1}{\Gamma} \mathcal{R}\left(\frac{\beta}{\Gamma}, r\Gamma\right), \tag{21.49}$$

and the explicit solutions for the scaling functions are

$$\mathcal{P}(y, y') = \frac{2y'}{e^{2y'} - 1} \exp\left(-\frac{2yy'}{e^{2y'} - 1}\right),$$

$$\mathcal{R}(y, y') = \mathcal{P}(y, -y'). \tag{21.50}$$

Again, the reader can verify that (21.49) and (21.50) constitute exact solutions of (21.45). So we have available a family of probability distributions, parameterized by the single tuning parameter r, and there is a quantum critical point at $r = 0$ separating the magnetically ordered phase ($r < 0$) from the quantum paramagnetic phase ($r > 0$).

Let us look at the explicit predictions of the above results for the low-energy properties of the quantum paramagnet. For this we place ourselves in the paramagnet by fixing $r > 0$ and then access low energies by sending $\Gamma \to \infty$ (recall that this was the order of limits discussed in Section 21.2). Then we find the probability distribution of transverse fields to be given by

$$P(g) \sim g^{-1+2\delta}, \tag{21.51}$$

while all exchange constants are essentially at zero energy with $P(J) = \delta(J)$. The spins are therefore effectively decoupled; each site can be independently diagonalized and has two energy levels separated by $2g$. From this we can determine the leading low-energy behavior of the average local spectral density $\overline{\chi_L''(\omega)}$. A naive calculation, using the form (7.6), suggests that

$$\overline{\chi_L''(\omega)} \sim \omega^{-1+1/\tilde{z}}, \tag{21.52}$$

where we have used the notation suggested by (21.10), and the value of the exponent $\tilde{z}$ is given by

$$\tilde{z} = \frac{1}{2r}. \tag{21.53}$$

However, this result is not entirely correct. We also need to know the probability that any given original spin $\hat{\sigma}_i^z$ will be active in the set of effective spins upon which transverse fields given by (21.51) act (i.e. we must ensure that this spin has not been decimated in the renormalization group transformation). We do not have the information yet to compute this precisely (although it can be reconstructed from [142]), but it becomes clear from the analysis below that the only consequence of rectifying this omission is to change (21.52) by powers of logarithms, as in (21.10). So our result for $\overline{\chi_L''(\omega)}$ is consistent with the general arguments of the GM singularities. Further, we have found that the "dynamic exponent" $\tilde{z}$ *diverges* as we approach the critical point with $r \to 0$. This is also precisely the behavior found in (21.32) for the dilute quantum Ising models in $d > 1$ in Section 21.5. By analogy, we may then expect that the present model also exhibits activated dynamic scaling with the critical dynamic exponent $z = \infty$; further, at the critical point $r = 0$, we may expect from (21.52) that $\overline{\chi_L''(\omega)} \sim 1/\omega$ times powers of $\ln(1/\omega)$, as in (21.33).

To truly establish the existence of activated dynamic scaling, we need information about length scales. In particular, we need to know the average spacing between the spins when we have renormalized down to a characteristic energy scale $\omega \sim \Omega_I e^{-\Gamma}$. We can obtain this information by simply keeping track of the total number of spins, $\mathcal{N}(\Gamma)$, that have not been decimated at a scale Γ. From (21.43) and (21.42) we know that this quantity satisfies the differential equation

$$\frac{d\mathcal{N}(\Gamma)}{d\Gamma} = -(P(0, \Gamma) + R(0, \Gamma))\mathcal{N}(\Gamma). \tag{21.54}$$

Using the result (21.47) we can now conclude that at the critical point $r = 0$

$$\mathcal{N}(\Gamma) \sim \frac{1}{\Gamma^2}. \tag{21.55}$$

Thus the average spacing between the spins increases as Γ^2; we may identify this as the characteristic length scale, ℓ, and we have

$$\ell \sim \Gamma^2 \sim [\ln(1/\omega)]^2. \tag{21.56}$$

This is precisely the behavior characteristic of activated dynamic scaling, as exhibited in the scaling form (21.30). We can now also obtain the correlation length, ξ, as the system moves away from criticality. From the scaling (21.49), we know that $\Gamma \sim 1/r$, and therefore from (21.56) we have

$$\xi \sim \Gamma^2 \sim \frac{1}{r^2}. \tag{21.57}$$

This gives us an exponent $\nu = 2$, which saturates the bound (21.5) in $d = 1$. The length ξ actually sets the scale for the decay of the disorder-averaged correlation functions; typical spin correlations (i.e. the most probable), however, decay at a different length scale that diverges with exponent 1 [466]. For the dilute Ising model of Section 21.5, the typical spin correlations were simply zero, as two spins chosen at random typically belong to different clusters.

Fisher [142] has obtained far more precise information on the nature of the spatial correlations. We will not discuss the details of this here, but review the general strategy and indicate some further results. So far we have only kept track of the probability distributions of the coupling constants, but it is also possible to include additional information about the nature of the renormalized spins as the decimation proceeds. In particular, we can associate with each spin a magnetic moment m_i. Initially, all spins have $m_i = 1$. However, when we decimate a large $J_{i,i+1}$, the two spins at i and $i + 1$ combine to form a single effective spin with moment $m = 2$. So, in general, parallel with the recursion relation (21.40) for each bond decimation, we have the recursion relation for the magnetic moments

$$m = m_i + m_{i+1}. \tag{21.58}$$

In addition, we can also associate a length ℓ_B with each bond and a length ℓ_S with each spin. We begin with a spin chain with unit lattice spacing. Let us associate a length of $1/2$ with each spin and with each bond. Then, when we decimate the bond $J_{i,i+1}$ we get a new spin of length $3/2$. In general, the recursion relation corresponding to the decimation (21.40) is

$$\ell_S = \ell_{S,i} + \ell_{S,i+1} + \ell_{B,i,i+1}. \tag{21.59}$$

Similarly, along with the decimation of the spin with transverse field g_i in (21.39) we have the length recursion

$$\ell_B = \ell_{B,i-1,i} + \ell_{B,i,i+1} + \ell_{S,i}. \tag{21.60}$$

We can transcribe (21.58)–(21.60) into renormalization group flow equations, just as we mapped (21.39) and (21.40) into (21.45). For this, we generalize the earlier distributions for the couplings $P(\zeta; \Gamma)$ and $R(\beta; \Gamma)$ into *joint* probability distributions $P(\zeta, \ell_B; \Gamma)$ and $R(\beta, \ell_S, m; \Gamma)$. The joint distributions account for the fact that the length or moment of any given spin is certainly correlated with the size of the transverse field acting on it – a spin with a very weak transverse field must have been obtained after substantial decimation and is thus more likely to be longer and have a larger moment; similarly for the bonds. However, the couplings, lengths, and bonds of neighboring spins remain uncorrelated, as they have been obtained by independent decimation steps. The transformations (21.39), (21.40), (21.58), (21.59), and (21.60) imply flow equations for $P(\zeta, \ell_B; \Gamma)$ and $R(\beta, \ell_S, m; \Gamma)$ that are very similar to (21.45); we will not write them out explicitly but note that the first two terms on the right-hand sides have essentially the same form (the distributions only have additional obvious arguments), while the last term has additional integrals over ℓ and/or m along with delta functions imposing (21.58)–(21.60).

A thorough analysis of these new flow equations has been carried out by Fisher [142]. Here we simply note that in the limit $\Gamma \to \infty$ the distribution functions satisfy scaling forms that generalize (21.49):

$$P(\zeta, \ell_B; \Gamma) = \frac{1}{\Gamma^3} \mathcal{P}\left(\frac{\zeta}{\Gamma}, \frac{\ell_B}{\Gamma^2}, r\Gamma\right),$$

$$R(\beta, \ell_S, m; \Gamma) = \frac{1}{\Gamma^{3+\mu}} \mathcal{R}\left(\frac{\beta}{\Gamma}, \frac{\ell_S}{\Gamma^2}, \frac{m}{\Gamma^\mu}, r\Gamma\right). \tag{21.61}$$

The prefactor of the power of Γ can be deduced simply from the requirement that P and R are normalized probability distributions. The scaling $\ell \sim \Gamma^2$ was already obtained in (21.56), but it also follows from an analysis of the present flow equations. Finally, there is a nontrivial exponent μ, which controls the scaling of m; it differs from that of ℓ because of the difference in the structure of (21.58) from that of (21.59) and (21.60); it was shown by Fisher that $\mu = (\sqrt{5} + 1)/2$, the golden mean.

We can use these results to analyze the response to a uniform field h coupling to $\hat{\sigma}^z$, as was also done for the dilute Ising model in $d > 1$ below (21.35). Consider (at $T = 0$) the magnetization $M(h, r)$ of the system as a function of external applied field h. In the presence of a magnetic field h, the energy levels of an otherwise-free cluster of magnetic moment m split into two with an energy splitting $E_h = 2mh$. We stop the renormalization when the maximum coupling $\Omega \sim E_h$. The extreme broadness of the distribution implies that almost all the clusters that have already been eliminated have transverse fields considerably bigger than E_h while almost all that are yet to be eliminated have transverse fields considerably smaller than E_h. Therefore, an asymptotically exact expression for $M(h, r)$ is obtained by aligning all the remaining clusters at $\Omega = E_h$ in the direction of the magnetic field. Thus

$$M(h, r) = \bar{m} \times (\text{total number of active spins at scale } \Gamma_h = \ln(D_h/h) + \cdots ,$$

(21.62)

where $\bar{m}$ and D_h are nonuniversal constants. This total number is easily reconstructed from the probability distributions, and we therefore have the scaling form

$$\begin{aligned}
M(h, r) &= \bar{m}\mathcal{N}(\Gamma_h) \int d\zeta \, d\ell \, dm \frac{m}{\Gamma^{3+\mu}} \mathcal{R}\left(\frac{\zeta}{\Gamma}, \frac{\ell}{\Gamma^2}, \frac{m}{\Gamma^{\mu}}, r\Gamma\right) \\
&= \bar{m}\Gamma_h^{\mu-2}\bar{\Phi}_M(r\Gamma_h) \\
&= \frac{\bar{m}}{(\ln(D_h/h))^{2-\mu}} \Phi_M\left(\frac{\ln(D_h/h)}{\xi^{1/2}}\right).
\end{aligned}$$

(21.63)

We see that this scaling form is identical in structure to (21.37) of the $d > 1$ dilute Ising model. This is clearly another consequence of activated dynamic scaling. Using standard scaling arguments, we can obtain the following results from (21.63): as we approach the transition from the ordered side, the spontaneous magnetization vanishes as $N_0 \sim |r|^{\beta}$ with $\beta = 2 - \mu$; right at the critical point, $M_{cr}(h) \sim (\ln(D_h/h))^{\mu-2}$. Both forms are identical to those in Section 21.5.3.

Similar arguments can be made to obtain the exact scaling forms for the T dependence of the linear susceptibility χ_h. At the critical point, $\chi_h(T) \sim 1/T(\ln(1/T))^{2-2\mu}$ while it has power-law T dependence in the ordered and disordered phases, reflecting the GM singularities.

21.7 Discussion

We have met two rather distinct scenarios for quantum critical points in random Ising/rotor models in this chapter. Let us review their main properties in turn.

The first was discussed in Section 21.3. For the most part, the scaling structure of the quantum critical point was similar to those discussed in Part II for clean systems. Dynamic scaling was conventional, with characteristic length (ℓ) and frequency (ω) scales at the critical point obeying $\omega \sim \ell^{-z}$, with z the usual dynamic critical exponent. The phases flanking the critical point exhibited Griffiths–McCoy singularities in their low-energy behavior. For $N \geq 2$ these were only very weak essential singularities. However, they were stronger for $N = 1$ and led to a power-law divergence in the low-energy density of states, which we characterized by the exponent $\tilde{z}$. The value of $\tilde{z}$ varied continuously in the phases, and it remains an open question whether it approaches z as we move toward the critical point. Note that there is no obvious mathematical inconsistency with the two values remaining different, as they characterize regions of the spectrum reached by distinct orders of limits.

The second scenario of *activated dynamic scaling* was realized in two solvable models in Sections 21.6. This is a special property of the $N = 1$ case and has been argued to occur for the generic $N = 1$ quantum transition in $d = 1$ (Section 21.6); it was also found for a rather special dilute Ising model in all $d > 1$ in Section 21.5 but is only expected to occur for the generic transition for low values of d, possibly $d = 2$. The characteristic property of activated dynamic scaling is that the diverging scales ℓ and $1/\omega$ of the critical point are related by $\ln(1/\omega) \sim \ell^{z_a}$, where z_a is now the universal dynamic exponent. There were very strong power-law GM singularities on either side of the transition, and the exponent $\tilde{z}$ diverged as $\tilde{z} \sim \xi^{z_a}$ upon the approach to the critical point.

It is interesting that both solvable models belonged to the second class showing activated behavior. We believe that this is not an accident, and the activated scaling is a simplifying physical property that leads to the solvability. In particular, there is a clear separation of scales at which the predominant effects of quantum and disorder-induced fluctuations appear. At any given energy scale, the underlying quantum mechanics mainly serves to separate the system into mutually decoupled clusters of "active" spins. The subsequent physical properties are then determined by the random geometry and statistics of these active clusters. The spins in each cluster are tightly coupled and each contributes a term of order unity to the magnetization. As we approach the critical point, the contribution of the active spins to the magnetization does not go to zero (as it would if quantum mechanics were playing a more central role); rather the vanishing of the magnetization at the critical point is due to the vanishing of the *number* of active spins at the lower energy scales.

Further progress in this field would be greatly aided by solvable models with disorder and interactions that exhibit conventional dynamic scaling.

21.8 Applications and extensions

The exact results for the random quantum Ising chain in Section 21.6 have been very successfully compared with numerical computations [143, 550, 552]. Closely related methods have also been applied to other one-dimensional random spin models, including $S = 1/2$ Heisenberg and XY spin chains [141, 212, 222], Potts and clock models [460], $S = 1$ antiferromagnetic spin chains [231, 232], and the experimentally realizable case of chains with mixed ferromagnetic and antiferromagnetic exchange [157, 159, 540].

Turning to higher dimensions, the quantum transition in the random Ising model in $d = 2$ has been studied in sophisticated Monte Carlo simulations [233, 379], and there are indications that the activated dynamic scaling behavior is generic. As we noted earlier, the large-N limit of the random quantum rotor model was studied in [268] by a renormalization group analysis, but no stable fixed point was found; the quantum phase transition of this model has been studied numerically [199] and by an alternative renormalization group defined directly on the saddle-point equations [208].

Useful reviews of theoretical and experimental studies of random metallic systems near magnetic ordering transitions can be found in the work of Vojta [517, 530, 531].

In three dimensions, random Heisenberg antiferromagnets have been studied by the renormalization group of Section 21.6 and applied to properties of doped semiconductors [46].

Random versions of the boson models of Chapters 9 and 16 have also been studied [148, 167, 317, 484, 535] and are of considerable experimental importance.

In this chapter, we want to move beyond the simplest disordered models considered in Chapter 21 and consider systems that have magnetically ordered states rather more complicated than those in which the average moments are in a regular arrangement, as in (21.3). In the context of the Ising/rotor models, such states can be obtained by relaxing the constraint $J_{ij} > 0$ and allowing the J_{ij} to fluctuate randomly over both negative and positive values (we can always choose the g_i to be positive by a local redefinition of the spin orientations, and we assume this is the case below). In particular, we are interested here in the magnetically ordered "spin-glass" state in which orientation of the spontaneous moment varies randomly from site to site, with a vanishing average over sites, $\overline{\langle \hat{\sigma}_i^z \rangle} = 0$ (or $\overline{\langle \hat{\mathbf{n}}_i \rangle} = 0$); such states are clearly special to disordered systems. For classical spin systems, that is models (21.1) and (21.2) at $g_i = 0$, such ordered states have been reviewed at length elsewhere [48, 138, 551]. The structure of the ordered spin-glass phases of quantum models is very similar, and so this is not the focus of our interest here. Rather, we are interested in the quantum phase transition from the spin glass to a quantum paramagnet, and in the nature of the finite-temperature crossovers in its vicinity, where quantum mechanics plays a more fundamental role.

The quantum Ising/rotor models of Part II also form the basis of much of our discussion of quantum spin glasses. However, in parallel, we also consider the appearance of spin-glass order in the metallic systems of Chapter 18. One of our interests is the transition from a paramagnetic Fermi liquid to a *spin density glass* state. Such a state is characterized by the analog of the order parameter defined in Section 18.3 for the ordinary spin density wave state, but now the orientation and magnitude of ϕ_α vary randomly in space and there are random phase offsets in the cosine.

We begin by introducing the order parameter that characterizes a spin-glass phase [48, 138] using, for now, the familiar terrain of the quantum Ising/rotor models. While a spin glass has no magnetic moment when averaged over all sites, its characteristic property is that each spin has a definite orientation whose memory it retains for all time. We can use this long-time memory to introduce the Edwards–Anderson order parameter, q_{EA}, defined, for $N = 1$, by

$$q_{EA} = \lim_{t \to \infty} \overline{\langle \hat{\sigma}_i^z(t) \hat{\sigma}_i^z(0) \rangle}, \tag{22.1}$$

and similarly using rotor variables for $N > 1$. For each site i the long-time limit gives the *square* of the local static moment; this is nonnegative, and so q_{EA} has a nonzero average in the spin-glass phase.

One of the primary objectives of the theory of quantum spin glasses is to understand the nature of dynamics of spin fluctuations in the vicinity of the quantum critical point where

q_{EA} vanishes. As in Chapter 21, we expect Griffiths–McCoy singularities to appear in both the spin-glass and quantum paramagnetic phases. Reliable information on these and the critical singularities for low-dimensional systems with short-range interactions is so far only available through numerical simulations. A great deal of work has also been done on simplified models with infinite-range interactions that display spin-glass phases [51, 74, 173, 276, 335, 361, 519, 548], and the solution of classical infinite-range models was an important step in the development of spin-glass theory [48, 138]. Here, we restrict our attention to the development of a mean-field theory of the quantum critical point (and its vicinity) between a spin glass and a paramagnet in the systems noted earlier. The physical properties of the mean-field theory are closely related to those of the models with infinite-range interactions, but the former also offers a formalism for understanding fluctuations in systems with shorter-range interactions; initial attempts at understanding such fluctuations have been made [400, 436], but these are not discussed here.

We present a derivation [400, 548] of the effective action controlling quantum fluctuations of the spin-glass order parameter in Section 22.1. The mean-field solution of this effective action and its physical properties then follow in Section 22.2.

22.1 The effective action

We begin by considering, for definiteness, the appearance of spin-glass order in the quantum rotor Hamiltonian H_{Rd} in (21.2); the results also apply to the Ising case simply by restricting the $O(N)$ vector indices $\alpha, \beta \ldots$ to just one value. The extension to the metallic systems of Chapter 18 follows in Section 22.1.1.

We set all the $g_i = g$ and take the J_{ij} to be distributed independently according to the Gaussian probability

$$P(J_{ij}) \sim \exp\left(-\frac{J_{ij}^2}{2J^2}\right). \tag{22.2}$$

We average over this distribution using the replica method, which was introduced briefly in Section 21.3 (see [138] for a more complete treatment). The averaged, replicated partition function becomes

$$\bar{Z}^n = \int \mathcal{D}n_{i\alpha a}\delta\left(n_{i\alpha a}^2 - 1\right)\exp\left[-\frac{1}{2g}\int d\tau \sum_{i,a}\left(\frac{\partial n_{i\alpha a}}{\partial \tau}\right)^2\right.$$
$$\left. -\frac{J^2}{2}\sum_{\langle ij\rangle}\int d\tau_1 d\tau_2 \sum_{ab} n_{i\alpha a}(\tau_1)n_{j\alpha a}(\tau_1)n_{i\beta b}(\tau_2)n_{j\beta a}(\tau_2)\right], \tag{22.3}$$

where i, j are site indices, α, β are $O(N)$ vector indices, and a, b are replica indices; we employ the usual summation convention over repeated $O(N)$ vector indices, but all other summations are explicitly noted. We now want to manipulate this into a form in which the on-site spin correlations responsible for the spin-glass order in (22.1) are somehow related to a primary "order parameter" field; to this end we use the Hubbard–Stratanovich transformation, which we first met in Section 9.3, to decouple the quartic term in (22.3). Among the several possible transformations, we choose to decouple the four spin-operators by picking out the one which emphasizes the correlations appearing in (22.1); this gives us

$$\bar{Z}^n = \int \mathcal{D}Q_{i\alpha\beta}^{ab} \exp\left(-\int \frac{d\tau_1 d\tau_2}{2J^2} \sum_{ijab} Q_{i\alpha\beta}^{ab}(\tau_1, \tau_2) K_{ij}^{-1} Q_{j\alpha\beta}^{ab}(\tau_1, \tau_2)\right) \prod_i Z_i[Q_i],$$

(22.4)

where K_{ij} is the connectivity matrix of the lattice (its matrix elements are unity for sites i, j that were coupled by a random exchange, and zero otherwise), and the subsidiary partition function Z_i is a functional of the values of the field $Q_{i\alpha\beta}^{ab}(\tau_1, \tau_2)$ only on the site i. It is obtained after a functional integral only over the site i quantum field $n_{i\alpha a}$:

$$Z_i[Q_i] = \int \mathcal{D}n_{\alpha a} \delta\left(n_{\alpha a}^2 - 1\right) \exp\left[-\frac{1}{2g} \int d\tau \sum_a \left(\frac{\partial n_{\alpha a}}{\partial \tau}\right)^2\right.$$
$$\left. - \int d\tau_1 d\tau_2 \sum_{ab} Q_{i\alpha\beta}^{ab}(\tau_1, \tau_2) n_{\alpha a}(\tau_1) n_{\beta b}(\tau_2)\right].$$

(22.5)

We have dropped the dummy site index on **n** as this field is integrated over. Note that the functional integral in $Z_i[Q_i]$ is closely related to those considered in Chapter 3 in our study of classical $d = 1$ spin chains. The latter models are exactly soluble, and their known correlators can be used to construct an expansion for $Z_i[Q_i]$ in powers of Q_i for an arbitrary time-dependent Q_i. It should be kept in mind that there are n decoupled copies of the classical chain here, and this does lead to an interesting and important structure in the resulting action. After evaluating $Z_i[Q_i]$ in this manner, we take the spatial continuum limit and obtain our spin-glass partition function, which we write schematically in the form

$$Z_{sg} = \int \mathcal{D}Q_{\alpha\beta}^{ab}(x, \tau_1, \tau_2) \exp(-\mathcal{S}_{sg}[Q]).$$

(22.6)

Now the focus of our attention is the field $Q_{\alpha\beta}^{ab}(x, \tau_1, \tau_2)$, which plays the role of an order parameter for the quantum spin glass. Before turning our attention to the structure of the action $\mathcal{S}_{sg}[Q]$, we discuss the physical interpretation of Q. From the structure of the Hubbard–Stratanovich transformation it is clear that we have the correspondence

$$Q_{\alpha\beta}^{ab}(x_i, \tau_1, \tau_2) \sim n_{i\alpha a}(\tau_1) n_{i\beta b}(\tau_2),$$

(22.7)

where the symbol $\sim$ indicates that correlators of Q are closely related to the corresponding correlators of the right-hand side; for simplicity we assume the proportionality constant is unity and replace (22.7) by an equality. From (22.7) we see that the replica *diagonal* components have the mean value

$$\lim_{n \to 0} \frac{1}{n} \sum_a \left\langle\!\left\langle Q_{\alpha\beta}^{aa}(x_i, \tau_1, \tau_2)\right\rangle\!\right\rangle = \overline{\langle n_{i\alpha}(\tau_1) n_{i\beta}(\tau_2)\rangle}$$
$$= \delta_{\alpha\beta} \overline{\chi_L(\tau_1 - \tau_2)},$$

(22.8)

where the double angular brackets represent averages taken with the translationally invariant replica action $\mathcal{S}_{sg}$ in (22.6) (recall that single angular brackets represent thermal/quantum averages for a fixed realization of randomness, and overlines represent averages over disorder). So the mean value of Q contains information on the entire time (or frequency) dependence of the average dynamic local susceptibility, which we also considered earlier in (21.7); in a sense, it is the time-dependent χ_L which is the "order parameter functional" for the quantum spin glass. From (22.1) and (22.8), we see that the Edwards–Anderson order parameter, q_{EA}, can be extracted for the replica diagonal components of Q by taking the long-time limit of (22.8) for *real* times t. Precisely at $T = 0$, this long-time limit can also be taken along the imaginary time axis (for $T > 0$, $\chi_L(\tau)$ is a periodic function with period $1/T$, and so the long-time limit is only defined for real times), and we have

$$q_{EA} = \lim_{\tau \to \infty} \lim_{n \to 0} \frac{1}{n} \sum_a \left\langle\!\left\langle Q^{aa}(x, \tau_1 = 0, \tau_2 = \tau) \right\rangle\!\right\rangle, \qquad T = 0. \tag{22.9}$$

Turning to the replica off-diagonal components, we see by a standard application of replica technology [48, 138] that

$$\lim_{n \to 0} \frac{1}{n(n-1)} \sum_{a \neq b} \left\langle\!\left\langle Q_{\alpha\beta}^{ab}(x_i, \tau_1, \tau_2) \right\rangle\!\right\rangle = \overline{\langle n_{i\alpha}(\tau_1)\rangle\langle n_{i\beta}(\tau_2)\rangle}$$

$$= q_{EA}. \tag{22.10}$$

The thermal average in the second step leads to time-independent values, and so the expectation value of the off-diagonal components is independent of both τ_1 and τ_2. In the last step, we have assumed that the thermal ensemble has the "clustering" property, which demands that the long-time limit of the correlator in (22.1) is simply the square of the static magnetic moment on the site, as discussed in Section 1.4: although it is certainly possible to construct states that do not obey clustering, imposing a suitable infinitesimal external field on each site will select the ensemble that does obey (22.10). In (22.10) we have also ignored subtleties that may arise as a consequence of the intricate phenomenon known as "replica symmetry breaking." In the simple mean-field theory we consider below, replica symmetry breaking does not occur; however, it does appear when additional higher order couplings are included [400], but fortunately the structure and analysis of replica symmetry breaking in the spin-glass phase turn out to be essentially identical to those discussed elsewhere in the classical case [48, 138]. For these reasons, and also because our interest is primarily in the spin fluctuations in the paramagnetic phase, we do not consider this phenomenon here further.

Returning to our determination of the form of $\mathcal{S}_{sg}[Q]$, recall that we noted that the only realistic option was an expansion in powers of Q. It is worthwhile to ponder a bit on the validity of such an expansion. In the vicinity of the quantum critical point, we expect q_{EA} to become small; in the spirit of Landau theory, it would then certainly be appropriate to expand in powers of q_{EA}. However, for the quantum transition we need the full time-dependent Q, and not just its long-time limit. For very short $|\tau_1 - \tau_2|$, the local on-site spin correlations will certainly be of order unity, and so Q will not be small for these times.

What we need to do is to "subtract out" the uninteresting short-time part of Q and focus on only its long-time part for which a Landau-like expansion could possibly be valid. To do this we consider the following transformation:

$$Q_{\alpha\beta}^{ab}(x, \tau_1, \tau_2) \rightarrow Q_{\alpha\beta}^{ab}(x, \tau_1, \tau_2) - C\delta^{ab}\delta(\tau_1 - \tau_2), \qquad (22.11)$$

where C is a constant, and the delta function $\delta(\tau_1 - \tau_2)$ is a schematic for a function that decays rapidly to zero on a short microscopic time. The value of C should be adjusted so that the resulting Q contains only the interesting long-time physics. (At this point, it is not clear how this can be done, but we see shortly that a simple constraint on the effective action allows us to do this quite easily.)

Let us consider the expansion of $Z_i[Q_i]$ in (22.5) in powers of Q: we discuss the nature of the low-order terms explicitly, and from these the principles that restrict the structure of the general term emerge.

The first term is one linear in Q. It is multiplied by a two-point correlator of n, which is nonzero only if both replica indices are the same. Further, the subsequent replica-diagonal average correlates the two time arguments in Q, and we get an expression such as

$$\int d^d x \, d\tau_1 d\tau_2 \, Q_{\alpha\alpha}^{aa}(x, \tau_1, \tau_2)\chi_L^0(\tau_1 - \tau_2), \qquad (22.12)$$

where the superscript 0 on the local susceptibility reminds us that this is a bare susceptibility, evaluated without accounting for intersite correlations. Now an important property of $\chi_L^0(\tau)$ (and all other multipoint correlations of n) is that it decays rapidly to zero over a time τ of order $1/g$. In frequency space, we have in the low-frequency limit

$$\chi_L^0(\omega_n) \sim (\omega_n^2 + \Delta_0^2)^{-1} \sim \Delta_0^{-2} - \omega_n^2\Delta_0^{-4} + \cdots, \qquad (22.13)$$

where $\Delta_0 \sim g$ is the gap of the classical chain model $Z_i[Q = 0]$ studied in Chapter 3. If we just take the leading frequency-independent term in (22.13), we have effectively replaced χ_L^0 by a constant and set $\tau_1 = \tau_2$. This is an important principle, which applies also to higher order terms: an even number of replica indices can take the same value, but then the associated time "indices" must also be set equal, as they can be correlated by quantum fluctuations of the underlying rotors. Subleading corrections involve derivatives of the difference in times, and it turns out to be necessary to retain the additional ω_n^2 dependence in (22.13) only for the linear term in (22.12).

Moving on to higher order terms, we see that the number of allowed terms proliferates very rapidly. In particular, at nth order there are terms that can have between 1 and n independent replica indices summed over; associated with each independent replica index is a time "index," which is integrated over in the action. Thus the terms have a variable number of time integrations, and it turns out that most important are those with a maximum number of independent time (and replica) indices. This is not difficult to see from a renormalization group perspective, because each additional time integration increases the scaling dimension of the associated coupling constant.

Proceeding in this manner, we can assert the following results for $\mathcal{S}_{sg}$ for the quantum Ising/rotor spin glass; we have used the benefit of hindsight and retained only the terms necessary to obtain the leading critical singularities within mean-field theory:

$$
\begin{aligned}
\mathcal{S}_{sg} = \frac{1}{w} \int d^d x \Bigg\{ & \frac{1}{\kappa} \int d\tau \sum_a \left[\frac{\partial}{\partial \tau_1} \frac{\partial}{\partial \tau_2} + r \right] Q^{aa}_{\alpha\alpha}(x, \tau_1, \tau_2) \Bigg|_{\tau_1 = \tau_2 = \tau} \\
& + \frac{1}{2} \int d\tau_1 d\tau_2 \sum_{ab} \left[\nabla Q^{ab}_{\alpha\beta}(x, \tau_1, \tau_2) \right]^2 \\
& - \frac{\kappa}{3} \int d\tau_1 d\tau_2 d\tau_3 \sum_{abc} Q^{ab}_{\alpha\beta}(x, \tau_1, \tau_2) Q^{bc}_{\beta\rho}(x, \tau_2, \tau_3) Q^{ca}_{\rho\alpha}(x, \tau_3, \tau_1) \\
& + \frac{1}{2} \int d\tau \sum_a \Big[u\, Q^{aa}_{\alpha\beta}(x, \tau, \tau) Q^{aa}_{\alpha\beta}(x, \tau, \tau) \\
& + v\, Q^{aa}_{\alpha\alpha}(x, \tau, \tau) Q^{aa}_{\beta\beta}(x, \tau, \tau) \Big] \Bigg\} \\
& - \frac{1}{2w^2} \int d^d x \int d\tau_1 d\tau_2 \sum_{ab} Q^{aa}_{\alpha\alpha}(x, \tau_1, \tau_1) Q^{bb}_{\beta\beta}(x, \tau_2, \tau_2). \quad (22.14)
\end{aligned}
$$

There are seven terms in the action, but only five coupling constants: w, r, κ, u, and v. Rescaling of space and time coordinates has allowed us to absorb the other two. The first two terms are linear in Q and are clearly a transcription of the two terms retained explicitly in (22.13). As the notation suggests, the coupling r turns out to be the relevant tuning parameter that moves the system between its two phases, and we are interested in the phase diagram in the r–T plane. The spatial gradient term arises from the K^{-1} coupling in (22.4), which couples different sites. This last coupling, and also the expansion of $Z_i[Q]$, also allow the simple quadratic term

$$
\int d^d x\, d\tau_1 d\tau_2 \sum_{ab} \left[Q^{ab}_{\alpha\beta}(x, \tau_1, \tau_2) \right]^2, \quad (22.15)
$$

which we have not included in $\mathcal{S}_{sg}$; instead we have chosen the freedom allowed by the transformation (22.11) to demand that the coefficient of this term be exactly zero. At the moment, this appears just as a convenient choice, but it is seen later to be exactly the criterion required to focus on only the interesting low-frequency behavior of Q. The quadratic terms proportional to u and v have only a single replica index and account for the nonlinear, quantum mechanical interactions of the quantum rotors. We have retained only a single cubic term, proportional to κ/w, the one with the maximum possible three time integrations; other allowed cubic terms are not as important.

Finally, the last term, proportional to $1/w^2$, actually does not appear in the expansion for $\mathcal{S}_{sg}$ as we have chosen to explicitly generate it. To obtain it, we have to allow for *on-site* disorder in the value of g_i as can be schematically seen in a "soft-spin" approach where randomness in g corresponds to a random mass multiplying $\phi^2 \sim Q^{aa}$; averaging over the random mass then leads to the last term in (22.14). However, even in the present model with g fixed, the $1/w^2$ term is generated upon any renormalization with the remaining couplings in $\mathcal{S}_{sg}$. In any case, this $1/w^2$ term plays no role in the mean-field theory to follow, and so we will not discuss it further. It is, however, important to retain it in any analysis of fluctuations.

22.1.1 Metallic systems

Let us consider extension of the analysis of phase transition of Fermi liquids in Chapter 18 to the case of a "spin density glass" [368, 369, 409, 436, 457]. For this we generalize to a model with a random exchange interaction

$$H_{\text{sdg}} = \int \frac{d^d k}{(2\pi)^d} (\varepsilon_k - \mu) c^\dagger_{\vec{k}\mu} c_{\vec{k}\mu} - \sum_{\langle ij \rangle} J_{ij} \hat{\mathbf{S}}_i \cdot \hat{\mathbf{S}}_j, \tag{22.16}$$

where

$$\hat{\mathbf{S}}_i \equiv \frac{1}{2} \sum_{\mu\nu} c^\dagger_{i\mu} \vec{\sigma}_{\mu\nu} c_{i\nu} \tag{22.17}$$

are the electron spin operators on site i. As was the case for the Ising/rotor models above, we take the J_{ij} to be independent Gaussian random variables. We refer the reader to a review by the author [423] for the arguments motivating (22.16) as an appropriate low-energy model for a large class of disordered metallic systems; a discussion of the strong Griffiths–McCoy singularities in such models [45, 338] may also be found there.

Our analysis of H_{sdg} follows closely the steps presented above for the Ising/rotor models. The field $\mathbf{S}$ replaces $\mathbf{n}$ and so now we have

$$Q^{ab}_{\alpha\beta}(x_i, \tau_1, \tau_2) \sim S_{i\alpha a}(\tau_1) S_{i\beta b}(\tau_2), \tag{22.18}$$

replacing (22.7). Also, the on-site action $(\partial n/\partial\tau)^2/(2g)$ is replaced by the first kinetic energy term in (22.16). All other steps are the same, and we obtain an expression identical to (22.4), with the modifications just noted in the definition of $Z_i[Q_i]$. The steps in the derivation of $\mathcal{S}_{sg}$ are also the same, except the functional integral over the metallic electrons leads to differences in the time dependence of the terms. In particular, from the arguments just above (18.33) we see that the expression (22.13) for the local susceptibility is replaced by

$$\chi_L(\omega_n) \sim A_1 - A_2 |\omega_n| + \cdots, \tag{22.19}$$

for some constants A_1 and A_2. This turns out to be the only significant change in $\mathcal{S}_{sg}$. Hence the final result takes exactly the form (22.14) except that the single time-derivative term (with coefficient $1/(w\kappa)$) is replaced by (after a Fourier transform of (22.19))

$$-\frac{1}{\pi w\kappa} \int d^d x \, d\tau_1 d\tau_2 \sum_a \frac{Q^{aa}_{\alpha\alpha}(x, \tau_1, \tau_2)}{(\tau_1 - \tau_2)^2}. \tag{22.20}$$

This change in the time-derivative term is completely analogous to the change between (14.2) and (18.33) for the case of regular magnetic order.

22.2 Mean-field theory

We now analyze the action $\mathcal{S}_{sg}$ in (22.14), and its metallic extension modification (22.20), in a simple mean-field theory. An analysis of the rather complex structure of fluctuations about this mean field has been attempted [400, 436], but we do not discuss it here as the results are quite inconclusive. The mean-field theory is useful in that it gives a simple picture of the quantum critical point and the finite-temperature crossovers in its vicinity, which should serve as a starting point for more sophisticated analyses.

Our strategy is to obtain saddle points of $\mathcal{S}_{sg}$ over variations in a mean-field value of the field $Q(x, \tau_1, \tau_2)$. We expect the saddle point to be invariant under translation in space and time, which implies that Q is independent of x and a function only of $\tau_1 - \tau_2$. Fourier transforming to Matsubara frequencies by

$$Q_{\alpha\beta}^{ab}(x, \omega_{n1}, \omega_{n2}) = \int_0^{1/T} d\tau_1 \int_0^{1/T} d\tau_2 \, Q_{\alpha\beta}^{ab}(x, \tau_1, \tau_2) e^{-i(\omega_{n1}\tau_1 + \omega_{n2}\tau_2)} \tag{22.21}$$

motivates the following saddle-point ansatz:

$$Q_{\alpha\beta}^{ab}(x, \omega_{n1}, \omega_{n2}) = \frac{q_{EA}}{T^2}\delta_{\omega_{n1},0}\delta_{\omega_{n2},0}\delta_{\alpha\beta} + \frac{\chi_L(\omega_{n1})}{T}\delta^{ab}\delta_{\omega_{n1}+\omega_{n2},0}\delta_{\alpha\beta}. \tag{22.22}$$

The first term is independent of the replica indices, and it therefore has been parameterized in terms of the Edwards–Anderson order parameter by (22.10). The second replica diagonal term is related to the local susceptibility by (22.8) (we have dropped the overline representing the disorder average because it is always implied in the present context). Quite independent of these physical interpretations it is clear that (22.22) is the most general replica-symmetric ansatz for Q in terms of the parameters q_{EA} and $\chi_L(\omega_n)$. We insert (22.21) and (22.22) into (22.14) and (22.20) and obtain for the mean-field free energy density per replica, $\mathcal{F}/n$:

$$\frac{\mathcal{F}}{Nn} = \frac{T}{w}\sum_{\omega_n}\left[\left(\frac{M(\omega_n)+r}{\kappa}\right)\chi_L(\omega_n) - \frac{\kappa}{3}\chi_L^3(\omega_n)\right]$$

$$+ \frac{u+Nv}{2w}\left[q_{EA} + T\sum_{\omega_n}\chi_L(\omega_n)\right]^2 + \frac{q_{EA}}{w}\left(\frac{r}{\kappa} - \kappa\left[\chi_L(0)\right]^2\right), \tag{22.23}$$

where

$$M(\omega_n) = \begin{cases} \omega_n^2 & \text{for the Ising/rotor models,} \\ |\omega_n| & \text{for the metallic system.} \end{cases} \tag{22.24}$$

There should also be an additional term in (22.23) coming from the last term in (22.14), but it is proportional to n and therefore does not contribute in the replica limit $n \to 0$. Under these circumstances, the coupling $1/t$ appears only as a prefactor in front of the total free energy, and so the value of w therefore plays no role in the mean-field theory. The replica limit $n \to 0$ has also been taken to simplify terms arising from the cubic coupling in $\mathcal{S}_{sg}$. Also, because we are considering a metallic system with Heisenberg symmetry, we should set $N = 3$.

We now determine the saddle point of (22.23) with respect to variations in q_{EA} and $\chi_L(\omega_n)$ for every ω_n; the resulting expressions can be written in the form

$$\chi_L(\omega_n) = -\frac{1}{\kappa}\sqrt{M(\omega_n) + \Upsilon},$$

$$q_{EA}\sqrt{\Upsilon} = 0, \tag{22.25}$$

where Υ is an intermediate parameter satisfying the equation

$$\Upsilon = r + (u + Nv)\left(\kappa q_{EA} - T\sum_{\omega_n}\sqrt{M(\omega_n) + \Upsilon}\right). \tag{22.26}$$

Equations (22.25) and (22.26) clearly have two distinct types of solution. The first corresponds to the paramagnetic phase in which the spin-glass order parameter vanishes, and so

$$q_{EA} = 0, \tag{22.27}$$

$$\chi_L(\omega_n) = -\frac{1}{\kappa}\sqrt{M(\omega_n) + \Upsilon}, \tag{22.28}$$

$$\Upsilon = r - (u + Nv)T\sum_{\omega_n}\sqrt{M(\omega_n) + \Upsilon}; \tag{22.29}$$

the parameter $\Upsilon > 0$ is then determined from the solution of the nonlinear equation (22.29). The second solution is that of the spin-glass phase in which $\Upsilon = 0$, and so

$$\chi_L(\omega_n) = -\frac{1}{\kappa}\sqrt{M(\omega_n)},$$

$$q_{EA} = -\frac{r}{\kappa(u + Nv)} + \frac{T}{\kappa}\sum_{\omega_n}\sqrt{M(\omega_n)}. \tag{22.30}$$

It is clear that for sufficiently large $r > 0$ the paramagnetic solution is the only physically sensible one, and it has a large $\Upsilon > 0$. As we decrease r at fixed T, the value of Υ decreases, and we have phase transition into the spin-glass phase where Υ first vanishes; this happens at $r = r_c(T)$, which is determined by setting $\Upsilon = 0$ in (22.29):

$$r_c(T) \equiv (u + Nv)T\sum_{\omega_n}\sqrt{M(\omega_n)}. \tag{22.31}$$

The spin-glass phase therefore exists for $r < r_c(T)$. It should be clear from this discussion that r plays the role of the relevant tuning parameter for the quantum transition, and this notation is consistent with that of Chapter 14. As in that chapter, it is convenient to shift variables by defining

$$s \equiv r - r_c(0), \tag{22.32}$$

so that the quantum critical point is precisely at $s = 0$, $T = 0$; at $T = 0$ the system is paramagnetic for $s > 0$ and a spin glass for $s < 0$. For $T > 0$ we have a phase boundary at $s = s_c(T) < 0$ whose precise shape will be determined shortly below. These considerations lead to the phase diagram shown in Fig. 22.1.

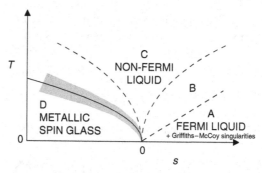

Mean-field phase diagram of a metallic spin glass as a function of the ground state tuning parameter s and temperature T. The $T = 0$ state is a metallic spin glass for $s < 0$ and a disordered, paramagnetic Fermi liquid for $s > 0$. The solid line is the only thermodynamic phase transition and is at $s = s_c(T)$ or $T = T_c(s)$ given in (22.44). The quantum critical point is at $s = 0, T = 0$. The dashed lines denote crossovers between different finite-T regions of the quantum field theory (22.14): the low-T regions are A, B (on the paramagnetic side) and D (on the ordered side), while the high-T region (C) displays "non-Fermi liquid" behavior. The crossovers on either side of C, and the spin-glass phase boundary $T_c(s)$, all scale as $T \sim |s|^{2/3}$; the boundary between A and B obeys $T \sim s$. The shaded region has classical critical fluctuations described by theories of the type discussed by [48] and [138].

Let us briefly discuss the physical properties of the phases found here in mean-field theory. In the paramagnetic phase, the local spectral density of the Ising/rotor models (with $M(\omega_n) = \omega_n^2$) is given by

$$\chi_L''(\omega) = \text{sgn}(\omega)\frac{\sqrt{\omega^2 - \Upsilon}}{\kappa}\theta(|\omega| - \sqrt{\Upsilon}); \qquad (22.33)$$

hence there is an energy gap, and spectral density increases with a square-root threshold above this gap. Clearly, we can expect that this gap will be filled in at $T = 0$ by Griffiths–McCoy singularities once fluctuation effects are included; for $T > 0$ ordinary thermal fluctuations are adequate to destroy the gap. The mean-field spectrum becomes gapless precisely at the critical point where $\Upsilon = 0$ and the spectral density vanishes linearly with frequency. The spectral density of the paramagnetic phase of the metallic systems is quite different; now we have $M(\omega_n) = |\omega_n|$, and this leads to

$$\chi_L''(\omega) = \frac{1}{\sqrt{2}\kappa}\frac{\omega}{\sqrt{\Upsilon + \sqrt{\omega^2 + \Upsilon^2}}}; \qquad (22.34)$$

now there is no gap, but the spectral density is linear, $\sim\omega$, for frequencies smaller than Υ, and a square root, $\sim\sqrt{\omega}$, for larger frequencies. We make some further remarks on the physical interpretation of this spectral density below.

Turning to the spin-glass phase, it is clear from (22.31) and (22.32) that the Edwards–Anderson order parameter is given by

$$q_{EA} = \frac{1}{\kappa(u + Nv)}\left[s_c(T) - s\right]. \qquad (22.35)$$

The spectral density remains pinned at the $\Upsilon = 0$ case of (22.33) and (22.34) in the entire spin-glass phase in the present mean-field theory.

An interesting property of the above solutions is that the low-frequency limit of the function $\chi_L(\omega_n)$ becomes small as one approaches the phase boundary (for real ω both the real and imaginary parts of χ_L become small), which indicates that expanding in powers of Q was appropriate. This smallness is actually a consequence of the shift (22.11) used to eliminate the term (22.15) from the action. If we had instead included (22.15) in the present mean-field analysis, we would have found a very similar solution, but the resulting χ_L would have an additional frequency-independent contribution to its real part, which remained of order unity at the phase boundary. Such a regular frequency-independent term does not modify the interesting long-time correlations or the low-frequency spectral weight, which actually remain as we have found them here. This is then the promised a-posteriori justification for the expansion employed in obtaining $\mathcal{S}_{sg}$.

Let us discuss the nature of the finite-temperature crossovers within the paramagnetic phase of the metallic system, as shown in Fig. 22.1; the behavior of the Ising/rotor system is closely related and details may be found elsewhere [400]. The spectral density is given everywhere by (22.34), which depends solely on the energy scale Υ, to be determined by the solution of (22.29). We present a complete derivation of the universal T and s dependence of Υ in the vicinity of the quantum critical point $T = 0$, $s = 0$; despite the seemingly simple equation (22.29) to be solved, a great deal of structure emerges, including some nontrivial crossover functions. We begin by combining (22.29), (22.31), and (22.32) into

$$\Upsilon + (u + Nv)T\sqrt{\Upsilon} = s - (u + Nv)\left(T\sum_{\omega_n \neq 0}\sqrt{|\omega_n| + \Upsilon} - \int \frac{d\omega}{2\pi}\sqrt{|\omega|}\right). \quad (22.36)$$

For convenience, we have chosen to move the $\omega_n = 0$ term in the frequency summation from the right-hand to the left-hand side. To leading order in $u + Nv$, this equation has the simple solution $\Upsilon = s$. To improve this result it is adequate to simply set $\Upsilon = s$ on the right-hand side of (22.36) since the minimum value of ω_n in the summation is $2\pi T$ and this always turns out to be much larger than Υ in the interesting universal region, as becomes clear from the analysis below. This strategy of separating the $\omega_n = 0$ and $\omega_n \neq 0$ terms, and of treating the $\omega_n = 0$ term with more care, is reminiscent of the approach applied in Chapter 14 for finite-T crossovers; we see below that the resulting crossovers are very similar to those found in Section 14.2.2 for the case when the clean Ising/rotor model was above its upper-critical dimension. After making the noted approximation, we can further manipulate (22.36) into

$$\begin{aligned}
\Upsilon + (u + Nv)T\sqrt{\Upsilon} = {} & s + (u + Nv)T\sqrt{s} \\
& - (u + Nv)\left(T\sum_{\omega_n}\sqrt{|\omega_n| + s} - \int \frac{d\omega}{2\pi}\sqrt{|\omega| + s}\right) \\
& - (u + Nv)\int \frac{d\omega}{2\pi}\left(\sqrt{|\omega| + s} - \sqrt{|\omega|} - \frac{s}{2\sqrt{|\omega|}}\right) \\
& - (u + Nv)\int \frac{d\omega}{2\pi}\frac{s}{2\sqrt{|\omega|}}. \quad (22.37)
\end{aligned}$$

The manipulations above are similar to those discussed below (11.47) and used extensively in Chapter 14: we always subtract from the summation over Matsubara frequencies of any function, the integration of precisely the same function; the difference is then convergent in the ultraviolet, and such a procedure leads naturally to the universal crossover functions [422]. We now manipulate (22.37) into a form where it is evident that Υ is analytic as a function of s at $s = 0$ for $T > 0$. This analyticity is of course closely related to that discussed in Sections 14.2.1 and 14.2.2 and is due to the absence of any thermodynamic singularity for $T > 0, s = 0$ (see Fig. 22.1). We use the identity

$$\int_0^\infty \sqrt{\alpha}d\alpha \left(\frac{1}{\alpha + a} - \frac{1}{\alpha + b} \right) = \pi \left(\sqrt{b} - \sqrt{a} \right) \tag{22.38}$$

to rewrite (22.37) as

$$\Upsilon + (u + Nv)T\sqrt{\Upsilon} = s \left(1 - \frac{(u + Nv)\Lambda_\omega^{1/2}}{\pi} \right) + (u + Nv)T\sqrt{s} + \frac{(u + Nv)}{\pi}$$
$$\times \int_0^\infty \sqrt{\alpha}d\alpha \left(T \sum_{\omega_n} \frac{1}{\alpha + |\omega_n| + s} - \int \frac{d\omega}{2\pi} \frac{1}{\alpha + |\omega| + s} \right)$$
$$+ \frac{(u + Nv)}{\pi} \int_0^\infty \sqrt{\alpha}d\alpha \int \frac{d\omega}{2\pi} \left(\frac{1}{\alpha + |\omega| + s} \right.$$
$$\left. - \frac{1}{\alpha + |\omega|} + \frac{r}{(\alpha + |\omega|)^2} \right), \tag{22.39}$$

where Λ_ω is an upper cutoff for the frequency. We evaluate the frequency summation by expressing it in terms of the digamma function ψ, and we perform all frequency integrals exactly. After some elementary manipulations (including use of the identity $\psi(s + 1) = \psi(s) + 1/s$), we obtain our final result for Υ, in the form of a solvable quadratic equation for $\sqrt{\Upsilon}$:

$$\Upsilon + (u + Nv)T\sqrt{\Upsilon} = s \left(1 - \frac{(u + Nv)\Lambda_\omega^{1/2}}{\pi} \right) + (u + Nv)T^{3/2}\Phi_{\text{sdg}} \left(\frac{s}{T} \right), \tag{22.40}$$

where the universal crossover function of the spin-density glass $\Phi_{\text{sdg}}(y)$ is given by

$$\Phi_{\text{sdg}}(y) = \frac{1}{\pi^2} \int_0^\infty \sqrt{\alpha}d\alpha \left[\log \left(\frac{\alpha}{2\pi} \right) - \psi \left(1 + \frac{\alpha + y}{2\pi} \right) + \frac{\pi + y}{\alpha} \right]. \tag{22.41}$$

Notice the similarity in the structure of the above results to that of (14.32) and (18.83): in all cases we have universal crossover functions for the characteristic energy scale in the vicinity of a quantum critical point; indeed, for accidental reasons the universal function $\Phi_{\text{sdg}}(y)$ is proportional to the universal function L in (18.84) in $d = 3$. Also note that the crossovers in (22.40) depend upon the magnitude of the microscopic couplings u and v, which represent the quantum mechanical interactions between the rotors or Ising spins.

This is also a feature of (14.32) and (18.83), and by analogy we may conclude that the couplings u and v are formally irrelevant at the quantum critical point but are nevertheless crucial in constructing the crossovers at nonzero temperatures (i.e. they are *dangerously irrelevant*). This expectation is verified by an explicit renormalization group analysis of $\mathcal{S}_{sg}$ [436] ,which we shall not discuss here.

The above expression for $\Phi_{\text{sdg}}(y)$ is clearly analytic for all $y \geq 0$, including $y = 0$, as we hoped to achieve. As was the case for (14.32) and (14.33), we can use the above result for $y < 0$ until we hit the first singularity at $y = -2\pi$, which is associated with singularity of the digamma function $\psi(s)$ at $s = 0$. However, this singularity is of no physical consequence, as it occurs within the spin-glass phase (Fig. 22.1), where the above solution is not valid; as shown below, the transition to the spin-glass phase occurs for $y \sim -(u + Nv)T^{1/2}$, which is well above -2π. For our subsequent analysis, it is useful to have the following limiting results, which follow from (22.41):

$$\Phi_{\text{sdg}}(y) = \begin{cases} \sqrt{1/2\pi}\,\zeta(3/2) + \mathcal{O}(y), & y \to 0, \\ (2/3\pi)y^{3/2} + y^{1/2} + (\pi/6)y^{-1/2} + \mathcal{O}(y^{-3/2}), & y \to \infty. \end{cases} \tag{22.42}$$

The expression (22.34), combined with the results (22.40) and (22.41), completely specifies the s and T dependence of the dynamic susceptibility in the paramagnetic phase and allow us to obtain the phase diagram shown in Fig. 22.1, whose details we now discuss. There is a quantum critical point at $s = 0$, $T = 0$, and the characteristic energy scale Υ vanishes *linearly* upon approach to this point at $T = 0$:

$$\Upsilon \sim s, \qquad \text{for } s > 0, T = 0. \tag{22.43}$$

There is a line of finite-temperature phase transitions, denoted by the full line in Fig. 22.1, which separates the spin-glass and paramagnetic phases; this line is determined by the condition $\Upsilon = 0$ and is at $r = s_c(T)$(or $T = T_c(s)$), with

$$s_c(T) = -(u + Nv)\Phi(0)T^{3/2} \quad \text{or} \quad T_c(s) = [-s/(u + Nv)\Phi(0)]^{2/3}. \tag{22.44}$$

The crossovers within the paramagnetic phase are similar to those found in Sections 14.2.2 and 18.4, and we discuss below the characteristics of the different limiting regimes.

(A, B) Low-T Paramagnetic Fermi Liquid, $T < [s/(u + Nv)]^{2/3}$

This is the "Fermi liquid" region, where the leading contribution to the characteristic energy scale Υ is its $T = 0$ value $\Upsilon(T) \sim \Upsilon(0) = s$. The leading temperature-dependent correction to Υ is, however, different in two subregions. In the lowest T region A, $T < s$, we have a Fermi liquid-like T^2 power law

$$\Upsilon(T) - \Upsilon(0) = \frac{(u + Nv)\pi T^2}{6\sqrt{s}} \qquad \text{(region Ia)}. \tag{22.45}$$

At higher temperatures, in region B, $s < T < [s/(u + Nv)]^{2/3}$, we have an anomalous temperature dependence

$$\Upsilon(T) - \Upsilon(0) = (u + Nv)\Phi(0)T^{3/2} \qquad \text{(region Ib and II)}. \tag{22.46}$$

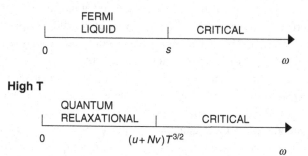

Fig. 22.2 Crossovers as a function of frequency, ω, in the regions of Fig. 22.1 of the metallic spin glass. The low-T Fermi liquid region is on the paramagnetic side ($s > 0$).

It is also interesting to consider the properties of regions A and B as a function of observation frequency, ω, as sketched in Fig. 22.2. At large frequencies, $\omega \gg s$, the local dynamic susceptibility behaves like $\chi_L'' \sim \text{sgn}(\omega)\sqrt{|\omega|}$, which is the spectrum of critical fluctuations; at the $T = 0$, $s = 0$ critical point, this spectrum is present at all frequencies. At low frequencies, $\omega \ll s$, there is a crossover (Fig. 22.2) to the characteristic Fermi liquid spectrum of local spin fluctuations $\chi_L'' \sim \omega/\sqrt{s}$. Upon consideration of fluctuations beyond mean field, one finds the appearance of Griffiths–McCoy singularities in this region, as discussed in related contexts in Chapter 21; these are quite important for experimental comparisons at low temperatures and are further discussed in [24, 45, 69, 118, 119, 282, 284, 338–340, 423].

(C) High-T Region, $T > [|s|/(u + Nv)]^{2/3}$
Here temperature-dependent contributions to Υ dominate over those due to the deviation of the coupling d from its critical point, $d = 0$. Therefore thermal effects are dominant, and the system behaves as if its microscopic couplings are at those of the critical ground state. The T dependence in (22.46) continues to hold, as we have already noted, with the leading contribution being

$$\Upsilon \approx (u + Nv)\Phi(0)T^{3/2}. \tag{22.47}$$

Note that the characteristic energy scale now does not scale simply as $\sim T$, as it did for the high-T region of the models in Part II with $d < 3$. Instead, all T-dependent corrections arise from the irrelevant coupling u, which leads to the anomalous power law in (22.47). As in the low-T paramagnetic region, it is useful to consider properties of this region as a function of ω (Fig. 22.2). For large ω ($\omega \gg (u + Nv)T^{3/2}$) we again have the critical behavior $\chi_L'' \sim \text{sgn}(\omega)\sqrt{|\omega|}$; this critical behavior is present at large enough ω in all the regions of the phase diagram. At small ω ($\omega \ll (u+Nv)T^{3/2}$), thermal fluctuations quench the critical fluctuations, and we have relaxational behavior with $\chi_L'' \sim \omega/(u+Nv)^{1/2}T^{3/4}$.

(D) Low-T Region above Spin Glass, $T < [-s/(u + Nv)]^{2/3}$, $s < 0$
Effects due to the formation of a static moment are now paramount. As one approaches the spin-glass boundary (22.44) from above, the system enters a region of purely classical

thermal fluctuations, $|T - T_c(s)| \ll (u + Nv)^{2/3} T_c^{4/3}(s)$ (shown shaded in Fig 22.1), where

$$\Upsilon = \left(\frac{s - s_c(T)}{T(u + Nv)} \right)^2 . \tag{22.48}$$

Notice that Υ depends on the *square* of the distance from the finite-T classical phase transition line, in contrast to its linear dependence at $T = 0$ in (22.43). It turns out that (22.48), when inserted into the static correlation functions, reproduces precisely the critical singularities of the theory of classical spin glasses [400]. Indeed, the reader is invited to show by the methods of Part II that the fluctuations in the shaded region of Fig. 22.1 are described by precisely the classical critical theories of [48] and [138].

The above results for the local dynamic susceptibility can be extended to a number of other experimentally important observables: as the basic methods are similar to those already developed here, we refer the interested reader to the literature [436, 457].

22.3 Applications and extensions

Much of the theoretical interest in quantum spin glasses has been driven by experiments performed by the group led by Rosenbaum and Aeppli [541, 542] on the insulating, dipolar Ising spin glass $LiHo_x Y_{1-x} F_4$ in a transverse field. These clearly show a crossover between thermal and quantum fluctuation dominated regimes, but the nature of the quantum critical point remains unclear. The vicinity of the spin-glass phase is dominated by real-time glassy dynamics, which drives the system out of equilibrium. On the theoretical side, we have already noted the work on the quantum Ising spin glass in infinite-range models. Models with finite-range interactions have been studied in imaginary-time computer simulations [186, 403], which yield information on thermodynamic properties and critical exponents. However, it is clear that an understanding of the experiments requires a theory of the real-time dynamics of quantum spin glasses. We discussed the real-time, nonzero temperature, physics near nonrandom quantum critical points in Part II, but there are no corresponding results for the random case. Recent steps toward understanding the real-time dynamics include the droplet model picture of Thill and Huse [503] and the infinite-range model studies of Rozenberg and Grempel [412].

The experiments also display ferromagnetic phases; these appear in models with a nonzero mean J_{ij} and have been studied [103] in the mean-field framework discussed here.

A significant application of the concepts discussed here on metallic spin glasses has been in the "heavy fermion" series of compounds. The Griffiths–McCoy singularities of the Fermi liquid phase have been the subject of much attention; discussions along with comparisons with experiments may be found in [24, 69, 118, 119, 339, 340, 423]. The anomalous power laws at high T in the vicinity of the quantum critical point [436, 457] have also been examined in experimental studies [489].

We have not commented in this chapter on insulating quantum spin glasses of Heisenberg spins: such models would generalize those of Chapter 19 to the case of random exchange interactions. These turn out to be considerably more complicated; some work on infinite-range models may be found in [57, 177, 275, 441, 456].

As we noted earlier, fluctuation corrections to the mean-field theory presented in this chapter have been considered in the literature [400, 436, 456, 457]. The first two works [400, 436] focused on spatial fluctuations in the spin-glass order parameter, although the last two [456, 457] considered the quantum fluctuations in the on-site "quantum impurity" model. These latter works argued that although the mean-field theory of this chapter is an adequate starting point for metallic electronic systems with Ising symmetry, those with full Heisenberg symmetry appear to be controlled by the critical quantum paramagnetic state found in [441] in the study of insulating Heisenberg spin models. The critical state of [441] also appears in a most interesting analysis by Parcollet and Georges [371] of a doped Mott insulator with random exchange interactions. The temperature and frequency crossovers found in this insulator are closely related to those observed in the high-temperature superconductors. Recent numerical work [178] on infinite range models of metallic spin glasses with Ising symmetry has obtained results in good agreement with the analytical predictions discussed here.

References

[1] Abanov, Ar., and Chubukov, A. (2000) *Phys. Rev. Lett.* **84**, 5608.

[2] Abanov, Ar., and Chubukov, A. (2004) *Phys. Rev. Lett.* **93**, 255702.

[3] Abanov, Ar., Chubukov, A., and Schmalian, J. (2003) *Advances in Physics* **52**, 119.

[4] Aeppli, G., Broholm, C., Di Tusa, J. F., *et al.* (1997) *Physica B* **237–238**, 30.

[5] Aeppli, G., Mason, T. E., Hayden, S. M., Mook, H. A., and Kulda, J. (1998) *Science* **278**, 1432.

[6] Affleck, I. (1985) *Nucl. Phys. B* **257**, 397.

[7] Affleck, I. (1985) *Nucl. Phys. B* **265**, 409.

[8] Affleck, I. (1990) in *Fields, Strings, and Critical Phenomena, Les Houches Summer School Proceedings, Vol. 49*, ed. E. Brézin and J. Zinn-Justin (North-Holland, Amsterdam).

[9] Affleck, I. (1990) *Phys. Rev. B* **41**, 6697.

[10] Affleck, I. (1991) *Phys. Rev. B* **43**, 3215.

[11] Affleck, I. (1998) *J. Phys. A: Math. Gen.* **31**, 4573.

[12] Affleck, I., and Haldane, F. D. M. (1987) *Phys. Rev. B* **36**, 5291.

[13] Affleck, I., Kennedy, T., Lieb, E. H., and Tasaki, H. (1987) *Phys. Rev. Lett.* **59**, 799.

[14] Affleck, I., Kennedy, T., Lieb, E. H., and Tasaki, H. (1988) *Commun. Math. Phys.* **115**, 477.

[15] Aifer, A. H., Goldberg, B. B., and Broido, D. A. (1996) *Phys. Rev. Lett.* **76**, 680.

[16] Ain, M., Lorenzo, J. E., Regnault, L. P., *et al.* (1997) *Phys. Rev. Lett.* **78**, 1560.

[17] Akhiezer, A. I., Bar'yakhtar, V. G., and Peletminskii, S. V. (1968) *Spin Waves*, Chapter 8 (North-Holland, Amsterdam).

[18] Altshuler, B. L. (1985) *Pis'ma Zh. Éksp. Fiz.* **51**, 530 (*JETP. Lett.* **41**, 648).

[19] Altshuler, B. L., Lee, P. A., and Webb, R. A. eds. (1991) *Mesoscopic Phenomena in Solids* (North Holland, Amsterdam).

[20] Ambegaokar, V., Eckern, U., and Schön, G. (1982) *Phys. Rev. Lett.* **48**, 1745.

[21] Amico, L., Falci, G., Fazio, R., and Giaquinta, G. (1997) *Phys. Rev. B* **55**, 1100.

[22] Anderson, P. W. (1987) *Science* **235**, 1196.

[23] Ando, Y., Segawa, K., Komiya, S., and Lavrov, A. N. (2002) *Phys. Rev. Lett.* **88**, 137005.

[24] Andrade, M. C. de, Chau, R., Dickey, R. P., *et al.* (1998) *Phys. Rev. Lett.* **81**, 5620.

[25] Andraka, B., and Tsvelik, A. M. (1991) *Phys. Rev. Lett.* **67**, 2886.

[26] Angelucci, A. (1991) *Int. J. Mod. Phys. B* **5**, 659.

[27] Arovas, D. P., and Auerbach, A. (1988) *Phys. Rev. B* **38**, 316.

[28] Ashcroft, N. W., and Mermin, N. D. (1976) *Solid State Physics* (Saunders College, Philadelphia).

[29] Auerbach, A. (1994) *Interacting Electrons and Quantum Magnetism* (Springer-Verlag, New York).

[30] Augier, D., Poilblanc, D., Sorensen, E., and Affleck, I. (1998) *Phys. Rev. B* **58**, 9110.

[31] Azaria, P., Delamotte, B., Delduc, F., and Jolicoeur, Th. (1993) *Nucl. Phys. B* **408**, 485.

[32] Azaria, P., Lecheminant, P., and Mouhanna, D. (1995) *Nucl. Phys. B* **455**, 648.

[33] Azuma, M., Hiroi, Z., Takano, M., Ishida, K., and Kitaoka, Y. (1994) *Phys. Rev. Lett.* **73**, 3463.

[34] Babelon, O., and Bernard, D. (1992) *Phys. Lett. B* **288**, 113.

[35] Bao, W., Broholm, C., Aeppli, G., *et al.* (1998) *Phys. Rev. B* **58**, 12727.

[36] Barouch, E., and McCoy, B. M. (1971) *Phys. Rev. A* **3**, 786.

[37] Barrett, S. E., Dabbagh, G., Pfeiffer, L. N., West, K. W., and Tycko, R. (1995) *Phys. Rev. Lett.* **74**, 5112.

[38] Baskaran, G., Zou, Z., and Anderson, P. W. (1987) *Solid State Commun.* **63**, 973.

[39] Batrouni, G. G., Scalettar, R. T., and Zimanyi, G. T. (1990) *Phys. Rev. Lett.* **65**, 1765.

[40] Batrouni, G. G., Scalettar, R. T., Zimanyi, G. T., and Kampf, A. P. (1995) *Phys. Rev. Lett.* **74**, 2527.

[41] Baym, G., Blaizot, J.-P., Holzmann, M., Laloë, F., and Vautherin, D. (1999) *Phys. Rev. Lett.* **83**, 1703.

[42] Baym, G., Blaizot, J.-P., Zinn-Justin, J. (2000) *Europhys. Lett.* **49**, 150.

[43] Beal-Monod, M. T., and Maki, K. (1975) *Phys. Rev. Lett.* **34**, 1461.

[44] Beard, B. B., Birgeneau, R. J., Greven, M., and Wiese, U.-J. (1998) *Phys. Rev. Lett.* **80**, 1742.

[45] Bhatt, R. N., and Fisher, D. S. (1992) *Phys. Rev. Lett.* **68**, 3072.

[46] Bhatt, R. N., and Lee, P. A. (1982) *Phys. Rev. Lett.* **48**, 344.

[47] Bijlsma, M., and Stoof, H. T. C. (1996) *Phys. Rev. A* **54**, 5085.

[48] Binder, K., and Young, A. P. (1986) *Rev. Mod. Phys.* **58**, 801.

[49] Bitko, D., Rosenbaum, T. F., and Aeppli, G. (1996) *Phys. Rev. Lett.* **77**, 940.

[50] Bödeker, D., McLerran, L., and Smilga, A. (1995) *Phys. Rev. D* **52**, 4675.

[51] Boechat, B., dos Santos, R. R., and Continentino, M. A. (1994) *Phys. Rev. B* **49**, 6404.

[52] Bose, I., and Mitra, P. (1991) *Phys. Rev. B* **44**, 443.

[53] Bourges, P. (1998) in *The Gap Symmetry and Fluctuations in High Temperature Superconductors*, ed. J. Bok, G. Deutscher, D. Pavuna, and S. A. Wolf (Plenum, New York); cond-mat/9901333.

[54] Boyanovsky, D., and Cardy, J. L. (1982) *Phys. Rev. B* **26**, 154.

[55] Braaten, E., and Platter, L. (2008) *Phys. Rev. Lett.* **100**, 205301.

[56] Braaten, E., Kang, D., and Platter, L. (2010) arXiv:1001.4518.

[57] Bray, A. J., and Moore, M. A. (1980) *J. Phys. C* **13**, L655.

[58] Brézin, E. (1982) *J. de Physique* **43**, 15.

[59] Brézin, E., Le Guillou, J. C., and Zinn-Justin, J. (1976) in *Phase Transitions and Critical Phenomena*, **6**, ed. C. Domb and M. S. Green (Academic Press, London).

[60] Brézin, E., and Wallace, D. J. (1973) *Phys. Rev. B* **7**, 1967.

[61] Brézin, E., and Zinn-Justin, J. (1976) *Phys. Rev. B* **14**, 3110.

[62] Brézin, E., and Zinn-Justin, J. (1985) *Nucl. Phys. B* **257**, 867.

[63] Broholm, C. (1998) in *Dynamical Properties of Unconventional Magnetic Systems, NATO Advanced Study Institute, Series B: Physics*, ed. A. Skjeltorp and D. Sherrington (Kluwer Academic, Norwell MA).

[64] Bruce, A. D., and Wallace, D. J. (1976) *J. Phys. A* **9**, 1117.

[65] Buragohain, C., and Sachdev, S. (1999) *Phys. Rev. B* **59**, 9285.

[66] Calabrese, P., Parruccini, P., Pelissetto, A., and Vicari, E. (2004) *Phys. Rev. B* **70**, 174439.

[67] Capel, H. W., and Perk, J. H. H. (1977) *Physica A* **87**, 211.

[68] Cardy, J. L. (1984) *J. Phys. A* **17**, L385.

[69] Castro Neto, A. H., Castilla, G. E., and Jones, B. A. (1998) *Phys. Rev. Lett.* **81**, 3531.

[70] Cha, M.-C., Fisher, M. P. A., Girvin, S. M., Wallin, M., and Young, A. P. (1991) *Phys. Rev. B* **44**, 6883.

[71] Chaboussant, G., Crowell, P. A., Levy, L. P., *et al.* (1997) *Phys. Rev. B* **55**, 3046.

[72] Chaboussant, G., Julien, M.-H., Fagot-Revurat, Y., *et al.* (1997) *Phys. Rev. Lett.* **79**, 925.

[73] Chaboussant, G., Fagot-Revurat, Y., Julien, M.-H., *et al.* (1998) *Phys. Rev. Lett.* **80**, 2713.

[74] Chakrabarti, B. K., Dutta, A., and Sen, P. (1996) *Quantum Ising Phases, and Transitions in Transverse Ising Models* (Springer, Berlin).

[75] Chakravarty, S., Halperin, B. I., and Nelson, D. R. (1989) *Phys. Rev. B* **39**, 2344.

[76] Chakravarty, S., Kivelson, S., Zimanyi, G. T., and Halperin, B. I. (1987) *Phys. Rev. B* **35**, 7256.

[77] Chakravarty, S., and Orbach, R. (1990) *Phys. Rev. Lett.* **64**, 224.

[78] Chalker, J. T., and Coddington, P. D. (1988) *J. Phys. C* **21**, 2665.

[79] Chamati, H., Pisanova, E. S., and Tonchev, N. S. (1998) *Phys. Rev. B* **57**, 5798.

[80] Chandra, P., and Doucot, B. (1988) *Phys. Rev. B* **38**, 9335.

[81] Chayes, J. T., Chayes, L., Fisher, D. S., and Spencer, T. (1986) *Phys. Rev. Lett.* **57**, 2999.

[82] Chitra, R., and Giamarchi, T. (1997) *Phys. Rev. B* **55**, 5816.

[83] Choi, J., and José, J. V. (1989) *Phys. Rev. Lett.* **62**, 1904.

[84] Chubukov, A. V. (1991) *Phys. Rev. B* **44**, 392.

[85] Chubukov, A. V., and Sachdev, S. (1993) *Phys. Rev. Lett.* **71**, 169.

[86] Chubukov, A. V., Sachdev, S., and Ye, J. (1994) *Phys. Rev. B* **49**, 11919.

[87] Chubukov, A. V., Sachdev, S., and Senthil, T. (1994) *Nucl. Phys. B* **426**, 601.

[88] Chubukov, A. V., Sachdev, S., and Sokol, A. (1994) *Phys. Rev. B* **49**, 9052.

[89] Chubukov, A. V., and Starykh, O. A. (1996) *Phys. Rev. B* **53**, R14729.

[90] Coldea, R., Tennant, D. A., Wheeler, E. M., *et al.* (2010) *Science* **327**, 177.

[91] Coleman, P. (1999) *Physica B*, **259–261**, 353.

[92] Combescot, R., Alzetto, F., and Leyronas, X. (2009) *Phys. Rev. A* **79**, 053640.

[93] Continentino, M. A. (1994) *Phys. Rep.* **239**, 179.

[94] Continentino, M. A. (1996) *Z. Phys. B* **101**, 197.

[95] Cowley, R. A. (1998) in *Dynamical Properties of Unconventional Magnetic Systems, NATO Advanced Study Institute, Series B: Physics*, ed. A. Skjeltorp and D. Sherrington (Kluwer Academic, Norwell MA).

[96] Creswick, R. J., and Wiegel, F. W. (1983) *Phys. Rev. A* **28**, 1579.

[97] Cross, M. C., and Fisher, D. S. (1979) *Phys. Rev. B* **19**, 402.

[98] Crowell, P. A., van Keuls, F. W., and Reppy, J. D. (1995) *Phys. Rev. Lett.* **75**, 1106.

[99] Crowell, P. A., van Keuls, F. W., and Reppy, J. D. (1995) *Phys. Rev. B* **55**, 12620.

[100] d'Adda, A., Luscher, M., and Di Vecchia, P. (1978) *Nucl. Phys. B* **146**, 63.

[101] Dagotto, E., and Moreo, A. (1989) *Phys. Rev. Lett.* **63**, 2148.

[102] Dagotto, E., and Rice, T. M. (1996) *Science* **271**, 618.

[103] Dalidovich, D., and Phillips, P. (1999) *Phys. Rev. B,* **59**, 11925.

[104] Damle, K. (1998) Ph.D. thesis, Yale University.

[105] Damle, K., and Sachdev, S. (1996) *Phys. Rev. Lett.* **76**, 4412.

[106] Damle, K., and Sachdev, S. (1997) *Phys. Rev. B* **56**, 8714.

[107] Damle, K., and Sachdev, S. (1998) *Phys. Rev. B* **57**, 8307.

[108] Daou, R., Chang, J., LeBoeuf, D., *et al.* (2010) *Nature* **463**, 519.

[109] Dasgupta, C., and Halperin, B. I. (1981) *Phys. Rev. Lett.* **47**, 1556.

[110] Dasgupta, C., and Ma, S.-k. (1980) *Phys. Rev. B* **22**, 1305.

[111] Das Sarma, S., Sachdev, S., and Zheng, L. (1997) *Phys. Rev. Lett.* **79**, 917.

[112] Das Sarma, S., Sachdev, S., and Zheng, L. (1998) *Phys. Rev. B* **58**, 4672.

[113] Del Maestro, A., Rosenow, B., Müller, M., and Sachdev, S. (2008) *Phys. Rev. Lett.* **101**, 035701.

[114] Del Maestro, A., Rosenow, B., and Sachdev, S. (2009) *Annals of Physics* **324**, 523.

[115] Devreux, F., and Boucher, J. P. (1987) *J. de Physique* **48**, 1663.

[116] Dhar, D. (1983) in *Stochastic Processes: Formalism and Applications*, ed. G. S. Agarwal and S. Dutta Gupta (Springer-Verlag, Berlin).

[117] Dhar, D., and Barma, M. (1980) *J. Stat. Phys.* **22**, 259.

[118] Dobrosavljevic, V., Kirkpatrick, T. R., and Kotliar, G. (1992) *Phys. Rev. Lett.* **69**, 1113.

[119] Dobrosavljevic, V., and Kotliar, G. (1993) *Phys. Rev. Lett.* **71**, 3218.

[120] Dombre, T., and Read, N. (1989) *Phys. Rev. B* **39**, 6797.

[121] Doniach, S. (1981) *Phys. Rev. B* **24**, 5063.

[122] Dorogovstev, S. N. (1980) *Phys. Lett.* **76A**, 169.

[123] dos Santos, R. R. (1982) *J. Phys. C* **15**, 3141.

[124] Dyson, F. J. (1956) *Phys. Rev.* **102**, 1230.

[125] Eckern, U., Schön, G., and Ambegaokar, V. (1984) *Phys. Rev. B* **30**, 6419.

[126] Eggert, S., Affleck, I., and Takahashi, M. (1994) *Phys. Rev. Lett.* **73**, 332.

[127] Elliott, R. J., Pfeuty, P., and Wood, C. (1970) *Phys. Rev. Lett.* **25**, 443.

[128] Elstner, N., Glenister, R. L., Singh, R. R. P., and Sokol, A. (1995) *Phys. Rev. B* **51**, 8984.

[129] Elstner, N., and Singh, R. R. P. (1998) *Phys. Rev. B* **57**, 7740.

[130] Elstner, N., and Singh, R. R. P. (1998) *Phys. Rev. B* **58**, 11484.

[131] Emery, V. J., and Kivelson, S. A. (1998) *J. Phys. Chem. Solids* **59**, 1705.

[132] Engel, L. W., Shahar, D., Kurdak, C., and Tsui, D. C. (1993) *Phys. Rev. Lett.* **71**, 2638.

[133] Fateev, V. A., Onofri, E., and Zamolodchikov, A. B. (1993) *Nucl. Phys. B* **406**, 521.

[134] Fazio, R., and Zappala, D. (1996) *Phys. Rev. B* **53**, R8883.

[135] Fertig, H. A., Brey, L., Côté, R., and MacDonald, A. H. (1994) *Phys. Rev. B* **50**, 11018.

[136] Fetter, A. L., and Walecka, J. D. (1971) *Quantum Theory of Many-Particle Systems* (McGraw-Hill, New York).

[137] Figueirido, F., Karlhede, A., Kivelson, S., *et al.* (1990) *Phys. Rev. B* **41**, 4619.

[138] Fischer, K. H., and Hertz, J. A. (1991) *Spin Glasses* (Cambridge University Press, Cambridge).

[139] Fisher, D. S., and Hohenberg, P. C. (1988) *Phys. Rev. B* **37**, 4936.

[140] Fisher, D. S. (1992) *Phys. Rev. Lett.* **69**, 534.

[141] Fisher, D. S. (1994) *Phys. Rev. B* **50**, 3799.

[142] Fisher, D. S. (1995) *Phys. Rev. B* **51**, 6411.

[143] Fisher, D. S., and Young, A. P. (1998) *Phys. Rev. B* **58**, 9131.

[144] Fisher, M. E. (1964) *Am. J. Phys.* **32**, 343.

[145] Fisher, M. P. A. (1990) *Phys. Rev. Lett.* **65**, 923.

[146] Fisher, M. P. A., and Grinstein, G. (1988) *Phys. Rev. Lett.* **60**, 208.

[147] Fisher, M. P. A., Grinstein, G., and Girvin, S. M. (1990) *Phys. Rev. Lett.* **64**, 587.

[148] Fisher, M. P. A., Weichman, P. B., Grinstein, G., and Fisher, D. S. (1989) *Phys. Rev. B* **40**, 546.

[149] Fong, H. F., Keimer, B., Dogan, F., and Aksay, I. A. (1997) *Phys. Rev. Lett.* **78**, 713.

[150] Forster, D. (1975) *Hydrodynamic Fluctuations, Broken Symmetry, and Correlation Functions* (Benjamin/Cummings, Reading, MA).

[151] Fradkin, E., and Kivelson, S. A. (1990) *Mod. Phys. Lett. B* **4**, 225.

[152] Fradkin, E., and Susskind, L. (1978) *Phys. Rev. D* **17**, 2637.

[153] Freericks, J. K., and Monien, H. (1994) *Europhys. Lett.* **26**, 545.

[154] Freericks, J. K., and Monien, H. (1996) *Phys. Rev. B* **53**, 2691.

[155] Frey, E., and Balents, L. (1997) *Phys. Rev. B* **55**, 1050.

[156] Friere, F., O'Connor, D., and Stephens, C. R. (1994) *J. Stat. Phys.* **74**, 219.

[157] Frischmuth, B., and Sigrist, M. (1997) *Phys. Rev. Lett.* **79**, 147.

[158] Fritz, L., and Sachdev, S. (2009) *Phys. Rev. B* **80**, 144503.

[159] Furusaki, A., Sigrist, M., Westerberg, E., *et al.* (1994) *Phys. Rev. Lett.* **73**, 2622.

[160] Gebhard, F. (1997) *The Mott Metal-Insulator Transition*, Springer Tracts in Modern Physics **137**, (Springer-Verlag, Berlin).

[161] Gelfand, M. P. (1990) *Phys. Rev. B* **42**, 8206.

[162] Gelfand, M. P., Singh, R. R. P., and Huse, D. A. (1989) *Phys. Rev. B* **40**, 10801.

[163] Gelfand, M. P., Zheng, W., Hamer, C. J., and Oitmaa, J. (1998) *Phys. Rev. B* **57**, 392.

[164] Gemelke, N., Zhang, X., Hung, C.-L., and Chin, Cheng (2009) *Nature* **460**, 995.

[165] Georges, A., Kotliar, G., Krauth, W., and Rozenberg, J. (1996) *Rev. Mod. Phys.* **68**, 13.

[166] Ghosal, A., Randeria, M., and Trivedi, N. (1998) *Phys. Rev. Lett.* **81**, 3940.

[167] Giamarchi, T., and Schulz, H. J. (1988) *Phys. Rev. B* **37**, 327.

[168] Girvin, S. M., and Mahan, G. D. (1979) *Phys. Rev. B* **20**, 4896.

[169] Glauber, R. J. (1963) *J. Math. Phys.* **4**, 294.

[170] Glazman, L. I., and Larkin, A. I. (1997) *Phys. Rev. Lett.* **79**, 3736.

[171] Goff, J. P., Tennant, D. A., and Nagler, S. E. (1995) *Phys. Rev. B* **52**, 15992.

[172] Goldenfeld, N. (1992) *Lectures on Phase Transitions and the Renormalization Group* (Addison-Wesley, Reading, MA).

[173] Goldschmidt, Y. Y., and Lai, P.-Y. *Phys. Rev. Lett.* **64**, 2467.

[174] Granroth, G. E., Meisel, M. W., Chaparala, M., *et al.* (1996) *Phys. Rev. Lett.* **77**, 1616.

[175] Greiner, M., Mandel, O., Esslinger, T., Hänsch, T. W., and Bloch, I. (2002) *Nature* **415**, 39.

[176] Grempel, D. R. (1988) *Phys. Rev. Lett.* **61**, 1041.

[177] Grempel, D. R., and Rozenberg, M. J. (1998) *Phys. Rev. Lett.* **80**, 389.

[178] Grempel, D. R., and Rozenberg, M. J. (1999) *Phys. Rev. B* **60**, 4702.

[179] Greven, M., Birgeneau, R. J., Endoh, Y., *et al.* (1994) *Phys. Rev. Lett.* **72**, 1096.

[180] Griffiths, R. B. (1969) *Phys. Rev. Lett.* **23**, 17.

[181] Gros, C., and Werner, R. (1988) *Phys. Rev. B* **58**, R14677.

[182] Grosche, F. M., Lister, S. J. S., Carter, F. V., *et al.* (1997) *Physica B* **239**, 62.

[183] Grüter, P., Ceperley, D., and Laloë, F. (1997) *Phys. Rev. Lett.* **79**, 3549.

[184] Gruzberg, I. A., Read, N., and Sachdev, S. (1997) *Phys. Rev. B* **55**, 10593.

[185] Gruzberg, I. A., Read, N., and Sachdev, S. (1997) *Phys. Rev. B* **56**, 13218.

[186] Guo, M., Bhatt, R. N., and Huse, D. A. (1994) *Phys. Rev. Lett.* **72**, 4137.

[187] Guo, M., Bhatt, R. N., and Huse, D. A. (1996) *Phys. Rev. B* **54**, 3336.

[188] Hagiwara, M., Katsumata, K., Affleck, I., Halperin, B. I., and Renard, J. P. (1990) *Phys. Rev. Lett.* **65**, 3181.

[189] Haldane, F. D. M. (1981) *Phys. Rev. Lett.* **47**, 1840.

[190] Haldane, F. D. M. (1982) *Phys. Rev. B* **25**, 4925.

[191] Haldane, F. D. M. (1982) *J. Phys. C* **15**, L831.

[192] Haldane, F. D. M. (1983) *Phys. Rev. Lett.* **50**, 1153.

[193] Haldane, F. D. M. (1988) *Phys. Rev. Lett.* **61**, 1029.

[194] Halperin, B. I., Hohenberg, P. C., and Ma, S.-k. (1972) *Phys. Rev. Lett.* **29**, 1548.

[195] Halperin, B. I., Hohenberg, P. C., and Ma, S.-k. (1974) *Phys. Rev. B* **10**, 139.

[196] Halperin, B. I., and Saslow, W. M. (1977) *Phys. Rev. B* **16**, 2154.

[197] Hamer, C. J., Kogut, J. B., and Susskind, L. (1979) *Phys. Rev. D* **19**, 3091.

[198] Hammar, P. R., Reich, D. H., Broholm, C., and Trouw, F. (1998) *Phys. Rev. B* **57**, 7846.

[199] Hartman, J. W., and Weichman, P. B. (1995) *Phys. Rev. Lett.* **74**, 4584.

[200] Harris, A. B. (1974) *J. Phys. C* **7**, 1671.

[201] Harris, A. B. (1974) *J. Phys. C* **7**, 3082.

[202] Hardy, W. N., Kamal, S., Liang, R., *et al.* (1996) in *Proceedings of the 10th Anniversary HTS Workshop on Physics, Materials and Applications*, ed. B. Batlogg *et al.* (World Scientific, Singapore), p. 223.

[203] Hartnoll, S. A., Kovtun, P. K., Müller, M., and Sachdev, S. (2007) *Phys. Rev. B* **76**, 144502.

[204] Hasegawa, H., and Moriya, T. (1974) *J. Phys. Soc. Jpn.* **36**, 1542.

[205] Hasenfratz, P., Maggiore, M., and Niedermayer, F. (1990) *Phys. Lett. B* **245**, 522.

[206] Hasenfratz, P., and Niedermayer, F. (1990) *Phys. Lett. B* **245**, 529.

[207] Hasenfratz, P., and Niedermayer, F. (1991) *Phys. Lett. B* **268**, 231.

[208] Hastings, M. B. (1998) *Phys. Rev. B* **60**, 9755.

[209] Haussmann, R., Punk, M., and Zwerger, W. (2009) *Phys. Rev. A* **80**, 063612.

[210] Hayden, S. M., Aeppli, G., Mook, H., Rytz, D., Hundley, M. F., and Fisk, Z. (1991) *Phys. Rev. Lett.* **66**, 821.

[211] Hebard, A. F. (1994) in *Strongly Correlated Electronic Materials*, ed. K. S. Bedell, Z. Wang, D. E. Meltzer, and E. Abrahams (Addison-Wesley, Reading, MA).

[212] Henelius, P., and Girvin, S. M. (1998) *Phys. Rev. B* **57**, 11457.

[213] Henelius, P., Sandvik, A. W., Timm, C., and Girvin, S. M. (1999) *Phys. Rev. B* **61**, 364.

[214] Henkel, M., and Hoeger, C. (1984) *Z. Phys. B* **55**, 67.

[215] Herbut, I. F. (1997) *Phys. Rev. Lett.* **79**, 3502.

[216] Herbut, I. F. (1998) *Phys. Rev. Lett.* **81**, 3916.

[217] Herbut, I. F., Juričić, V., and Roy, B. (2009) *Phys. Rev. B* **79**, 085116.

[218] Hertz, J. A. (1976) *Phys. Rev. B* **14**, 1165.

[219] Herzog, C. P., Kovtun P., Sachdev, S., and Son, D. T. (2007) *Phys. Rev. D* **75**, 085020.

[220] Heuser, K., Scheidt, E.-W., Schreiner, T., and Stewart, G. R. (1998) *Phys. Rev. B* **57**, R4198.

[221] Hinkov, V., Haug, D., Fauqué, B., *et al.* (2008) *Science* **319**, 597.

[222] Hirsch, J. E., and José, J. V. (1980) *Phys. Rev. B* **22**, 5355.

[223] Hohenberg, P. C., and Halperin, B. I. (1977) *Rev. Mod. Phys.* **49**, 435.

[224] Holzmann, M., Baym, G., Blaizot, J.-P., and Laloë, F. (2001) *Phys. Rev. Lett.* **87**, 120403.

[225] Hoyos, J. A., and Vojta, T. (2008) *Phys. Rev. Lett.* **100**, 240601.

[226] Hoyos, J. A., Kotabage, C., and Vojta, T. (2007) *Phys. Rev. Lett.* **99**, 230601.

[227] Huckenstein, B. (1995) *Rev. Mod. Phys.* **67**, 357.

[228] Huh, Y., Fritz, L., and Sachdev, S. (2010) *Phys. Rev. B* **81**, 144432.

[229] Huo, R. E., Hetzel, R. E., and Bhatt, R. N. (1993) *Phys. Rev. Lett.* **70**, 481.

[230] Huse, D. A., and Elser, V. (1988) *Phys. Rev. Lett.* **60**, 2531.

[231] Hyman, R. A., Yang, K., Bhatt, R. N., and Girvin, S. M. (1996) *Phys. Rev. Lett.* **76**, 839.

[232] Hyman, R. A., and Yang, K. (1997) *Phys. Rev. Lett.* **78**, 1783.

[233] Ikegami, T., Miyashita, S., and Rieger, H. (1998) *J. Phys. Soc. Jpn.* **67**, 2761.

[234] Imai, T., Slichter, C. P., Yoshimura, K., and Kosuge, K. (1993) *Phys. Rev. Lett.* **70**, 1002.

[235] Imai, T., Slichter, C. P., Yoshimura, K., Katoh, M., and Kosuge, K. (1993) *Phys. Rev. Lett.* **71**, 1254.

[236] Ioffe, L. B., and Millis, A. J. (1995) *Phys. Rev. B* **51**, 16151.

[237] Irkhin, V. Yu, and Katanin, A. A. (1998) *Phys. Rev. B* **58**, 5509.

[238] Ising, E. (1925) *Z. Phys.* **31**, 253.

[239] Itoh, Y., and Yasuoka, H. (1997) *J. Phys. Soc. Jpn.* **66**, 334.

[240] Its, A. R., Izergin, A. G., and Korepin, V. E. (1991) *Physica D* **53**, 187.

[241] Its, A. R., Izergin, A. G., Korepin, V. E., and Novokshenov, V. Ju. (1990) *Nucl. Phys. B* **340**, 752.

[242] Its, A. R., Izergin, A. G., Korepin, V. E., and Varzugin, G. G. (1992) *Physica D* **54**, 351.

[243] Itzhaki, N., Maldacena, J. M., Sonnenschein, J., and Yankielowicz, S. (1998) *Phys. Rev. D* **58**, 046004.

[244] Itzykson, C., and Drouffe, J.-M. (1989) *Statistical Field Theory* (Cambridge University Press, Cambridge).

[245] Itzykson, C., and Zuber, J. B. (1980) *Quantum Field Theory* (McGraw-Hill, New York).

[246] Jalabert, R., and Sachdev, S. (1991) *Phys. Rev. B* **44**, 686.

[247] Jeon, S., and Yaffe, L. G. (1996) *Phys. Rev. D* **53**, 5799.

[248] Jevicki, A., and Papanicolaou, N. (1979) *Ann. Phys.* **120**, 107.

[249] Jepsen, D. W. (1965) *J. Math. Phys.* **6**, 405.

[250] Jolicoeur, Th., and Golinelli, O. (1994) *Phys. Rev. B* **50**, 9265.

[251] Jordan, P., and Wigner, E. (1928) *Z. Phys.* **47**, 631.

[252] José, J. V., Novotny, M. A., and Goldman, A. M. (1988) *Phys. Rev. B* **38**, 4562.

[253] Joyce, G. S. (1967) *Phys. Rev.* **155**, 478.

[254] Juričić, V., Herbut, I. F., and Semenoff, G. W. (2009) *Phys. Rev. B* **80**, 081405 (R).

[255] Julian, S. R., Carter, F. V., Grosche, F. M., *et al.* (1998) *J. Magnetism, and Magnetic Matt.* **177–181**, 265.

[256] Kaganov, M. I., and Chubukov, A. V. (1987) *Usp. Fiz. Nauk* **153**, 537 (*Sov. Phys. Usp.* **30**, 1015).

[257] Kageyama, H., Yoshimura, K., Stern, R., *et al.* (1999) *Phys. Rev. Lett.* **82**, 3168.

[258] Kambe, S., Raymond, S., Regnault, L.-P., *et al.* (1996) *J. Phys. Soc. Jpn.* **65**, 3294.

[259] Kapusta, J. I. (1993) *Finite-Temperature Field Theory* (Cambridge University Press, Cambridge).

[260] Kärkkäinen, L., Lacaze, R., Lacock, P., and Petersson, B. (1994) *Nucl. Phys. B* **415**, 781.

[261] Kasner, M., and MacDonald, A. H. (1996) *Phys. Rev. Lett.* **76**, 3204.

[262] Kasner, M., Palacios, J. J., and MacDonald, A. H. (2000) *Phys. Rev. B* **62**, 2640.

[263] Keimer, B., Belk, N., Birgeneau, R. J., *et al.* (1992) *Phys. Rev. B* **46**, 14034.

[264] Khveshchenko, D. V., and Paaske, J. (2001) *Phys. Rev. Lett.* **86**, 4672.

[265] Kikuchi, J., Yamauchi, T., and Ueda, Y. (1997) *J. Phys. Soc. Jpn.* **66**, 1622.

[266] Kim, E.-A., Lawler, M. J., Oreto, P., *et al.* (2008) *Phys. Rev. B* **77**, 184514.

[267] Kim, J.-K., and Troyer, M. (1997) *Phys. Rev. Lett.* **80**, 2705.

[268] Kim, Y.-B., and Wen, X. G. (1994) *Phys. Rev. B* **49**, 4043.

[269] Kim, Y.-J., Greven, M., Wiese, U.-J., and Birgeneau, R. J. (1998) *Eur. Phys. J. B* **4**, 291.

[270] Kirkpatrick, T. R., and Belitz, D. (1996) *Phys. Rev. Lett.* **76**, 2571 (**78**, 1197 (E)).

[271] Kivelson, S. A., Rokhsar, D. S., and Sethna, J. P. (1987) *Phys. Rev. B* **35**, 8865.

[272] Klauder, J., and Skagerstam, B. eds. (1985) *Coherent States* (World Scientific, Singapore).

[273] Kolomeisky, E. B., and Straley, J. P. (1992) *Phys. Rev. B* **46**, 11749.

[274] Kolomeisky, E. B., and Straley, J. P. (1992) *Phys. Rev. B* **46**, 13942.

[275] Kopeć, T. K. (1995) *Phys. Rev. B* **52**, 9590.

[276] Kopeć, T. K. (1997) *Phys. Rev. Lett.* **79**, 4266.

[277] Kopietz, P., and Chakravarty, S. (1989) *Phys. Rev. B* **40**, 4858.

[278] Korepin, V. E., Bogoliubov, N. M., and Izergin, A. G. (1993) *Quantum Inverse Scattering Method and Correlation Functions* (Cambridge University Press, Cambridge).

[279] Korepin, V. E., and Slavnov, N. A. (1990) *Commun. Math. Phys.* **129**, 103.

[280] Kosevich, Y. A., and Chubukov, A. V. (1986) *Zh. Eksp. Teor. Fiz.* **91**, 1105 (*Sov. Phys. JETP* **64**, 654).

[281] Kovtun, P. K., Son, D. T., and Starinets, A. (2005) *Phys. Rev. Lett.* **94**, 11601.

[282] Lakner, M., von Löhneysen, H., Langenfeld, A., and Wolfle, P. (1994) *Phys. Rev. B* **50**, 17064.

[283] Lammert, P. E., Rokhsar, D. S., and Toner, J. (1993) *Phys. Rev. Lett.* **70**, 1650.

[284] Langenfeld, A., and Wolfle, P. (1995) *Ann. Physik* **4**, 43.

[285] Laughlin, R. B. (1998) *Phys. Rev. Lett.* **80**, 5188.

[286] Laughlin, R. B. (1998) *Adv. in Phys.* **47**, 943.

[287] Lawrie, I. D. (1978) *J. Phys. C* **11**, 3857.

[288] Lebowitz, J. L., and Percus, J. K. (1969) *Phys. Rev.* **188**, 487.

[289] Leclair, A., Lesage, F., Saleur, H., and Sachdev, S. (1996) *Nucl. Phys. B* **482**, 579.

[290] Lee, H.-L., Carini, J. P., Baxter, D. V., and Gruner, G. (1998) *Phys. Rev. Lett.* **80**, 4261.

[291] Lee, P. A., and Stone, A. D. (1985) *Phys. Rev. Lett.* **55**, 1622.

[292] Lee, S. S. (2009) *Phys. Rev. B* **80**, 165102.

[293] Leggett, A. J., Chakravarty, S., Dorsey, A. T., *et al.* (1987) *Rev. Mod. Phys.* **59**, 1.

[294] Lenard, A. (1966) *J. Math. Phys.* **7**, 1268.

[295] Lesage, F., and Saleur, H. (1997) *Nucl. Phys. B* **493**, 613.

[296] Leung, P. W., and Lam, N.-w. (1995) *Phys. Rev. B* **53**, 2213.

[297] Leung, P. W., Chiu, K. C., and Runge, K. J. (1996) *Phys. Rev. B* **54**, 12938.

[298] Li, M.-R., Hirschfeld, P. J., and Wölfle, P. (2001) *Phys. Rev. B* **63**, 054504.

[299] Lieb, E., Schultz, T., and Mattis, D. (1961) *Ann. Phys.* **16**, 406.

[300] Lister, S. J. S., Grosche, F. M., Carter, F. V., *et al.* (1997) *Z. Phys. B* **103**, 263.

[301] Liu, Y., and Goldman, A. M. (1994) *Mod. Phys. Lett. B* **8**, 277.

[302] Löhneysen, H. v., Rosch, A., Vojta, M., and Wölfle, P. (2007) *Rev. Mod. Phys.* **79**, 1015.

[303] Loinaz, W., and Willey, R. S. (1998) *Phys. Rev. D* **58**, 076003.

[304] Lonzarich, G. G. (1997) in *Electron*, Chap. 6, ed. M. Springford (Cambridge University Press, Cambridge).

[305] Lonzarich, G. G. and Taillefer, L. (1985) *J. Phys. C* **18**, 4339.

[306] Lukyanov, S. (1998) *Nucl. Phys. B* **522**, 533.

[307] Luscher, M. (1982) *Phys. Lett. B* **118**, 391.

[308] Luther, A., and Emery, V. J. (1974) *Phys. Rev. Lett.* **33**, 589.

[309] Luther, A., and Peschel, I. (1974) *Phys. Rev. B* **9**, 2911.

[310] Ma, S.-k. (1972) *Phys. Rev. Lett.* **29**, 1311.

[311] Ma, S.-k. (1973) *Rev. Mod. Phys.* **45**, 589.

[312] Ma, S.-k. (1976) *Modern Theory of Critical Phenomena* (Benjamin Cummings, Reading, MA).

[313] Ma, S.-k., Dasgupta, C., and Hu, C.-K. (1979) *Phys. Rev. Lett.* **43**, 1434.

[314] Mahan, G. D. (1990) *Many-Particle Physics* (Plenum, New York).

[315] Majumdar, C. K., and Ghosh, D. K. (1970) *J. Phys. C* **3**, 911.

[316] Maki, K. (1981) *Phys. Rev. B* **24** 335.

[317] Makivic, M., Trivedi, N., and Ullah, S. (1993) *Phys. Rev. Lett.* **71**, 2307.

[318] Maleev, S. (1957) *Zh. Eksp. Teor. Fiz.* **33**, 1010.

[319] Manfra, M. J., Aifer, E. H., Goldberg, B. B., *et al.* (1996) *Phys. Rev. B* **54**, R17327.

[320] Mason, T. E., Schröder, A., Aeppli, G., Mook, H. A., and Hayden, S. M. (1996) *Phys. Rev. Lett.* **77**, 1604.

[321] Mathur, N. D., Grosche, F. M., Julian, S. R., *et al.* (1998) *Nature* **394**, 39.

[322] Matsuda, T., and Hida, K. (1990) *J. Phys. Soc. Jpn.* **59**, 2223.

[323] Matsushita, Y., Gelfand, M. P., and Ishii, C. (1997) *J. Phys. Soc. Jpn.* **66**, 3648.

[324] Matsushita, Y., Gelfand, M. P., and Ishii, C. (1998) *J. Phys. Soc. Jpn.* **68**, 247.

[325] Mattis, D. C., and Lieb, E. H. (1965) *J. Math. Phys.* **6**, 375.

[326] Mazin, I. I., Singh, D. J., Johannes, M. D., and Du, M.-H. (2008) *Phys. Rev. Lett.* **101**, 057003.

[327] McCoy, B. M. (1968) *Phys. Rev.* **173**, 531.

[328] McCoy, B. M. (1969) *Phys. Rev. Lett.* **23**, 383.

[329] McCoy, B. M. (1969) *Phys. Rev.* **188**, 1014.

[330] McCoy, B. M., Barouch, E., and Abraham, D. B. (1971) *Phys. Rev. A* **4**, 2331.

[331] McCoy, B. M., Perk, J. H. H., and Shrock, R. E. (1983) *Nucl. Phys. B* **220**, 35.

[332] McCoy, B. M., and Wu, T. T. (1973) *The Two-Dimensional Ising Model* (Harvard University Press, Cambridge, MA).

[333] Metlitski, M., and Sachdev, S. (2010) *Phys. Rev. B* **82**, 075127.

[334] Metlitski, M., and Sachdev, S. (2010) arXiv:1005.1288.

[335] Miller, J., and Huse, D. A. (1993) *Phys. Rev. Lett.* **70**, 3147.

[336] Millis, A. J. (1993) *Phys. Rev. B* **48**, 7183.

[337] Millis, A. J., and Monien, H. (1994) *Phys. Rev. B* **50**, 16606.

[338] Milovanovic, M., Sachdev, S., and Bhatt, R. N. (1989) *Phys. Rev. Lett.* **63**, 82.

[339] Miranda, E., Dobrosavljevic, V., and Kotliar, G. (1996) *J. Phys. Cond. Matt.* **8**, 9871.

[340] Miranda, E., Dobrosavljevic, V., and Kotliar, G. (1997) *Phys. Rev. Lett.* **78**, 290.

[341] Mishra, S. G., and Ramakrishnan, T. V. (1978) *Phys. Rev. B* **18**, 2308.

[342] Mishra, S. G., and Sreeram, P. A. (1998) *Phys. Rev. B* **57**, 2188.

[343] Miyahara, S., and Ueda, K. (1999) *Phys. Rev. Lett.* **82**, 3701.

[344] Miyake, K., Schmitt-Rink, S., and Varma, C. M. (1986) *Phys. Rev. B* **34**, 6554.

[345] Mook, H. A., Yehiraj, M., Aeppli, G., Mason, T. E., and Armstrong, T. (1993) *Phys. Rev. Lett.*, **70**, 3490.

[346] Moreo, A., Dagotto, E., Jolicoeur, Th., and Riera, J. (1990) *Phys. Rev. B* **42**, 6283.

[347] Moriya, T., and Kawabata, A. (1973) *J. Phys. Soc. Jpn.* **34**, 639.

[348] Moriya, T., and Kawabata, A. (1973) *J. Phys. Soc. Jpn.* **35**, 669.

[349] Moriya, T. (1985) *Spin Fluctuations in Itinerant Electron Magnetism* (Springer Verlag, Berlin).

[350] Moriya, T., and Takimoto, T. (1995) *J. Phys. Soc. Jpn.* **64**, 960.

[351] Motrunich, O. I., and Vishwanath, A. (2004) *Phys. Rev. B* **70**, 075104.

[352] Murthy, G., Arovas, D., and Auerbach, A. (1997) *Phys. Rev. B*, **55**, 3104.

[353] Murthy, G., and Sachdev, S. (1990) *Nucl. Phys. B* **344**, 557.

[354] Myers, R. C., Sachdev, S., and Singh, A. (2010) arXiv:1010.0443.

[355] Nagler, S. E., Buyers, W. J. L., Armstrong, R. L., and Briat, B. (1983) *Phys. Rev. B* **28**, 3873.

[356] Narumi, Y., Hagiwara, M., Sato, R., *et al.* (1998) *Physica B* **246–247**, 509.

[357] Negele, J. W., and Orland, H. (1988) *Quantum Many-Particle Systems* (Addison-Wesley, Redwood City, CA).

[358] Nelson, D. R., and Kosterlitz, J. M. (1977) *Phys. Rev. Lett.* **39**, 1201.

[359] Nelson, D. R., and Pelcovits, R. (1977) *Phys. Rev. B* **16**, 2191.

[360] Nelson, D. R., and Rudnick, J. (1975) *Phys. Rev. Lett.* **35**, 178.

[361] Nieuwenhuizen, T. M., and Ritort, F. (1997) *Physica A* **250**, 8.

[362] Nikolic, P., and Sachdev, S. (2007) *Phys. Rev. A* **75**, 033608.

[363] Nishida, Y., and Son, D. T. (2007) *Phys. Rev. A* **75**, 063617.

[364] Niyaz, P., Scalettar, R. T., Fong, C. Y., and Batrouni, G. G. (1994) *Phys. Rev. B* **50**, 362.

[365] Normand, B., and Rice, T. M. (1996) *Phys. Rev. B* **54**, 7180.

[366] Normand, B., and Rice, T. M. (1997) *Phys. Rev. B* **56**, 8760.

[367] O'Connor, D., and Stephens, C. R. (1994) *Int. J. Mod. Phys. A* **9**, 2805.

[368] Oppermann, R., and Binderberger, M. (1994) *Ann. Physik.* **3**, 494.

[369] Oppermann, R., and Mueller, A. (1993) *Nucl. Phys. B* **401**, 507.

[370] Oshikawa, M., Yamanaka, M., and Affleck, I. (1997) *Phys. Rev. Lett.* **78**, 1984.

[371] Parcollet, O., and Georges, A. (1999) *Phys. Rev. B* **59**, 5341.

[372] Patashinskii, A. Z., and Pokrovski, V. L. (1974) *Sov. Phys. JETP* **37**, 733.

[373] Pellegrini, V., Pinczuk, A., Dennis, B. S., *et al.* (1997) *Phys. Rev. Lett.* **78**, 310.

[374] Pellegrini, V., Pinczuk, A., Dennis, B. S., *et al.* (1998) *Science* **281**, 799.

[375] Pépin, C., and Lavagna, M. (1999) *Phys. Rev. B* **59**, 2591.

[376] Perk, J. H. H., and Capel, H. W. (1977) *Physica A* **89**, 265.

[377] Perk, J. H. H., Capel, H. W., Quispel, G. R. W., and Nijhoff, F. W. (1984) *Physica A* **123**, 1.

[378] Pfeuty, P. (1970) *Ann. of Phys.* **57**, 79.

[379] Pich, C., Young, A. P., Rieger, H., and Kawashima, N. (1998) *Phys. Rev. Lett.* **81**, 5916.

[380] Pinczuk, A., Dennis, B. S., Heiman, D., *et al.* (1992) *Phys. Rev. Lett.* **68**, 3623.

[381] Pisarski, R. D., and Tytgat, M. (1997) *Phys. Rev. Lett.* **78**, 3622.

[382] Plischke, M., and Bergersen, B. (1989) *Equilibrium Statistical Physics* (Prentice Hall, Englewood Cliffs, NJ).

[383] Poilblanc, D., Gagliano, E., Bacci, S., and Dagotto, E. (1991) *Phys. Rev. B* **43**, 10970.

[384] Polyakov, A. M. (1975) *Phys. Lett. B* **59**, 87.

[385] Polyakov, A. M. (1987) *Gauge Fields and Strings* (Harwood, London).

[386] Popov, V. N. (1972) *Teor. Mat. Phys.* **11**, 354.

[387] Popov, V. N. (1983) *Functional Integrals in Quantum Field Theory and Statistical Physics* (D. Reidel, Dordrecht).

[388] Prokof'ev, N., Ruebenacker, O., and Svistunov, B. (2001) *Phys. Rev. Lett.* **87**, 270402.

[389] Rajaraman, R. (1982) *Solitons and Instantons: An Introduction to Solitons and Instantons in Quantum Field Theory* (North-Holland, Amsterdam).

[390] Ramakrishnan, T. V. (1974) *Solid State Commun.* **14**, 449.

[391] Ramakrishnan, T. V. (1974) *Phys. Rev. B* **10**, 4014.

[392] Ramond, P. (2001) *Field Theory: A Modern Primer* (Westview Press, Boulder, CO).

[393] Randeria, M., Sethna, J. P., and Palmer, R. G. (1985) *Phys. Rev. Lett.* **54**, 1321.

[394] Rasolt, M., Stephen, M. J., Fisher, M. E., and Weichman, P. B. (1984) *Phys. Rev. Lett.* **53**, 798.

[395] Read, N., and Sachdev, S. (1989) *Nucl. Phys. B* **316**, 609.

[396] Read, N., and Sachdev, S. (1989) *Phys. Rev. Lett.* **62**, 1694.

[397] Read, N., and Sachdev, S. (1990) *Phys. Rev. B* **42**, 4568.

[398] Read, N., and Sachdev, S. (1991) *Phys. Rev. Lett.* **66**, 1773.

[399] Read, N., and Sachdev, S. (1995) *Phys. Rev. Lett.* **75**, 3509.

[400] Read, N., Sachdev, S., and Ye, J. (1995) *Phys. Rev. B* **52**, 384.

[401] Reger, J. D., and Young, A. P. (1988) *Phys. Rev. B* **37**, 5978.

[402] Reppy, J. D. (1984) *Physica B* **126**, 335.

[403] Rieger, H., and Young, A. P. (1994) *Phys. Rev. Lett.* **72**, 4141.

[404] Rieger, H., and Young, A. P. (1996) *Phys. Rev. B* **54**, 3328.

[405] Rimberg, A. J., Ho, T. R., Kurdak, C., *et al.* (1997) *Phys. Rev. Lett.* **78**, 2632.

[406] Rokhsar, D., and Kivelson, S. A. (1988) *Phys. Rev. Lett.* **61**, 2376.

[407] Rosch, A. (1999) *Phys. Rev. Lett.* **82**, 4280.

[408] Rosch, A., Schröder, A., Stockert, O., and von Löhneysen, H. (1997) *Phys. Rev. Lett.* **79**, 159.

[409] Rosenow, B., and Oppermann, R. (1996) *Phys. Rev. Lett.* **77**, 1608.

[410] Rosenstein, B., Warr, B. J., and Park, S. H. (1991) *Phys. Rep.* **205**, 59.

[411] Rossat-Mignod, J., Regnault, L. P., Vettier, C., *et al.* (1991) *Physica C* **185–189**, 86.

[412] Rozenberg, M. J., and Grempel, D. R. (1998) *Phys. Rev. Lett.* **81**, 2550.

[413] Rudnick, J., Guo, H., and Jasnow, D. (1985) *J. Stat. Phys.* **41**, 353.

[414] Ruegg, C., Normand, B., Matsumoto, M., *et al.* (2008) *Phys. Rev. Lett.* **100**, 205701.

[415] Sachdev, S. (1992) *Phys. Rev. B.* **45**, 12377.

[416] Sachdev, S. (1994) *Phys. Rev. B* **50**, 13006.

[417] Sachdev, S. (1994) *Z. Phys. B* **94**, 469.

[418] Sachdev, S. (1995) in *Low Dimensional Quantum Field Theories for Condensed Matter Physicists*, ed. Y. Lu, S. Lundqvist, and G. Morandi (World Scientific, Singapore); cond-mat/9303014.

[419] Sachdev, S. (1996) in *Proceedings of the 19th IUPAP International Conference on Statistical Physics, Xiamen, China*, ed. B.-L. Hao (World Scientific, Singapore); cond-mat/9508080.

[420] Sachdev, S. (1996) *Nucl. Phys. B* **464**, 576.

[421] Sachdev, S. (1997) in *Strongly Correlated Magnetic and Superconducting Systems*, ed. G. Sierra and M. A. Martin-Delgado (Springer-Verlag, Berlin).

[422] Sachdev, S. (1997) *Phys. Rev. B* **55**, 142.

[423] Sachdev, S. (1998) *Phil. Trans. R. Soc. London A* **356**, 173.

[424] Sachdev, S. (1998) in *Dynamical Properties of Unconventional Magnetic Systems*, NATO ASI Series E: Applied Sciences, Vol. 349, ed. A. Skjeltorp and D. Sherrington (Kluwer Academic, Dordrecht); cond-mat/9705266.

[425] Sachdev, S. (1998) *Phys. Rev. B* **57**, 7157.

[426] Sachdev, S. (1999) *Phys. Rev. B* **59**, 14054.

[427] Sachdev, S. (1998) in *Highlights in Condensed Matter Physics*, ed. Y. M. Cho and M. Virasoro (World Scientific, Singapore); cond-mat/9811110.

[428] Sachdev, S. (1999) *Phys. World* **12**, No. 4, p. 33.

[429] Sachdev, S. (2004) in *Quantum Magnetism*, U. Schollwock, J. Richter, D. J. J. Farnell, and R. A. Bishop eds, Lecture Notes in Physics, (Springer, Berlin); cond-mat/0401041.

[430] Sachdev, S. (2008) *Exotic phases and quantum phase transitions: model systems and experiments*, Rapporteur talk at the 24th Solvay Conference on Physics, *Quantum Theory of Condensed Matter*, Brussels, Oct 2008, arXiv:0901.4103

[431] Sachdev, S. (2010) *Physica Status Solidi B* **247**, 537 (*Preprint* arXiv:0907.0008).

[432] Sachdev, S., Chubukov, A. V., and Sokol, A. (1995) *Phys. Rev. B* **51**, 14874.

[433] Sachdev, S., and Damle, K. (1997) *Phys. Rev. Lett.* **78**, 943.

[434] Sachdev, S., and Jalabert, R. (1990) *Mod. Phys. Lett. B* **4**, 1043.

[435] Sachdev, S., and Read, N. (1991) *Int. J. Mod. Phys. B* **5**, 219.

[436] Sachdev, S., Read, N., and Oppermann, R. (1995) *Phys. Rev. B* **52**, 10286.

[437] Sachdev, S., and Senthil, T. (1996) *Ann. Phys.* **251**, 76.

[438] Sachdev, S., Senthil, T., and Shankar, R. (1994) *Phys. Rev. B* **50**, 258.

[439] Sachdev, S., and Starykh, O. A. (2000) *Nature* **405**, 322.

[440] Sachdev, S., and Ye, J. (1992) *Phys. Rev. Lett.* **69**, 2411.

[441] Sachdev, S., and Ye, J. (1993) *Phys. Rev. Lett.* **70**, 3339.

[442] Sachdev, S., and Young, A. P. (1997) *Phys. Rev. Lett.* **78**, 2220.

[443] Sandvik, A. W., and Scalapino, D. J. (1994) *Phys. Rev. Lett.* **72**, 2777.

[444] Sandvik, A. W., Chubukov, A. V., and Sachdev, S. (1995) *Phys. Rev. B* **51**, 16483.

[445] Sawada, A., Ezawa, Z. F., Ohno, H., *et al.* (1998) *Phys. Rev. Lett.* **80**, 4534.

[446] Scalapino, D. J., Loh, E., and Hirsch, J. E. (1986) *Phys. Rev. B* **34**, 8190.

[447] Scalettar, R. T., Trivedi, N., and Huscroft, C. (1999) *Phys. Rev. B* **59**, 4364.

[448] Schneider, W., and Randeria, M. (2010) arXiv:0910.2693.

[449] Schneider, W., Shenoy, V. B., and Randeria, M. (2009) *Phys. Rev. A* **81**, 021601(R).

[450] Schön, G., and Zaikin, A. D. (1990) *Phys. Rep.* **198**, 237.

[451] Schröder, A., Aeppli, G., Bucher, E., Ramazashvili, R., and Coleman, P. (1998) *Phys. Rev. Lett.* **80**, 5623.

[452] Schulz, H. J. (1986) *Phys. Rev. B* **34**, 6372.

[453] Schulz, H. J. (1995) in *Mesoscopic Quantum Physics, Les Houches Summer School Proceedings, Vol. 61*, ed. E. Akkermans, G. Montambaux, J. Pichard, and J. Zinn-Justin (North-Holland, Amsterdam).

[454] Schulz, H. J., and Ziman, T. A. L. (1992) *Europhys. Lett.* **18**, 355.

[455] Schulz, H. J., Ziman, T. A. L., and Poilblanc, D. (1996) *J. Physique I* **6**, 675.

[456] Sengupta, A. M. (2000) *Phys. Rev. B* **61**, 4041.

[457] Sengupta, A. M., and Georges, A. (1995) *Phys. Rev. B* **52**, 10295.

[458] Senthil, T., Balents, L., Sachdev, S., Vishwanath, A., and Fisher, M. P. A. (2004) *Phys. Rev. B* **70**, 144407.

[459] Senthil T., and Fisher, M. P. A. (2000) *Phys. Rev. B* **62**, 7850.

[460] Senthil, T., and Majumdar, S. (1996) *Phys. Rev. Lett.* **76**, 3001.

[461] Senthil, T., and Sachdev, S. (1996) *Phys. Rev. Lett.* **77**, 5295.

[462] Senthil, T., Vishwanath, A., Balents, L., Sachdev, S., and Fisher, M. P. A. (2004) *Science* **303**, 1490.

[463] Seo, Kangjun, Bernevig, B. A., and Hu, Jiangping (2008) *Phys. Rev. Lett.* **101**, 206404.

[464] Shankar, R. (1994) *Principles of Quantum Mechanics*, 2nd ed. (Plenum, New York).

[465] Shankar, R. (1994) *Rev. Mod. Phys.* **66**, 129.

[466] Shankar, R., and Murthy, G. (1987) *Phys. Rev. B* **36**, 536.

[467] Shankar, R., and Read, N. (1990) *Nucl. Phys. B* **336**, 457.

[468] Shastry, B. S., and Sutherland, B. (1981) *Phys. Rev. Lett.* **47**, 964.

[469] Shastry, B. S., and Sutherland, B. (1981) *Physica* **108B**, 1069.

[470] Shenker, S. H., and Tobochnik, J. (1980) *Phys. Rev. B* **22**, 4462.

[471] Shiramura, W., Takatsu, K., Kurniawan, B., *et al.* (1998) *J. Phys. Soc. Jpn.* **67**, 1548.

[472] Si, Q., and Smith, J. L. (1996) *Phys. Rev. Lett.* **77**, 3391.

[473] Singh, K. K. (1975) *Phys. Rev. B* **12**, 2819.

[474] Singh, K. K. (1978) *Phys. Rev. B* **17**, 324.

[475] Singh, S., and Pathria, R. K. (1985) *Phys. Rev. B* **31**, 4483.

[476] Smith, J. L., and Si, Q. (1999) *Europhys. Lett.* **45**, 228.

[477] Smith, J. L., and Si, Q. (2000) *Phys. Rev. B* **61**, 5184.

[478] Sokol, A., Glenister, R. L., and Singh, R. R. P. (1994) *Phys. Rev. Lett.* **72**, 1549.

[479] Son, D. T., and Thompson, E. G. (2010) *Phys. Rev. A* **81**, 063634.

[480] Sondhi, S. L., Karlhede, A., Kivelson, S. A., and Rezayi, E. H. (1993) *Phys. Rev. B* **47**, 16419.

[481] Sondhi, S. L., Girvin, S. M., Carini, J. P., and Shahar, D. (1997) *Rev. Mod. Phys.* **69**, 315.

[482] Sorensen, E. S., and Affleck, I. (1993) *Phys. Rev. Lett.* **71**, 1633.

[483] Sorensen, E. S., and Roddick, E. (1995) *Phys. Rev. B* **53**, 8867.

[484] Sorensen, E. S., Wallin, M., Girvin, S. M., and Young, A. P. (1992) *Phys. Rev. Lett.* **69**, 828.

[485] Stanley, H. E. (1968) *Phys. Rev.* **176**, 718.

[486] Starykh, O. A., Singh, R. R. P., and Sandvik, A. W. (1997) *Phys. Rev. Lett.* **78**, 539.

[487] Starykh, O. A., Sandvik, A. W., and Singh, R. R. P. (1997) *Phys. Rev. B* **55**, 14953.

[488] Stauffer, D., and Aharony, A. (1992) *Introduction to Percolation Theory* (Taylor and Francis, New York).

[489] Steglich, F., Buschinger, B., Gegenwart, P., *et al.* (1996) *J. Phys. Cond. Matt.* **8**, 9909.

[490] Steiner, M., Breznay, N., and Kapitulnik, A. (2008) *Phys. Rev. B* **77**, 212501.

[491] Stinchcombe, R. B. (1981) *J. Phys. C* **14**, L263.

[492] Stockert, O., von Löhneysen, H., Rosch, A., Pyka, N., and Loewenhaupt, M. (1998) *Phys. Rev. Lett.* **80**, 5627.

[493] Stone, M. ed. (1994) *Bosonization* (World Scientific, Singapore).

[494] Stoner, E. C. (1936) *Proc. Roy. Soc. London A* **154**, 656.

[495] Strongin, M., Thompson, R. S., Kammerer, O. F., and Crow, J. E. (1970) *Phys. Rev. B* **1**, 1078.

[496] Suzuki, M. (1976) *Prog. Theor. Phys.* **56**, 1454.

[497] Takahashi, M., Nakamura, H., and Sachdev, S. (1996) *Phys. Rev. B* **54**, R744.

[498] Takigawa, M., Asano, T., Ajiro, Y., Mekata, M., and Uemura, Y. J. (1996) *Phys. Rev. Lett.* **76**, 2173.

[499] Takigawa, M., Starykh, O. A., Sandvik, A. W., and Singh, R. R. P. (1997) *Phys. Rev. B* **56**, 13681.

[500] Tan, S. (2008) *Annals of Physics* **323**, 2952.

[501] Tan, S. (2008) *Annals of Physics* **323**, 2971.

[502] Taylor, J. R. (1972) *Scattering Theory: The Quantum Theory of Non-Relativistic Collisions* (Wiley, New York).

[503] Thill, M. J., and Huse, D. A. (1995) *Physica A* **15**, 321.

[504] Timm, C., Girvin, S. M., Henelius, P., and Sandvik, A. W. (1998) *Phys. Rev. B* **58**, 1464.

[505] Tomonaga, S. (1950) *Prog. Theor. Phys.* **5**, 544.

[506] Tranquada, J. M., Shirane, G., Keimer, B., Shamoto, S., and Sato, M. (1989) *Phys. Rev. B* **40**, 4503.

[507] Tranquada, J. M., Shirane, G., Keimer, B., Shamoto, S., and Sato, M. (1992) *Phys. Rev. B* **46**, 5561.

[508] Trivedi, N., Scalettar, R. T., and Randeria, M. (1996) *Phys. Rev. B* **54**, 3756.

[509] Troyer, M., and Sachdev, S. (1998) *Phys. Rev. Lett.* **81**, 5418.

[510] Troyer, M., Tsunetsugu, H., and Rice, T. M. (1996) *Phys. Rev. B* **53**, 251.

[511] Troyer, M., Zhitomirsky, M. E., and Ueda, K. (1997) *Phys. Rev. B* **55**, R6117.

[512] Tsvelik, A. M. (1987) *Zh. Eksp. Teor. Fiz.* **93**, 385 (*Sov. Phys. JETP* **66**, 221).

[513] Tsvelik, A. M. (1990) *Phys. Rev. B* **42**, 10499.

[514] Tsvelik, A. M., and Reizer, M. (1993) *Phys. Rev. B* **48**, 9887.

[515] Tyc, S., Halperin, B. I., and Chakravarty, S. (1989) *Phys. Rev. Lett* **62**, 835.

[516] Tyc, S., and Halperin, B. I. (1990) *Phys. Rev. B* **42**, 2096.

[517] Ubaid-Kassis, S., Vojta, T., and Schroeder, A. (2010) *Phys. Rev. Lett.* **104**, 066402.

[518] Uemura, Y. J., *et al.* (1989) *Phys. Rev. Lett.* **62**, 2317.

[519] Usadel, K. D., Büttner, G., and Kopeć, T. K. (1991) *Phys. Rev. B* **44**, 12583.

[520] van Otterlo, A., and Wagenblast, K.-H. (1994) *Phys. Rev. Lett.* **72**, 3598.

[521] van Otterlo, A., Wagenblast, K.-H., Baltin, R., *et al.* (1995) *Phys. Rev. B* **52**, 16176.

[522] van Otterlo, A., Wagenblast, K.-H., Fazio, R., and Schön, G. (1993) *Phys. Rev. B* **48**, 3316.

[523] Varma, C. M., Littlewood, P. B., Schmitt-Rink, S., Abrahams, E., and Ruckenstein, A. E. (1989) *Phys. Rev. Lett.* **63**, 1996.

[524] Varma, C. M. (1997) *Phys. Rev. B* **55**, 14554.

[525] Veillette, M. Y., Sheehy, D. E., and Radzihovsky, L. (2007) *Phys. Rev. A* **75**, 043614.

[526] Veillette, M. Y., Moon, E. G., Lamacraft, A., *et al.* (2008) *Phys. Rev. A* **78**, 033614.

[527] Vojta, M., Zhang, Y., and Sachdev, S. (2000) *Phys. Rev. Lett.* **85**, 4940; Erratum (2008) *Phys. Rev. Lett.* **100**, 089904(E).

[528] Vojta, M., Zhang, Y., and Sachdev, S. (2000) *Int. J. Mod. Phys. B* **14**, 3719.

[529] Vojta, T. (2003) *Phys. Rev. Lett.* **90**, 107202.

[530] Vojta, T. (2006) *J. Phys. A* **39**, R143.

[531] Vojta, T., and Hoyos, J. A. (2008) in *Recent Progress in Many-Body Theories*, ed. J. Boronat, G. Astrakharchik and F. Mazzanti (World Scientific, Singapore).

[532] von Löhneysen, H., Neubert, A., Schröder, A., *et al.* (1998) *Eur. Phys. J. B* **5**, 447.

[533] Voss, D. (1998) *Science* **282**, 221.

[534] Wagenblast, K.-H., van Otterlo, A., Schön, G., and Zimanyi, G. T. (1997) *Phys. Rev. Lett.* **79**, 2730.

[535] Wallin, M., Sorensen, E. S., Girvin, S. M., and Young, A. P. (1994) *Phys. Rev. B* **49**, 12115.

[536] Weichman, P. B., Rasolt, M., Fisher, M. E., and Stephen, M. J. (1986) *Phys. Rev. B* **33**, 4632.

[537] Weiss, U. (1993) *Quantum Dissipative Systems* (World Scientific, Singapore).

[538] Wen, X. G., and Zee, A. (1990) *Int. J. Mod. Phys. B* **4**, 437.

[539] Wenzel, S., and Janke, W. (2009) *Phys. Rev. B* **79**, 014410.

[540] Westerberg, E., Furusaki, A., Sigrist, M., and Lee, P. A. (1997) *Phys. Rev. B* **55**, 12578.

[541] Wu, W., Ellman, B., Rosenbaum, T. F., Aeppli, G., and Reich, D. H. (1991) *Phys. Rev. Lett.* **67**, 2076.

[542] Wu, W., Bitko, D., Rosenbaum, T. F., and Aeppli, G. (1993) *Phys. Rev. Lett.* **71**, 1919.

[543] Xu, C., and Sachdev, S. (2009) *Phys. Rev. B* **79**, 064405.

[544] Xu, G., DiTusa, J. F., Ito, T., *et al.* (1996) *Phys. Rev. B* **54**, R6827.

[545] Xu, G., Broholm, C., Soh, Yeong-Ah, *et al.* (2007) *Science* **317**, 1049.

[546] Ye, J. (1998) *Phys. Rev. B* **58**, 9450.

[547] Ye, J., and Sachdev, S. (1998) *Phys. Rev. Lett.* **80**, 5409.

[548] Ye, J., Sachdev, S., and Read, N. (1993) *Phys. Rev. Lett.* **70**, 4011

[549] Yoshizawa, H., Hirakawa, K., Satija, S. K., and Shirane, G. (1981) *Phys. Rev. B* **23**, 2298.

[550] Young, A. P. (1997) *Phys. Rev. B* **56**, 11691.

[551] Young, A. P. ed. (1997) *Spin Glasses and Random Fields* (World Scientific, Singapore).

[552] Young, A. P., and Rieger, H. (1996) *Phys. Rev. B* **53**, 8486.

[553] Zamolodchikov, A. B., and Zamolodchikov, A. B. (1979) *Ann. of Phys.* **120**, 253.

[554] Zamolodchikov, A. B., and Zamolodchikov, A. B. (1992) *Nucl. Phys. B* **379**, 602.

[555] Zheng, L., Radtke, R. J., and Das Sarma, S. (1997) *Phys. Rev. Lett.* **78**, 2453.

[556] Zheng, W., and Sachdev, S. (1989) *Phys. Rev. B* **40**, 2704.

[557] Zinn-Justin, J. (1993) *Quantum Field Theory and Critical Phenomena* (Oxford University Press, Oxford).

[558] Zülicke, U., and Millis, A. J. (1995) *Phys. Rev. B* **51**, 8996.

[559] Zwerger, W. (2004) *Phys. Rev. Lett.* **92**, 027203.

Index

Printed in the United States
By Bookmasters